Ionic Polymer Metal Composites (IPMCs)

Smart Multi-Functional Materials and Artificial Muscles

Volume 2

RSC Smart Materials

Series Editors:
Professor Hans-Jörg Schneider, *Saarland University,* Germany
Professor Mohsen Shahinpoor, *University of Maine,* USA

Titles in this Series:

1: Janus Particle Synthesis, Self-Assembly and Applications
2: Smart Materials for Drug Delivery: Volume 1
3: Smart Materials for Drug Delivery: Volume 2
4: Materials Design Inspired by Nature
5: Responsive Photonic Nanostructures: Smart Nanoscale Optical Materials
6: Magnetorheology: Advances and Applications
7: Functional Nanometer-Sized Clusters of Transition Metals: Synthesis, Properties and Applications
8: Mechanochromic Fluorescent Materials: Phenomena, Materials and Applications
9: Cell Surface Engineering: Fabrication of Functional Nanoshells
10: Biointerfaces: Where Material Meets Biology
11: Semiconductor Nanowires: From Next-Generation Electronics to Sustainable Energy
12: Supramolecular Materials for Opto-Electronics
13: Photocured Materials
14: Chemoresponsive Materials: Stimulation by Chemical and Biological Signals
15: Functional Metallosupramolecular Materials
16: Bio-Synthetic Hybrid Materials and Bionanoparticles: A Biological Chemical Approach Towards Material Science
17: Ionic Polymer Metal Composites (IPMCs): Smart Multi-Functional Materials and Artificial Muscles, Volume 1
18: Ionic Polymer Metal Composites (IPMCs): Smart Multi-Functional Materials and Artificial Muscles, Volume 2

How to obtain future titles on publication:
A standing order plan is available for this series. A standing order will bring delivery of each new volume immediately on publication.

For further information please contact:
Book Sales Department, Royal Society of Chemistry, Thomas Graham House, Science Park, Milton Road, Cambridge, CB4 0WF, UK
Telephone: +44 (0)1223 420066, Fax: +44 (0)1223 420247
Email: booksales@rsc.org
Visit our website at www.rsc.org/books

Ionic Polymer Metal Composites (IPMCs)

Smart Multi-Functional Materials and Artificial Muscles

Volume 2

Edited by

Mohsen Shahinpoor
University of Maine, Orono, Maine, USA
Email: shah@maine.edu

ROYAL SOCIETY OF CHEMISTRY

THE QUEEN'S AWARDS FOR ENTERPRISE: INTERNATIONAL TRADE 2013

RSC Smart Materials No. 18

Print ISBN: 978-1-78262-721-0
PDF eISBN: 978-1-78262-723-4
ISSN: 2046-0066

A catalogue record for this book is available from the British Library

Published by The Royal Society of Chemistry,
Thomas Graham House, Science Park, Milton Road,
Cambridge CB4 0WF, UK

Registered Charity Number 207890

For further information see our web site at www.rsc.org

Printed in the United Kingdom by CPI Group (UK) Ltd, Croydon, CR0 4YY, UK

Preface

It is a great honor to serve as editor for this historic volume on electroactive ionic polymers and in particular ionic polymer metal composites or IPMCs as smart multi-functional polymeric actuators, sensors and energy harvesters, among others. I can proudly proclaim that I have brought together in this volume the leading researchers in the world on various aspects of this amazing biomimetic robotic electronic material that play quite a role in the future of electroactive polymers and smart multi-functional materials.

IPMC is a class of electroactive polymers that can be both actuator and sensor. By applying voltage it exhibits large deformation, which is why it is known as artificial muscle and, on the other hand, it acts as a smart material that can sense mechanical bending by creating proportional voltage. While other strain sensors require a power source to work, the IPMC sensors not only do not require power, but also they can create voltage and power that makes them a potential candidate for battery-less sensors. Conversely, an applied small voltage or electric field can induce an array of spectacularly large deformation or actuation behaviors in IPMCs, such as bending, twisting, rolling, twirling, steering and undulating.

My vision of the future of IPMC artificial muscles may be summarized below in terms of both medical and industrial applications. Note that IPMCs are excellent sensors that generate huge outputs in terms of millivolts, which can be employed for the sensing, transduction and harvesting of energy from wind or ocean waves. These unique materials work perfectly well in a wet environment and thus they are excellent candidates for medical applications. These might range from endovascular steerers and stirrers to enable navigation within the human vasculature; use as deep brain stimulators or employed in flat diaphragm micropumps for precision drug delivery, glaucoma and hydrocephalus; artificial muscles for the surgical correction of

RSC Smart Materials No. 18
Ionic Polymer Metal Composites (IPMCs): Smart Multi-Functional Materials and Artificial Muscles, Volume 2
Edited by Mohsen Shahinpoor

Published by the Royal Society of Chemistry, www.rsc.org

ptosis (drooping eyelid syndrome); ophthalmological and vision improvement applications; artificial muscles to assist a failing heart; correction of facial paralysis, facioscapulohumeral and other applications in muscular dystrophy; to mediate the control of drainage or flow within the human body; and myriad additional purposes. On the industrial side, due to the fact that the IPMCs are excellent sensors and low-voltage actuators, they can be used for both sensing and simultaneous actuation in many engineering applications. In the sensing mode they have a very good bandwidth to sense low as well as high frequencies, in contrast to piezoelectric materials such as PZT (Lead Zirconate Titanate) or lithium niobate, which are only suitable for high-frequency sensing. Two emerging visions of the future are to see IPMCs heavily utilized in atomic force microscopes as novel and dynamic probes in scanning probe microscopy, as well as robotic surgery to facilitate the conveyance of specific haptic, force, tactile and impedance feedback to surgeons. IPMCs as active substrate and micro-pillars may be used to monitor nano-bio and cellular dynamics in real time.

These two volumes on IPMCs provide a broad coverage of the state of the art and recent advances in the field with detailed information on the characteristics and applications of these materials by some of the world's leading experts on various characterizations and modeling of IPMCs. This volume contains 27 chapters to present a thorough coverage of all properties and characteristics of IPMCs. Chapter 1 covers the fundamentals of IPMCs, Chapter 2 covers optimal manufacturing of IPMCs, Chapter 3 discusses graphene-based IPMCs, Chapter 4 describes what happens to IPMC electrode interfaces and their effects on actuation and sensing, Chapter 5 presents step-by-step modeling of IPMCs using the multiphysics package of Comsol, Chapter 6 describes IPMCs with electrochemical electrodes, Chapter 7 presents electromechanical distributed modeling of IPMCs while Chapter 8 discusses modeling for engineering design of IPMC devices and Chapter 9 covers electric energy storage using flexible IPMC capacitors, Chapter 10 models the environmental dependency of IPMCs' actuation and sensing dynamics, Chapter 11 discusses the precision feedback/feedforward control of IPMC dynamics while Chapter 12 covers the design, testing and micromanipulation of IPMC microgrippers, Chapter 13 discusses the phenomenon of spatially growing waves of snake-like robots and natural generation of biomimetic swimming motions. Volume 1 of the two volumes on IPMCs ends here with Chapter 13 and Volume 2 starts with Chapter 14. Chapter 14 covers underwater sensing of impulsive loading of IPMCs, Chapter 15 presents a design of a micropump for drug delivery employing IPMCs, Chapter 16 presents the modeling and characterization of IPMC transducers, Chapter 17 discusses IPMCs as postsilicon transducers for the realization of smart systems and Chapter 18 covers micromachined IPMC actuators for biomedical applications, Chapter 19 presents recent advances in IPMC self-sensing while Chapter 20 describes the continuum multiphysics theory for IPMCs, Chapter 21 covers multiphysics modeling of nonlinear plates made with IPMCs, Chapter 22 describes the applications of

IPMCs to dexterous manipulation and haptic feedback/tactile sensors for minimally invasive robotic surgery, Chapter 23 covers IPMCs as soft biomimetic robotic artificial muscles, Chapter 24 describes a family of ionic electroactive actuators with giant electromechanical responses while Chapter 25 describes the multiphysics modeling and simulation of dynamics sensing in IPMCs with applications to soft robotics, and finally Chapter 26 presents a comprehensive review on electroactive paper actuators.

I am hoping that the collection of these chapters by the leading authorities on IPMCs will appeal to readers from chemistry, materials science, engineering, physics and medical communities interested in both IPMC-related materials and their applications.

Mohsen Shahinpoor
Orono, Maine, USA

Contents

Volume 1

RSC Smart Materials No. 18
Ionic Polymer Metal Composites (IPMCs): Smart Multi-Functional Materials and Artificial Muscles, Volume 2
Edited by Mohsen Shahinpoor

Published by the Royal Society of Chemistry, www.rsc.org

Volume 2

CHAPTER 14

Energy Exchange between Coherent Fluid Structures and Ionic Polymer Metal Composites, toward Flow Sensing and Energy Harvesting

SEAN D. PETERSON*[a,b] AND MAURIZIO PORFIRI[b]

[a] Department of Mechanical and Mechatronics Engineering, University of Waterloo, 200 University Avenue West, Waterloo, ON N2L 3G1, Canada;
[b] Department of Mechanical and Aerospace Engineering, New York University Polytechnic School of Engineering, Six MetroTech Center, Brooklyn, NY 11201, USA
*Email: peterson@mme.uwaterloo.ca

14.1 Introduction

The growing integration of small-scale electronics into modern life, ranging from consumer to research-grade devices, provides continuous impetus for the investigation of smart material-based energy harvesting to extend battery lifetime and enhance functionalities.[1–5] Beyond well-studied structural vibrations, energy sources that have been considered in the design of miniature energy harvesting systems include human and animal locomotion,[6–8] jaw movements,[9] and heartbeats,[10] to name a few. While these studies widely

RSC Smart Materials No. 18
Ionic Polymer Metal Composites (IPMCs): Smart Multi-Functional Materials and Artificial Muscles, Volume 2
Edited by Mohsen Shahinpoor

Published by the Royal Society of Chemistry, www.rsc.org

differ in their target applications, they are all based upon energy extraction from mechanical deformation of a solid body with which they interact. Recently, research efforts have been focused on expanding the range of energy harvesting sources to encompass fluid flows.

The majority of these studies have employed piezoelectric transducers to convert small-scale fluid energy into a usable electrical form. The first instance of energy conversion from a fluid flow is the so-called harvesting eel,[11,12] which utilizes the Kármán vortex street shed from a cylinder in a cross flow to deform a piezoelectric element. Building on this concept, higher Reynolds numbers and different geometric configurations have been explored over subsequent years.[13–16] Fluid–structure interactions associated with flutter instabilities from a mean flow and vibrations from a moving base have also been considered.[17–20] In addition, hydraulic fluctuations and hydroelastic impacts have been demonstrated to be viable means for energy extraction.[21,22]

To afford large structural deformations and lower the operation frequency, recent studies have proposed the integration of ionic polymer metal composites (IPMC)[23–28] into fluid energy harvesting. Upon mechanical deformation, a complex cascade of chemo-electro-mechanical processes takes place within the ionomer core and especially at the ionomer–electrode interfaces, resulting in an electric potential sensed across the electrodes.[29–33] By shunting the IPMC electrodes with an external resistor, a small fraction of the mechanical energy is converted into electrical form and then dissipated as heat.[19,34–36] Thus, IPMCs have been used to harvest energy from the flutter instability of a heavy flag,[37] hydroelastic impact,[38] fluid-induced buckling,[39] hydroelastic interactions in arrays of flexible structures,[40] fluid–structure interactions from a compliant miniature turbine,[41] and the beating of an artificial fish tail.[8]

Coherent fluid structures are persistent, organized collections of rotational fluid parcels that are ubiquitous in nature, examples of which include vortex rings,[42,43] Kármán vortex streets,[44] and hairpin vortices.[45] These fluid structures arise from such diverse sources as an undulating fish tail,[46] the propulsion of some invertebrate marine animals,[47] the shear layer of a jet,[48] the oscillation of a sharp-edged structure in a quiescent fluid,[49] the wake of a bluff body,[44] and in turbulent boundary layers.[45,50] The energy associated with these coherent structures is regulated by salient geometric features and local flow circulation.[42,43,51] As an example, the energy of a vortex ring is a function of the vortex core structure, ring radius, and circulation.[42,43]

The advection and diffusion of coherent fluid structures are influenced by interactions with neighboring vortices and solid boundaries, comprising confining walls and nearby compliant bodies.[52] For example, in the absence of viscosity, potential flow can be used to predict the kinematics of a pair of vortices approaching a rigid wall of infinite extent,[53] including deformation of the core.[54] The role of viscosity on impact dynamics has been subsequently elucidated through computational fluid dynamics (CFD).[55] The inclusion of viscosity results in complex vortex dynamics due to the generation of secondary and tertiary structures on no-slip or porous boundaries.[56–58] Scaling laws for the generation of vorticity on no-slip boundaries have been determined, along with the role of the angle of impact.[59–61] The three-dimensional (3D) impact of a vortex ring with a wall has been addressed both numerically and experimentally.[62,63]

Here, we summarize recent work exploring the energy exchange between coherent fluid structures and IPMCs during impact events.[64–68] We specifically focus on impulsive loading of compliant IPMC strips and annuli from self-propagating vortex rings. Beyond experimental evidence in favor of the potential of sensing and energy extraction during short-duration fluid–structure interactions, we propose potential flow and CFD solutions to offer insight into the physics of the impact.

14.2 Experiments

We present two experimental demonstrations of IPMC-based energy harvesting from coherent fluid structures. Specifically, we first review bending deformation of a cantilevered IPMC strip,[65] followed by axisymmetric bending of an IPMC annulus, due to the impact of a vortex ring.[64] Particle image velocimetry (PIV) is utilized in both studies to resolve the attendant flow field. The IPMCs are fabricated in-house from Nafion-117 membranes plated with platinum electrodes,[23] which are held short-circuited to offer evidence for the possibility of energy conversion.

14.2.1 Impulsive Loading of a Cantilever Strip

The vortex is generated by a piston plunged through a cylinder submersed in the water, as shown in Figure 14.1. The vortex ring has circulation Γ, velocity V_{vr}, and diameter a. The vortex ring is launched so that it impacts the IPMC of length L orthogonally, with its center at the tip. A Cartesian coordinate system is introduced, with the x-axis along the IPMC and the y-axis aligned with the direction of the vortex propagation (see Figure 14.1). The circulation

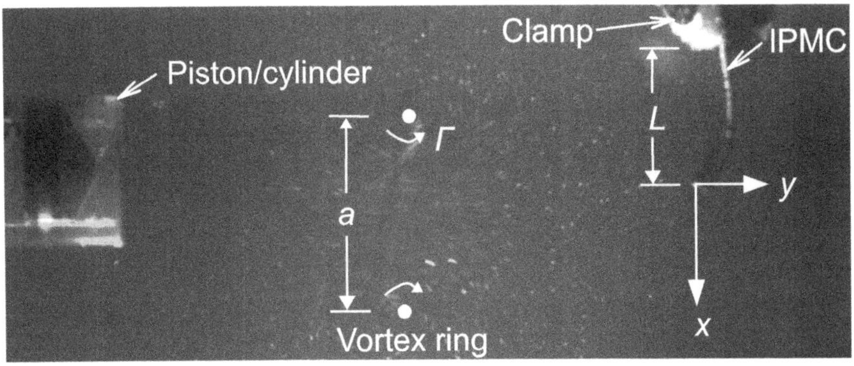

Figure 14.1 Experimental configuration with overlaid coordinate system and variable definitions.
Reproduced from Peterson and Porfiri.[65]

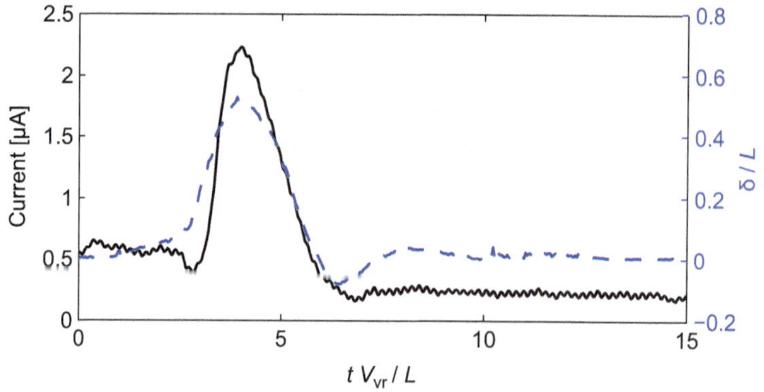

Figure 14.2 Scaled IPMC tip deflection (dashed line) as a function of the non-dimensional time tV_{vr}/L and short-circuit current (solid line) through the IPMC.
Reproduced from Peterson and Porfiri.[65]

Γ is computed from the line integral of the velocity around each of the two vortices in the 2D slice captured by PIV.

As the piston plunges into the cylinder, water is ejected, forming a vortex ring. At the end of the plunger stroke, the vortex ring pinches off and advects towards the IPMC at a nearly constant speed. Once the vortex ring is approximately within 1.5 L of the IPMC, the IPMC bends away from it with negligible influence on the flow physics. Upon impact, the vortex ring breaks down and the IPMC suddenly deflects as energy is transferred from the ring to the structure.

The IPMC tip deflection δ as a function of time t superimposed with the short-circuit current is presented in Figure 14.2 for a representative instance of impact. The maximum deflection of the IPMC is as large as 0.6 L and is attained after vortex ring breakup. Specifically, when the vortex ring reaches the initial position of the IPMC, the deflection is on the order of 0.2 L, after which the IPMC tip speed increases rapidly, implying that breakdown of the vortex ring is a determinant of energy exchange. Such an exchange is revealed by the concomitant increase in the short-circuit current, which closely follows IPMC deformation. An estimate of the energy conversion is obtained from the strain and electrical energy in the IPMC, which amounts to roughly 0.001%.

14.2.2 Impulsive Loading of an Annulus

The problem of a vortex ring advecting through a placid fluid and orthogonally impacting a co-axial IPMC annulus is explored. The experimental setup consists of a custom vortex ring facility comprising a mechanically actuated piston/cylinder vortex generator immersed in a water tank (see Figure 14.3a). Rings of varying circulation are formed by controlling the piston stroke. The IPMC annulus is clamped along its outer radius through a custom fixture with embedded electrodes (see Figure 14.3b).

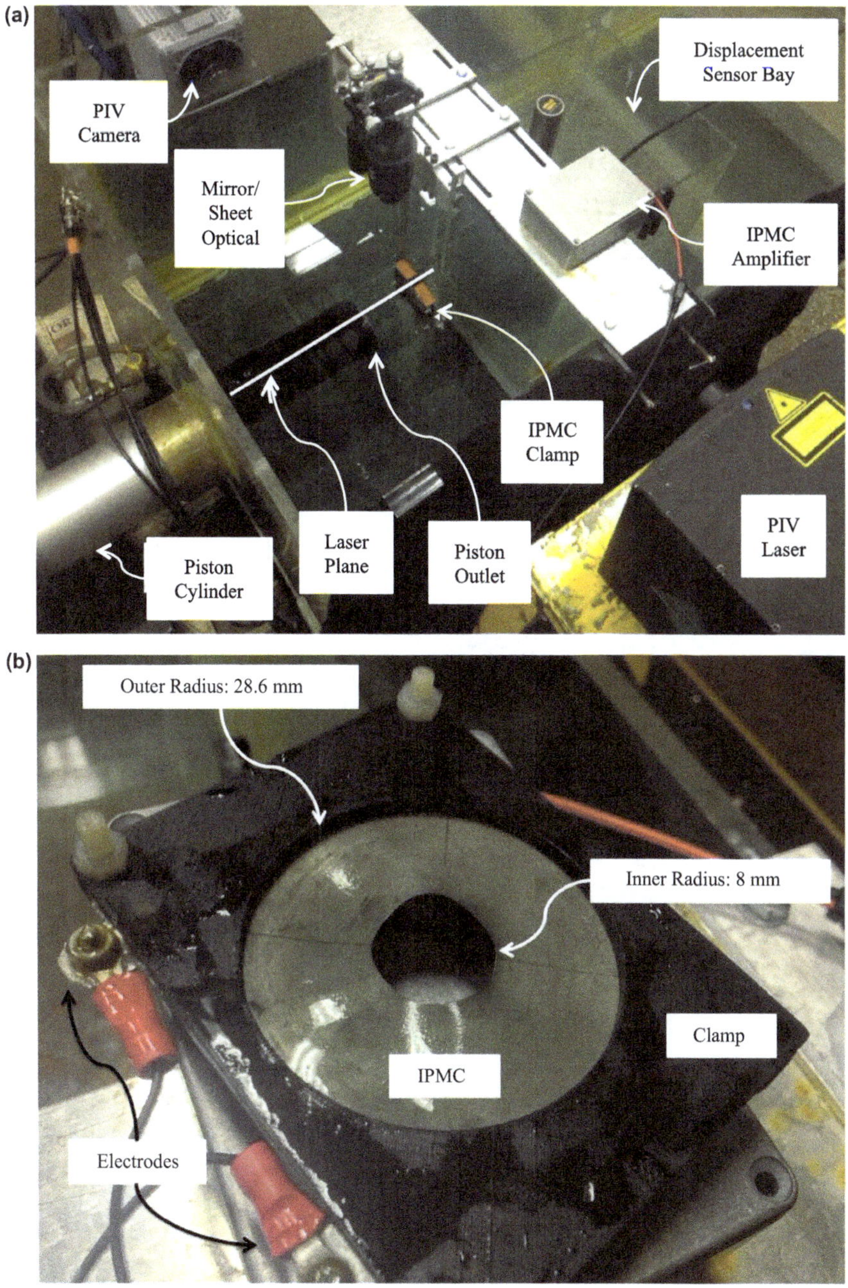

Figure 14.3 (a) Image of the experimental setup for the axisymmetric study, and (b) close-up of the annular IPMC harvester.
Reproduced from Hu *et al.*[64]

The velocity and vorticity fields of the advecting vortex ring pre- and post-impact are presented in Figure 14.4 for a representative experiment. The gray region in each frame represents the clamped IPMC. The vortex ring core is

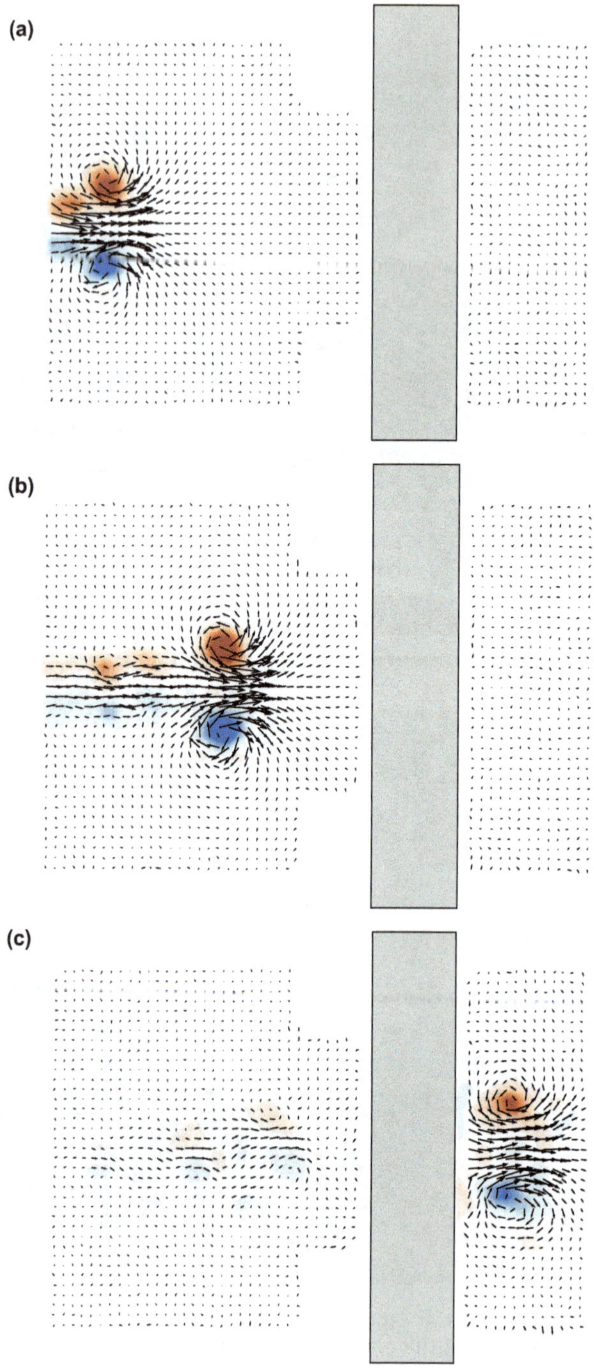

Figure 14.4 Velocity and vorticity fields (a,b) pre- and (c) post-impact. Velocity is presented as vectors with vorticity contours overlaid. The gray blocks indicate the location of the IPMC clamp, which is not optically clear. Reproduced from Hu et al.[64]

evident in the contours of vorticity, appearing as regions of red and blue, indicting opposite signs of vorticity. As in the previous study, the vortex ring advects towards the IPMC through self-induction.

Trailing vorticity is found behind the vortex ring as a result of the formation process.[51] As seen in Figure 14.4c, a vortex ring exists post-impact, suggesting that the fluid passing through the annulus rolls up into another coherent structure of lower circulation.

The time traces of the displacement and current follow those presented in the previous study,[65] whereby just after impact the IPMC attains its maximum deflection, which is accompanied by a substantial increase in the short-circuit current. Likely due to the different impact geometries, we observe a modest deflection of the IPMC towards the approaching ring prior to impact. This was not observed in the previous study due to the relatively high stagnation pressure at the tip of the IPMC strip. A further difference between the response of the annulus and the strip is the drastically reduced deformation levels experienced by the annular structure, due to the increased stiffness associated with the 2D axisymmetric bending. Despite the differences in deformation, the energy conversion is within the same order of magnitude.

14.3 Insights from Modeling and Simulation

Here, we employ analytical modeling techniques and CFD to obtain further insights into the physics of the impact and associated energy transfer. Specifically, we report on a fully coupled potential flow-based fluid–structure interaction model for energy harvesting design[66] and a discussion of its associated limitations. To address some of these drawbacks, we develop a CFD framework for refined analysis of the vortex dynamics.[67,68]

14.3.1 Potential Flow Modeling

To gain insight into the impact dynamics of a vortex ring with a deformable structure, we develop a modeling framework in 2D in which the vortex ring is represented as a pair of point vortices and the IPMC strip is modeled as a Kirchhoff–Love plate undergoing cylindrical bending in an ideal fluid. As such, the circulation of the vortices remains constant during the impact, and no new coherent fluid structures are shed from the strip. Using the potential flow model, the plate is assimilated to a vortex sheet with zero net circulation, and the pressure field is expressed in terms of the configuration of the impacting vortices and the distributed vorticity of the sheet. The pressure field is in turn utilized to compute the impulsive hydrodynamic loading on the strip, which determines the strip deformation. As the structure deforms, the boundary condition on the fluid changes, thus influencing the flow physics, namely the propagation of the free vortices and the vorticity distribution on the strip.

The coupled partial differential equations are solved by projecting the strip deflection onto a basis of Chebyshev polynomials, which is afforded through the introduction of Lagrange multipliers to enforce kinematic boundary conditions on the strip. A similar basis is used to project the vorticity distribution on the strip and ultimately transform the problem into a set of coupled ordinary differential equations in the complex domain.

Snapshots of the interaction between the strip and the vortex pair are detailed in Figure 14.5. In this figure, the left column depicts the vortex positions (dots) and the configuration of the strip (black line). The center column displays a magnified view of the strip shape, while the right column shows the hydrodynamic loading on the strip, which is computed from the pressure jump across the two faces. Similarly to our experimental findings,[65] the stagnation pressure induced by the approaching vortices results in deflection of the strip away from the vortices. When the pair is in close proximity to the strip, the magnitude of the hydrodynamic loading dramatically increases. The model not only anticipates the drastic temporal variation in the loading prior to impact, but also predicts a distinct spatial pressure localization. Specifically, the hydrodynamic loading can alternate sign along the strip, with the strip being pushed away from the vortices in the region between the pair and being sucked towards them in the vicinity of their cores. Unlike the experimental observations, where vorticity can be generated along the IPMC and thus influence the vortex dynamics, resulting in the vortex ring breakdown, the vortex pair in the potential flow model simply passes around the plate and continues to advect away indefinitely, albeit possibly at a different speed.

The maximum tip deflection is shown in Figure 14.6a *versus* circulation of the vortices. The amplitude of vibration increases monotonically with circulation due to the larger induced pressure field associated with higher circulation vortices. For lower circulation strengths, the strip vibration is primarily along the fundamental mode shape of the cantilever, though higher modes are excited at higher circulation values. The strip energy, including kinetic and strain energies, is presented in dimensionless form in Figure 14.6b for various values of vortex circulation as a function of time. As the vortices approach, the energy of the strip increases due to the transfer of energy through the hydrodynamic loading, and as such an increase is modulated by the vortex circulation. As the vortex circulation increases, the strip undergoes larger deflection, which results in more energy transfer. After the interaction, the strip continues to vibrate in the fluid, maintaining a constant mean energy level, due to the absence of fluid viscosity and structural damping.

We note that in comparison with experiments,[65] the potential flow model matches well in the early stages of the impact, where interaction between the advecting vortices and the vortex sheet co-located with the strip is relatively weak. However, viscous effects become important when the vortices are near the plate, and thus the potential flow model assumptions lose accuracy.

A similar framework has been proposed to model the experimental study of the interaction between a vortex ring and an annulus.[64] However, due to

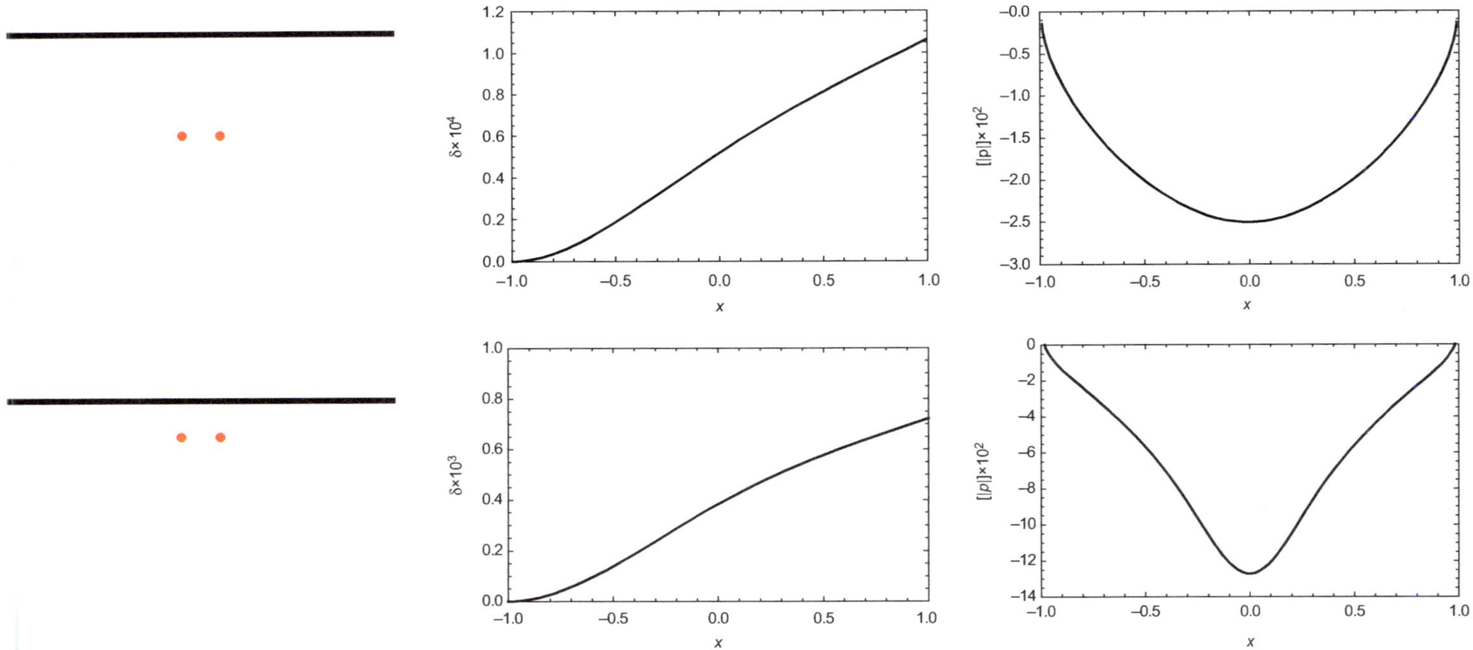

Figure 14.5 Snapshots of the vortex pair location with respect to the strip (left column), the non-dimensional strip shape $\delta(x,t)$ (middle column), and the non-dimensional pressure difference across the strip $[|p|](x,t)$ (right column). Time is increasing moving down the rows.
Reproduced from Peterson and Porfiri[66] (details on the non-dimensionalization process can be found therein).

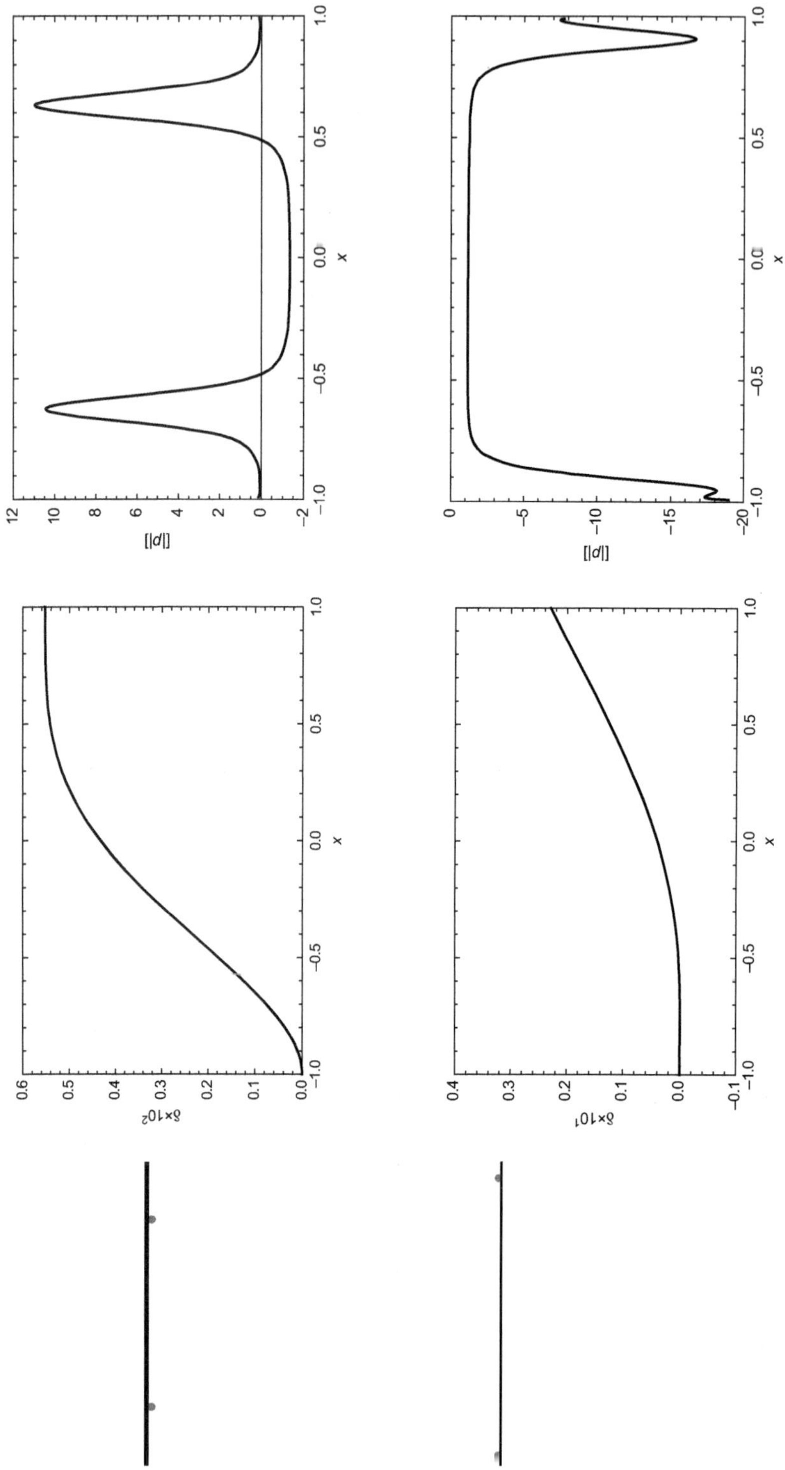

Figure 14.5 (Continued)

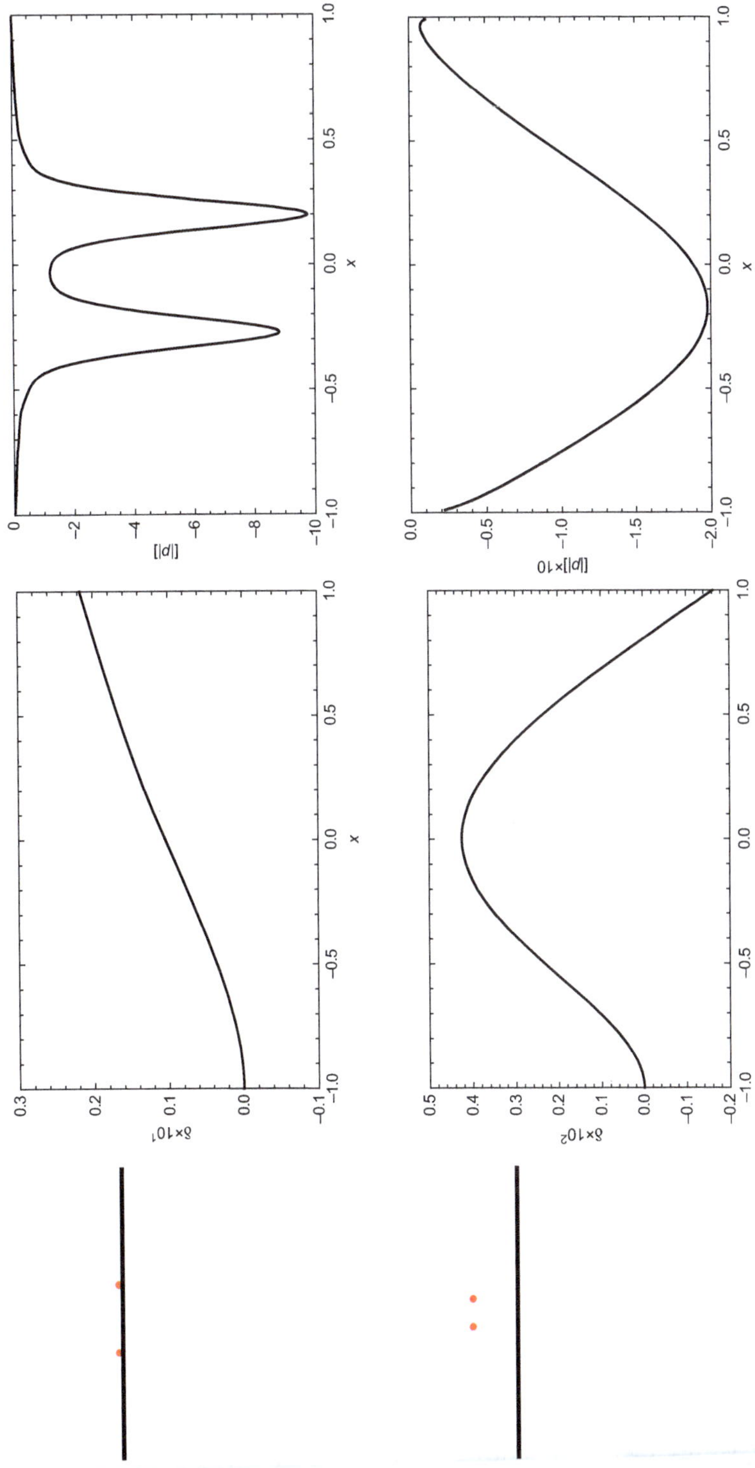

Figure 14.5 (Continued)

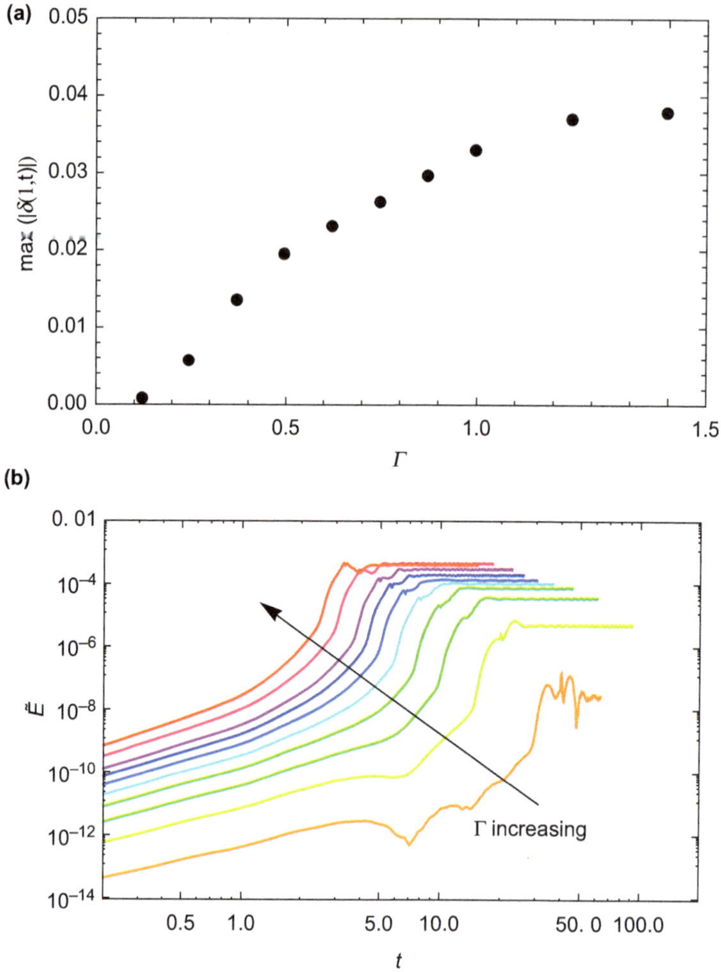

Figure 14.6 (a) Maximum non-dimensional strip deflection and (b) non-dimensional strip energy for various values of vortex strength. Reproduced from Peterson and Porfiri.[66]

the additional complexity of the coherent fluid structure and the minute deflection of the annulus observed in the experiments, the bi-directional coupling between the fluid flow and the annulus deformation has been neglected. Thus, vortex propagation is studied assuming a rigid annulus and the resulting hydrodynamic loading is independently imposed to predict the short-circuit current through the IPMC electrodes.

14.3.2 CFD

In an effort to overcome the limitations of the potential flow solution in resolving vortex dynamics during and after impact, we develop a CFD modeling framework.[67,68] While the point vortex model is amenable to analytical treatment through the potential flow solution, it contains a singularity in the velocity field at the vortex core, hampering CFD. As such, we replace the vortex pair with a Lamb dipole[69] for numerical analysis. Simulations are performed using the icoFoam incompressible flow solver in the open-source CFD package OpenFOAM, which solves the 3D Navier–Stokes equations using the finite volume method. Structural deformations of the plate are modeled by discretizing Kirchhoff–Love plate equations using a central finite difference approximation.

The vortex dynamics of a dipole impacting the tip of a rigid semi-infinite plate are presented in Figure 14.7 for a dipole Reynolds number of $Re = 1500$, where the Reynolds number is based upon the initial dipole convection speed and radius. Initially, all of the vorticity in the domain is contained within the dipole. Prior to impact, positive vorticity is observed at the plate tip, while negative vorticity is generated in the thin boundary layer on the plate face. This induced vorticity is a viscous effect not captured in the potential flow model. The importance of this induced vorticity is evident during the impact, in which the top half of the initial dipole pairs with the induced vorticity along the wall, resulting in the formation of a secondary dipole. Similarly, the lower half of the initial dipole pairs with the vorticity shed from the tip, forming another secondary dipole. These secondary dipoles follow circular trajectories that result in subsequent impacts with the wall. Viscous dissipation slowly diffuses the vorticity within the domain, in contrast with the potential flow model in which vortices advect away from the plate, maintaining their circulation.

To elucidate the hydrodynamic loading on the wall, we compute the time-dependent tip load by integrating the pressure jump in a region that extends from the free end to one dipole initial radius. This resultant force is presented in Figure 14.8 for three dipole Reynolds numbers. Initially, as the dipole approaches the wall, the force increases slowly, which is in agreement with the potential flow model. Upon impact with the wall, the resultant force rapidly increases, attaining a maximum value when the dipole center reaches the wall. During this primary impact, we observe that the force is relatively insensitive to the dipole Reynolds number. Subsequent deviations of the resultant force for varying Reynolds numbers after primary impact are a result of the complex vortex dynamics shown in Figure 14.7.

This vortex dipole/rigid wall impact study is extended to fully coupled fluid–structure interaction simulations by replacing the semi-infinite wall with a finite length Kirchhoff–Love plate undergoing cylindrical bending.[68] As the dipole approaches the plate, it deflects slowly, followed by a rapid deflection upon impact, which corresponds to a rapid increase in the strain energy. Similar to the rigid wall case, the impact results in the formation of two secondary dipoles, which return for subsequent interactions with the plate. These secondary impacts result in additional peaks in the deflection and strain energy. In comparison with the rigid wall case, the vorticity induced along the plate during impact is reduced due to the plate compliance.

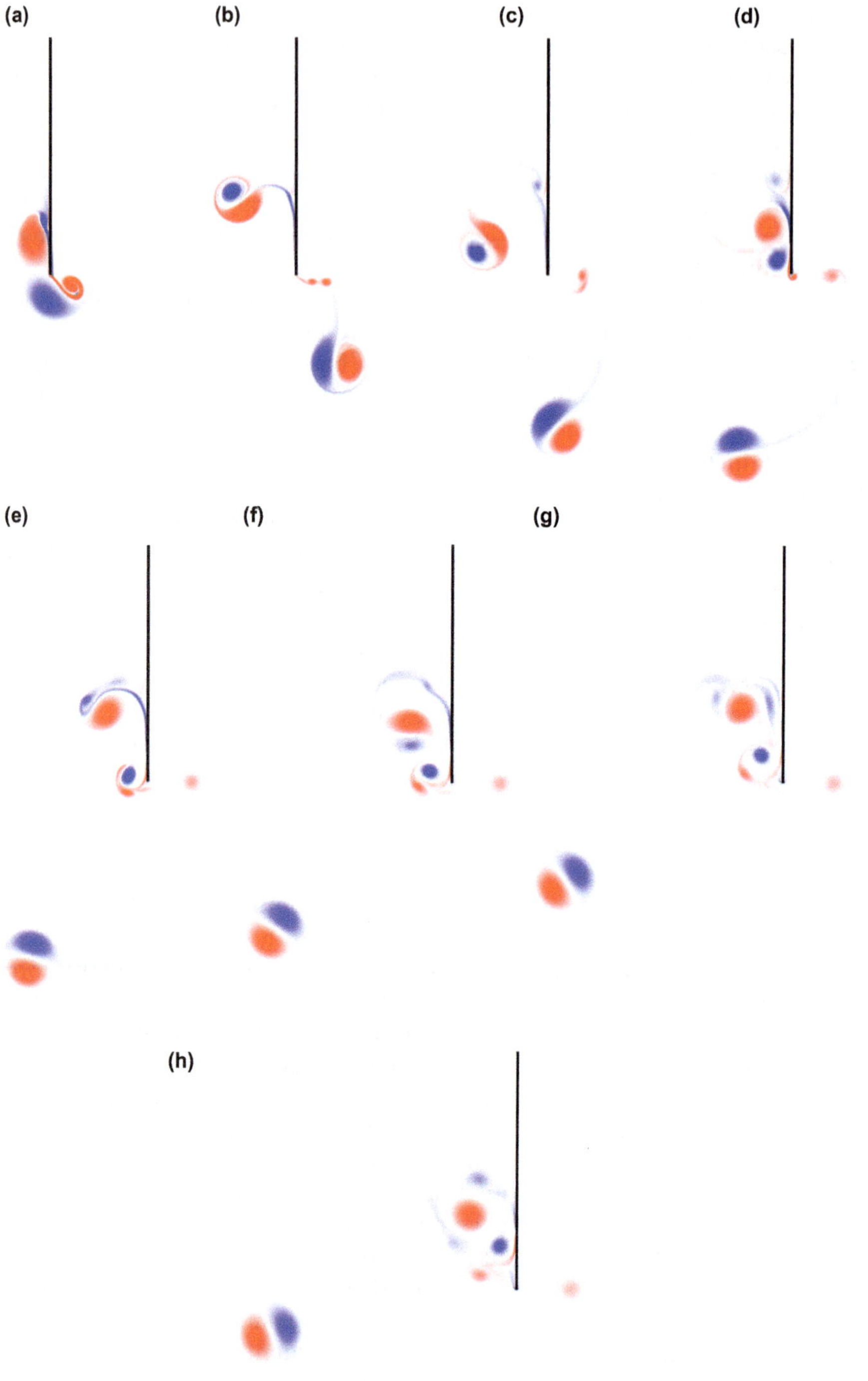

Figure 14.7 Vortex dipole evolution during impact with the tip of a semi-infinite rigid plate. Red and blue contours show positive and negative vorticity, respectively. Time is increasing from (a) to (h).
Reproduced from Peterson and Porfiri.[67]

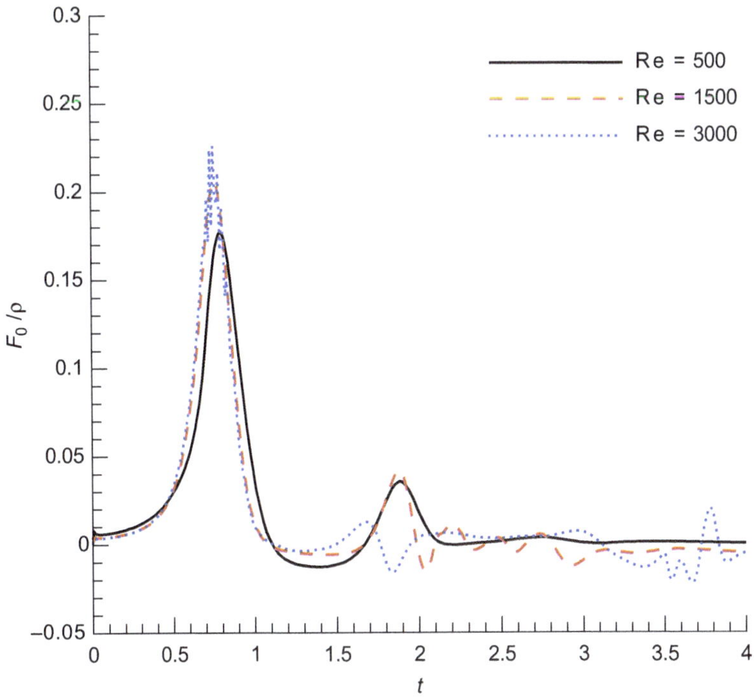

Figure 14.8 Force per unit depth on the tip of the rigid plate *versus* time for three dipole Reynolds numbers.
Reproduced from Peterson and Porfiri.[67]

Consequently, the secondary dipoles remain much closer to the plate and have a stronger influence on the hydrodynamic loading. The primary impact is largely independent of the Reynolds number, as found in the rigid wall case; however, the interaction with the secondary dipoles and the associated plate dynamics and strain energy are highly dependent on the Reynolds number.

14.4 Summary and Conclusions

In this chapter, we have examined the impulsive loading of IPMCs by coherent fluid structures for the design of small-scale fluid energy harvesters. Dedicated experimental setups for simultaneous mechanical, fluid, and electrical measurements have been integrated with analytical and numerical schemes to elucidate the physics of the energy exchange.

We have demonstrated the feasibility of extracting energy from self-propagating vortex rings in aqueous environments through impact with IPMC strips and annuli. PIV measurements indicate that most of the energy exchange takes place as the vortex ring reaches the IPMC, which results in a sudden deflection of the IPMC and simultaneous vortex breakdown. For the annular geometry, the IPMC deflections tend to be modest due to the nature of the mechanical deformation and fluid passing through the hole in the

IPMC rolling up into a second vortex ring of lower circulation that propagates away. For the strip geometry, large deflections are observed, along with complex vortex dynamics in the vicinity of the tip. The short-circuit current through the IPMCs is found to consistently follow the rapid deformation, occurring over a few milliseconds.

The fully coupled potential flow solution accurately predicts the early stage of the fluid–structure interaction, where the coupling between the impacting coherent fluid structure and induced vorticity on the IPMC is negligible. The model describes the impulsive hydrodynamic loading on the IPMC as a function of the locations of the vortices and the IPMC vibration. We predict the energy stored in the IPMC as the vortex pair advects away post-interaction and dissect the role of the vortex pair strength. CFD allows for resolving the vortex dynamics during the impact, which leads to the formation of secondary and tertiary coherent fluid structures that significantly contribute to the hydrodynamic loading experienced by the IPMC.

While we acknowledge that there is a significant gap between the power requirements of existing miniature electronic devices and the energy transduction capacity of IPMCs, our results offer compelling evidence for the feasibility of IPMC use in highly unsteady flow sensing applications, along with an experimental and theoretical framework to inform the design of energy harvesters.

Acknowledgements

This research was supported by the Natural Sciences and Engineering Research Council of Canada under grant number 386282-2010, the National Science Foundation under grant numbers CMMI-0745753, CMMI-0926791, and CBET-1332204, the Office of Naval Research under grant N00014-10-1-0988, and the Mitsui USA Foundation. The authors would like to thank Dr Youngsu Cha, Mr Jia Cheng Hu, Mr Eugene Zivkov, and Dr Serhiy Yarusevych, who have contributed to the research efforts summarized in this chapter.

References

1. L. Mateu and F. Moll, in *VLSI Circuits and Systems II, Pts 1 and 2*, ed. J. F. Lopez, F. V. Fernandez, J. M. LopezVillegas and J. M. DelaRosa, 2005, vol. 5837, pp. 359–373.
2. H. A. Sodano, D. J. Inman and G. Park, *Shock Vibration Digest*, 2004, **36**, 197–205.
3. B. E. White Jr., *Nat. Nanotechnol.*, 2008, **3**, 71–72.
4. A. Harb, *Renewable Energy*, 2011, **36**, 2641–2654.
5. J. W. Matiko, N. J. Grabham, S. P. Beeby and M. J. Tudor, *Meas. Sci. Technol.*, 2014, **25**.
6. N. S. Shenck and J. A. Paradiso, *IEEE Micro*, 2001, **21**, 30–42.
7. M. Renaud, P. Fiorini, R. Van Schaijk and C. Van Hoof, *Smart Mater. Struct.*, 2009, **10**, 16.

8. Y. Cha, M. Verotti, H. Walcott, S. D. Peterson and M. Porfiri, *Bioinspir. Biomim.*, 2013, **8**, 036003.
9. A. Delnavaz and J. Voix, *Smart Mater. Struct.*, 2014, **23**, 105020.
10. G.-T. Hwang, H. Park, J.-H. Lee, S. Oh, K.-I. Park, M. Byun, H. Park, G. Ahn, C. K. Jeong, K. No, H. Kwon, S.-G. Lee, B. Joung and K. J. Lee, *Adv. Mater.*, 2014, **26**, 4880.
11. J. J. Allen and A. J. Smits, *J. Fluids Struct.*, 2001, **15**, 629–640.
12. G. W. Taylor, J. R. Burns, S. M. Kammann, W. B. Powers and T. R. Welsh, *IEEE J. Oceanic Eng.*, 2001, **26**, 539–547.
13. H. D. Akaydin, N. Elvin and Y. Andreopoulos, *Exp. Fluids*, 2010, **49**, 291–304.
14. H. D. Akaydin, N. Elvin and Y. Andreopoulos, *J. Intell. Mater. Syst. Struct.*, 2010, **21**, 1263–1278.
15. H. D. Akaydin, N. Elvin and Y. Andreopoulos, *Smart Mater. Struct.*, 2012, **21**, 025007.
16. O. Goushcha, N. Elvin and Y. Andreopoulos, *Appl. Phys. Lett.*, 2014, **104**, 021919.
17. S.-D. Kwon, *Appl. Phys. Lett.*, 2010, **97**, 164102.
18. S. Michelin and O. Doare, *J. Fluid Mech.*, 2013, **714**, 489–504.
19. Y. Cha, L. Shen and M. Porfiri, *Smart Mater. Struct.*, 2013, **22**, 055027.
20. A. Erturk and G. Delporte, *Smart Mater. Struct.*, 2011, **20**, 125013.
21. K. A. Cunefare, E. A. Skow, A. Erturk, J. Savor, N. Verma and M. R. Cacan, *Smart Mater. Struct.*, 2013, **22**, 025036.
22. R. Panciroli and M. Porfiri, *Int. J. Impact Eng.*, 2014, **66**, 18–27.
23. K. J. Kim and M. Shahinpoor, *Smart Mater. Struct.*, 2003, **12**, 65–79.
24. M. Shahinpoor, Y. Bar-Cohen, J. O. Simpson and J. Smith, *Smart Mater. Struct.*, 1998, **7**, R15–R30.
25. M. Shahinpoor and K. J. Kim, *Smart Mater. Struct.*, 2001, **10**, 819.
26. M. Shahinpoor and K. J. Kim, *Smart Mater. Struct.*, 2004, **13**, 1362–1388.
27. M. Shahinpoor and K. J. Kim, *Smart Mater. Struct.*, 2005, **14**, 197–214.
28. C. Jo, D. Pugal, I. K. Oh, K. J. Kim and K. Asaka, *Prog. Polym. Sci.*, 2013, **38**, 1037–1066.
29. Y. Cha, F. Cellini and M. Porfiri, *Phys. Rev. E: Stat., Nonlinear, Soft Matter Phys.*, 2013, **88**, 062603.
30. Y. Cha and M. Porfiri, *J. Mech. Phys. Solids*, 2014, **71**, 156–178.
31. S. Nemat-Nasser, *J. Appl. Phys.*, 2002, **92**, 2899–2915.
32. T. Wallmersperger, A. Horstmann, B. Kroplin and D. J. Leo, *J. Intell. Mater. Syst. Struct.*, 2009, **20**, 741–750.
33. G. Del Bufalo, L. Placidi and M. Porfiri, *Smart Mater. Struct.*, 2008, **17**, 045010.
34. J. Brufau-Penella, M. Puig-Vidal, P. Giannone, S. Graziani and S. Strazzeri, *Smart Mater. Struct.*, 2008, **17**, 015009.
35. R. Tiwari, K. J. Kim and S. M. Kim, *Smart Struct. Syst.*, 2008, **4**, 549–563.
36. M. Aureli, C. Prince, M. Porfiri and S. D. Peterson, *Smart Mater. Struct.*, 2010, **19**, 015003.
37. A. Giacomello and M. Porfiri, *J. Appl. Phys.*, 2011, **109**, 084903.
38. Y. Cha, C. N. Phan and M. Porfiri, *Appl. Phys. Lett.*, 2012, **101**, 094103.

39. F. Cellini, Y. Cha and M. Porfiri, *J. Intell. Mater. Syst. Struct.*, 2014, **25**, 1496–1510.
40. F. Cellini, C. Intartaglia, L. Soria and M. Porfiri, *Smart Mater. Struct.*, 2014, **23**, 045015.
41. F. Cellini, J. Pounds, S. D. Peterson and M. Porfiri, *Smart Mater. Struct.*, 2014, **23**, 085023.
42. I. S. Sullivan, J. J. Niemela, R. E. Hershberger, D. Bolster and R. J. Donnelly, *J. Fluid Mech.*, 2008, **609**, 319–347.
43. K. Shariff and A. Leonard, *Annu. Rev. Fluid Mech.*, 1992, **24**, U235–U279.
44. C. H. K. Williamson, *Annu. Rev. Fluid Mech.*, 1996, **28**, 477–539.
45. M. R. Head and P. Bandyopadhyay, *J. Fluid Mech.*, 1981, **107**, 297–338.
46. F. E. Fish and G. V. Lauder, *Annu. Rev. Fluid Mech.*, 2006, **38**, 193–224.
47. J. O. Dabiri, *Annu. Rev. Fluid Mech.*, 2009, **41**, 17–33.
48. P. S. Krueger and M. Gharib, *Phys. Fluids*, 2003, **15**, 1271–1281.
49. M. Aureli, M. E. Basaran and M. Porfiri, *J. Sound Vibration*, 2012, **331**, 1624–1654.
50. S. J. Kline, W. C. Reynolds, F. A. Schraub and P. W. Runstadl, *J. Fluid Mech.*, 1967, **30**, 741–773.
51. M. Gharib, E. Rambod and K. Shariff, *J. Fluid Mech.*, 1998, **360**, 121–140.
52. S. G. L. Smith and R. J. Nagem, *Regular Chaotic Dyn.*, 2013, **18**, 194–201.
53. L. M. Milne-Thomson, *Theoretical Hydrodynamics*, Dover Publishers, New York, 2011.
54. P. G. Saffman, *J. Fluid Mech.*, 1979, **92**, 497–503.
55. P. Orlandi, *Phys. Fluids A*, 1990, **2**, 1429–1436.
56. W. Kramer, H. J. H. Clercx and G. J. F. van Heijst, *Phys. Fluids*, 2007, **19**, 1–13.
57. A. K. Hinds, I. Eames, E. R. Johnson and N. R. McDonald, *J. Geophys. Res.: Oceans*, 2009, **114**, C06006.
58. M. Cheng, J. Lou and T. T. Lim, *Phys. Fluids*, 2014, **26**, 103602.
59. A. R. Cieslik, R. A. D. Akkermans, L. P. J. Kamp, H. J. H. Clercx and G. J. F. van Heijst, *Eur. J. Mech. B Fluids*, 2009, **28**, 397–404.
60. H. J. H. Clercx and C. H. Bruneau, *Comput. Fluids*, 2006, **35**, 245–279.
61. G. H. Keetels, W. Kramer, H. J. H. Clercx and G. J. F. van Heijst, *Theor. Comput. Fluid Dyn.*, 2011, **25**, 293–300.
62. P. Orlandi and R. Verzicco, *J. Fluid Mech.*, 1993, **256**, 615–646.
63. G. Arevalo, R. H. Hernandez, C. Nicot and F. Plaza, *Phys. Fluids*, 2010, **22**, 053604.
64. J. C. Hu, Y. S. Cha, M. Porfiri and S. D. Peterson, *Smart Mater. Struct.*, 2014, **23**, 074014 .
65. S. D. Peterson and M. Porfiri, *Appl. Phys. Lett.*, 2012, **100**, 114102.
66. S. D. Peterson and M. Porfiri, *J. Intell. Mater. Syst.Struct.*, 2012, **23**, 1485–1504.
67. S. D. Peterson and M. Porfiri, *Phys. Fluids*, 2013, **25**, 093103.
68. E. Zivkov, S. Yarusevych, M. Porfiri and S. D. Peterson, *J. Fluid. Struct.*, 2015, in press.
69. H. Lamb, *Hydrodynamics*, Dover Publishers, New York, 1945.

CHAPTER 15

Miniature Pump with Ionic Polymer Metal Composite Actuator for Drug Delivery

JIAQI WANG, ANDREW McDAID, RAJNISH SHARMA, WEI YU AND KEAN C. AW*

Mechanical Engineering, University of Auckland, New Zealand
*Email: k.aw@auckland.ac.nz

15.1 Introduction

Over the last few decades, the fast development in micro-technologies has led to the growing importance of microfluidics. Miniature pumps are the key components in microfluidic systems. They are widely used in many applications such as lab-on-a-chip, micro-total analysis systems and microdosage systems. The first miniature pump was fabricated in the 1980s.[1] Since then, a number of miniature pumps have been presented and many are still under development. The most important part for miniature pumps is the actuating mechanism because it is directly related to factors such as flow rate, efficiency, driving source and cost. Currently, the most common driving mechanisms of miniature pumps are piezoelectric, electromagnetic, thermopneumatic, electrostatic and so on. They all have their own advantages and disadvantages. However, the majority of pumps are relatively expensive; the fabrication processes are generally quite complicated.[2] To obtain a cost-effective and fabrication-optimized design, a simple valveless pump driven

RSC Smart Materials No. 18
Ionic Polymer Metal Composites (IPMCs): Smart Multi-Functional Materials and Artificial Muscles, Volume 2
Edited by Mohsen Shahinpoor

Published by the Royal Society of Chemistry, www.rsc.org

by ionic polymer metal composites (IPMCs) and controlled with the iterative feedback tuning (IFT) technique will be presented in this chapter.

15.2 IPMC Fundamentals

IPMCs, also known as ionic conducting polymer gel films, are a type of electroactive polymer (EAP) that are lightweight, soft, flexible and known as "artificial muscles" because EAPs convert electric energy into mechanical energy.[3] With low excitation voltages (typically less than 5 V), IPMC strips can produce relatively large displacements but with limited force outputs and slow reactions. Its properties make it a promising candidate for building artificial devices such as biologically inspired robots.[4] IPMC actuators are best suited for reciprocating motion because they can apply force in one direction for only a short period. There are also many other issues concerning the use of IMPC actuators in real-world applications, which need to be taken into account. Some of their properties may restrict their performance in one application but might be feasible for another. In order to utilize IPMC actuators, one has to find a way to benefit from their unique shape and behaviour.[4]

An IPMC is typically a thin elastic strip made from either Nafion® or Flemion®, which has both sides coated with highly conducting metal (normally platinum or gold). These metals act as the contacts or, more precisely, electrodes. Currently, the two main suppliers for Nafion® and Flemion® are DuPont, USA, and Asahi Glass, Japan, respectively.[5] Taking Nafion® as an example, it is a perfluorinated polymer containing hydrogen (or metal) sulfonate functional groups.[6] It is commonly used as a polymer matrix in IPMC applications. A short operation time and a low generative blocking force are regarded as the major drawbacks of conventional Nafion-based IPMCs. The short operation time of IPMCs in dry condition is due to the loss of the inner water through natural evaporation, leakage due to surface expansion and continuous electrolysis during operation. Various attempts have been made to improve the performance of IPMCs in terms of durability as well as blocking force.[7] When electric voltage is applied to the both sides of the IPMC, the strip bends toward the anode (+). This is due to the expansion along one side and contraction along the other the side of the polymer membrane. By applying an electric field across the hydrated membrane, mobile cations are attracted to the cathode, which in turn attracts water molecules trapped within the membrane. The accumulation of water on one side of the polymer causes expansion, therefore producing the bending stress towards the anode.[3,5,6]

15.3 Advantages of IPMCs and Current Applications

In general, EAPs are novel actuation and sensing materials, which have promising applications in robotics and biomedical systems.[9] IPMC materials have a number of unique properties that make them more favourable than traditional actuators, as well as other EAPs:[1]

- *Lightweight and small size*. A typical IPMC is a thin film between 0.2 and 2 mm thick. Its small size and good scalability are perfect in miniature and micro-applications.
- *Low actuation voltages*. These are typically between 1 and 5 V, which is much lower than conventional actuators that are commonly used in biological applications.
- *Large displacements*. A bending angle greater than 360° is achievable with sufficient length.
- *Biocompatibility*. This is an important feature for IPMCs and makes it possible for human implants, biomimetics and other similar products.
- *Flexible and soft*. Since the structural material is a polymer gel, it has the basic properties of a polymer.
- *Environmentally friendly*. Operation is possible in water, at low temperatures and in vacuum.
- *Noiseless*. Unlike electric motors or pneumatics, silent actuation makes IPMCs very useful in many applications in which quiet operation is required.
- *Direct energy conversion*. No components other than a pair of electrodes are required to convert electrical energy into mechanical motion.
- *Bi-directional operation*. By changing the polarity of the applied voltage, the deflection can be controlled in both directions.

With these useful properties of IPMCs, much research has been done to utilize them in real-world applications such as biomedical devices, flapping robots, scout robots, grippers and micromanipulators, dust wipers, vibration dampers and linear actuators.[4]

15.4 IPMC Control Techniques

Although there is currently no general model to define IPMC actuators, the knowledge of electrochemical mechanical behaviour allows the control of their tip force and displacement. However, the material properties of IPMCs are highly dependent on factors such as hydration level, manufacture, actuation time and clamping status. This means that the controller must be robust and self-adaptive.[8] A number of techniques have been put into practice to control the output displacement of an IPMC actuator. Open-loop (OL) controllers have been implemented based on the models for the quasi-static response of the materials.[9] The performance was improved by using the inverse Preisach operator to model quasi-static hysteresis separately from the linear dynamics.[9] However, due to the highly nonlinear behaviour of IPMC properties described above, performance under OL control is still very unstable and thus not desirable for most applications. Closed-loop control is considered to be the best method to control IPMCs.

Some classical proportional, proportional-integral (PI) or proportional-integral-derivative (PID) controllers have been successfully implemented using linear approximate models of the actuator[10–13] or linear quadratic regulator theory with more optimized results.[13] An H-infinity controller that seemed to be better than PI control was also developed using a model optimized for control purposes.[14] In addition, force control was achieved using hybrid or impedance controllers.[10–12] With the feedback from a laser sensor or precision force sensor, these controllers were capable of controlling the tip displacement of an IPMC strip for both step and periodic inputs. However, because of the time-varying characteristics of IMPCs, these controllers are still not good enough for long operation applications as they do not continuously tune themselves.

15.4.1 Development of Miniature Pump Technology

Microelectromechanical systems (MEMS) technologies have brought to the world various types of sensors and actuators by allowing nonelectrical components onto microchips. In the early years of MEMS development, fluidic devices were the first components that were realized in microscale using silicon technology.[1] Microfluidics consists of the design and development of miniature devices, which manipulate small amounts of fluids (10^{-9}–10^{-18} litres) using small channels. By miniaturizing the size, the channels can be scaled down to just a few micrometres, which facilitates controlling and transferring tiny volumes of fluids.[15] Although studies on microfluidic devices, fabricated with MEMS technology, has its origin almost 40 years ago, it is still a very actively research topic.[16] Over the last decades, the fast development of genomics, proteomics, life sciences and the discovery of new drugs led to the growing importance of microfludic systems.[1,17] For example, micro-total analysis systems (μTAS) in chemical and biological analysis and detection, implantable drug delivery systems for high-precision flow control, microchips integrated in computers to circulate coolant for cooling, micro-pumping systems in portable fuel cells, as well as impeller systems for blood flow regulation all need a fluid delivery unit to transfer the fluid from a reservoir to the target place with accuracy and reliability.[15] Numerous microdevices were developed such as filters, mixers, reactors and separators.[1] Different functions of microfluidic operations, such as pumping and mixing, have to be integrated into a single lab-on-a-chip.[17] New effects such as electrokinetic effects, acoustic streaming, magneto-hydrodynamic effects, electrochemical effects and more, which previously were neglected in macroscopic applications, now gain their importance at the microscale.[1,15]

The most common components in microfluidic devices are flow sensors, microvalves and miniature pumps. According to the definition of "MEMS", miniaturized pumping devices fabricated by micromachining technologies are called "miniature pumps".[18] As it was said that the miniature pump is the beating heart of microfluidics, miniature pumps serve as the actuation source and have been the subject of a number of research projects since the early years of MEMS development.[15] One of the very first miniaturized pumps was patented back in 1975 and was fabricated using conventional techniques, and it was not until 1984 that a miniature pump based on silicon microfabrication technologies was patented by Smits[10] and published later in 1990.[10] The miniature pump designed by Smits was a peristaltic pump consisting of three active valves actuated by piezoelectric discs. The device was primarily developed for use in controlled insulin delivery systems.[18]

In contrast to other MEMS devices, miniature pumps are components with a large variety of operating principles. Similar to other MEMS applications, the first approach made by researchers was to move away from the well-known principles in the macroscale by micromachining.[2] Generally, miniature pumps can be classified into two categories—mechanical and non-mechanical—which is based on whether the pump converts external mechanical energy or non-mechanical energy directly into kinetic energy.[2,15,18] Electrical, magnetic, thermal, optical and acoustic energies have been used for transfer into mechanical energy in miniature pump applications.[15] Mechanical pumps usually contain moving parts such as check valves, oscillating membranes or turbines for delivering a certain amount of fluid in each pump cycle, while non-mechanical pumps achieve pumping by adding momentum to the fluid, which is done by converting another energy form into the kinetic energy.[1] Under this system, mechanical pumps can be further divided into displacement pumps (also known as "reciprocating miniature pumps")[2] and dynamic pumps according to whether the mechanical energy is added intermittently to increase pressure to move the fluid or constantly to increase the velocity of the fluid.[15]

Displacement pumps are the most common type of mechanical miniature pump that consist of a pump chamber closed with a flexible diaphragm. A schematic illustration of diaphragm type mechanical miniature pump is shown in Figure 15.1.

Fluid flow is achieved by the oscillatory movement of the actuator diaphragm, which creates under- and over-pressure transients (Δp) in the pump chamber. During the so-called "supply phase", under-pressure in the pump chamber causes fluid to flow into the pump chamber through the inlet valve. During the following "pump phase", over-pressure in the pump chamber transfers the fluid out of the pump chamber through the outlet valve.[2,18] The valves at the inlet and outlet are in place to prevent reverse flow in the respective pump phases. The medium (*i.e.* liquid or gas) is obviously delivered in a series of discrete amount. The amount depends on the actuator stroke volume ΔV (*i.e.* its net volumetric displacement during one cycle).[2]

Based on the understanding of new fluid motion phenomena, other non-mechanical miniature pumps have also been developed. However, their operation and performance are dependent on the properties of the pumping fluids and surface material. In most cases, non-mechanical miniature pumps cannot generate high flow rates, driving pressures and fast response

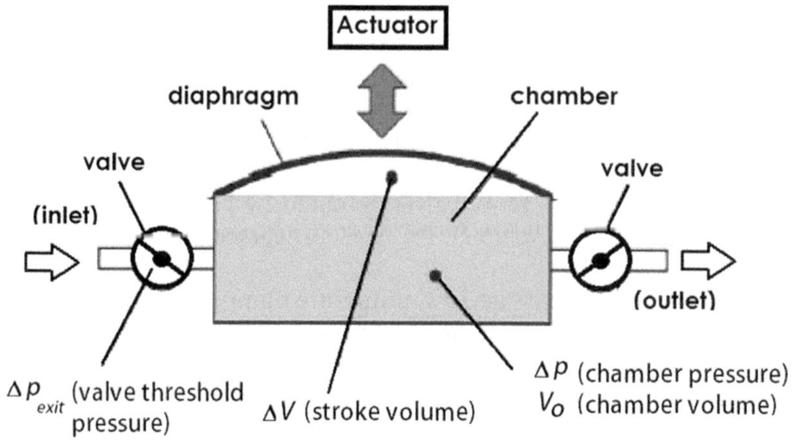

Figure 15.1 Schematic illustration of a diaphragm type miniature pump.

times, but they are very suitable for small regulation and precise control in the μTAS.[15] In terms of inlet/outlet mechanisms, categories of miniature pumps are also divided into with-valve miniature pumps and without-valve (or valveless) miniature pumps.[19] Early development of miniature pumps always involved check valves. The leak rate of these check valves must be small compared to the pump flow rate (which means in the nl/min range). However, long-standing problems related to sedimentation or wear should also be considered.[16] On the other hand, valveless miniature pumps, using nozzle/diffuser elements, are easily created in small sizes and can avoid the wear and fatigue of moving parts.[19] Figure 15.2 shows a miniature pump with conical nozzle/diffuser elements. This is similar to other membrane pumps with a pump chamber in which a certain pressure and volume stroke can be generated. The check valves, however, are replaced by diffuser/nozzle elements having no movable parts, but yet showing fluid rectification.[16] As shown in Figure 15.2, the diaphragm bends upward in the pump mode, and downward in the supply mode. In the pump mode, the large and small solid arrows represent outlet and inlet flows, respectively, while in the supply mode, the large and small dotted arrows illustrate flows through the inlet and outlet parts, respectively.[19]

Miniature pumps are very attractive devices since they can be used for dispensing therapeutic agents, cooling microelectronic systems, development of μTAS, propelling microspacecraft, *etc.* For such a variety of applications, many types of miniature pumps have been developed. At the same time, drawbacks of the various miniature pump concepts have been revealed. The overall commercialization process is still in its infancy. Low cost and simple fabrication designs and good performance reproducibility are definitely the next goals to be achieved. There is still much to learn about the fundamentals of miniature pumps and other microfluidic components and systems.[2,19]

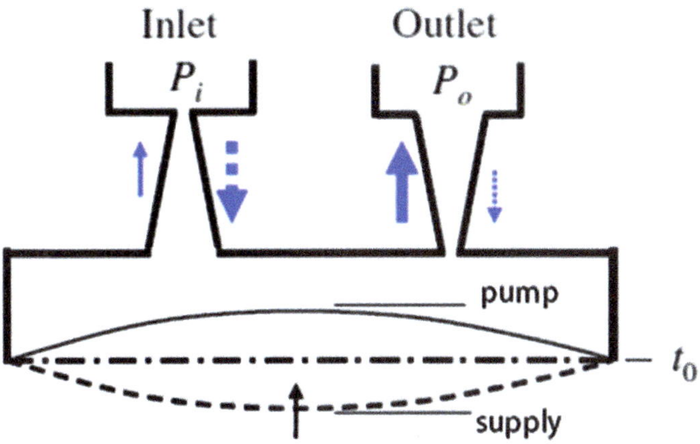

Figure 15.2 A schematic diagram of a membrane miniature pump with nozzle/diffuser elements.

15.4.2 Overview and Discussion of Miniature Pump Actuation Mechanisms

Ideally, actuators are easy to construct and can provide large force, have large strokes, have fast response times, consume little power and have low thermal losses. The choice of whether to use an integrated or an external actuator depends on the specific requirements of the application. For example, to achieve large strokes for higher flow rates, the large force and displacement of external actuators is desirable. However, the trade-off of using an external actuator is the large physical size. During the last decade, in attempting to meet these various requirements, such as working liquid, maximum pumping rate or back pressure and power consumption limitations, a number of miniature pumps have been designed and implemented.[15]

Piezoelectric actuation was the first and by far the most popular mechanism for miniature pump applications.[2,20] It is a very attractive concept, as it provides a simple structure, a comparatively high actuation force and a fast mechanical response. Moreover, commercial PZT (lead zirconate titanate) material is readily available. However, the fabrication process of integrating PZT materials with silicon is complicated. The high actuation voltage and small stroke are also regarded as the major drawbacks.[2,15,19] The other actuation mechanisms seem to be less popular in miniature pump applications due to disadvantages such as the long response time in shape memory alloys, the large size of the electromagnetic and pneumatic actuations, the high temperatures and high power consumption in thermos-pneumatic actuations and the low output force and small stroke in electrostatic actuations,[20,21] whereas some others effects, like bimetallic, magnetostrictive effects or optical effects, are rarely found.[2]

The excellent properties of EAP materials have drawn much attention from researchers when developing miniature pumps. IPMCs are one of the most popular polymer composites. When applied to miniature pumps, they have advantages such as a low driving voltage, reasonable responses and biocompatibility. In addition, they can work in aqueous environments. Much work has been done recently to develop IPMC miniature pumps in the known literature. Guo *et al.* designed an IPMC miniature pump with active valves made of IPMCs.[22] A maximum flow rate of 37.8 μl/min was recorded. However, the IPMC valves of this miniature pump are considered to be very complicated in terms of fabrication and control. The sealing of the IPMC valves might be another problem due to their rough surfaces and the hysteresis behaviour of IPMC material. Pak *et al.* presented a valveless IPMC miniature pump, which reached a maximum flow rate of 9.97 μl/min.[23] They also reported drawbacks such as durability and sealing of the IPMC diaphragm. In 2006, Lee *et al.*[19] proposed a design and flow rate estimation of a valveless IPMC miniature pump. The IPMC diaphragm of this miniature pump was designed to be made from Nafion®. Later, Nguyen *et al.*[21] reported a flap valve miniature pump with a multilayered IPMC as the actuator. Experiments showed a flow rate of up to 760 μl/min at a 3 V actuation voltage. The miniature pump was tested and found to be suitable for running for longer than 1 hour at an applied voltage of 1.5 V. Unlike Lee *et al.*, this study used an IPMC diaphragm fixed by a flexible polydimethylsiloxane (PDMS) polymer instead of Nafion®. This design was aimed at increasing the degrees of freedom of movement of the diaphragm while covered entirely with electrode. With Nguyen *et al.*'s design, the IPMC material was integrated into the PDMS layer, thus restricting the total diaphragm stroke.

More recently, Santos *et al.*[24] introduced a new concept of IPMC diaphragms in a miniature pump design. The electrodes were covered at the centre of the diaphragm. A maximum flow rate of 481.2 μl/min was measured. However, this design obviously decreased the durability of the IPMC material, as most of the diaphragm was covered by the electrodes. The size of the miniature pump was also as large as 80 mm. Overall, these designs of IPMC-driven miniature pumps require the IPMC to be an actuator as well as a diaphragm that is fixed or attached to the chamber. At least one side of the actuator is not in contact with liquid. This can cause water loss from the IPMC and therefore decrease its durability, rendering it incapable of running continuously over an extended period. Moreover, fabrication of IPMC membrane actuators is quite complex and relatively expensive.[15,18,20]

15.4.3 Advantages of IPMCs for Drug Delivery Miniature Pumps

Miniature pumps for drug delivery applications must meet basic requirements, which are biocompatibility, actuation safety, desired and controllable flow rate, small chip size and low power consumption. The biocompatibility of MEMS based miniature pumps is becoming increasingly important and is regarded as a key requirement for drug delivery systems. Biocompatibility is defined as "the ability of a material to perform with an appropriate host response in a specific application". As miniature pumps in drug delivery systems can be implanted inside the human body, therefore the materials used for fabrication must strictly meet biocompatibility and bio-stability requirements.[18]

Recently, IPMCs have attracted considerable attention due to their large deformation at low applied voltages, low power consumption and relatively short response time characteristics. Their small size and biocompatibility make them capable of human implantation. Even though IPMCs can operate well in dry environments when fabricated as self-contained encapsulated actuators, they can perform extremely well in a humid environment. Thus, IPMCs are considered to be promising candidates for miniature pump applications, particularly for use in a humid environment such as inside a drug delivery miniature pump.[20]

15.4.4 Design and Fabrication of Miniature Pumps

The fabrication processes of current miniature pumps are relatively complex, requiring special equipment and techniques, hence increasing the cost of these devices. For these reasons, a low-cost miniature pump was designed and prototyped based on a simple fabrication technique structure. The ultimate aim is to produce a miniature pump using a simple fabrication technique with an IPMC actuator that can be reliably controlled.

The most widely used miniature pump actuator is piezoelectric based. It usually produces high actuation forces and fast mechanical responses, but they generate relatively small stroke volumes and need very high actuation voltages,[25] which is energy consuming and not safe for some applications, such as a human wearable devices. Thermo-pneumatically actuated diaphragms requires low input voltages while generating higher pump rates; however, high power consumption and long thermal time constants are the restrictions to their wide application. Electrostatically actuated diaphragms have good performance in terms of response times, MEMS compatibility and low power consumption, but they also have drawbacks of small stroke volumes, performance degradation and high input voltages. Electromagnetically actuated diaphragms come with fast response times, but they have disadvantages of bad MEMS compatibility and high power consumption.[25] IPMC-driven diaphragms have merits of low input voltages, large stroke volumes and low power consumption, which make them promising candidates for miniature pump application.

In most pumps, mechanical check valves were usually used for the inlet and outlet ports, either with membranes or flaps. The process of designing and fabricating this type of valve is very complex. The pump itself may suffer from problems such as high pressure drop across the valves.[26] Some crucial properties like backward flow and switching speed have to be controlled precisely in order to achieve a working miniature pump. Moreover, wear and fatigue can be critical issues, especially in polymer-fabricated devices.[2] This may result in reduced lifetimes and reliabilities.[26] There is also the risk of valve blocking by even small particles, which instantly degrade the pumping performance or cause damage to sensitive fluids. This limits the application range of most valve-based miniature pumps to filtered media.[2] Therefore, there was a need for miniature pumps with no movable parts.

15.4.5 Valveless Miniature Pumps

The "valveless" miniature pump can eliminate these problems. The device was first introduced by Stemme and Stemme.[26] It uses "diffuser/nozzle" elements with flow-rectifying properties to provide similar function to a check valve. According to,[26] a maximum achievable forward–backward flow ratio of 2.23 was calculated for this type of "valve", which is sufficient for a pump. The working principle of this valveless miniature pump was shown earlier in Figure 15.2. A diffuser is a diverging duct and a nozzle is a converging duct. These ducts can be geometrically designed to have a lower pressure loss in the diffuser direction than in the nozzle direction for the same flow velocity.[26] When chamber volume increases (the supply mode), the inlet element acts as a diffuser with a lower flow restriction than the outlet element, which acts as a nozzle. Therefore, a larger volume is transported through the inlet into the chamber than through the outlet. On the other hand, during decreasing chamber volume (the pump mode), the outlet element acts as a diffuser with a lower flow restriction than the inlet element, which acts as a nozzle. This means that a larger volume is transported through the outlet out of the chamber than through the input. As a result, a net volume is transported (*i.e.* pumped) from the inlet side to the outlet side. Therefore, a complete pump cycle is achieved, despite the fact that the fluid travelled in both directions of the diffuse/nozzle element.

15.4.6 Miniature Pump Design

Figure 15.3 shows the final design for the valveless miniature pump. It consists of four Perspex layers (each 2 mm thick) and a latex diaphragm. The top layer is the cover. It also has a micro-channel inlet, which can be connected to an external reservoir. The second layer is the most critical part that contains the pump's micro-channel outlet and two nozzles/diffusers for the inlet and outlet. It requires precision machining to ensure the diffuser and nozzle dimensions are correct for proper operation. The pump chamber third is located in the Perspex layer. A latex film is directly attached (glued) to the third layer to cover the pump chamber at the bottom. A small piece of strong double-sided tape (1 mm×1 mm) of about 1 mm thickness is placed underneath the latex film at the centre. The IPMC tip is now attached to the latex diaphragm.

The contact area of the diaphragm to the IPMC is small, amplifying the exerting pressure, since the IPMC is only capable of providing limited force,

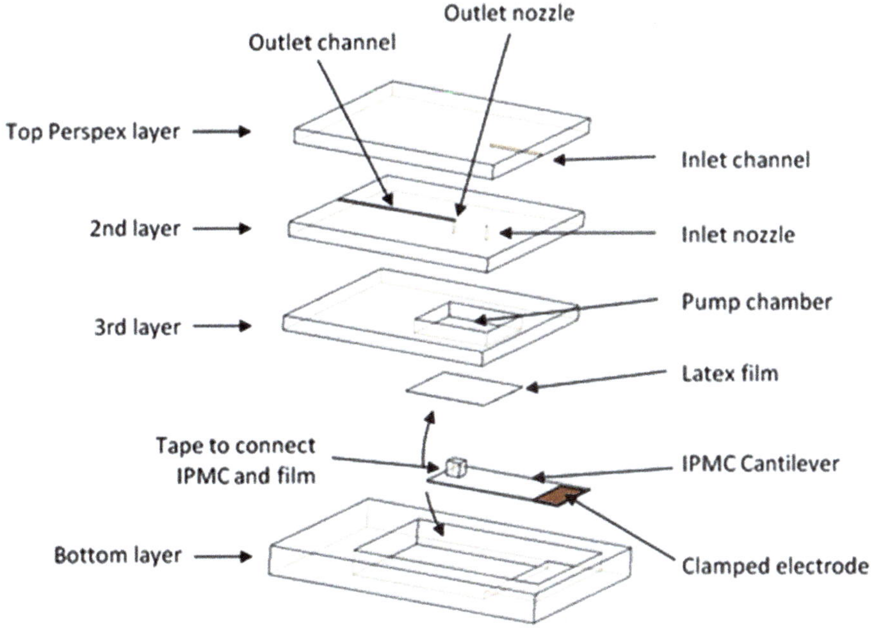

Figure 15.3 Exploded view of a miniature pump designed for dispensing drugs.[37]

typically up to 10 gf. To overcome the dehydration problem, the IPMCs should always be kept in a liquid. A bottom layer with a cavity to hold water is added. This layer is thicker than the other layers (4 mm instead of 2 mm) so that the cavity is sufficient to hold water while providing enough space for the IPMC strip to deflect. With this bottom layer, the clamping part is also submerged in water; thus continuous operation is now possible. In this prototype, the bottom layer is slightly larger than the rest of the pump so that the IPMC clamped section can also be completely submerged in water. A picture of the assembled pump prototype is shown in Figure 15.4.

Since this is a prototype only, the pump was not made to be very small. The overall size is 30 × 40 × 10 mm (length × width × height). The pump chamber is an 11 × 11 mm square of 2 mm in depth. The actuator used in this prototype is a Nafion®-based IPMC strip 18 mm long and 10 mm wide, with an average thickness of 0.22 mm. A pair of copper plates as electrodes was used to clamp the two sides of the IPMC strip, providing the stimulus electrical signal to the IPMC. The clamping length is 5 mm. The conical nozzle/diffuser dimensions are shown in Figure 15.5 and it has an area ratio of 64 and, with a length of 2 mm, equates to a half angle, θ, of 10°.

15.4.7 Simulation of the Pump

Figure 15.6 shows the fluid model of the valveless miniature pump created for ANSYS® CFX, which includes a pump chamber, two nozzle/diffuser

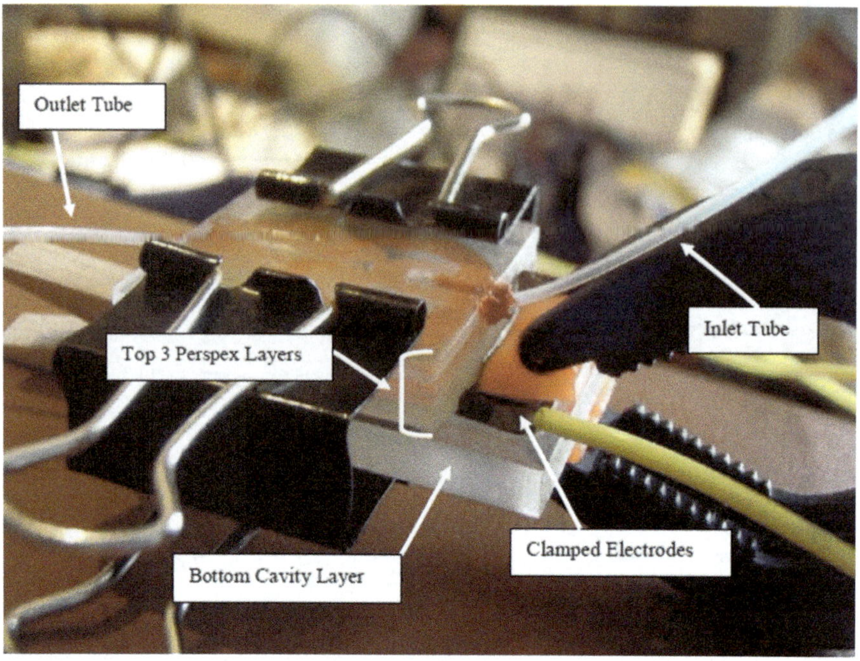

Figure 15.4 An assembled pump prototype.[37]

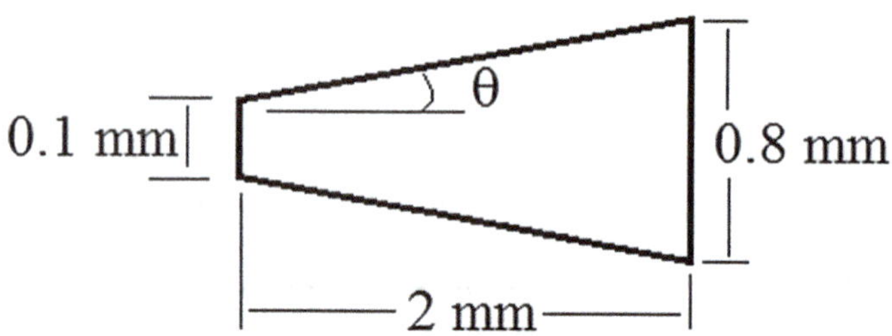

Figure 15.5 The dimensions of the nozzle/diffuser used in the miniature pump.

elements and partial inlet and outlet channels. Because the inlet and outlet allow the fluid to move in both directions through them, the boundary conditions at both the inlet and the outlet were set to be "opening" with zero static pressure. A "no-slip wall" boundary condition was imposed on the solid wall boundary of the fluid model. Since there would be motions along the solid–fluid interface, a moving mesh boundary condition was imposed along it, with an arbitrary formulation. The flow was assumed to be laminar, and the fluid material was set to be water at 25 °C and 1 atmospheric pressure. These properties were assumed to have no dependence on the temperature change.

When a sinusoidal input voltage signal is applied to the IPMC actuator, the IMPC strip starts to oscillate and cause the latex film to move up and

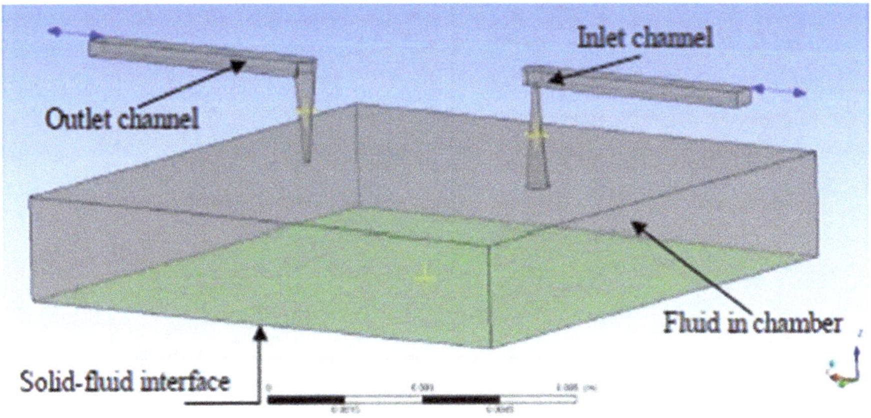

Figure 15.6 Fluid domain model of the valveless miniature pump.

down. This action can be regarded as a pressure signal being applied to the diaphragm. It was assumed that the pressure imposed had the following form of a sine function:

$$f(x) = P \sin(\omega t)$$

where P is the amplitude of the pressure on the diaphragm and ω is the driven frequency of applied voltage on the IPMC. Different values of P and ω were simulated to analyse their effects.

Figure 15.7 shows the simulation results for a complete pumping cycle with the conditions of $P=1200$ Pa and $\omega=0.2$ Hz. The period for one pumping cycle is 5 s, thus making 50 time steps.

At $t=0.01$ s in Figure 15.7(a), the fluid starts to flow out from the outlet, but little fluid is flowing out through the inlet channel. This is correct because the diaphragm is being pushed upwards. It is similar to the "pump mode" of the nozzle/diffuser elements. At $t=0.625$ s, as the pressure approaches the maximum value, more fluid is flowing out from the outlet channel. At the same time, some fluid is also flowing out through the inlet, but at a much lower amount. Turbulences are also built up at the inlet nozzle position, as shown in Figure 15.7(b). The miniature pump is still under pump mode. At $t=1.25$ s, the pressure has reached its maximum and the diaphragm is at its upmost position. The miniature pump is now changing from pump mode to supply mode. Fluids are now flowing through both the inlet and outlet. However, turbulences start to build up inside the pump chamber at the outlet channel, which resists fluid flow in from the outlet channel (Figure 15.7(c)). Then, at $t=1.875$ s, as shown in Figure 15.7(d), as the diaphragm is moving downwards, more fluid flows from the inlet channel to the pump chamber than the outlet channel. The miniature pump has now entered supply mode. The diaphragm reaches the middle point at $t=2.5$ s in Figure 15.7(e), which is half of the period. The miniature pump is now in the equilibrium position. However, it is still in supply mode so that

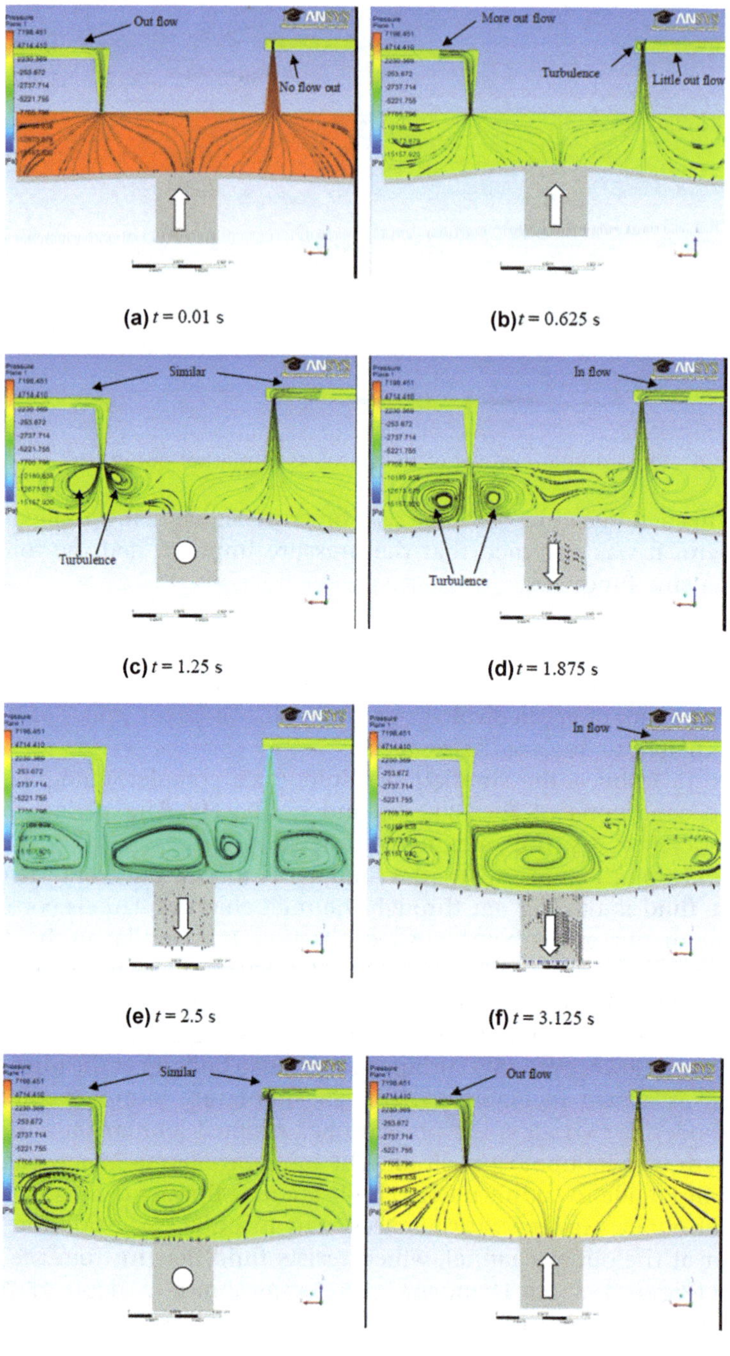

(a) $t = 0.01$ s **(b)** $t = 0.625$ s **(c)** $t = 1.25$ s **(d)** $t = 1.875$ s **(e)** $t = 2.5$ s **(f)** $t = 3.125$ s **(g)** $t = 3.75$ s **(h)** $t = 4.375$ s

Figure 15.7 Simulation of a complete pumping cycle at different time steps (a) 0.01 s, (b) 0.625 s, (c) 1.25 s, (d) 1.875 s, (e) 2.5 s, (f) 3.125 s, (g) 3.75 s and (h) 4.375 s. The wide arrows indicate the diaphragm direction of motion using ANSYS® CFX.

fluids are flowing into the chamber. At $t = 3.125$ s, the diaphragm has passed the middle point and continues to move downwards. As shown in Figure 15.7(f), more fluid tends to flow into the pump chamber from the inlet channel. The diaphragm reaches the lowest position at $t = 3.75$ s, as shown in Figure 15.7(g). The miniature pump is now at the transition from supply mode to pump mode. Equal flows are observed through the inlet and the outlet. Figure 15.7(h) shows the diaphragm moving upwards again in the pump mode. More fluid flows out of the outlet, while hardly any fluid flows out through the inlet channel. Once it passes the middle point again, the pumping cycle is complete. This simulation clearly shows that fluid is being pumped from the inlet to the outlet.

15.5 Control of IPMC Actuators

Because there is no appropriately accurate model available for IPMC actuators, it is essential to develop an adaptive control method that does not rely heavily on the availability and accuracy of the plant model. One of the most suitable solutions for solving this kind of control problem is a tuning method called IFT, which was proposed in 1998 by Hjalmarsson *et al.*[27] It is a tuning method that iteratively identifies and updates the controller to obtain optimized results. In this case, it uses an iterative gradient descent algorithm and the objective is to minimize the cost function or gradient of a controller design criterion directly from an unknown system; in this case, a least-squares fit of the tracking error. The method does not require a system model to get an unbiased estimate of the gradient; instead, it uses the error data gathered from the normal operation of the system with the current controller. Then, a special experiment, fed with the output signal of the system operating in normal conditions as the reference input to the control system, is performed to obtain additional data. By using the information of the time domain response obtained from this additional test, controller parameters can be updated in the negative gradient direction at the start of each iteration. This means that the controller parameter updates are only based on the experimental data from the actual system. Knowledge of the system's model is therefore not required. Although the assumptions for this method are that the system is linear and time-invariant, it has been shown that IFT can improve the performance of such a system.[27]

There have already been a number of existing IFT applications in both research fields and industry fields. The implementation of IFT has shown good results in applications such as profile cutting machine control systems,[28] speed and position control of servo drives,[29] temperature regulation in a distillation column[26] and photo-resistant film thickness control.[30] The traditional approach of tuning the system with the IFT method is "offline", or prior to the standard operation. This is due to the fact that a "special" gradient experiment is required to calculate the updated parameters; however, its trajectory may diverge far from the trajectory of the normal experiment. In this research, the IFT algorithm used has a modification from the standard version, which can keep the trajectory close to the normal reference in the gradient experiment. It will be a truly "online" tuning method, where the PID controller is continuously updated at a predetermined time. Below is an overview of the key theory of IFT and an explanation to the modification of the standard IFT algorithm to implement an "online" IFT.

15.5.1 IFT Algorithm

The miniature pump system with a PID controller shown in Figure 15.8 will be tuned using an IFT algorithm so that the IPMC actuator can be accurately controlled over a continuous operating period. IFT is a time domain approach whose objective is to obtain an optimized system performance by minimizing the cost function or design criterion based on the controller performance. Various design criteria based on tracking error and control effort can be considered. Weight filters can also be incorporated to the control effort depending on the application requirements. For example, to ensure more emphasis is placed on the transient or steady state, a time domain weighting filter could be added, or the system performance can be more flexible by placing a frequency-weighted filter.

For this proposed IPMC-driven miniature pump, the design criterion $J(\rho)$ consists of a quadratic function based on a least squares fit of the tracking error, $\tilde{y}_t$, as shown below in eqn (15.1).

$$J(\rho) = \frac{1}{2N}\sum_{t=1}^{N}\tilde{y}(\rho)^2 \tag{15.1}$$

where ρ is a vector of the controller parameters to be tuned, N is the total number of samples for a given experiment, t is the discrete time step index and $\tilde{y}_t(\rho) = y_t(\rho) - r$ represents the desired output minus the system output produced by the currently tuned controller (*i.e.* the system error) at time step t. The objective of the IFT tuning scheme is to make the gradient equal to 0 in order to find a minimum for the design criterion, J; in this case, finding the minimum tracking error over the entire experiment. To obtain the gradient, eqn (15.1) needs to be differentiated. By making it equal to 0, as shown in eqn (15.2), and finding the solution to it, the minimum of the criterion can be found.

$$\frac{\partial J(\rho)}{\partial \rho} = \frac{1}{N}\sum_{t=1}^{N}\left(\tilde{y}_t(\rho)\frac{\partial y_t(\rho)}{\partial \rho}\right) = 0 \tag{15.2}$$

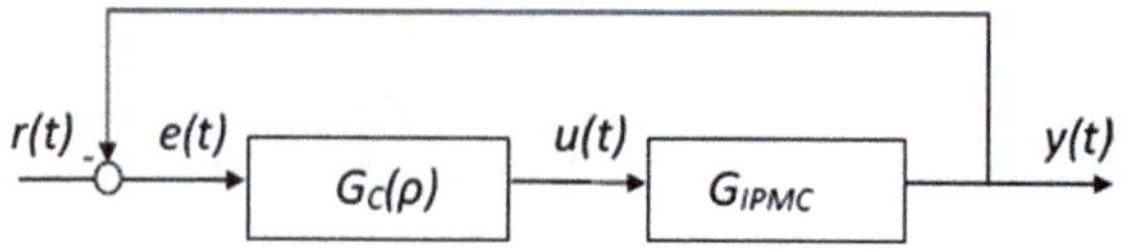

Figure 15.8 A PID control system.[37]

Applying the iterative algorithm in eqn (15.3), eqn (15.1) and (15.2) can be solved to update the controller gains for the next iteration. The solution for ρ can be found to obtain the minimum error for the system. This is a gradient search algorithm.

$$\rho_{i+1} = \rho_i - \gamma R_i^{-1} \frac{\partial J(\rho_i)}{\partial \rho} \tag{15.3}$$

γ is a positive real scalar that decides the step size. R_i is an appropriate positive definite matrix that controls the search direction for the optimization. i represents the iteration number. Appling the identity matrix to R_i gives a negative gradient direction. Many researchers have suggested that the controller performance can be improved by using the Gauss–Newton approximation of the Hessian of $J(\rho)$ for R_i.[27,28,31] This becomes more important when the time step size is small. The Hessian is given below in eqn (15.4).

$$R_i = \frac{1}{N} \sum_{t=1}^{N} \left(\frac{\partial y_t(\rho_i)}{\partial \rho} \left[\frac{\partial y_t(\rho_i)}{\partial \rho} \right]^T \right) \tag{15.4}$$

In order to get the updated controller gains, two signals, $\tilde{y}_t(\rho)$ and $\frac{\partial y_t(\rho)}{\partial \rho}$, are required to be solved in eqn (15.2). These two signals must be unbiased from each other, thus they needed to be solved independently.[31] In the standard IFT algorithm, these two signals are found over two consecutive experiments that are independent of each other, as follows:

1) In the first experiment, the input reference signal r is applied to the control system under normal operating conditions and the output y is recorded. $\tilde{y}_t(\rho)$ can then be found from $\tilde{y}_t = y_t(\rho) - r$.
2) Then, a second "special" experiment is performed. This is the same as the first experiment, except that this time the input signal r is the error $\tilde{y}_t(\rho)$ from the first experiment. From this operation, the gradient $\frac{\partial y_t(\rho)}{\partial \rho}$ can be found.

After these two experiments, $\tilde{y}_t(\rho)$ can be obtained directly from the first experiment, while $\frac{\partial y_t(\rho)}{\partial \rho}$ is calculated by the following process.

For the PID controlled system (Figure 15.8) used to control the IPMC actuator, the closed-loop output is defined as:

$$y(\rho_i) = \frac{G_c(\rho_i) G_{\mathrm{IPMC}}}{1 + G_c(\rho_i) G_{\mathrm{IPMC}}} r \tag{15.5}$$

Then, by differentiating the output, $\frac{\partial y_t(\rho)}{\partial \rho}$ can be found as:

$$\frac{\partial y(\rho_i)}{\partial \rho} = \frac{1}{G_c(\rho_i)} \frac{\partial G_c(\rho_i)}{\partial \rho} \left[\frac{G_c(\rho_i) G_{\mathrm{IPMC}}}{1 + G_c(\rho_i) G_{\mathrm{IPMC}}} (r - y_i(\rho_i)) \right] \tag{15.6}$$

From eqn (15.5) and (15.6), we can see that the term in the square brackets in eqn (15.6) is the result of injecting the error from the first experiment through the closed-loop system. Thus, the output collected in the second test gives the term in the square brackets in eqn (15.6). By differentiating the controller itself, the two terms $\frac{1}{G_c(\rho_i)}\frac{\partial G_c(\rho_i)}{\partial \rho}$ can be easily found. The gradient $\frac{\partial y(\rho_i)}{\partial \rho}$ can now be established. By putting this result back into eqn (15.4), the Hessian can be calculated and $\frac{\partial J(\rho)}{\partial \rho}$ can be found as well. By now, the new updated controller parameters ρ_{i+1} are available, which will give an improved controller for the system. Depending on the application requirement, this procedure is repeated for the desired number of iterations, or until the desired system performance is achieved.

15.5.2 Online IFT Tuning

As discussed in the last section, tuning a one degree-of-freedom PID controller with the standard IFT algorithm requires two independent experiments. This means that with this procedure, the system is tuning for half the time during operation; as a result, the target input, which is the error signal, will vary far from the normal or desired trajectory. This also means that the gradient experiment differs far from the normal experiment, which in practice is not acceptable for online tuning. However, it is possible to improve performance by using a different second experiment to calculate the gradient, and more detail of this is available in.[31] This reference proposes an IFT algorithm that offers a solution for dealing with the production waste caused by the second gradient experiment. The reference used for the gradient experiment is "switched" to $y_t(\rho)$ from the first experiment (instead of $(r - y_t(\rho))$, then eqn (15.6) for calculating $\frac{\partial y(\rho_i)}{\partial \rho}$ can be modified to:

$$\frac{\partial y(\rho_i)}{\partial \rho} = \frac{1}{G_c(\rho_i)}\frac{\partial G_c(\rho_i)}{\partial \rho}[(y_1 - y_2)] \tag{15.7}$$

This modification is acceptable because when the system is tracking the trajectory at a reasonable accuracy, then the reference for the gradient experiment, $y_t(\rho)$, should be equal to r for the normal experiment, with some small tracking error. However, this substitution also brings another problem that is $\tilde{y}_t(\rho)$, which will be used for calculating $\frac{\partial J(\rho)}{\partial \rho}$ as in eqn (15.2), is based on $\frac{\partial y(\rho_i)}{\partial \rho}$. This is because the first and second experiments are used to calculate $\frac{\partial y(\rho_i)}{\partial \rho}$ directly, and the second experiment is dependent on the first

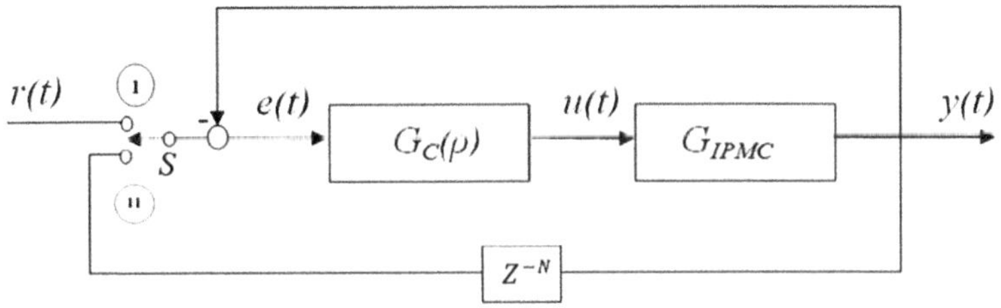

Figure 15.9 Block diagram of the control system for the online IFT.[37]

experiment. To overcome this problem, a third experiment must be conducted using the same conditions as in the first experiment.

Figure 15.9 is a block diagram for the modified "online" IFT-tuned control system. In this implementation, a switch S is added to switch the reference input to the system during three experiments. In the first and third experiment, the switch is set to (I), which is the reference signal r, and for the second gradient experiment, the switch is set to (II), which feeds back the output of the first experiment, as there is an N sample delay.

However, if the system is under-damped, there will be a problem because the reference for the gradient experiment, $y_t(\rho)$, will contain the exact resonance frequencies of the system. This is not preferred for the gradient experiment, which may cause the system to become heavily excited.[31] Therefore, with the modified Ziegler–Nicholas model-based tuning method stated previously, the damping should be large enough (over-damped or critically damped) so that any unwanted oscillations can be eliminated.

For a stable initial control system, if the step size is small enough and the data set is relatively large, the search algorithm of online IFT will always be in the negative direction, which can ensure convergence to the local minimum of the design criteria.[27] Since the online tuning method has a modified experiment every third period to calculate the error, it may cause some deviation from the desired reference, so it is expected to have some overshoots in every third cycle. To minimize this problem, the total number of tuning experiments should be reduced during the entire system operation.

15.5.3 Experimental Results

It has been demonstrated that the IFT algorithm can work over a number of different frequencies and target displacements and so now the IPMC and pump are assembled for testing on the real system. The IPMC is attached to the diaphragm and will push against the diaphragm to pump the fluid. The laser sensor has been positioned so that it can measure the diaphragm displacement. Figure 15.10 shows the comparison of the diaphragm's displacement with OL control, PID control and PID control with the online IFT for a targeted diaphragm displacement of 25 μm at 0.1 Hz. From this it can be seen that OL control cannot accurately track a desired displacement and the displacement drifts over time, which will cause the fluid flow rate to be

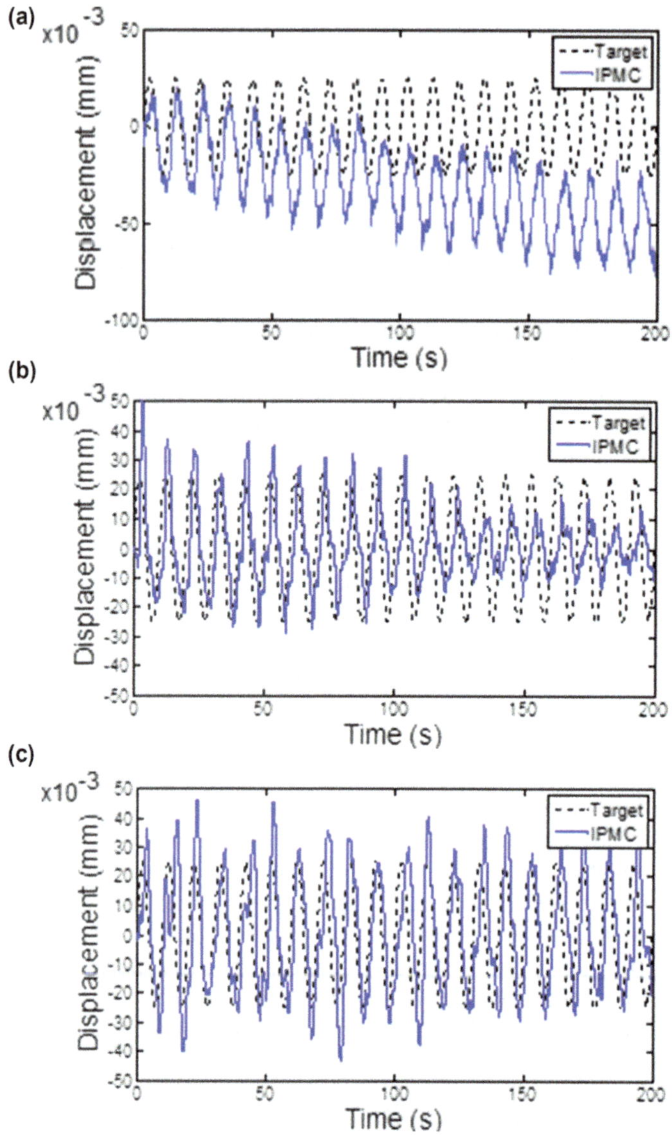

Figure 15.10 The displacement of the pump's diaphragm with (a) OL control, (b) PID control and (c) PID control with online IFT.[32]

inconsistent. The standard PID control does maintain the overall mean displacement and prevents the drift that is seen in the OL control. The PID controller, however, is still not very accurate throughout the operation as the controller parameters are not optimally tuned. The PID controller with the new online IFT algorithm is able to control the diaphragm's displacement more consistently over time than a standard PID controller. It does overshoot at the start, similarly to the standard PID controller, but it can adjust

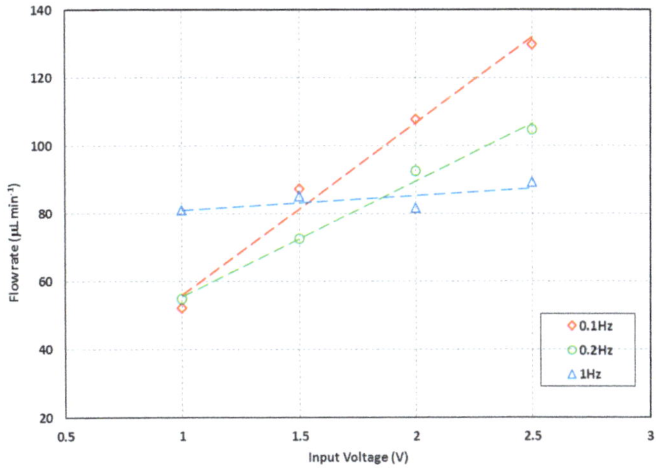

Figure 15.11 Average flow rate for varying amplitude voltage sinusoids at different frequencies.[37]

its gains over the period of operation to maintain a more accurate displacement and hence flow rate.

To realize a continuous pumping rate, a sinusoid IPMC displacement is required. The performance of the pump was tested with sine wave voltages of varying amplitudes and frequencies were applied to the IPMC and the average flow rate is plotted in Figure 15.11. The flow rate is higher at lower frequencies (<1 Hz). This suggests that at lower frequencies a higher IPMC displacement is achieved, causing a higher flow rate. As the IPMC is continually hydrated, it can be continuously operated at these higher voltages. From the experiments it was found that a frequency of around 0.1 Hz gives the highest pumping rate for this system. This is in line with,[33] which shows that a frequency of 0.1 Hz gives good pumping operation. In fact, low pumping rates for general drug dispensing will be required. As such, the frequencies of interest for the pump will be between 0.05 Hz and 0.1 Hz.

Similar conditions to the above experiment are use in the ANSYS® CFX simulation. In this simulation (0.1 Hz with IPMC deflection at 2.5 V), the flux in one cycle is 146 μl/min, which is close to the experimental result of 130 μl/min. However, the simulated diffuser efficiency (η) is only 1.38 and only occurs with a Reynolds number (Re) of 24. The simulated outlet velocity (m s^{-1}) *versus* time (s) is shown in Figure 15.12. Note that the input is the pump diaphragm pushing and pulling into the chamber according to the sinusoidal voltage applied to the IPMC, which in turn varies the fluid velocity accordingly.

15.5.4 Performance Optimization of Valveless Pumps

The valveless pump needs to be further miniaturized and therefore there needs to be a reduction in the area ratio (AR) of the diffuser/nozzle element in actual application for drug delivery so that the pump can be wearable and

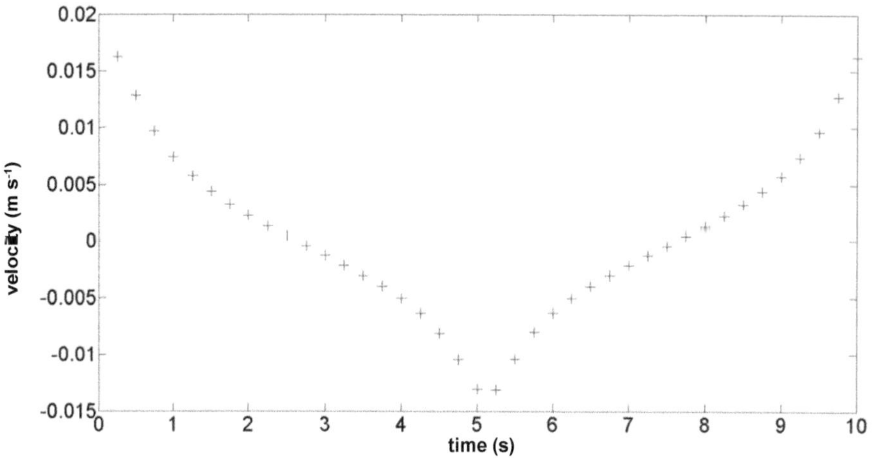

Figure 15.12 Outlet velocity *vs.* time.

portable. Hence, it is important that the valveless pump is still efficient at a smaller scale. Although many former researchers have investigated different diffuser/nozzle structures, few of them were concerned about the flow characteristics in low *Re* ranges (<100).[34] Under the trend of minimization and high-viscosity working fluid (especially in drug delivery), there comes the need to understand the detailed performance of valveless miniature pumps in this *Re* range.

In this study, the flow characteristics of diffuser/nozzle elements with *AR* values of 4, 6.25 and 9 were studied numerically using the commercial computational fluid dynamics (CFD) software package ANSYS® CFX.[35] The prototyped valveless pump has diffuser/nozzle diameters of 100 μm/800 μm and $AR = 64$, and with smaller *AR* values of 4, 6.25 and 9, the diameters are now 100 μm/200μm, 100 μm/250 μm and 100 μm/300 μm, respectively. The schematic of a diffuser/nozzle element with a sudden contraction inlet condition and a sudden expansion outlet condition is illustrated in Figure 15.13. The diameter of the sections to which the diffuser/nozzle element is connected to is set to ten times the diameter of the widest section of the diffuser/nozzle element ($D_1/d_2 = D_2/d_2 = 10$), while the lengths of these sections are similar to that of the diffuser/nozzle element ($l_1 \approx l_2 \approx l$). This kind of structure can eliminate the error caused by different sudden expansion or contraction ratios and at the same time incorporate the entrance and exit pressure losses, allowing it to be directly coupled into the full pump model.

Diffuser efficiency η (defined as the ratio of the pressure loss coefficient for the nozzle direction to that of the diffuser direction) is recognized as the flow-directing ability of diffuser/nozzle element[36] and is numerically investigated and plotted in Figure 15.14 for different θ and *AR* values. There is a clear trend that the optimum θ, where maximum η occurs, decreases with *Re*, as the θ determines the length of the diffuser/nozzle element.

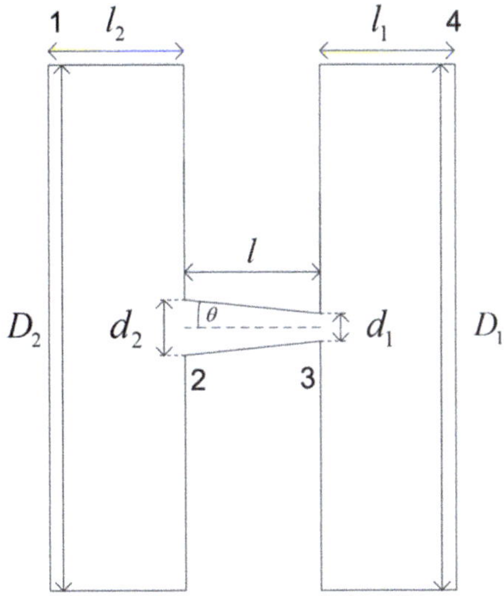

Figure 15.13 Diffuser/nozzle element structure used for ANSYS® CFX simulation.

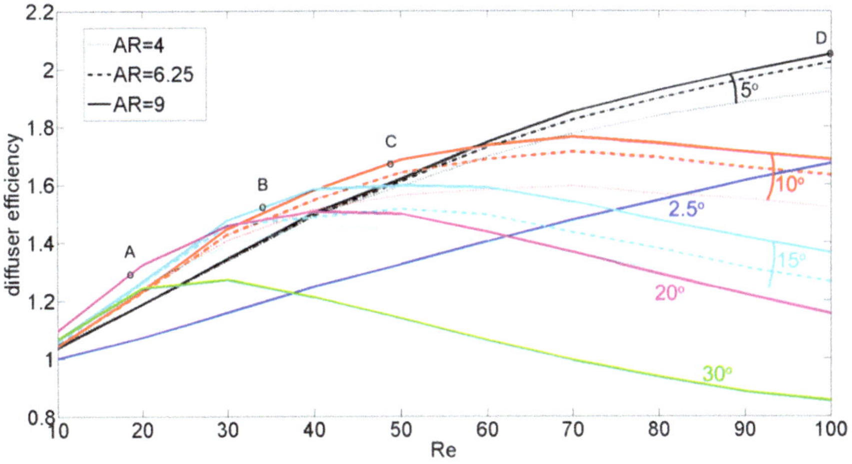

Figure 15.14 Diffuser efficiency of the diffuser/nozzle element.

From these simulations, even with further miniaturization, the η can be optimized between ~1.3 and 2.2 depending on the *Re* it is operating at. Several representative points in Figure 15.14 are selected to fit an "optimum half angle curve" shown in Figure 15.15. This conclusion can be regarded as the design guideline for the drug delivery valveless miniature pump, which works with low *Re* but still requires a certain amount of flux in limit dimensions.

Since the diffuser/nozzle model in this study can be directly coupled into a whole-pump model, Figure 15.15 can be used as a guideline, or as a design

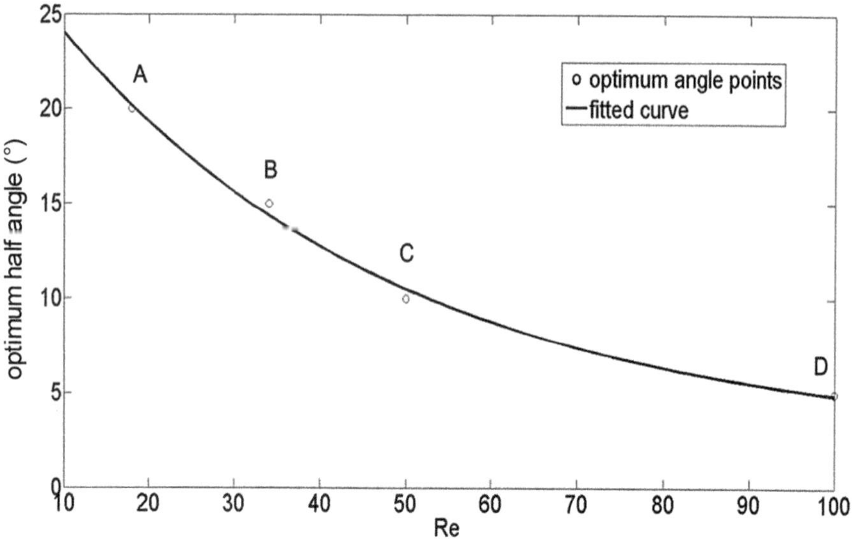

Figure 15.15 Fitted curves for the optimum half angle.

chart, in the selection of conical diffuser/nozzle elements for a valveless miniature pump at different *Re* and hence achieving the maximum flow-directing capability of the valveless pump. In general, the θ for the diffuser/nozzle needs to be increased as the *Re* decreases in order to achieve the optimal flow directivity performance of the valveless pump.

15.6 Conclusion

This study presents a prototype valveless pump used for drug delivery. Further numerical investigations of the flow characteristics of the diffuser/nozzle elements in the low *Re* range (<100) have been carried out. The results show that optimum half angle, θ of the diffuser/nozzle element at which maximum efficiency η occurs, is found to vary with *Re*. The highlights from these results could serve as a design guideline for diffuser/nozzle element selection of valveless micropumps operating at low *Re*.

References

1. N. T. Nguyen, X. Huang and T. K. Chuan, MEMS-micropumps: A review, *J. Fluids Eng. Trans. ASME*, 2002, **124**(2), 384–392.
2. P. Woias, Micropumps – Past, progress and future prospects, *Sens. Actuators, B*, 2005, **105**(1), 28–38.
3. M. Shahinpoor, K. J. Kim and M. Mojarrad, *Artificial Muscles: Applications of Advanced Polymeric Nanocomposites*. 2007, Taylor & Francis: New York.

4. M. Anton, A. Punning, A. Aabloo and M. Kruusmaa, *Validating usability of ionomeric polymer-metal composite actuators for real world applications*, in Proceeding of the IEEE/RSJ International Conference on Intelligent Robots and Systems, 2006, pp. 5441–5446.
5. M. Shahinpoor and K. J. Kim, Ionic polymer-metal composites: I. Fundamentals. , *Smart Mater. Struct.*, 2001, **10**(4), 819–833.
6. K. J. Kim, Ionic Polymer-Metal Composite as a New Actuator and Transducer Material, in *Electroactive Polymers for Robotic Applications: Artificial Muscles and Sensors*, ed. K. J. Kim and S. Tadokoro, Springer, Berlin; London, 2007, pp. 153–164.
7. V. K. Nguyen, J. W. Lee and Y. Yoo, Characteristics and performance of ionic polymer-metal composite actuators based on Nafion/layered silicate and Nafion/silica nanocomposites, *Sens. Actuators, B*, 2007, **120**(2), 529–537.
8. K. Mallavarapu and D. J. Leo, Feedback control of the bending response of ionic polymer actuators, *J. Intell. Mater. Syst. Struct.*, 2001, **12**(3), 143–155.
9. Z. Chen, X. Tan and M. Shahinpoor, *Quasi-static positioning of ionic Polymer-metal composite (IPMC) actuators*, in Proceeding of IEEE/ASME International Conferenceon Advanced Intelligent Mechatronics, 2005, pp. 60–65.
10. N. Bhat and W. J. Kim, Precision force and position control of an ionic polymer metal composite, in Proceedings of the Institution of Mechanical Engineers, Part I: J. Syst. Control Eng., 2004, 218, pp. 421–432.
11. C. Bonomo, L. Fortuna, P. Giannone and S. Graziani, A sensor-actuator integrated system based on IPMCs, in Proceedings of the IEEE International Conference on Sensors, 2004, pp. 489–492.
12. R. C. Richardson, M. C. Levesley, M. D. Brown, J. A. Hawkes, K. Watterson and P. G. Walker, Control of ionic polymer metal composites, *IEEE/ASME Transactions on Mechatronics*, 2003, **8**(2), 245–253.
13. K. S. Yun, *A novel three-finger IPMC gripper for microscale applications*, in PhD Thesis, Mechanical Engineering, 2006, A&M University, Texas.
14. Z. Chen and X. Tan, A control-oriented, physics-based model for ionic polymer-metal composite actuators, in Proceedings of the 46th IEEE Conference on Decision and Control, 2007, pp. 590–595.
15. F. Amirouche, Y. Zhou and T. Johnson, Current micropump technologies and their biomedical applications, *Microsyst. Technol.*, 2009, **15**(5), 647–666.
16. P. Gravesen, J. Branebjerg and O. S. Jensen, Microfluidics – A review, *J. Micromech. Microeng.*, 1993, **3**(4), 168–182.
17. S. S. Wang, X. Y. Huang and C. Yang, Valvelessmicropump with acoustically featured pumping chamber, *Microfluid. Nanofluid.*, 2009, 1–7.
18. A. Nisar, N. Afzulpurkar and B. Mahaisavariya, , AdisornTuantranont, *MEMS-based micropumps in drug delivery and biomedical applications*, *Sens. Actuators, B*, 2008, **130**(2), 917–942.

19. S. Lee and K. J. Kim, Design of IPMC actuator-driven valve-less micropump and its flow rate estimation at low Reynolds numbers, *Smart Mater. Struct.*, 2006, **15**(4), 1103–1109.
20. T. T. Nguyen, N. S. Goo, V. K. Nguyen, Y. Yoo and S. Park, Design, fabrication, and experimental characterization of a flap valve IPMC micropump with a flexibly supported diaphragm, *Sens. Actuators, A*, 2008, **141**(2), 640–648.
21. T. T. Nguyen, V. K. Nguyen, Y. Yoo and S. Park, A novel polymeric micropump based on a multilayered ionic polymer-metal composite, IEEE No. 4, 2006, pp. 4888–4892.
22. S. Guo, S. Hata, K. Sugumoto and T. Fukuda, *A new type of capsule micropump using ICPF actuator* in Proceedings of the International Symposium on Micromechatronics and Human Science, 1998, pp. 255–260.
23. J. J. Pak, J. Kim, S. W. Oh, J. H. Son, S. H. Cho, S. K. Lee; J. Y. Park and B. Kim, *Fabrication of ionic-polymer-metal-composite (IPMC) micropump using a commercial Nafion*, In Proceedings of SPIE – The International Society for Optical Engineering, 2004, 5385, pp. 272–280.
24. J. Santos, B. Lopes and P. J. C. Branco, Ionic polymer-metal composite material as a diaphragm for micropump devices, *Sens. Actuators, A*, 2010, **161**(1–2), 225–233.
25. S. Lee, K. J. Kim and H. C. Park, *Design and performance analysis of a novel IPMC-driven micropump*, in Proceedings of SPIE – The International Society for Optical Engineering, 2005, 5759, pp. 439–446.
26. E. Stemme and G. Stemme, A valveless diffuser/nozzle-based fluid pump, . *Sens. Actuators,*, 1993, **39**(2), 159–167.
27. H. Hjalmarsson, M. Gevers, S. Gunnarsson and O. Lequin, Iterative feedback tuning: Theory and applications Control Systems, IEEE, 1998, 18 (4), pp. 26–41.
28. A. E. Graham, A. J. Young and S. Q. Xie, Rapid tuning of controllers by IFT for profile cutting machines, *Mechatronics*, 2007, **17**, 121–128.
29. S. Kissling, Ph. Blanca, P. Myszkorowskib and I. Vaclavik, Application of iterative feedback tuning (IFT) to speed and position control of a servo drive, *Control Eng. Pract.*, 2009, **17**(7), 834–840.
30. A. Tay, W. K. Ho, J. Deng and B. K. Lok, Control of photoresist film thickness: Iterative feedback tuning approach, *Comput. Chem. Eng.*, 2006, **30**(3), 572–579.
31. H. Hjalmarsson, Iterative feedback tuning – an overview, *Int. J. Adaptive Control Signal Process.*, 2002, **16**(5), 373–395.
32. W. Yu, *The Development of an IPMC Actuated Micropump using PID Control with Iterative Feedback Tuning*, in Master's Thesis, Mechatronics Engineering, The University of Auckland, NZ, Auckland, 2011.
33. J. Santos, B. Lopes and P. J. C. Branco, Ionic polymer-metal composite material as a diaphragm for micropump devices, *Sens. Actuators,*, 2010, **161**(1–2), 225–233.

34. C. Sun and Z. H. Yang, Effects of the half angle on the flow rectification of a microdiffuser, *J. Micromech. Microeng.*, 2007, **17**(10), 2031–2038.
35. ANSYS CFX-Pre User's Guide, Canonsburg, PA, 2009.
36. T. Gerlach, M. Schuenemann and H. Wurmus, A new micropump principle of the reciprocating type using pyramidic micro flowchannels as passive valves, *J. Micromech. Microeng.*, 1995, **5**(2), 199–201.
37. K. C. Aw, W. Yu, A. J. McDaid and S. Q. Xie, *An IPMC driven micropump with adaptive on-line iterative feedback tuning*, In Proceedings of SPIE – The International Society for Optical Engineering, 2012, 8409, 84090I.

CHAPTER 16

Modelling and Characterisation of Ionic Polymer Metal Composite (IPMC) Transducers: From IPMC Infancy to Multiphysics Modelling

SALVATORE GRAZIANI

Dipartimento di Ingegneria Elettrica Elettronica e Informatica, Università degli Studi di Catania, Viale Andrea Doria 6, Catania, Italy
Email: salvatore.graziani@dieei.unict.it

16.1 Introduction

We are assisting in the deep interest in the development of intelligent systems intended to expand our action and sensing capabilities and, in such a way, helping people in everyday life.

The envisaged intelligent systems need to be capable of solving increasingly complex problems with little or no human intervention in strategic fields such as bio-inspired robotics, underwater applications, hazardous environment surveillance, aerospace, and medicine, to mention just a few.[1] Generally

RSC Smart Materials No. 18
Ionic Polymer Metal Composites (IPMCs): Smart Multi-Functional Materials and Artificial Muscles, Volume 2
Edited by Mohsen Shahinpoor

Published by the Royal Society of Chemistry, www.rsc.org

speaking, a huge number of applications can be foreseen in all those fields where human intervention may be either impossible or dangerous.

We are talking about, just to cite a few examples, bio-inspired robots that will perform repetitive tasks, active prostheses that will help with the rehabilitation of patients, smart textiles that can deliver drugs on the basis of well-established protocols, or even artificial tissues and organs that will eliminate the need to search for donors and the problem of rejection, in such a way as to improve the quality of life of the transplanted patient.

The realisation of such smart applications will be possible mainly because of new materials that are becoming available as a result of interdisciplinary research, including materials science, engineering, mechanics, and biology. These new materials are able to perceive external stimuli and react to them, for example by changing their mechanical or electrical properties, so that they are no longer passive objects or simple coating elements, instead cooperating to fulfil a task.

Electroactive polymers (EAPs) will play a main role in the realisation of such smart systems. Although the interest in EAPs has grown at an impressive rate in recent decades, the history of EAPs is much older. As a matter of fact, the first experiments on EAPs were performed by W. C. Röntgen, who in 1880 observed the mechanical reaction of natural rubber when it was activated using electrical stimuli.[2] A portrait of W. C. Röntgen is shown in Figure 16.1.

Figure 16.1 Portrait of Wilhelm Conrad Röntgen taken during the period he spent at the University of Wurzburg.

The portrait is from the same period as when Röntgen discovered X-rays, which earned him the Nobel Prize in Physics in 1901.

In that same year, experiments by the two brothers Jaques and Pierre Curie highlighted the piezoelectric effect on quartz crystals[3] that represents a further case of materials capable of reversible electro-mechanical coupling, although of course these are not polymeric materials.

Actually, the coincidence of the year when both phenomena were discovered is partly obscured by the fact that, since that time, piezoelectric non-polymeric materials have been investigated much more than EAPs. Nonetheless, the first piezoelectric device was not proposed until World War I, when underwater sonar for submarine detection was proposed by P. Langevin.

As is widely known, the increase in interest regarding EAPs, including ionic polymer metal composites (IPMCs), was more recent, beginning at the end of the last century.[2,4,5]

I have been studying IPMCs since 2002 and this is a considerable time interval for a relatively new technology, as is the case for IPMCs.

It is in any case a significant lapse of time for observing the changes over the years that have characterised the approach of the scientific community to the field. Of course, a corresponding evolution occurred in my research activity. In fact, over more than 10 years, new results emerged and the research focus shifted from one aspect to another.

Nevertheless, some constant aspects that I investigated since the beginning of my interest on IPMC transducers have emerged over time.

The very first characteristic that tantalised researchers, including myself, regarding IPMCs is that they are reversible materials, capable of working both as deformation sensors and actuators. This property makes them privileged candidates for realising some of the main functionalities required for smart systems. In fact, electric generating capabilities allow the use of IPMCs in both power generation and motion sensing, while the actuating capabilities can be exploited to realise motion actuators. No less important is their polymeric nature, since this allows their potential use in smart post-silicon devices, along with plastic electronics, whose performances are rapidly improving. Looking back at the very beginning of my activity, one of my first papers was on the realisation of a very simple prototype in which an integrated motion actuator–sensor system was proposed.[6] The system consisted of a couple of IPMC strips working as a sensor and as an actuator, respectively. It was meant to show the capability of IPMC transducers for realising smart devices, through the realisation of one system in which both of the working modes of IPMCs were exploited. More specifically, the prototype consisted on a hydrated strip for the actuator and a dehydrated strip for the sensor. Both strips were attached to a thin plastic sheet and pinned at one end in a cantilever configuration as shown in Figure 16.2.

During the experiments, the plastic supporting sheet was mechanically stimulated with impulsive inputs. The sensing IPMC was used to detect the sheet oscillations and to generate a corresponding voltage output. The

Figure 16.2 A picture of a simple IPMC-based actuator–sensor smart system. © 2004 IEEEE. Reprinted, with permission, from C. Bonomo, L. Fortuna, P. Giannone and S. Graziani, A sensor–actuator integrated system based on IPMCs, Proc. of IEEE Sensors, 3rd Conf. on Sens. 2004, 489–492.

sensing voltage was inverted and amplified to obtain a corresponding output voltage in the order of a few volts. This signal was used as the feedback to the actuator strip in a closed-loop scheme.

Though the controlling policy was very simple, it was able to reduce the disturbances induced on the sheet. This can be seen in Figure 16.3, where the sheet oscillations, caused by a mechanical impulse, as recorded by the IPMC sensor, are reported, both in open-loop and closed-loop conditions. The duration of the system oscillations proved to be reduced by using the IPMC-based actuator–sensor system.

A second aspect of my research activity on IPMCs is strictly linked to my academic education and my teaching classes. My research activity focus is on electronic measuring systems and I teach classes on this topic. For this reason, I realised and I strong believe that the research activity on IPMC-based transducers needs significant effort, both to improve the stability of their performance and to give a quantitative assessment of their uncertainty. Such knowledge, along with the corresponding models, is a mandatory aspect to be solved and investigated ahead of IPMCs being fully exploited in real-life applications and becoming more that laboratory-scale proof-of-concept prototypes.

In a broader sense, some constant topics can be identified in the research on IPMCs since they are considered mandatory investigation areas for the full exploitation of this new class of smart materials, including corresponding challenges that were and to some extent remain to be solved.

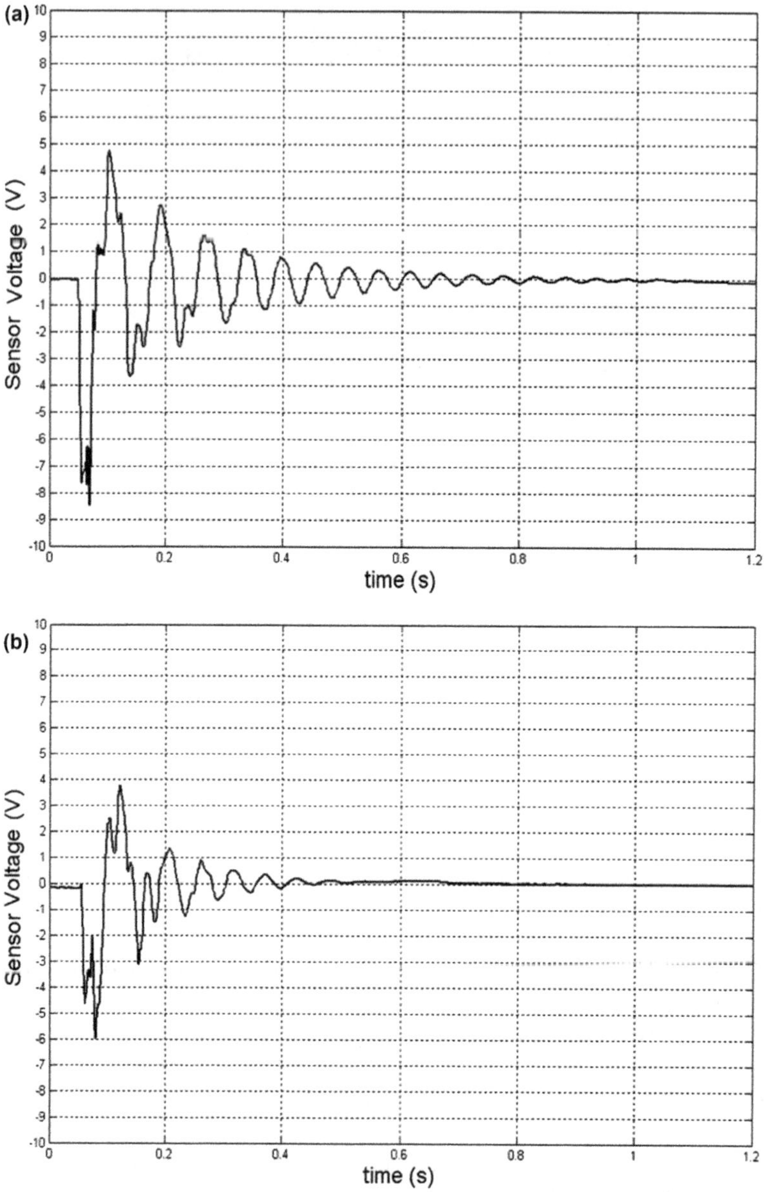

Figure 16.3 Sensor output recorded when a mechanical pulse is applied to the system shown in Figure 16.2. The open-loop response (a) and the closed loop response (b) are shown.
© 2004 IEEE. Reprinted, with permission, from C. Bonomo, L. Fortuna, P. Giannone and S. Graziani, A sensor-actuator integrated system based on IPMCs, Proc. of IEEE Sensors, 3rd Conf. on Sens. 2004, 489–492.

There is, in fact, a general consensus on the consideration that the main research activities in IPMCs deal with production technology, modelling, and application development and demonstration. Moreover, these research topics cannot be considered as independent fields; as an example, new production strategies, aimed at producing better materials, can be greatly helped if models are available for predicting, in advance, the effects of changes in the production protocols. Also, on the one hand, applications need to be described by suitable models, while on the other hand, new models can be experimentally verified by observing laboratory-scale prototypes. It can be said that new findings in each of the three mentioned fields influence the remaining two and that they evolve in a collaborative way.

In a similar way, it is not unusual for a researcher interested in IPMCs to face problems referring to each one of the three mentioned topics, without a sharp delineation between them. Nonetheless, for simplicity, it makes sense to describe research activities on IPMCs by respecting the above-mentioned taxonomy. In fact, this gives a clearer, although quite artificial, view of the addressed topics.

16.2 Modelling

In this chapter, for simplicity, the focus will be on my research activity in IPMC modelling, though the relative influences of the other aspects mentioned above will clearly emerge.

The full exploitation of a technology requires the capability to describe the involved phenomena by using suitable mathematical models. The investigation of such models is of interest in two different ways.

When a novel technology is discovered, hypotheses are introduced regarding the phenomena that rule that technology. Models are therefore proposed and compared with real observations on laboratory-scale prototypes. The agreement of model-estimated behaviour with corresponding real data gives evidence of the correctness of the hypothesised working model. In this sense, models can be a tool to confirm or refute the correctness of the introduced hypotheses and therefore to improve our level of knowledge about the investigated technology. This is very much true when transduction phenomena between different domains are considered and complex models are required, and is of course the case with IPMC transducers.

Once models have been successfully verified and tuned, they can be powerful tools for engineering applications, since they allow predicting, in advance, the behaviour of the device under design. Mathematical models can be used to describe the behaviour of the devices to be designed to a level of accuracy necessary for the intended applications. Such models are formidable tools for the user, since they allow predicting of the effects of any choice during the design step.

A quite general trend in system modelling starts from models that, on the basis of experimental investigations, try to fit the observed data to arrive at models that provide a higher level of generality. The second class of models are then able to predict the behaviour of devices other than the investigated ones, or used in working conditions different from those experimentally observed.

The modelling of IPMCs has followed this trend. Moreover, a significant improvement can be observed from the pioneering works to the latest models, which are characterised by an ever-growing level of complexity and accuracy.

Nowadays, tens of models of IPMC transduction phenomena are available in the literature, which can be grouped in three main classes: black-box or data-driven models; grey-box models; and first principle or white-box models. A good overview on the theory of system modelling can be found in Ljung.[7] Such a classification will be used in the remainder of the chapter to describe IPMC models that have been investigated so far.

16.2.1 Black-box Modelling

Different strategies are suitable for modelling real systems, mainly depending on the level of *a priori* knowledge of the involved phenomena. Models can be obtained either on the basis black-box or grey-box identification approaches or by using a white-box approach (also known as first principles analysis or mechanistic models).

Of course, the very first models proposed to describe IPMC transducing capabilities were black-box models. Moreover, since the most dramatic phenomenon that can be observed for IPMCs transducers is their large electrically induced deformation, this was the very first phenomenon to be modelled.

Black-box models use some universal approximators to fit experimental data and derive a macroscopic description of the device behaviours without any attempt to model their microscopic origin. In the considered case, data to be modelled represent the IPMC deformation, generally in a pinned beam configuration, as a consequence of an applied electrical stimulus. Applied stimuli were—and still generally are—either steps, if the time transience and regime deformation were investigated, or sine waves of varying frequencies, if the frequency dependence of the IPMC mechanical reaction was of interest.

When I started my research activity, such models were already considered as classic contributions that were widely referenced in the literature. A black-box model of IPMC actuators was proposed back in 1994 by Kanno and co-workers.[8] The proposed model is routinely referenced in papers on IPMCs[9,10] as one of the first attempts to model IPMC electromechanical transduction and, as a matter of fact, it was published no later than 2 years after IPMC actuation capabilities were reported.[11] It suffices here to say that in the contribution by Kanno *et al.*,[8] a pinned IPMC was considered and its behaviours were investigated under the effects of both step and sinusoidal signals. Eventually, a data-driven dynamic model of its deformation, under the effects of various input voltage steps, was identified by using the usual least squares method. In more details, the tip displacement was modelled with a number of exponential functions, whose free parameters were obtained by fitting the model estimations to experimental data. A corresponding fourth-order transfer function for the voltage-to-tip deflection was proposed. Though a linear model was identified, the acquired deformation revealed the nonlinear nature of the electromechanical transduction, since both the maximum displacement and the maximum absorbed current showed an exponential-like dependence on the applied voltage value.

When an electromechanical actuator is of interest, attention must be paid to its electrical behaviour, including the corresponding power consumption. This aspect can, in fact, influence practical applications of actuators and is particularly relevant when autonomous systems are of interest, as is the case for a number of robotics applications envisaged for IPMCs. It is therefore of great importance to study the relationship between applied voltages, absorbed currents, and produced mechanical reactions. This aspect was investigated by using a black-box approach in Bonomo *et al.*,[12] with the aim of modelling the nonlinearity of IPMC actuators' voltage to current conversion. This contribution is a good example of how good black-box models can be for solving the problem of complex phenomena modelling provided that "good-quality" data are available. In fact, though at that stage the authors were interested only in a quite general understanding of the involved phenomena, it was possible to approximate quite accurately behaviours observed under a number of different working conditions.

The model proposed in Bonomo *et al.*[12] is a possible electric circuit realisation of IPMC membrane nonlinear electrical behaviour and it represents an attempt to overtake the limitations imposed by linear models proposed in the literature.[13–16] The elements used in the different branches were introduced to mimic, by using a small set of lumped components, the expected behaviour of each current contribution, while the actual values of the introduced components were obtained by fitting the model-simulated output to the real absorbed current.

The total current through an IPMC actuator was considered to be the sum of the ionic, electronic, and displacement current contributions. The sum was modelled by combining two impedances and a nonlinear branch in parallel, each one representing the relative conduction phenomenon.

The proposed model is a nonlinear circuit consisting of lumped elements, whose scheme is shown in Figure 16.4. More specifically, the right-hand branch represents the ionic current due to the ionic transport phenomenon. The path on the left-hand side was introduced into the model to take into account the threshold-like behaviour observed for the current absorbed by IPMC actuators. In fact, the presence of the couple of diodes in anti-parallel connection allows the current flow only when the voltage applied is greater than a threshold value. The resistance R_L was eventually introduced into the model to take account of current leakage through the membrane. The central path, through capacitor C_1, was named as the displacement current and used to model the capacitive-like behaviour of the IPMC actuator. Finally,

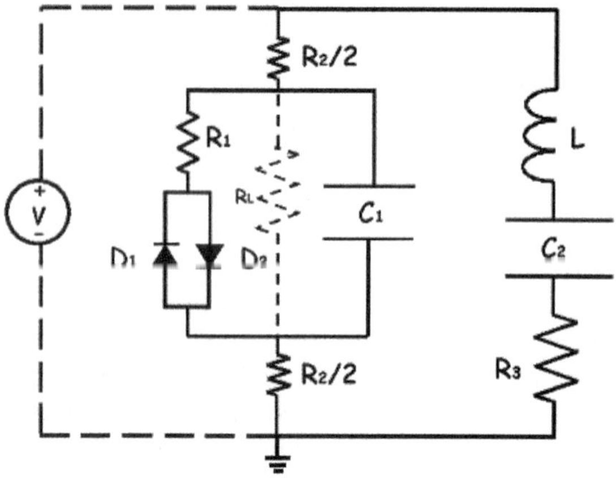

Figure 16.4 Scheme of the electric circuit used to model the current *versus* voltage behaviour of an IPMC. The right-hand path represents the ionic current, the left-hand path, through R_1, is the nonlinear contribution to the current, and the central path, through C_1, is the displacement current. The dotted path through R_L represents the leakage current.

the resistor R_2 was introduced to model the usual IPMC surface resistance due to the finite value of platinum electrode conductivity.

The values of the electrical components ruling the model were determined by using data acquired from experimental observations. More specifically, data were acquired on a Nafion® 117 (DuPont)-based IPMC, with platinum electrodes and Na^+ counterions. The considered actuator sizes were 35 mm×4 mm×200μm (length×width×thickness) and a photograph of the clamped device is reported in Figure 16.5.

This device was investigated under a number of experimental working conditions. More specifically, a survey was performed to outline the influence of the input level value on the voltage-to-current conversion. It was observed that, at low values of the input voltage, no significant nonlinear phenomena occurred, while by increasing the input voltage value, a significant waveform distortion occurred. This was considered to be a justification of the introduction of the left-hand side branch introduced in Figure 16.4, and an example of obtained polar plots is shown in Figure 16.6.

A second set of experiments was performed to investigate the dynamic nature of the device under test (DUT). To that aim, the most obvious survey is represented by a set of sinusoidal inputs of varying frequency. Experiments revealed that in the low-frequency range—from 10 mHz to 10.0 Hz in the considered experiments—a strong nonlinear behaviour characterised the voltage to current conversion. By increasing the input signal frequency—over 10.0 Hz and up to 500.0 Hz—the most relevant behaviour resulted in a

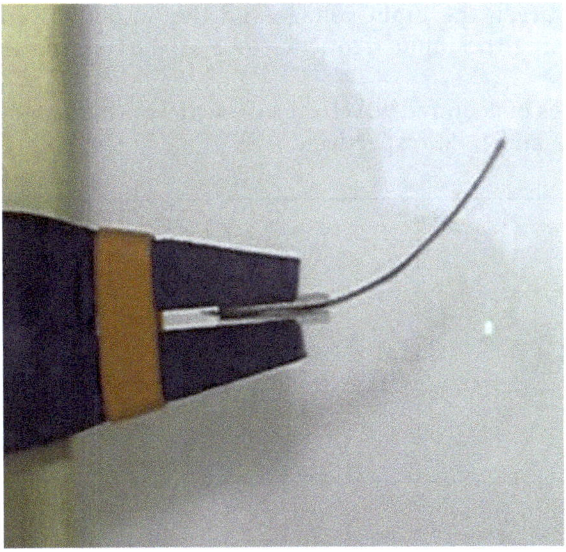

Figure 16.5 A photograph of the clamped actuator used to determine the voltage-to-current lumped component electrical realisation of the IPMC actuator. © 2006 IEEE. Reprinted, with permission, from C. Bonomo, L. Fortuna, P. Giannone and S. Graziani, A circuit to model the electrical behavior of an ionic polymer-metal composite, *IEEE Trans. Circuits Syst. I*, 2006, **53**(2), 338–350.

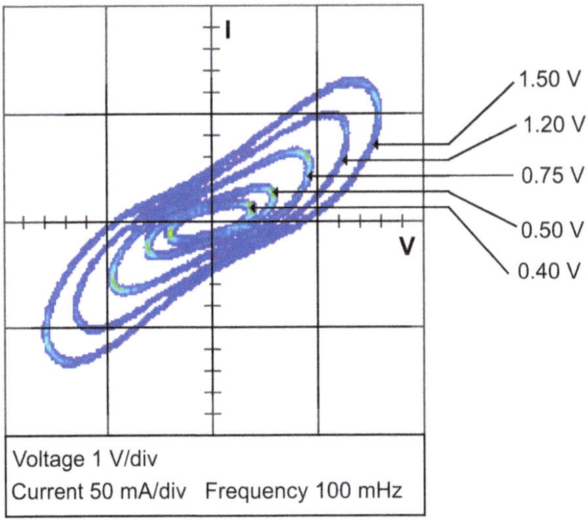

Figure 16.6 Experimental polar voltage-to-current plots of the IPMC-based actuator obtained by applying a sinusoidal input voltage at 100 mHz and varying the amplitude.
© 2006 IEEE. Reprinted, with permission, from C. Bonomo, L. Fortuna, P. Giannone, S. Graziani, A circuit to model the electrical behavior of an ionic polymer-metal composite, *IEEE Trans. Circuits Syst.* **I**, 2006, **53**(2), 338–350.

phase shift between the input voltage and the output current. Finally, by further increasing the input frequency (over 500.0 Hz), the IPMC behaved like a resistor.

Two examples of acquired waveforms in such working ranges are reported in Figures 16.7 and 16.8, respectively.

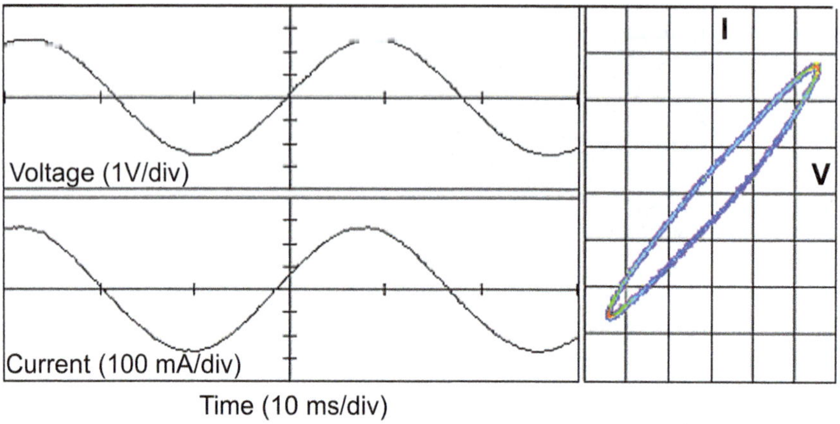

Figure 16.7 Capacitive behaviour of the IPMC at intermediate frequency values. The top image shows the sinusoidal input voltage signal at about 25 Hz and at an amplitude of 2.5 V. The bottom image shows the current response. The corresponding polar plot is also shown.

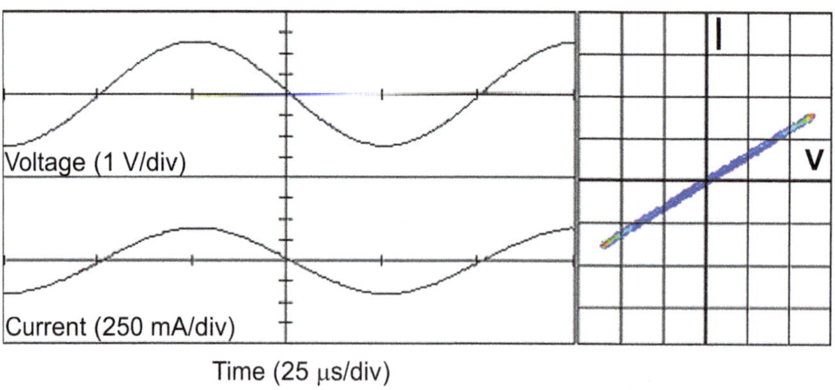

Figure 16.8 Resistive behaviour of the IPMC at high frequency values. The top image shows the sinusoidal input voltage signal at about 10.0 kHz and at an amplitude of 2.5 V. The bottom image shows the current response. The corresponding polar plot is also shown.

The lumped circuit introduced in Bonomo *et al.*,[12] while simple, is capable of modelling, for a suitable choice of component values, the complex voltage to current dependence experimentally observed for IPMC actuators and described so far, and this is why at that time we introduced the model.

As a matter of fact, the proposed circuit realisation is a black-box model and it needed to be tuned against real data. Experiments performed to this end were based on the step response of the IPMC under investigation and details are reported in the referred paper. Here it is worth noting that, after the circuit parameter identification was accomplished, we were able to simulate the voltage to current transduction for sinusoidal inputs and to compare simulated waveforms with acquired ones.

An enlightening example of the simulations we were able to obtain is reported in Figure 16.9(a) and (b), respectively. They show the effect of the input voltage frequency on the relative importance of the nonlinear and linear current contributions in the lumped circuit realisation. If a low-frequency forcing voltage signal is considered (250 mHz, 2.5 V in the reported case; see Figure 16.9(a)), the capacitance C_1 is close to an open circuit; in such conditions, a relevant part of the current flows through the nonlinear branch and this reflects the nonlinearity of the total current. The total current was measured by using a shunt resistor R_i and it is indicated in Figure 16.9 as $I(R_i)$. When a high-frequency voltage signal forces the circuit (25 Hz, 2.5 V; see Figure 16.9(b)), the capacitance C_1 is almost shorted and it represents a preferential path for the current. This reduces the nonlinear contribution to the total current.

It is worth remembering here that the realisation described so far is strictly linked to data used for its identification and no generalisation is possible of the circuit parameter values. This aspect is one of the main weaknesses of black-box models and of course also affects the described model. For this reason, research activity on IPMC models was aimed at finding more general models, as will be described below.

In any case, the proposed model has good estimation capabilities of the IPMC actuators' voltage to current transduction and its performance was considered by Cha *et al.*[17] to be comparable to other lumped models that have been become available in the literature and to a model proposed by the authors. An interesting table[17] compares the performances of such models though at low voltage levels, so that the nonlinearity introduced in Bonomo *et al.*[12] is not activated and the couple of diodes are reduced to open circuits. More specifically, a simple resistance-capacitor (RC) network linear model—consisting of a RC circuit and an additional resistor in parallel connection[15] and two parallel RC networks in an interconnected series with further resistance, respectively[16]—the nonlinear model proposed by Bonomo *et al.*,[12] and the model containing a Warburg impedance proposed by the authors were taken into account and compared *via* the R-square criterion. The results, reported in Cha *et al.*[17] for the modelling of a specific IPMC, are given in Table 16.1.

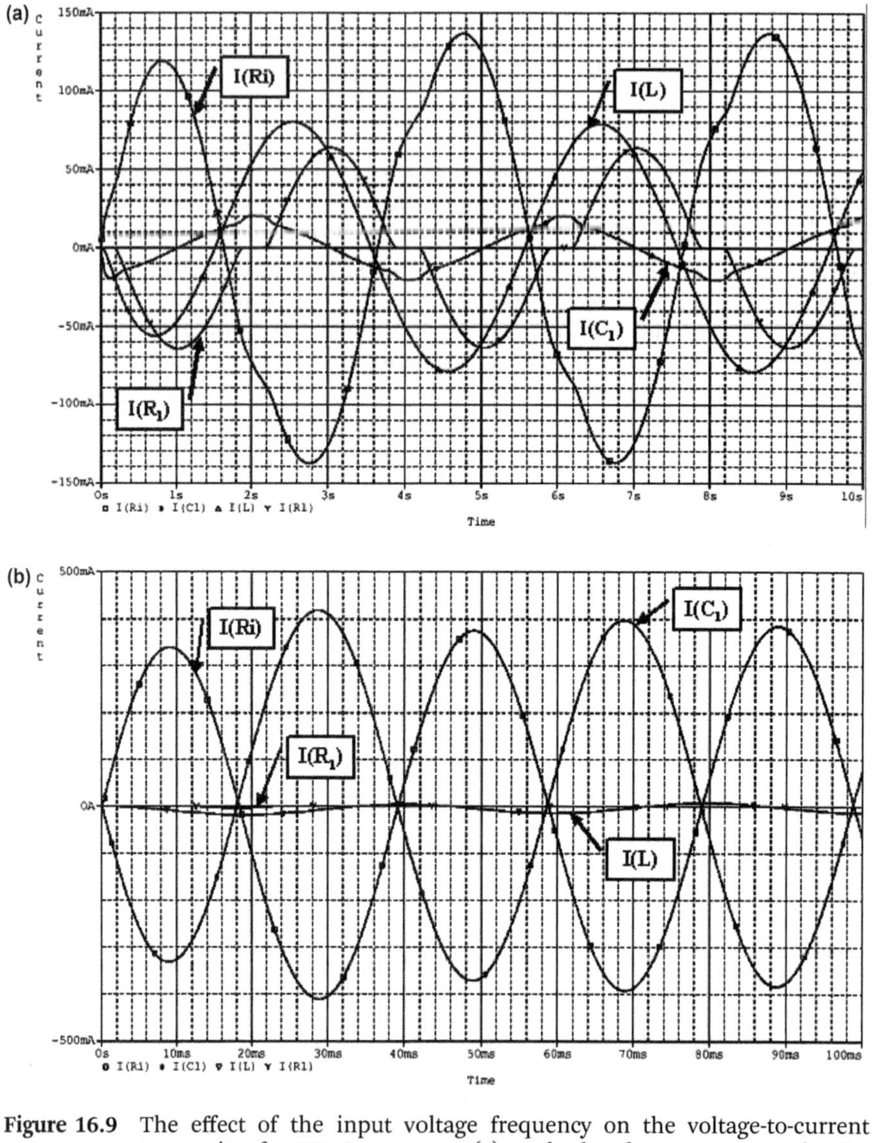

Figure 16.9 The effect of the input voltage frequency on the voltage-to-current conversion for IPMC actuators. (a) In the low-frequency range for the forcing voltage signal (250 mHz, 2.5 V), the relevant part of the current flows through the nonlinear branch and this reflects the nonlinearity of the total current, indicated as $I(R_i)$. (b) At higher-frequency voltage inputs (25 Hz, 2.5 V), the nonlinear contribution to the total current is negligible.

Table 16.1 Performances of models proposed in the literature to describe IPMC actuators' impedance (adapted from Cha *et al.*[17]).

Model	*R*-square
RC model	0.5023
Bao *et al.*[15]	0.8166
Paquette *et al.*[16]	0.8166
Bonomo *et al.*[12]	0.9429
Cha *et al.*[17]	0.9868

It is interesting to note here that, though 20 years separate the models compared in Table 16.1, they share a quite common scheme and behave in a roughly comparable way (not considering the case of the RC model that was, in any case, simply used for comparison). This looks to me to be evidence that, though our knowledge of the phenomena ruling IPMC electromechanical transduction has become increasingly detailed, the black-box models proposed during the "infancy" of the studies on IPMCs were successful in capturing their macroscopic behaviour and modelling the voltage to current conversion *via* lumped circuits.

The electrical realisation for modelling IPMC electrical behaviour by using lumped circuits has been a shared idea and it enables the breaking down of the global transduction phenomenon into a number of meaningful stages. More specifically, for the case of IPMC motion actuators, models have been introduced in time that transform the applied stimulus into charges, since there is a general consensus about the fact that the electromechanical transduction is produced because of charge accumulation in the proximity of the IPMC electrodes. Models are then completed by introducing some kind of electromechanical coupling term. The main differences between the proposed models lie in the complexity of the proposed circuit and/or in the nature of the circuit components.

The path described above has also been adopted for the case of black-box models. In Kanno *et al.*,[14] data were acquired on the voltage input applied to an IPMC actuator in a pinned configuration. Data on both the absorbed current and corresponding tip deflection were also recorded.

The experimental nonlinear electromechanical reactions of IPMC actuators were therefore fitted by an approximated linear dynamic model, consisting of an electrical lumped model that converts the applied voltage into the absorbed current.

A linear stress-generation stage, which accounts for the transduction of the absorbed current into an internal stress, and a mechanical stage, which accounts for the generation of a corresponding volumetric strain and produces the IPMC mechanical bending reaction, were further proposed by the same authors 1 year later.[18] The three stages were considered to be in series connection. Eventually, the authors suggested that the IPMC actuator nonlinearity was limited to the electrical stage, while the current to displacement transduction was modelled by a linear dynamic model.

In both Kanno *et al.*[14] and Kanno *et al.*,[18] the IPMC electrical model was obtained by dividing the actuator into 15 sections, in an attempt to approximate its distributed nature. All of the sections were 1 mm long, five sections were blocked between the powering electrodes, and ten were free to move under the effect of the applied voltage. The parameters in the model were identified by using the IPMC step response and the model of the electrical stage was tested against different waveforms, up to about 20 Hz. The parameters of both the stress-generation stage and the mechanical model were identified by fitting experimental data of the IPMC step input at a point close to the IPMC tip. A view of the proposed electrical stage is reported in Figure 16.10.[14]

The structure proposed by Kanno *et al.* was reconsidered in successive works by other authors who changed the basic assumptions in a number of ways, though maintaining the black-box modelling approach. More specifically, the black-box modelling approach was a powerful tool when the modelling efforts focused on IPMC nonlinearity and/or non-integer behaviour. In fact, though the voltage to current conversion of an IPMC can be roughly described as a capacitive phenomenon, a more accurate investigation reveals minute details that give evidence of a much complex system. It is generally observed that when a step input voltage is applied to an IPMC, the corresponding absorbed current shows a sharp increase

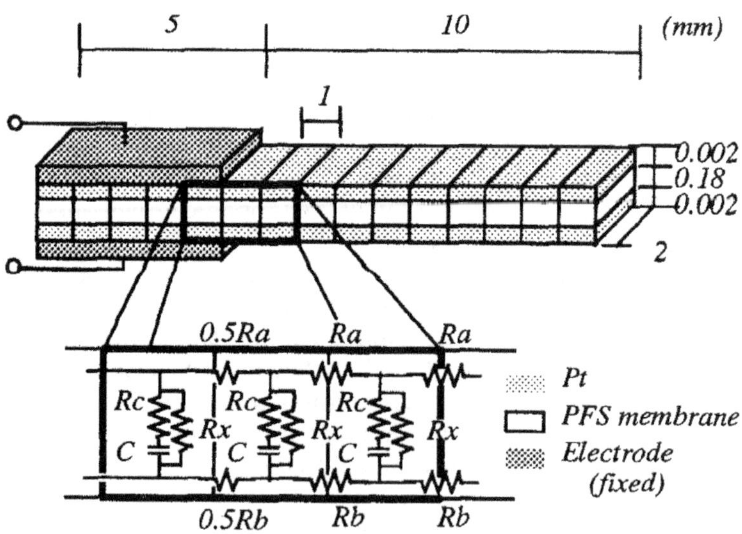

Figure 16.10 The lumped electrical model proposed for IPMC actuators. © 1995 IEEE. Reprinted, with permission, from R. Kanno, S. Tadokoro, M. Hattori and K. Oguro, Modeling of ICPF (Ionic Conductiong Polymer Film) Actuator – Modeling of Electrical Characteristics, Proc. of the 1995 IEEE IECON 21st International Conference on Industrial Electronics, Control, and Instrumentation, Orlando, FL, 1995, 913–918.

followed by a drop that cannot be described by simple capacitance: a faster decay of the time response is followed by slower and slower dynamics so that it looks like the system is increasing its capacitance value. Such a characteristic was observed by a number of different researchers, both for Nafion- and Flemion-based IPMCs.[15,19,20]

An approximate solution could be the introduction of a lumped circuit realisation of the IPMC consisting of a number of cells of RC series-connected elements, characterised by different time constants. This solution was proposed by Newbury and Leo,[19,20] who proposed the use of four parallel branches of RC groups connected in series in an attempt to model with an acceptable level of accuracy both the fast and slow dynamics of experimentally observed step responses for Nafion-based IPMCs.

In Bao *et al.*,[15] the authors compared the performances of Nafion- and Flemion-based IPMCs and then suggested that such behaviour is a consequence of the nature of the electrode–polymer interface.

Generally speaking, the metal–polymer interface is where double-layer capacitors are formed and mobile charges accumulate when external stimuli are applied. A further resistance can be introduced to take into account the effect of the electrolyte. Such models can be further completed by introducing a leakage resistance. Since the CRC branch can be simplified into an RC series, the simplest IPMC model is an RC circuit in parallel connection with the leakage resistance (*i.e.* a first-order system). Bao *et al.*[15] used such a tentative simple model to show its inadequacy for modelling the observed dynamics of the IPMC absorbed current, then a totally different electric circuit realisation was proposed.

It is widely known that electrodes of IPMCs are not just flat layers grown above the bulk polymer;[21] on the contrary, a complex structure can be observed and scanning electron microscopy (SEM) images revealed a quite complex structure. The authors modelled the real electrode structure observed for Flemion-based IPMCs with gold electrodes as a tree-like fractal structure in which each level has its own capacitance and is connected to the next one by a suitable resistance. A SEM view of a Flemion-based IPMC electrode is shown in Figure 16.11(a), while in Figure 16.11(b), the complex distribution of the platinum electrode into a Nafion-based IPMC is shown.[22]

The double-layer capacitance of the fractal electrode is therefore modelled by using a distributed RC line. The authors used such model to show that the absorbed current needs to vary with time as:

$$i(t) \propto t^{-0,5} \tag{16.1}$$

while they were able to accurately fit experimental data by using the expression:

$$i(t) = a + b(t - t_0)^{-0,45} \tag{16.2}$$

provided that experimental data were used to identify the parameters in the model.

In the described model, the fractal nature of the transduction process is limited to the electrical domain. After the current is predicted, a simple

(a)

(b)

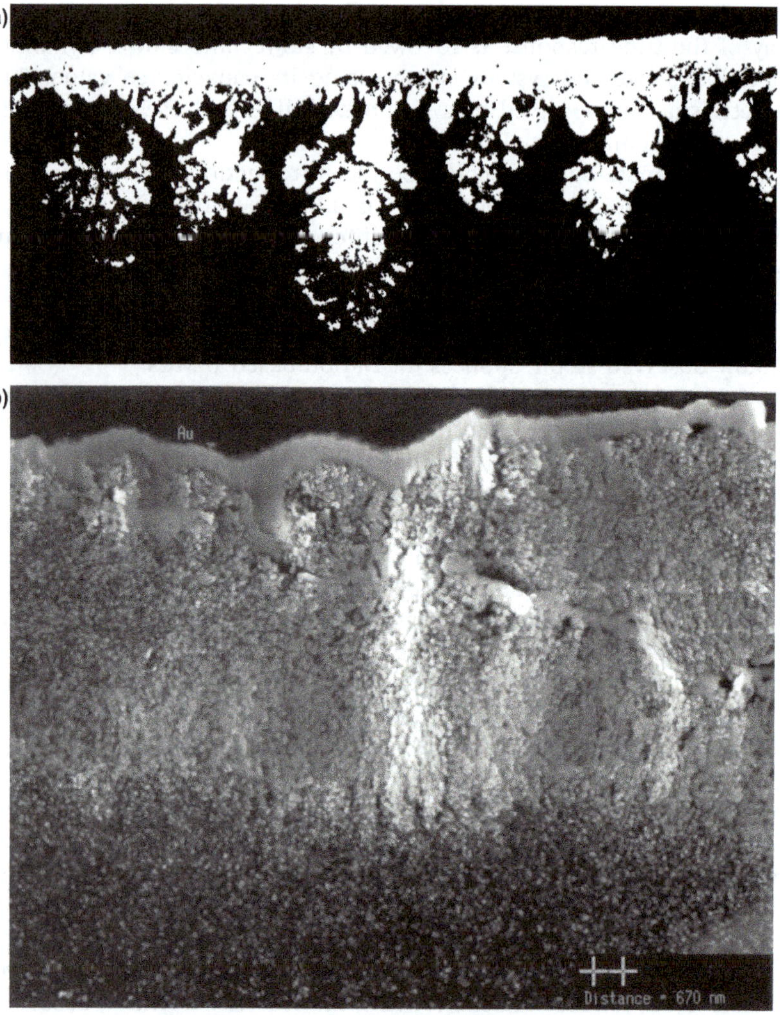

Figure 16.11 Examples of electrode–polymer interfaces. (a) A SEM photograph of an IPMC electrode revealing the fractal nature of the metal (gold in the reported case)–Flemion interface and the corresponding tree-like structure (adapted from ref. 15). Reprinted from X. Bao, Y. Bar-Cohen and S.-S. Lih, Measurements and Macro Models of Ionomeric Polymer-Metal Composites (IPMC), proc. SPIE EAPAD, San Diego, CA, 2002, 4695–4627. (b) Electrode structure for a gold–platinum electrode of a Nafion-based IPMC. Reprinted from S. Nemat-Nasser, Micromechanics of actuation of ionic polymer-metal composites, *J. Appl. Phys.*, 2002, **92**(5), 2899.

model is proposed for the mechanical reaction following a step application, both in the presence of the relaxation phenomenon and for IPMCs that did not show such behaviour. The proposed mechanical model can predict the IPMC tip motion only in cases of step input signals.

Based on the evidence described above, we investigated, using the black-box approach, the fractal nature of Nafion-based IPMC actuation performances. Though we started from the results reported in the literature, we preferred to investigate the frequency responses of actuators, as they are the tools that are generally adopted for studying non-integer dynamics and are considered helpful in the investigation and design of new applications. Of course, the particular nature of the investigated phenomena did not allow the use of the traditional tools adopted for frequency investigation of linear systems. On the contrary, we modelled IPMC actuators as non-integer-order systems.[23]

Also, bearing in mind the importance of actuator power consumption, especially for devices such as IPMC-based actuators, which have been considered promising devices for use in autonomous robotic systems, we focused on the non-integer modelling of both the electrical stage (*i.e.* the voltage to current transduction) and then of the following strain-generation stage (*i.e.* the current to deformation transduction).

Since we pursued a black-box modelling approach, experimental data were required. IPMC actuators were obtained by using Nafion® 117, plated with platinum electrodes that were realised by using the usual electroless plating procedure, and with sodium as the counterion. After the IPMC realisation, strips of 25 mm×3 mm were cut, while the sample thickness was, as usual, about 200 μm.

Measuring surveys were performed in air and an experimental setup was realised to acquire data on the applied voltage, absorbed current and the corresponding tip deflection. A view of the experimental setup is shown in Figure 16.12.

More specifically, the system was forced by a swept voltage sinusoidal signal with a peak-to-peak value equal to 6.0 V and with frequency values varying in the range from 0.5 Hz to 50.0 Hz as the interval in which relevant transduction capabilities of the IPMC were observed. The input signal was produced by using an Agilent Technologies 33 220 A, 20 MHz signal generator. The corresponding absorbed current was transduced into a voltage signal by using a shunt resistor. The IPMC tip deflection was measured by using a commercially available laser sensor (Baumer Electrics OADM12U6430). Signals were acquired at a sampling frequency equal to 1000 samples s^{-1} by using a National Instruments DAQ 6052E acquisition card trough a LabVIEW VI.

The system was approximated to a linear one and the corresponding transfer functions were estimated by using the acquired data. Three transfer functions were taken into account: the voltage to current, current to deflection and voltage to deflection transfer functions.

The corresponding transfer functions have been obtained by using the Matlab® function *ftestimate*. As an example, Figure 16.13 shows the obtained voltage to tip deflection Bode plots.

An inspection of reported Bode diagrams allows to conclude that the system presents non-integer behaviour,[24] since the module's graphs present a slope equal to $m\times 20$ dB per decade, and the corresponding phase diagrams present a phase lag equal to $m\times(\pi/2)$, with m being a real number.

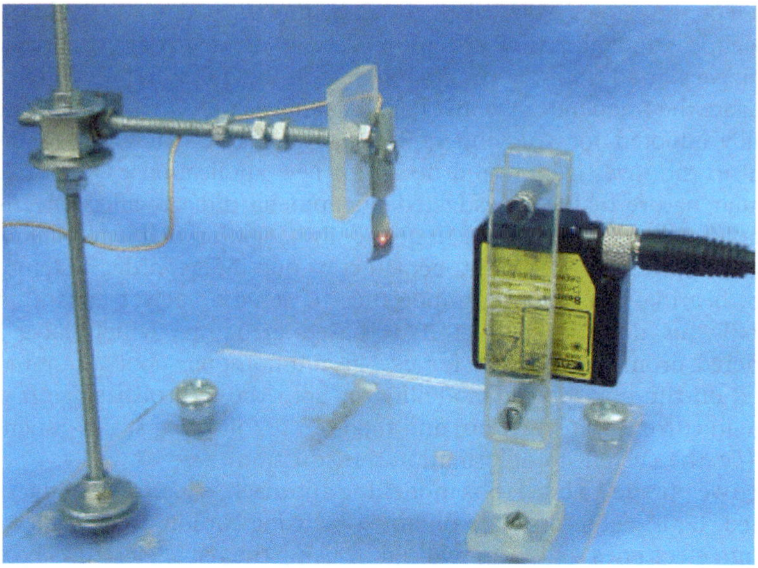

Figure 16.12 A view of the IPMC in cantilever configuration and of the laser sensor used to measure the IPMC tip deflection.
(Adapted from ref. 23). Reprinted from R. Caponetto, G. Dongola, L. Fortuna, A. Gallo and S. Graziani, *IPMC Actuators Non Integer Order Models, New Trends in Nanotechnology and Fractional Calculus Applications*, ed. D. Baleanu, Z. B. Guvenc and J. A. T. Machado, Springer, 2010, pp. 263–272.

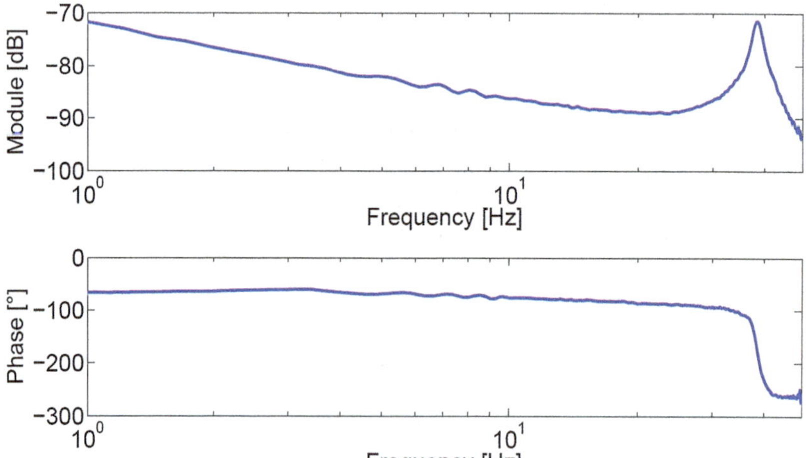

Figure 16.13 Bode plots of the voltage to tip deflection transduction.
Reprinted from R. Caponetto, G. Dongola, L. Fortuna, A. Gallo and S. Graziani, *IPMC Actuators Non Integer Order Models, New Trends in Nanotechnology and Fractional Calculus Applications*, ed. D. Baleanu, Z. B. Guvenc and J. A. T. Machado, Springer, 2010, pp. 263–272.

The next step was the identification of the transfer functions that approximate the reported Bode plots. It is worth noting that since both the pole and zero values and corresponding exponents are searched for, the identification problem is a nonlinear one and adequate minimisation algorithms are required.

The non-integer-order transfer functions were identified by using the Marquardt algorithm[25] and the results are reported in the following:

$$\frac{I(s)}{V(s)} = 0.5s^{0.09} \frac{\left(\frac{s}{0.01}+1\right)^{1.2}}{\left(\frac{s}{1.5}+1\right)^{1.2}} \tag{16.3}$$

$$\frac{D(s)}{I(s)} = \frac{680}{s^{0.876}\left(s^2+3.85s+5880\right)^{1.15}} \frac{\left(\frac{s}{1.5}+1\right)^{1.2}}{\left(\frac{s}{0.01}+1\right)^{1.2}} \tag{16.4}$$

$$\frac{D(s)}{V(s)} = \frac{340}{s^{0.786}\left(s^2+3.85s+5880\right)^{1.15}} \tag{16.5}$$

where $V(s)$, $I(s)$, and $D(s)$ are the Laplace representations of corresponding time-domain applied voltage $V(t)$, absorbed current $I(t)$, and produced tip deflection $\delta(t)$. Once the transfer functions were identified, it was possible to compare model estimations with the corresponding Bode diagrams obtained from experimental data manipulation.

Finally, the identified model was verified in the time domain by considering the swept sinusoidal input voltage and estimating the IPMC tip deflection as predicted by eqn (16.5). A comparison between the observed tip deflection and the corresponding estimation is given in Figure 16.14.

It is worth noting that the proposed model guaranteed an excellent estimation capability of the IPMC tip deflection, at least for the considered signal.

The model described so far was identified by using experimental data and a black-box approach. It is not surprising, therefore, that it cannot be applied to devices that are different to the considered DUT. In an attempt to improve the generality of the model, a further investigation was performed on a set of devices with different sizes.[26]

Actually, that paper was aimed at finding a relationship between the IPMC size and the structure of the non-integer transfer functions, in such a way as to make the obtained models closer to grey-box models as useful compromises between the competing requirements of simplicity, accuracy, and generality. In fact, we intended to demonstrate the possibility of fixing IPMC actuator geometry in advance according to the desired system dynamics, as a case of the role that models can have in the design stage of prototypes realised by using a new technology.

The same experimental setup shown in Figure 16.12 was used to investigate the DUTs, but a quite different value of the input signal was chosen.

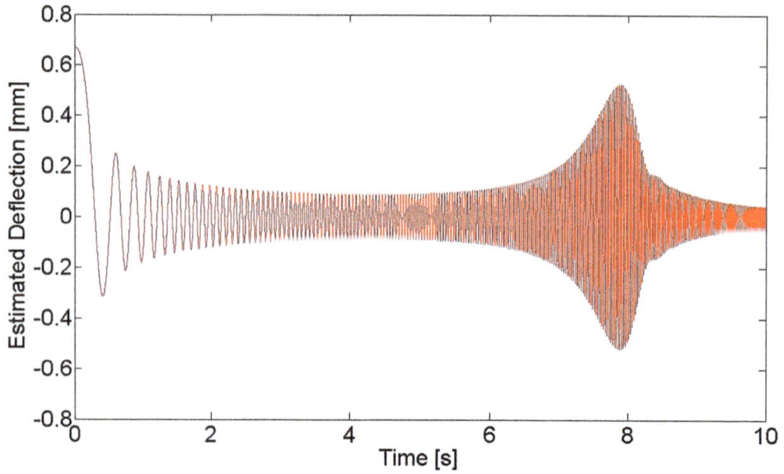

Figure 16.14 Acquired tip deflection for an IMPC stimulated by using a swept sinusoidal signal and corresponding estimation obtained by using the described non-integer transfer function.
Reprinted from R. Caponetto, G. Dongola, L. Fortuna, A. Gallo and S. Graziani, *IPMC Actuators Non Integer Order Models, New Trends in Nanotechnology and Fractional Calculus Applications*, ed. D. Baleanu, Z. B. Guvenc and J. A. T. Machado, Springer, 2010, pp. 263–272.

In fact, the investigated frequency range was 1.0–100.0 Hz. This choice was justified by the need to investigate a wider range of frequencies, since it was expected that the changes in the IPMC geometry would have a direct effect on the actuator frequency response. Moreover, in order to have longer acquisition intervals, the peak-to-peak value of the input signal was reduced to 3.0 V, to alleviate the adverse phenomenon of solvent (water) loss. Data were further processed to estimate the frequency responses.

According with already published findings,[23] frequency responses were approximated by using the non-integer transfer function:

$$F(s) = \frac{D(s)}{V(s)} = \frac{k_0}{s^n \left(\dfrac{s}{p_1} + 1\right)^m \left(\dfrac{s}{p_2} + 1\right)^m} \tag{16.6}$$

with the usual meaning of adopted symbols. By fixing $k = k_0(\alpha^2 + \beta^2)$, eqn (16.6) can be rewritten as:

$$F(s) = \frac{D(s)}{V(s)} = \frac{k}{s^n (s^2 + 2\alpha s + \alpha^2 + \beta^2)^m}. \tag{16.7}$$

The parameters α and β are further linked to the pole system values by the expressions:

$$p_1 = \alpha + j\beta \tag{16.8}$$

$$p_2 = \alpha - j\beta \tag{16.9}$$

The investigation was carried out on six DUTs, characterised by different lengths. Samples 15 mm, 18 mm, 20 mm, 25 mm, 27 mm, and 30 mm long were considered, while all the samples were 3 mm wide and about 200 μm thick.

The values of n and m were not allowed to vary and were held constant, according to the values reported above, while the values of k_0, α, and β were optimised by fitting the Bode plot of the model $\hat{F}(s)$ to the corresponding frequency responses, obtained directly from the acquired data. More specifically, a Matlab® script was used to minimise a cost function of the form:

$$J = J_{\mathrm{M}} + \lambda J_{\mathrm{PH}} \tag{16.10}$$

where:

$$J_{\mathrm{M}} = \frac{1}{N}\sum_{i=1}^{N}\left(|F(i)|^2 - |\hat{F}(i)|^2\right) \tag{16.11}$$

and:

$$J_{\mathrm{PH}} = \frac{1}{N}\sum_{i=1}^{N}(\measuredangle F(i) - \measuredangle \hat{F}(i))^2 \tag{16.12}$$

The sums in eqn (16.11) and (16.12) considered all of the available data and λ was a weighting factor introduced to adequately take into account the errors of both the transfer function modulus and phase.

The vectors of optimal α and β values were further interpolated by using both a linear regression and a second-order polynomial regression with respect to the transducer length. The obtained results are reported in Figure 16.15(a) and (b), respectively.

Actually, measurements from samples that were 18 mm and 27 mm long were not used in the model identification. They were instead used to validate the proposed non-integer model.

In Figure 16.16, the estimation capabilities of the proposed model are shown for the case of the IPMC actuator that was 27 mm long.

A number of works have outlined that IPMC transducers show quite evident nonlinearities. Though linear models have been accepted as useful approximations,[5,8] research efforts have been devoted to the explanation of such behaviour and to its modelling.

Of course, black-box modelling is a powerful tool for modelling nonlinear systems, especially when computing paradigms such as fuzzy logic and artificial neural networks are used. It fact, it is widely known that both artificial neural networks and fuzzy algorithms are universal approximators for a large class of functions and are particularly suited for multi-input models. Moreover, optimisation algorithms have been proposed for years that allow the fitting of both artificial neural networks and fuzzy algorithms to experimental data,[27,28] so that, after their optimisation, they can work as excellent data-driven models.[29,30] It is not surprising, therefore, that

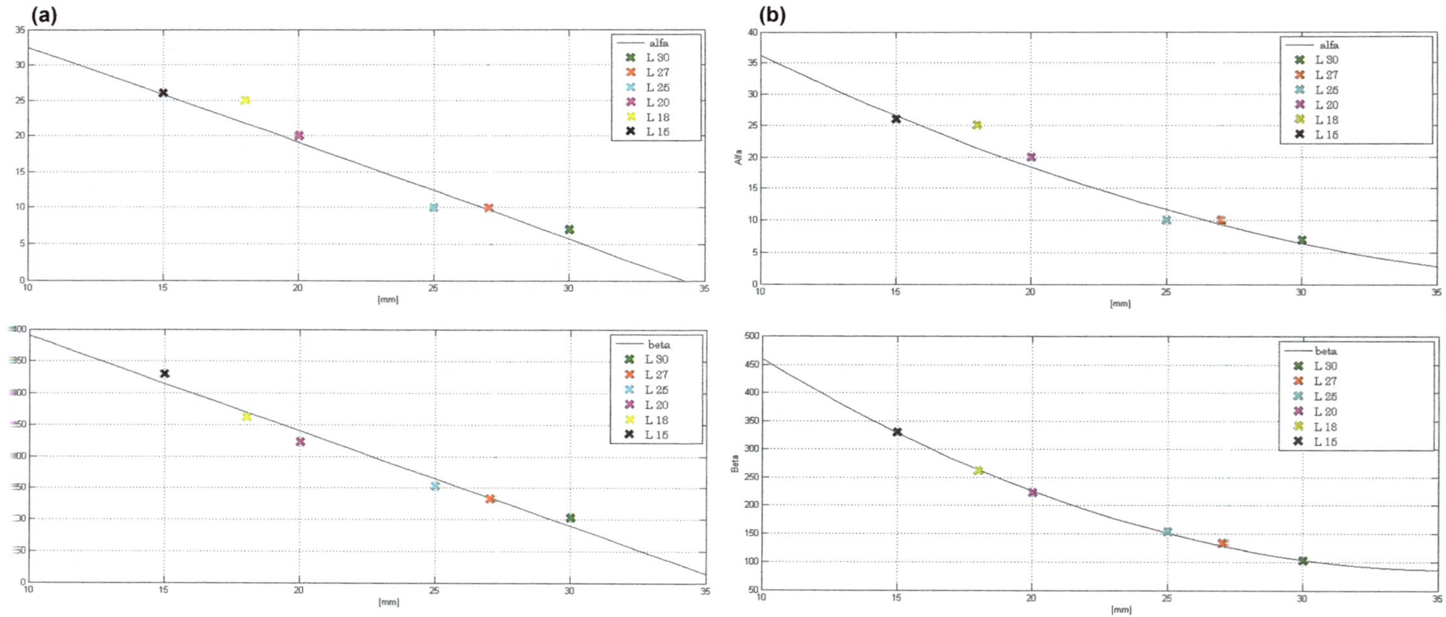

Figure 16.15 Interpolation of α and β values. (a) Linear regression; (b) second-order polynomial regression.
These figures were originally published in the Proceedings of the, IFAC 7th Vienna International Conference on Mathematical Modelling (MATHMOD 2012), Vienna, 2012, 593–596 IFAC-PapersOnLine © IFAC DOI: 2012 10.3182/20120215-3-AT-3016.00105.

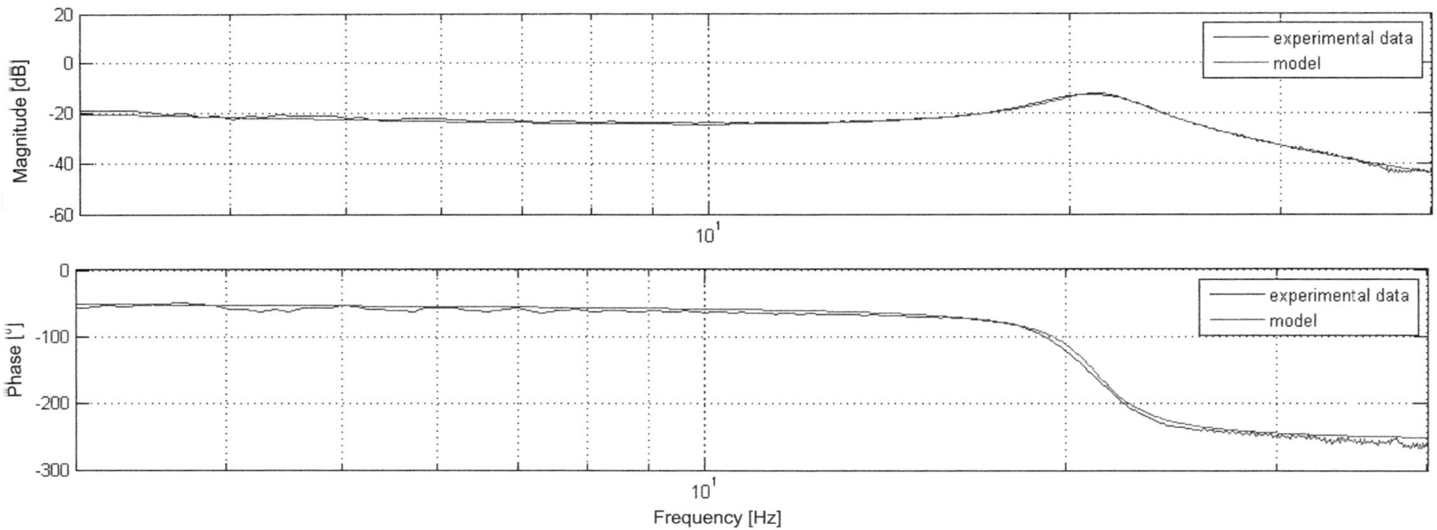

Figure 16.16 Experimental frequency response of the IPMC actuator that was 27 mm long and the corresponding estimation obtained by using the parameterised non-integer model.
This figure was originally published in the Proceedings of the, IFAC 7th Vienna International Conference on Mathematical Modelling (MATHMOD 2012), Vienna, 2012, 593–596 IFAC-PapersOnLine © IFAC DOI: 2012 10.3182/20120215-3-AT-3016.00105.

researchers focused recently on such tools to model the IPMC transducer behaviour.

Truong and Ahn[31] used a set of feedforward multilayer perceptrons trained by using experimental data acquired on an IPMC actuator in order to identify the neural model that best estimates the IPMC tip deflection as a function of the applied input voltage, the voltage signals measured at given points along the IPMC electrode surface and a suitable number of past samples of the IPMC tip deflection. According to the shown results,[31] the best neural model was a multilayer perceptron with one hidden layer and eight hidden neurons. This was chosen from among structures with one hidden layer and the number of hidden neurons spanning from 6 to 14. Also, models with no feedback and one and two feedbacks, respectively, were considered, and the best results were obtained by considering two delayed output values.

While during the network training phase the network inputs were built by considering data acquired from the real system, during the validation phase, the IPMC tip estimation was directly fed back to the model, as shown in Figure 16.17.

The training and testing phases of the investigated models were performed by using a square wave with amplitude equal to 4.2 V. The final validation phase was carried out by using both square waves of different amplitudes, sinusoidal signals, and pseudo-random binary (PRB) signals. An example of the estimation capabilities of the obtained neural nonlinear model is reported in Figure 16.18 for the case of PRB symmetric signals with a 3 V amplitude.

Annabestani and Naghavi[32] proposed an adaptive neuro-fuzzy inference system (*i.e.* a fuzzy algorithm whose structure has been implemented in such

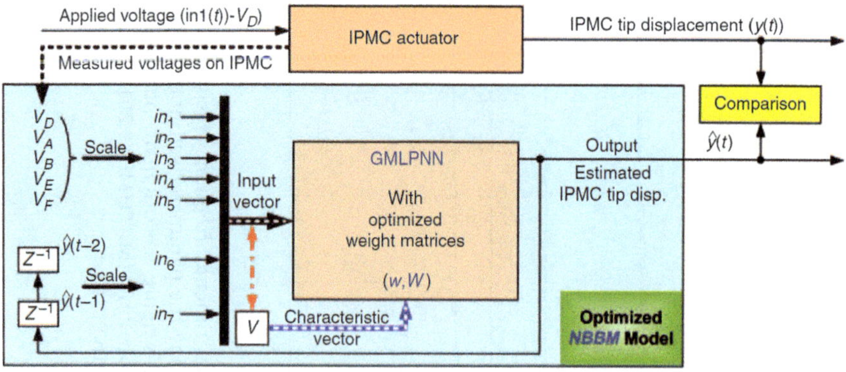

Figure 16.17 Structure of the neural (general multi-layer perceptron neural network – GMLPNN) dynamic nonlinear black-box model (NBBM) proposed in ref. 31.
Reprinted from D. Q. Truong and K. K. Ahn, Design and verification of a non-linear black-box model for ionic polymer metal composite actuators, *J. Intell. Mater. Syst. Struct.*, 2011, **22**, 253–269.

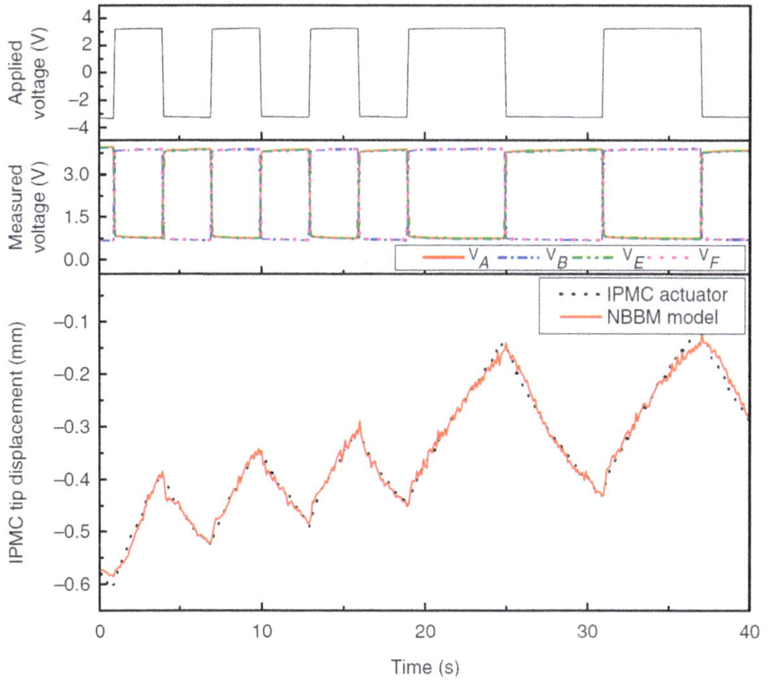

Figure 16.18 An example of the estimation capabilities of the nonlinear neural model proposed in ref. 31.
Reprinted from D. Q. Truong and K. K. Ahn, Design and verification of a non-linear black-box model for ionic polymer metal composite actuators, *J. Intell. Mater. Syst. Struct.*, 2011, **22**, 253–269.

a way that it can be fitted to experimental data by using in sequence the least square optimisation and the back propagation algorithms) to identify a nonlinear autoregressive with exogenous (NARX) input dynamic model of an IPMC actuator:[7]

$$\begin{gathered} y(k+1) = f(R(k)) \\ R(k) = [y(k), \ldots, y(k-n_y+1), u(k), \ldots, u(k-n_u+1)]^T \end{gathered} \tag{16.13}$$

More specifically, the authors show the results obtained with a first-order NARX model of the IPMC tip motion in the transversal direction. In this contribution, the proposed model estimates the current time discrete value of the tip motion as a function of the first past samples of the tip deflection itself and of the voltage input, respectively.

The authors concluded by showing the performances of the identified adaptive neuro-fuzzy inference system-NARX (ANFIS-NARX) model for a set of test signals, as reported in Figure 16.19.[32] More specifically, Figure 16.19(a)–(d) report testing cases obtained by using a sinusoidal, chirp, square wave, and sampled Gaussian noise voltage, respectively.

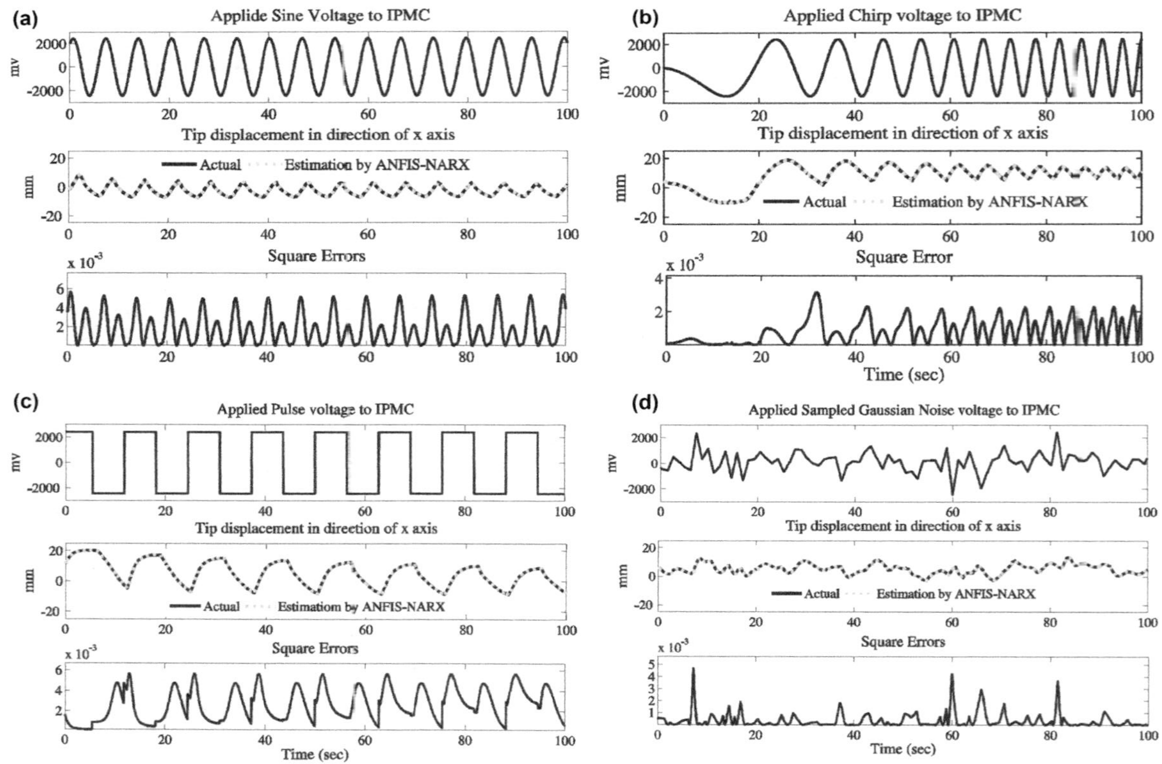

Figure 16.19 Performance of the IPMC ANFIS-NARX model for testing cases, obtained by using a sinusoidal (a), chirp (b), square wave (c), and sampled Gaussian noise (d) voltage, respectively.
Reprinted from M. Annabestani and N. Naghavi, Nonlinear identification of IPMC actuators based on ANFIS-NARX paradigm, *Sens. Actuators, A*, 2014, **209**, 140–148.

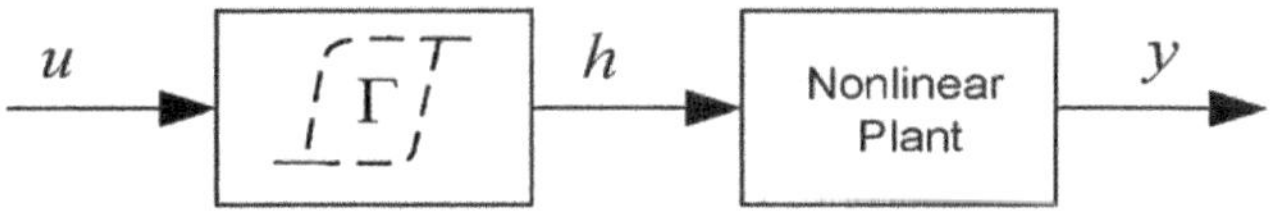

Figure 16.20 The cascaded structure proposed to model hysteresis and nonlinear dynamics in IPMC actuators.
Reprinted from D. N. C. Nam and K. K. Ahn, Identification of an ionic polymer metal composite actuator employing Preisach type fuzzy NARX model and Particle Swarm Optimization, *Sens. Actuators, A*, 2012, **183**, 105–114.

Eventually, a table was given that showed the superiority of the proposed model with respect to neural network-based and polynomial NARX models.

A further nonlinear black-box model was identified by Nam and Ahn[33] as an attempt to model both the nonlinearity and the hysteresis that characterise the IPMC actuator electromechanical transduction mechanism. The proposed model consists of the cascaded combination of a Preisach operator to model the rate-independent hysteresis behaviour, and a fuzzy NARX model to take into account the nonlinear dynamics of the IPMC. The block scheme of the proposed nonlinear model is shown in Figure 16.20.

More specifically, the fuzzy NARX model was obtained by the fuzzy manipulation of a number of linear time-invariant second-order ARX models, in the same way as already reported in eqn (16.13).

As usual for any black-box model, the authors realised an *ad hoc* test rig and acquired data on quantities relevant for the searched model (*i.e.* the voltage difference applied at one end of an IPMC strip and the corresponding free tip deflection).

Data were used to determine both the hysteretic and then dynamic parts of the model. The hysteresis is identified by forcing onto the IPMC the so-called first-order reversal input signal and then minimising the error between the hysteresis model and recorded IPMC output. The obtained results are reported in Figure 16.21.

After the Preisach operator was identified, the authors were also able to identify the dynamic part of the model by using a particle swarm optimisation approach to the fuzzy NARX model as an alternative to the genetic algorithm optimisation approach.

Finally, the validation of the proposed model was performed for a set of signals, and for comparison, the estimations of the proposed Preisach fuzzy NARX model were compared with the corresponding estimations obtained by using an optimised fuzzy NARX model. An example of the significant approximation capabilities obtained by using the proposed model can be seen in Figure 16.22, where the case of a very-low-frequency sinusoidal input (0.005 Hz) is considered as a type of signal capable of outlining the rate-independent behaviour that characterises a hysteretic phenomenon.

Lantada *et al.*[34] used artificial neural networks to model the time-varying nature of IPMC actuators, using water as the solvent, when working in dry environments (actually, experiments were run in a controlled humidity chamber with nominal relative humidity equal to 10%). More

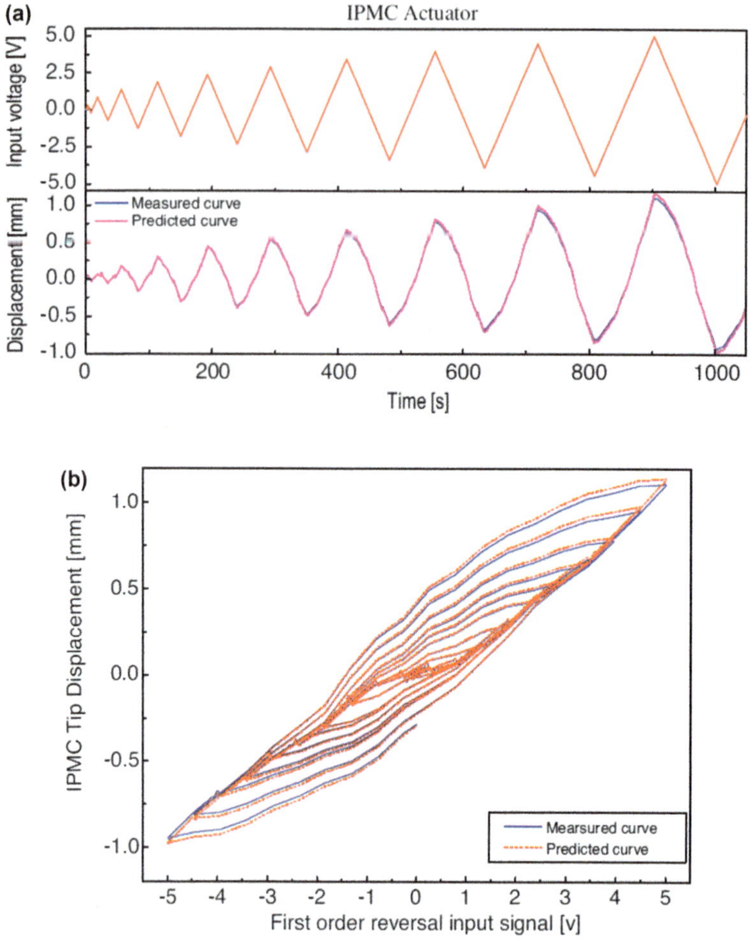

Figure 16.21 The IPMC actuator hysteresis identification. (a) IPMC time response to the first-order reversal input signal; (b) hysteresis curve.
Reprinted from D. N. C. Nam and K. K. Ahn, Identification of an ionic polymer metal composite actuator employing Preisach type fuzzy NARX model and Particle Swarm Optimization, *Sens. Actuators, A*, 2012, **183**, 105–114.

specifically, the authors investigated the possibility of estimating a suitable defined IPMC bending angle when a one-hidden-layer neural network was inputted with the time passed by the transducer out of water and the time since a square wave voltage signal was applied to the actuator. The neural network was intended to take into account the observed decay in the IPMC maximum deflection angle with time, as reported in Figure 16.23.

After a number of trials, the authors found that the best-performing neural network had ten hidden neurons, and such a neural structure was used to validate the proposed neural model. An example of the obtained results is shown in Figure 16.24.

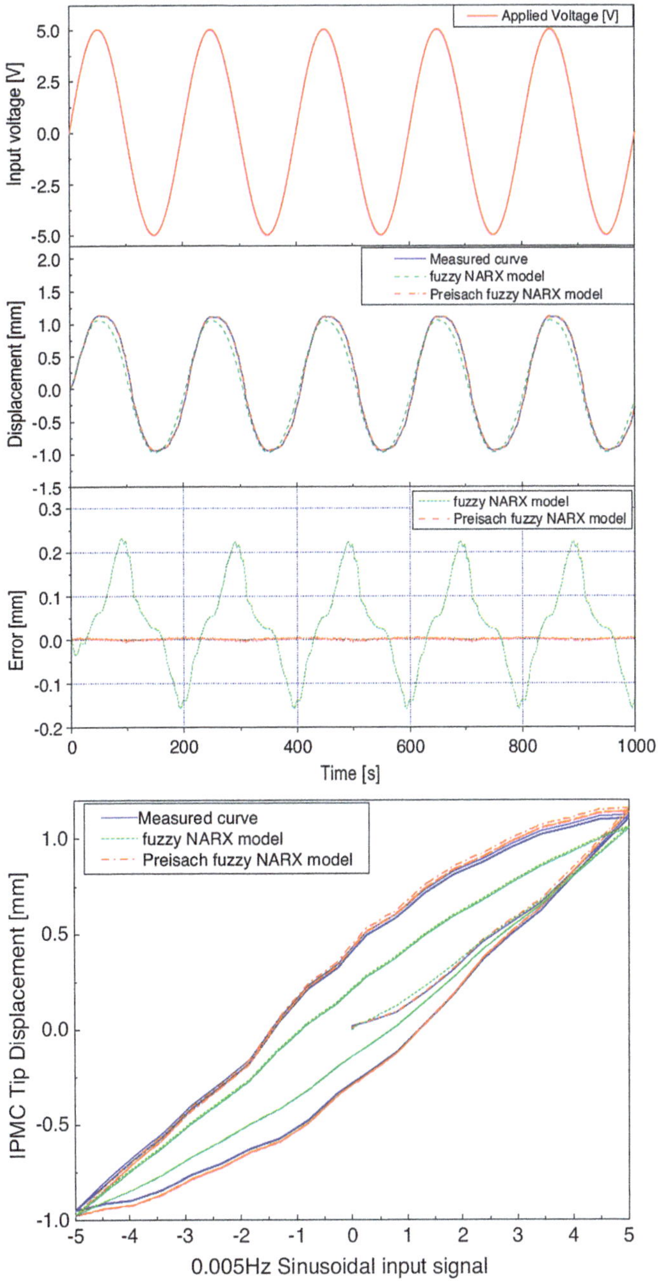

Figure 16.22 Estimation capabilities of the described Preisach fuzzy NARX model. Reprinted from D. N. C. Nam and K. K. Ahn, Identification of an ionic polymer metal composite actuator employing Preisach type fuzzy NARX model and Particle Swarm Optimization, *Sens. Actuators, A*, 2012, **183**, 105–114.

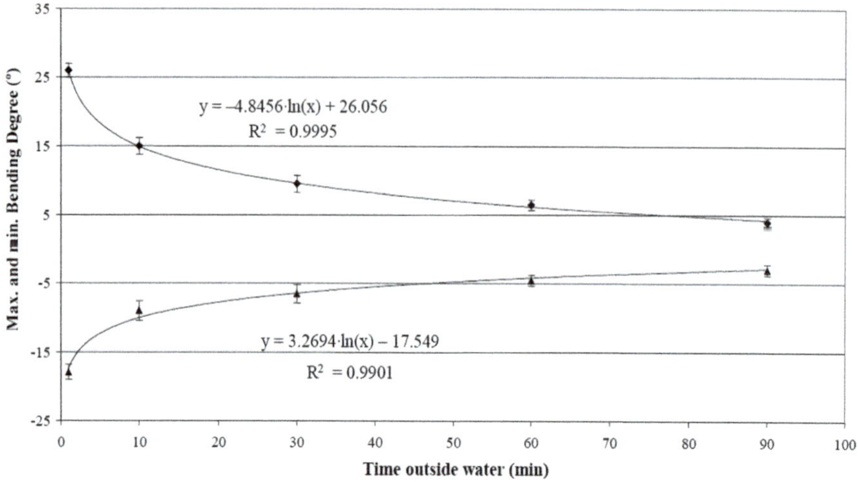

Figure 16.23 The decay of the IPMC bending angle with time spent out of water when a square wave input voltage is applied.
Reprinted from A. D. Lantada, P. L. Morgado, J. L. M. Sanz, J. M. M. Guijosa and J. E. Otero, Neural network approach to modelling the behaviour of ionic polymer-metal composites in dry environments, *J. Signal Information Process.*, 2012, **3**, 137–145.

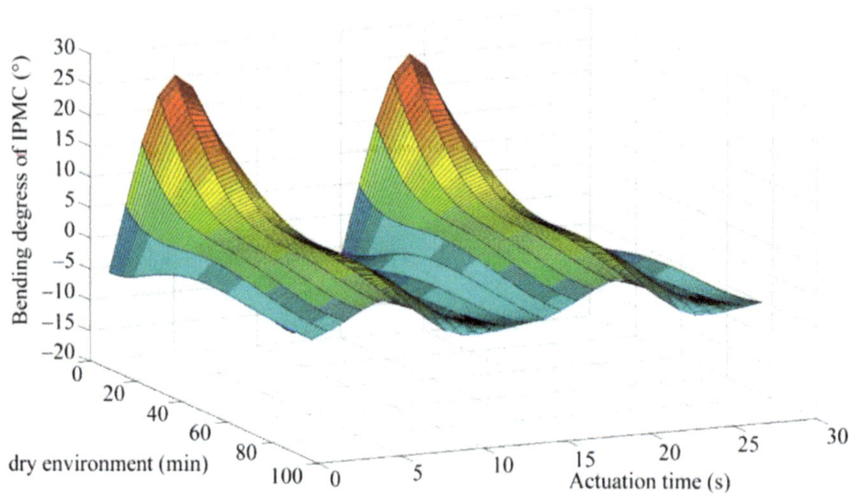

Figure 16.24 Neural model simulation of the IPMC deflection angle as a function of time spent out of water and of time since the input signal was applied.
Reprinted from A. D. Lantada, P. L. Morgado, J. L. M. Sanz, J. M. M. Guijosa and J. E. Otero, Neural network approach to modelling the behaviour of ionic polymer-metal composites in dry environments, *J. Signal Information Process.*, 2012, **3**, 137–145.

Since my research activity has also dealt with neural networks for data filtering and nonlinear dynamic system identification,[35] I focused on the black-box modelling of ionic polymeric transducer dynamic behaviour by using neural networks.

We investigated the possibility of using nonlinear black-box models, including artificial neural networks, to model the multi-input nature of polymeric transducers.[36] More specifically, the multi-input black-box modelling was pursued for a completely polymeric EAP: the ionic polymer–polymer composites (IP^{2}Cs).[37–39] It should not be a surprise that, while dealing with IPMC models, a novel class of EAPs is considered: on the one hand, research activity in this field is a multidisciplinary task and competencies in material sciences, physics, chemistry, and engineering interact to improve the state of technology, knowledge, and applications available. On the other hand, new findings in one of the mentioned fields represent a stimulus and raise new questions for the remaining fields.

Of course, this also occurs during my activity and I am part of a multidisciplinary team in which we share the awareness that improvements in the field of polymeric transducers requires continuous efforts in new materials synthesis, modelling, and meaningful applications. IP^{2}Cs are the result of such efforts and represent a novel class of polymeric transducers that we have investigated for a while. More details about IP^{2}C models we developed will be given in next sections of this chapter, while here the focus will be exclusively on their multi-input nonlinear black-box modelling.

IP^{2}Cs are direct descendants of IPMCs,[38] with replacement of metal electrodes by polymeric ones. While IPMCs combine the electromechanical properties of ionic polymer membranes with the conductivity of surface electrodes made of deposited noble metals (platinum or gold), IP^{2}Cs are manufactured through deposition of organic conductor electrodes directly onto an ionic polymer membrane. The completely polymeric nature of IP^{2}Cs maintains the electromechanical coupling capability, low required voltage, high compliance, lightness, and softness of IPMCs and requires simpler production procedures, lower production costs, and are suitable for the integration in envisaged post-silicon all-polymeric smart systems.

The IP^{2}Cs we used were manufactured through the direct polymerisation of poly(3,4-ethylenedioxythiophene):poly(styrenesulfonate) (PEDOT:PSS) on a Nafion membrane.[38]

As a matter of fact, we also proposed the modelling and characterisation of IPMC actuators with respect to humidity, obtained by using a grey-box approach,[40,41] and the obtained results will be described later in this chapter. Here, the fulfilment of the same objective, obtained by using the black-box approach with the aim of obtaining nonlinear neural models of IP^{2}Cs, will be described.[36]

Our interest in ionic polymeric transducer dependence on environmental humidity is twofold, as with anyone involved in transducer synthesis and characterisation. In fact, such an effect is generally considered to be an undesired phenomenon if the realisation of a motion transducer is of interest. It introduces a dependence of the system performance on an environmental parameter that is the source of an uncertainty in the system behaviour. In that case, the modelling of the undesired dependence is mandatory in order to cancel out the humidity effect.

The dependence of polymeric actuator vibration characteristics on relative humidity can represent, in any case, a possibility for using polymeric vibrating structures as humidity sensors. Ionic electroactive actuators in a pinned configuration behave, in fact, as second-order underdamped systems and the humidity values affect both the resonance frequency values and the corresponding resonance peak values. Both of these parameters could be used to estimate the environmental humidity value.

The IP^{2}C was used as a vibrating electromechanical transducer, mounted in the cantilever configuration: the IP^{2}C strip was pinned at one end and it bent when an electrical stimulus was applied across its thickness. The relative humidity changed the natural resonant frequency of the beam and the corresponding resonance peak. Actually, sensing devices based on resonant cantilevered beams are widely proposed in the literature to realise MEMS based on piezoelectric materials.[42] We were interested in the described model as a proof of concept of the possibility to use IP^{2}C for future realisations of polymer MEMS.[43,44]

The IP^{2}C actuator model was obtained by using neural networks. More specifically, the referred work was focused on the investigation of IP^{2}C dynamic behaviour in the frequency domain.

In order to develop a model for different values of humidity, a set of experimental data was acquired. Then three different neural network models (*i.e.* feed-forward neural network [FFNN], radial-basis neural network [RBNN], and recurrent neural network [RNN]) were developed. Eventually, the models were compared in terms of their performance.

The IP^{2}C was built by using the usual Nafion® 117 of thickness 180 μm, produced by DuPont, while the surface electrodes were made of PEDOT:PSS.

The measurement surveys were performed in a chamber realised *ad hoc* with the aim of controlling the relative humidity and measuring the environmental temperature. The experimental setup is shown in Figure 16.25. A swept sine-wave with a peak-to-peak amplitude of 4 V and a variable frequency in the range of 1 Hz to 45 Hz (V_{in}) was generated by a signal generator and fed into the IP^{2}C device through a conditioning circuit by a couple of copper contacts, which also allowed measuring of the absorbed current (I_{in}) over time. The tip displacement δ was measured using a distance laser sensor. The humidity range was 40–90%.

The estimated frequency responses as a function of the chamber relative humidity are seen in Figure 16.26. Data acquired during the measurement surveys were used to identify the model. More specifically, 70% of the data were used for the training phase, while 30% of the data were used for model validation. The frequency and the relative humidity were used as inputs to the neural networks and the frequency response as the output.

FFNN, RBNN, and RNN structures were investigated by using the mean square error (MSE) in the estimation of the frequency response as a

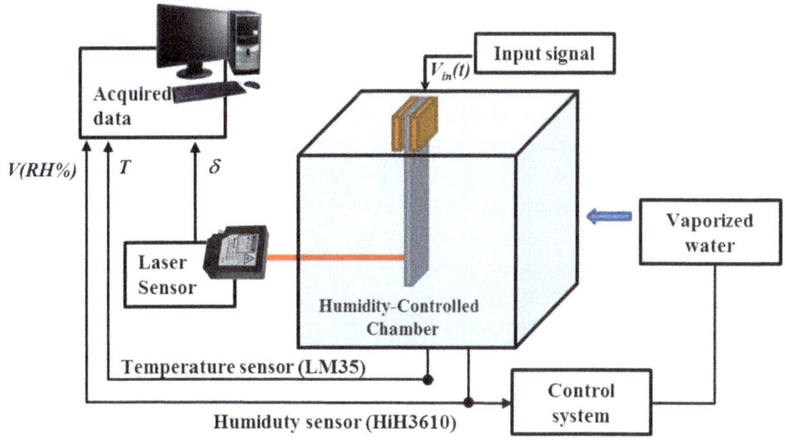

Figure 16.25 Experimental setup used for assessment of IP^{2}C dependence on relative humidity.

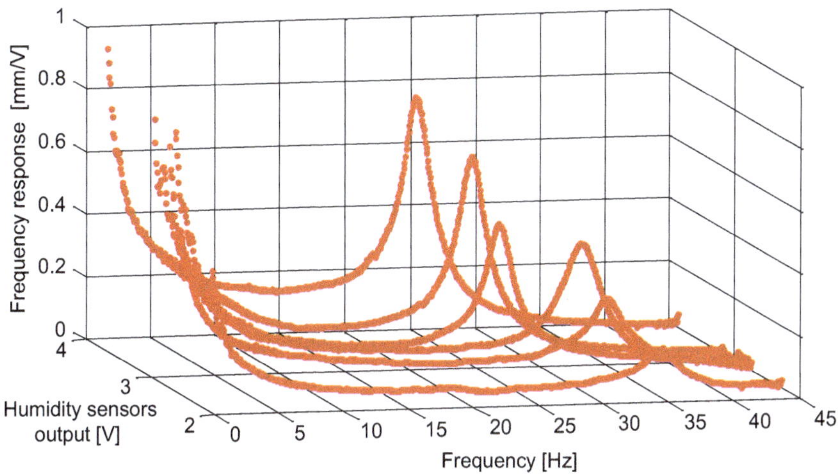

Figure 16.26 Estimated frequency responses as a function of the environmental relative humidity.

Table 16.2 Performance of the considered neural networks in the estimation of the IP^{2}C frequency response (adapted from De Luca *et al.*[36]).

	MSE	Nodes	Epochs	Correlation coefficient
FFNN	0.535×10^{-4}	37	411	0.9978
RBNN	0.00395	100	248	0.8904
RNN	0.628×10^{-4}	45	1000	0.9879

performance index. The obtained results are summarised in Table 16.2, showing that the best working neural network was the FFNN, both in terms of the MSE and of the corresponding correlation coefficient between the acquired data and the neural model estimations.

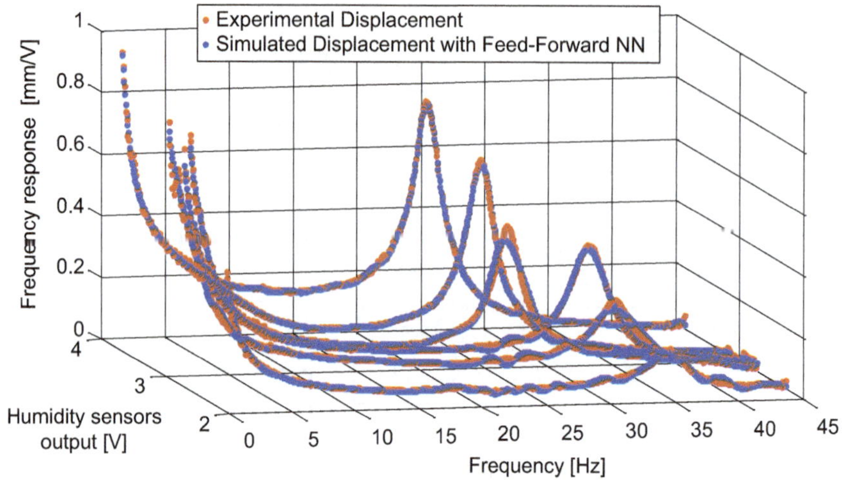

Figure 16.27 Performance of the FFNN in the estiamation of the IP^2C frequency response.

Finally, in Figure 16.27, the comparison between experimental and simulated data obtained using the FFNN is reported.

The models described so far show both the pros and cons of the black-box modelling approach. It can guarantee an excellent level of accuracy in the prediction of electromechanical transducer behaviour, even for quite complex soliciting signals (*e.g.* the described possibility of predicting either the influence of the input voltage[34] or even of the concurrent contribution of the electrical stimulus and of environmental parameters, such as relative humidity[36]). Such accuracy is an excellent tool when devices are used as measuring systems or are required for the realisation of accurate positioning systems (*e.g.* for the realisation of micro-robots or surgery aids).

Nevertheless, the proposed models do not give any explanation of the microscopic phenomena that produce the system reaction; in this sense, black-box models can improve only marginally our understanding of the processes they describe. This is also true for ionic polymeric transducers and is the reason why, with the exception of some niche topics, black-box models of IPMCs, which flourished during the infancy of this class of transducers, have been substituted over time with grey-box and white-box models, which will be the topics of the next sections.

16.2.2 Grey-box Modelling

Since the very first years of my research activity on IPMCs, I had an application-orientated approach to the field. There was a general consensus in the literature of the need to find "a dynamic model that is useful for the design of material systems that incorporate ionic polymer materials... a set of expressions which can be used to design sensors, actuators, or material systems that contain ionic polymer materials"[10] and that describes the transduction capabilities of the ionic polymer materials in terms of some material parameters and of the geometry of the transducer in such a way as to guarantee the scalability of the model. I quoted the sentences reported above because I believe that they are the real essence of what was asked of IPMC models during the first decade of this century, and it is the approach I followed in my research activity on IPMC modelling during the last decade.

The black-box models described so far fail to guarantee the capabilities described above. This can be clearly understood if another contribution by Leo and his coworkers is considered.[45] In that paper, a black-box approach was adopted to determine a dynamic linear model of an IPMC bender in a cantilever configuration. Nevertheless, I refer to it for matter of comparison, to outline the dramatic differences between this model and the grey-box or semi-empirical models that the same authors proposed shortly after.[19,20]

The model proposed by Newbury and Leo[45] is a coupled electromechanical transduction model of an IPMC bender that can be used with both IPMC-based sensors and actuators. More specifically, the IPMC transducer has been modelled by using a two-port equivalent electrical system as represented in Figure 16.28, where both an IPMC transducer mounted in cantilever configuration and the corresponding scheme are reported.

In the proposed model, v and i are the electrical variables (*i.e.* voltage and current at the IPMC electrodes), while f and $\dot{u}$ are the mechanical ones (*i.e.* external force and velocity of the point of the external force application L_d).

The authors stated that the choice was motivated at that time by the lack of adequate first-principle analysis models capable of describing the microscopic phenomena involved in IPMC transduction and by the desire to develop a model focused on the relationship that links variables relevant to the transducer user.

The equivalent model is a linear one and the ruling equations were written in the frequency domain by using a matrix of impedances:

$$\begin{bmatrix} v(\omega) \\ f(\omega) \end{bmatrix} = \begin{bmatrix} Z_{11} & Z_{12} \\ Z_{21} & Z_{22} \end{bmatrix} \begin{bmatrix} i(\omega) \\ \dot{u}(\omega) \end{bmatrix} \tag{16.14}$$

Measurement surveys were executed to identify the equivalent circuit impedances. More specifically, the model was identified by performing a large number of static and dynamic measurements on a Nafion-based 17 mm×5 mm×0.2 mm IPMC bender with golden electrodes, Li^+ as the

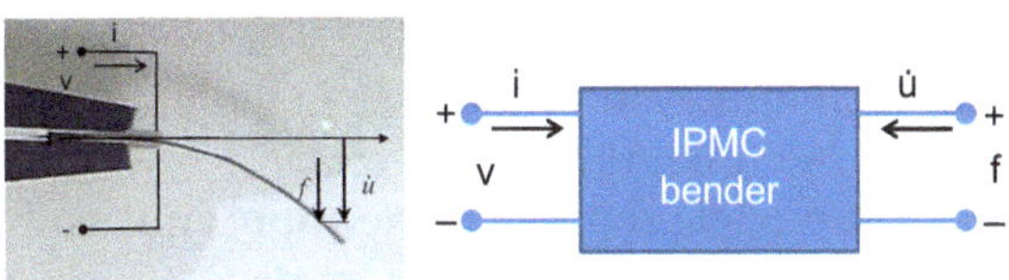

Figure 16.28 An IPMC mounted in cantilever configuration and the two-port model introduced by Newbury and Leo to model IPMC-based transducers.[45]

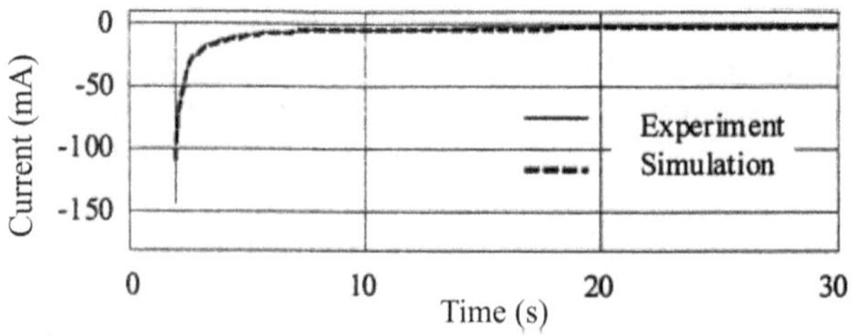

Figure 16.29 Simulated and experimental absorbed current in the blocked condition. Reprinted from K. M. Newbury, D. J. Leo, Electromechanical modeling and characterization of ionic polymer benders, *J. Intell. Mater. Syst. Struct.*, 2002, **13**, 51–60.

counterion, and water as the solvent. The observed time data were processed in order to identify the gains, zeros, and poles of the Z_{ij} in eqn (16.14).

After the impedances were identified, the authors validated the proposed model against experimental data. As an example, the experimental absorbed current in the blocked boundary condition, and the corresponding model estimation when a step voltage was applied, are shown in Figure 16.29. This experiment was used to confirm the relation:

$$\left(\frac{i}{v}\right)^{\dot{u}} = \frac{1}{Z_{11}} \tag{16.15}$$

which can be easily derived by the model in eqn (16.14). Similar tests were used to confirm the suitability of the proposed model and of the adopted Z_{ij} structures.

Though the proposed model presented excellent fitting capabilities, in the conclusion section, the authors state that since the parameters in the model were determined empirically, it is not possible to scale them when transducers with different geometries are considered. This is one of the key weaknesses of black-box models: they are valid only for the devices used during model identification.

The grey-box modelling approach can alleviate such a drawback. Grey-box models are obtained on the basis of some well-understood theories (*e.g.* the beam theory and an adaptation of the piezoelectric theory for the case of IPMC transducers) and are ruled by using parameters that can be estimated with a set of experiments. More specifically, data of relevant quantities are acquired and are processed by suitable optimisation algorithms that minimise some kind of cost function in order to fit model estimations to the observed data. The minimisation of the cost function allows estimating of the optimal values of the unknown parameters.

The evolution of the model proposed by Newbury and Leo[45] highlights the differences between black-box and grey-box models.

The IPMCs were still modelled as linear dynamic systems, and the Z impedance matrix of eqn (16.11) was reconsidered,[19,20] but this time the form of the Z_{ij} matrix entries depend only on the characteristics of the materials, so that the two-port model can be scaled on the basis of the size of the considered device.

More specifically, IPMCs were considered in a pinned configuration and the mechanical vibrations were modelled in the framework of the Euler–Bernoulli beam theory. Moreover, since IPMCs are viscoelastic materials (*i.e.* materials with both solid-like and liquid-like features), their Young's modulus cannot be assumed as a constant value. A complex formulation of the Young's modulus is required in the Laplace domain, where the real part, known as the storage modulus, models the linear elastic behaviour of the material, and the imaginary part takes into account the dissipative behaviour.

This is of particular relevance when dynamic phenomena are of interest, as for the case of IPMC deformations. The authors explicitly took into account this aspect by using the Golla–Hughes–McTavish (GHM) method.

The electrical behaviour of the IPMCs was modelled by considering the usual combination of resistive and capacitive elements, leading to the introduction of an equivalent material permittivity.

The electromechanical transduction phenomenon was modelled by using a suitable electromechanical coupling term, in a similar way as for the case of piezoelectric materials, but unlike this latter class of materials, a frequency-dependent coupling term was introduced and derived.

The transducer was represented by using an equivalent linear electric circuit, as reported in Figure 16.30 (see Figure 16.28 for the meaning of the external quantities).

The circuital elements were defined in accordance with the electrical, mechanical, and electromechanical properties of the material, represented by its dielectric permittivity, η, the viscoelastic modulus, Y, and an effective strain coupling term, d. The elements were scaled with the transducer geometry and size. This choice guarantees the scalability of the proposed model.

The electrical part of the model (left-hand part in Figure 16.30) was modelled considering that both at very low and very high frequencies, IPMCs behave with high value and low value resistance, respectively, while at intermediate frequencies, they behave like a capacitive load. Based on this

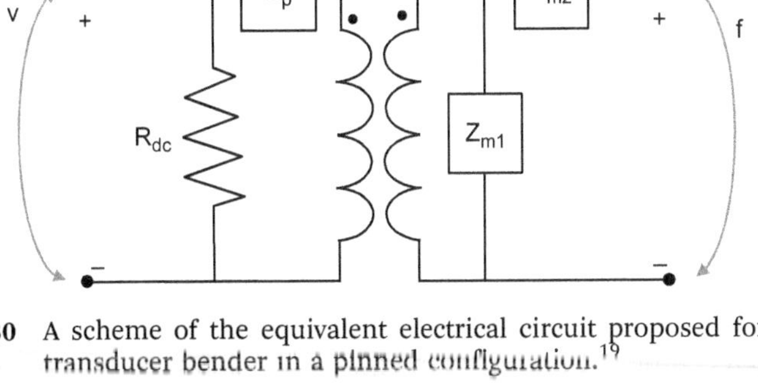

Figure 16.30 A scheme of the equivalent electrical circuit proposed for an IPMC transducer bender in a pinned configuration.[19]

consideration, the IPMC was modelled with a DC resistance, R_{dc}, and an impedance, Z_p, consisting of a number of RC groups in parallel connection, which essentially model the capability of the system to store charges.

The mechanical behaviour of the bender was modelled in the right-hand side of Figure 16.30. It consists of two mechanical impedances: Z_{m1}, used to model the system behaviour at very low frequencies; and Z_{m2}, introduced to model the bender behaviour at frequencies approaching the first mechanical resonant frequency.

The electromechanical coupling term was represented by the ideal transformer in the scheme of Figure 16.30 and assumes the form of a complex function of the applied signal frequency. Its form was obtained by adapting the piezoelectric equations that rule the well-known corresponding phenomena to the case of IPMC transduction.

The equivalent circuit in Figure 16.30 was therefore solved and an explicit form of eqn (16.14) was determined that describes the behaviour of a slender polymeric electroactive beam.

More specifically, the solution of the circuit shown in Figure 16.30 is represented by the following two linearly coupled equations:

$$\begin{bmatrix} v \\ f \end{bmatrix} = \begin{bmatrix} \dfrac{R_{dc}(N^2 Z_{m1} + Z_p)}{R_{dc} + N^2 Z_{m1} + Z_p} & \dfrac{N R_{dc} Z_{m1}}{R_{dc} + N^2 Z_{m1} + Z_p} \\ \dfrac{N R_{dc} Z_{m1}}{R_{dc} + N^2 Z_{m1} + Z_p} & \dfrac{(Z_{m1} + Z_{m2})(R_{dc} + Z_p) + N^2 Z_{m1} Z_{m2}}{R_{dc} + N^2 Z_{m1} + Z_p} \end{bmatrix} \begin{bmatrix} i \\ \dot{u} \end{bmatrix} \tag{16.16}$$

where:

$$Z_{m1} = \frac{1}{s}\frac{Y w t^3}{4 L_d^3} \tag{16.17}$$

$$Z_{m2} = s \frac{3 L_{free}^4 \rho_m w t}{L_d^3 \Gamma^4} \tag{16.18}$$

$$Z_p = \frac{1}{s L_t w} \frac{1}{\sum_{1=1}^{n} \left(\dfrac{\epsilon_i}{1 + s\epsilon_i \rho_i} \right)} \tag{16.19}$$

$$N = \frac{3 d L_d^2}{\sum_{1=1}^{n} \left(\dfrac{\epsilon_i^T}{1 + s\epsilon_i^T \rho_i^T} \right) L_t w t} = \frac{3 d L_d^2}{\eta^T L_t w t} \tag{16.20}$$

In the reported equations, L_t, L_{free}, w, and t are the total length, the free length, the width, and the thickness of the sample so that, unlike in eqn (16.14), each coefficient proposed in eqn (16.16) is now determined in terms of a set of basic material parameters and is scaled by transducer dimensions and geometry so that the model is scalable.

Moreover, the authors proposed[19] and demonstrated[20] that the reflected mechanical impedance is negligible with respect to electrical impedance, and this allowed for a simpler form of eqn (16.16). By applying suitable boundary conditions to such a simplified form of eqn (16.16) and considering a rectangular cross-section, the circuit reported in Figure 16.30 was solved, deriving equations useful for the description of typical IPMC transducer working conditions. More specifically, two relations relevant for the actuation were obtained. By imposing $\dot{u}=0$ (no motion is allowed), the blocked force is obtained, while if the condition $f=0$ (no force is applied) is set, the free tip deflection is determined as a consequence of the applied voltage. These expressions are, respectively:

$$\left(\frac{f}{v}\right)^{\dot{u}} = \frac{3dtwY^E}{4L_d} \tag{16.21}$$

$$\left(\frac{u}{v}\right)^{f} = -\frac{3dL_d^2}{\dfrac{12\rho_m L_f^4}{\Gamma^4 Y^E}s^2 + t^2} \tag{16.22}$$

When the sensing mode is considered, if $v=0$ is supposed, the short-circuit sensing current produced by the tip motion with a velocity $\dot{u}$ can be derived, which represents a commonly used sensor output working condition:

$$\left(\frac{i}{\dot{u}}\right)^{v} = -\frac{3dtwY^E}{4L_d} \tag{16.23}$$

Moreover, the same relation holds true when the accumulated charge with respect to the applied deformation is of interest.

The proposed electric realisation is valid up to the first mechanical mode of the system. Eventually, the model was identified and validated[20] by performing a number of experiments on both the actuating and sensing transduction.

Experiments were performed in air on an IPMC mounted in a cantilever configuration. The DUT was $L_t=33$ mm, $w=5$ mm, and $t=0.2$ mm. The free length was $L_{free}=25$ mm, while the driving point was $L_d=20$ mm away from the clamped edge. The DUT was investigated to identify the mechanical, electrical, and electromechanical coupling terms and such parameters were used to estimate the IPMC impedances. As a good practice, experiments were designed in such a way as to isolate the contribution of as small as possible a subset of the searched parameters (eventually just one of them) in order to make the inaccuracies of the correspondingly identified values smaller.

The three parameters relevant to the description of the IPMC transducing behaviour (*i.e.* the equivalent permittivity, the strain coupling term, and the viscoelastic modulus) were finally estimated.

The model was validated against experimental data. The transducer used for the model identification was also used for its validation. This was a necessary choice because of the lack of reproducibility that plagues IPMCs; nevertheless, working input–output relationships that were different from those used for the model identification were taken into account.

Models were verified for the case of the actuator (up to about 50 Hz for the case of the free deflection, and up to about 20 Hz for the blocked force) in the frequency domain and by using step voltage input in the time domain. For the case of the sensor, the charge-to-deflection transduction, in short-circuit conditions, was investigated up to about 20 Hz.

Finally, the scaling properties of the proposed model were verified by changing the length and the width of the transducer or the point of application of the external force. As an example, in Figure 16.31, the scaling effect on the blocked force and its dependence on the point of application of the external force are reported. More specifically, the estimated IPMC frequency response (Figure 16.31(a)) and the corresponding response after the scaling effect of the IPMC width, as predicted by the proposed model,[19] are cancelled out (Figure 16.31(b)). The good matching of the frequency response magnitude and phase was considered by the authors as evidence of the correctness of the proposed linear model.

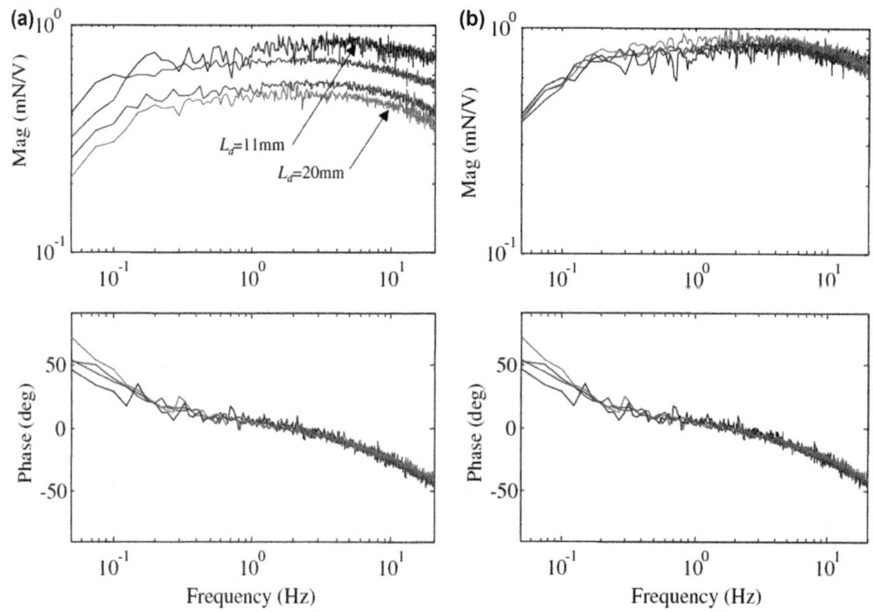

Figure 16.31 Validation of the scaling capabilities and the influence of the linear model proposed for IPMC trasnducers. The scaling effect of the point of application of the external applied force is verified.
Reprinted from K. M. Newbury and D. J. Leo, Linear electromechanical model of ionic polymer transducers – Part II: experimental validation, *J. Intell. Mater. Syst. Struct.*, 2003, 14, 343–357.

The model described so far was developed with the aim of providing a useful tool for people interested in the design and realisation of IPMC-based systems. Nevertheless, it was based on a number of simplifying assumptions, as recognised in Newbury and Leo,[20] where, for example, the impossibility to adequately fit both the step voltage absorbed current, step current-produced voltage, and the electrical impedance was considered as evidence that the dynamics of the transducer response were level dependent.

The same inadequacy was outlined when the electromechanical transduction was considered: for example, it was not possible to obtain, at the same time, good accuracies in the estimation of the blocked force produced by an applied voltage when both a step response and the frequency representation were considered.

Our interest was in the development of a model that, though being user orientated and scalable, was closer to the real behaviour of IPMC actuators.[40] More specifically, we exploited our previous experience in modelling the nonlinear behaviour of the IPMC voltage-to-current transduction[12] to propose a lumped electrical model of the electrical behaviour of an IPMC actuator. The model takes into account the finite resistance of the IPMC electrodes and the nonlinear voltage-to-current transduction nature.

As a first step, the model allows predicting of the absorbed current by using a nonlinear block (NLF). A fraction of this quantity is then used to model the electromechanical transduction in the case of both free deflection and blocked force, by using two further blocks indicated as LF, as reported in Figure 16.32.

A large number of IPMC samples were taken into account to obtain a comprehensive set of experimental data referring to different systems. More

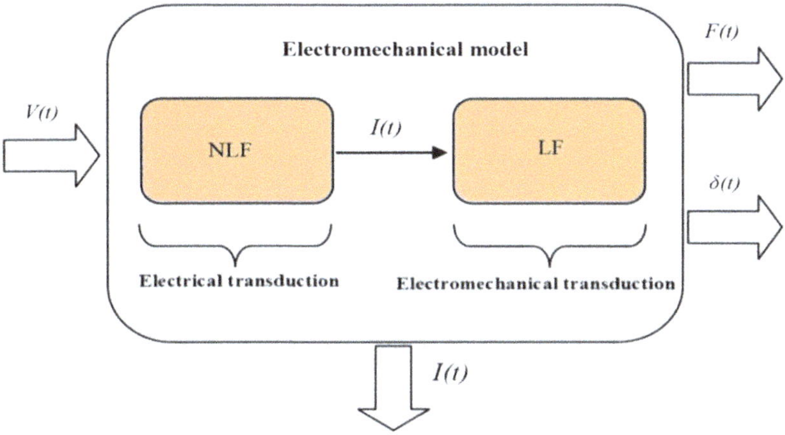

Figure 16.32 A scheme of the model used to describe the nonlinear IPMC actuator transduction.
Reprinted from C. Bonomo, L. Fortuna, P. Giannone, S. Graziani and S. Strazzeri, A nonlinear model for ionic polymer metal composites as actuators, *IOP Smart Mater. Struct.*, 2007, 16(1), 1.

specifically, the model was identified and verified for IPMC actuators realised both by using Nafion® 115 and Nafion® 117, whose nominal thicknesses are 127 μm and 183 μm, respectively, and with both Li^+ and Na^+ as the counterions. All experiments were performed in air, with the IPMC in a cantilever bender configuration.

An example of a voltage-to-current cyclic voltammogram is seen in Figure 16.33(a) for the case of a Nafion-based IPMC actuator using Li^+ as the counterion and water as the solvent. Figure 16.33(b) reports the equivalent electric circuit proposed[40] to model the IPMC nonlinear electric behaviour. The model is in the form of an electrical realisation and includes

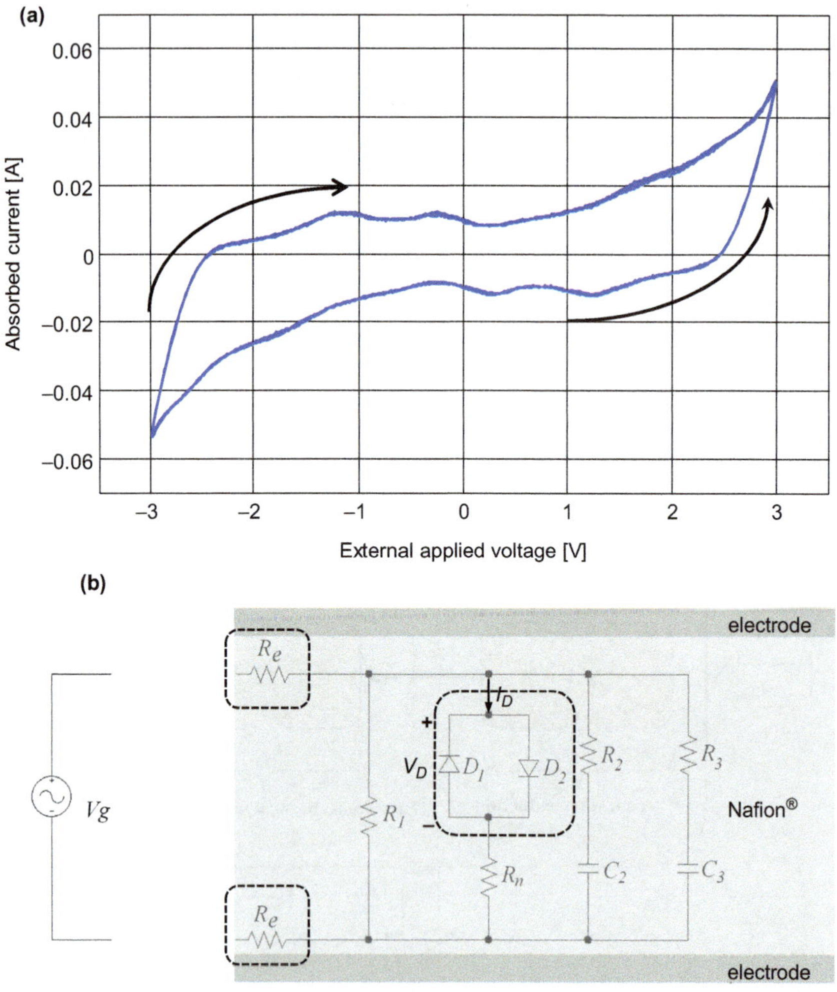

Figure 16.33 An example of a voltammogram obtained for a Nafion-based IPMC with Li^+ as the counterion and water as the solvent (a) and the nonlinear electric circuit realisation used to model IPMC actuator nonlinear behavior (b) (adapted from ref. 40).

the electrode resistance and the nonlinear elements, which were modelled by using two diode-like elements in anti-parallel connection. The newly introduced elements are outlined with the dashed boxes in Figure 16.33(b).

In Figure 16.33(b), R_e takes into account the resistance of the electrodes. R_1 is the equivalent bulk resistance of Nafion. R_n is the resistance associated with observed nonlinear phenomena. It has been introduced along with the two diodes D_1 and D_2 to model the nonlinearity observed in the absorbed current. The R_2C_2 and R_3C_3 branches reflect the capacitive nature of IPMCs. The electrical realisation is ruled by the following equations:

$$
\begin{aligned}
\dot{V}_{C2} &= \frac{V_D}{R_2C_2} + \frac{R_n}{R_2C_2} I_D - \frac{V_{C2}}{R_2C_2} \\
\dot{V}_{C3} &= \frac{V_D}{R_3C_3} + \frac{R_n}{R_3C_3} I_D - \frac{V_{C3}}{R_3C_3}
\end{aligned}
\tag{16.24}
$$

where V_{C2} and V_{C3} are the voltages across the capacitances C_2 and C_3, V_D is the voltage drop across the nonlinearity, and I_D is the current flowing through the nonlinearity.

The nonlinearity is modelled by using two diodes in anti-parallel connection and each diode is described by adapting the Shockley ideal diode equation:

$$
\begin{aligned}
I_{D1} &= Is_{nl}\left(e^{\frac{-V_D}{\gamma}} - 1\right) \\
I_{D2} &= Is_{nl}\left(e^{\frac{V_D}{\gamma}} - 1\right) \\
I_{nl} &= 2Is_{nl} \sinh\left(\frac{V_D}{\gamma}\right)
\end{aligned}
\tag{16.25}
$$

V_D is the solution of the non-linear equation:

$$
\begin{aligned}
&\frac{V_g}{2R_e} - V_D\left(\frac{1}{2R_e} + \frac{1}{R_1} + \frac{1}{R_2} + \frac{1}{R_3}\right) \\
&- 2Is_{nl} \sinh\left(\frac{V_D}{\gamma}\right)\left(\frac{R_n}{2R_e} + \frac{R_n}{R_1} + \frac{R_n}{R_2} + \frac{R_n}{R_3} + 1\right) + \frac{V_{C2}}{R_2} + \frac{V_{C3}}{R_3} = 0
\end{aligned}
\tag{16.26}
$$

Eqn (16.24) through (16.26) can be numerically solved and the current absorbed by the membrane can be finally estimated as:

$$
I = \frac{V_g - V_D - R_n 2Is_{nl} \sinh\left(\frac{V_D}{\gamma}\right)}{2R_e}
\tag{16.27}
$$

where V_g is the value of the applied voltage.

Electrical components in Figure 16.33(b) can be scaled according to the material properties and the device geometry, shown in Figure 16.34, where f

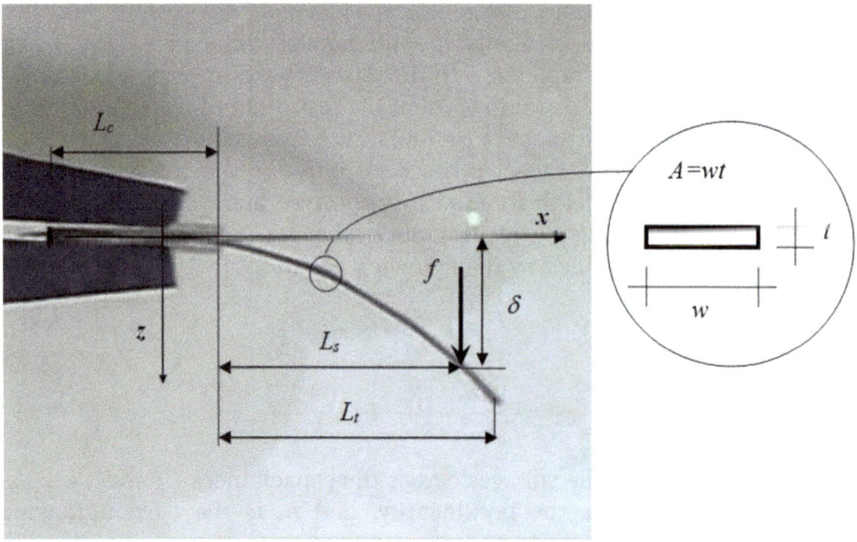

Figure 16.34 Geometry of the pinned IPMC beam.

is a force applied to (or developed by) the IPMC sample, δ is the beam deflection, L_c is the length of the clamped part of the IPMC, L_t is the total free length of the IPMC (without considering the length of the pinned part that will not contribute to the electrode resistance value), L_s is the point where the force is applied, and w and t are the dimensions of the IPMC cross-section.

By using the symbols introduced above, the following scaling relations were suggested for the circuit components in the static part of the electric IPMC realisation.

$$R_e = \frac{R_s L_t}{w} \tag{16.28}$$

where R_s is the resistance per unit length of the electrodes (for simplicity, the same value was assumed for both the electrodes of the IPMCs).

The resistance R_1, which was assumed to model the bulk resistance of the Nafion in DC conditions, was scaled as:

$$R_1 = \frac{\rho_1 t}{(L_t + L_{clamp})w} \tag{16.29}$$

In the same way, R_n, which is the equivalent bulk resistance of Nafion against species that induce non-linear phenomena conduction, was scaled. It was modelled as a function of both a Nafion resistivity ρ_{nl} in the nonlinear region and the geometrical dimensions of the samples:

$$R_n = \frac{\rho_{nl} t}{(L_t + L_{clamp})w} \tag{16.30}$$

Finally, according to the model of a diode, the current Is_{nl} used in eqn (16.25) was scaled as:

$$Is_{nl} = Js_{nl}\, w(L_t + L_{clamp}) \tag{16.31}$$

Dynamics phenomena were modelled as usual for IPMC transducers by using RC networks in parallel connection. The number of such branches is directly linked to the number of the free parameters that are available to fit model estimations to experimental data. In the proposed model,[40] two branches was considered to be a good compromise for modelling both the "slow" and "fast" dynamics observed for IPMC absorbed current.

An example reporting the experimentally observed current absorbed by a Nafion® 115-based IPMC in the case of a step voltage input, and the corresponding model estimation is reported in Figure 16.35(a). The contributions of the two RC branches are further reported in Figure 16.35(b).

The components in the two capacitive branches in the proposed model were also scaled as follows:

$$\begin{aligned} R_i &= \frac{\rho_i t}{(L_t + L_{clamp})w} \\ & \qquad\qquad\qquad i = \{2, 3\} \\ C_i &= \frac{\varepsilon_i (L_t + L_{clamp})w}{t} \end{aligned} \tag{16.32}$$

According to the general consensus that the current flowing through the IPMC and the corresponding charge accumulation are the causes of the electromechanical transduction phenomenon,[15,18,19] the current flowing through the two capacitive branches in the circuital realisation proposed was linked to the corresponding mechanical reaction. The transduction model proposed by Newbury and Leo[19] was adapted to the case of the nonlinear circuit in Figure 16.33(b). The following relationships between the current flowing through the capacitive branches and the free deflection or the blocking force were obtained:

$$\frac{\delta}{I_C} = \frac{\delta}{I_{C2} + I_{C3}} = \frac{1}{s} \frac{3d\, L_s^2}{\eta (L_t + L_{clamp}) wt} \left(\frac{1}{1 + s^2 \dfrac{12 L_t{}^4 \rho}{\Gamma^4 Y t^2}} \right) \tag{16.33}$$

$$\frac{f}{I_C} = \frac{f}{I_{C2} + I_{C3}} = \frac{1}{s} \frac{3\, Y\, d\, t^2}{\eta (L_t + L_{clamp}) 4 L_s} \tag{16.34}$$

where the adopted symbols have the same meanings as for the case of the corresponding equations given by Newbury and Leo.[19]

Since the model described so far is based on a grey-box modelling approach, the experimental measurement of some parameters and the estimation of the remaining parameters that were not possible to measure were required. More specifically, the identification of the parameters involved in

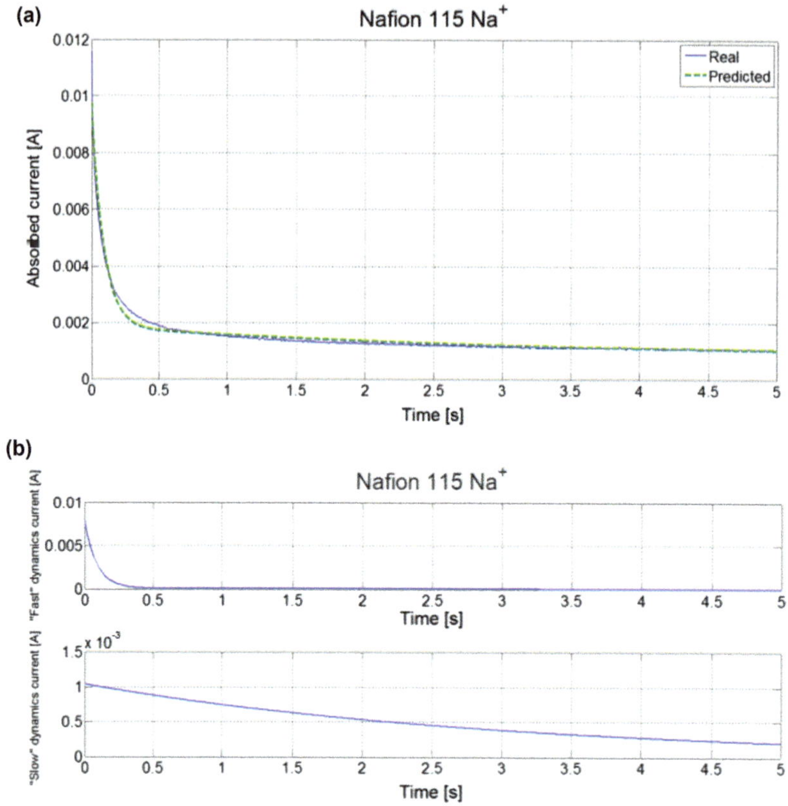

Figure 16.35 The contributions of the reactive current to the current absorbed by an IPMC for a step voltage input.

the electrical circuit and in both the complex functions d and Y were performed in different optimisation steps. As far as the value of ρ_1 was concerned, it was experimentally measured.

The parameters ρ_{nl}, γ, and Js_{nl}. were obtained considering the absorbed current as the result of two different terms: a static nonlinear one (flowing through R_1 and the nonlinear branch) and a dynamic one, due to the presence of the two RC branches.

The nonlinear branch was identified in the first step by static experiments. After the nonlinear term was identified, it was possible to proceed to the identification of the other parameters in the circuit. This second step was achieved by fitting the model-estimated current to the absorbed current when an *ad hoc* input signal is used. This signal consisted of a sequence of steps with different amplitudes and a sequence of samples of white noise with uniform distribution. This choice proved to be the most adequate for simulating a very wide range of the system dynamics.

At this point, it is worth noting that, as a constant aspect of research on IPMCs, a number of experimental setups needed to be realised from the beginning. These were, moreover, required to acquire data in conditions that were largely different from those usually encountered for other electromechanical transducers. As an example, think about the free deflection and the most widely investigated class of electromechanical transducers (*i.e.* piezoelectric transducers). It is quite obvious that even if a deformation is of interest in both cases, the IPMC deformation is orders of magnitude greater than that observed for the case of piezoelectric transducers. The opposite occurs for the blocked force, since for the case of IPMCs, forces that are much smaller than those produced by piezoelectric devices are developed.

It is then quite evident that an aspect relevant to the study of IPMC transducers is the *ad hoc* realisation of measurement facilities.

This occurred to me from the beginning, and in time, a number of different setups have been realised, improved, or even totally redesigned as a necessary and yet fascinating activity.

In Figure 16.36, the setup used is reported.[40] More specifically, the setup seen in Figure 16.36(a) was realised to measure the blocked force *via* a load cell, while in Figure 16.36(b), the setup was used to measure the free deflection, by using an infrared (IR) sensing system.

Examples of the signals involved in the identification of the electrical part of the described model are given in Figures 16.37 and 16.38. More specifically, in Figure 16.37, the experimental nonlinearity observed for an IPMC realised by using Nafion® 117 and Na^+ as the counterion is shown.

In Figure 16.38, the signals of interest to the dynamic components in the electrical model are reported, again for the case of an actuator based on Nafion® 117 with Na^+ as the counterion.

The input voltage consisted of a sequence of steps of various amplitudes followed by a sequence of samples of white noise with uniform amplitude.

Note that only a subset of the reported signals can be acquired, while remaining ones are the results of model estimation.

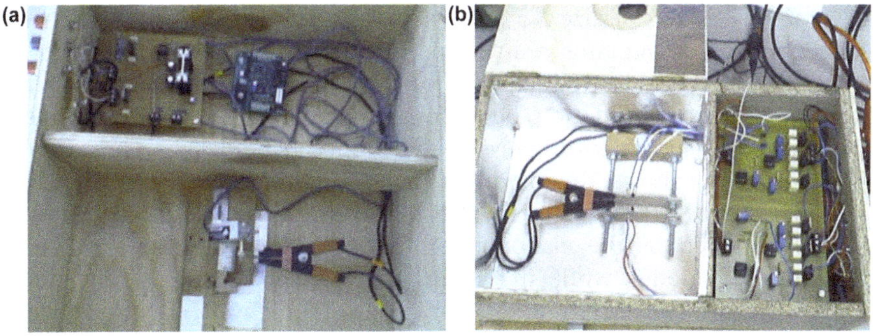

Figure 16.36 The system used to measure the blocked force (a) and the free deflection (b) produced by IPMC actuators in a cantilever configuration.
Reprinted from C. Bonomo, L. Fortuna, P. Giannone, S. Graziani and S. Strazzeri, A nonlinear model for ionic polymer metal composites as actuators, *IOP Smart Mater. Struct.*, 2007, **16**(1), 1.

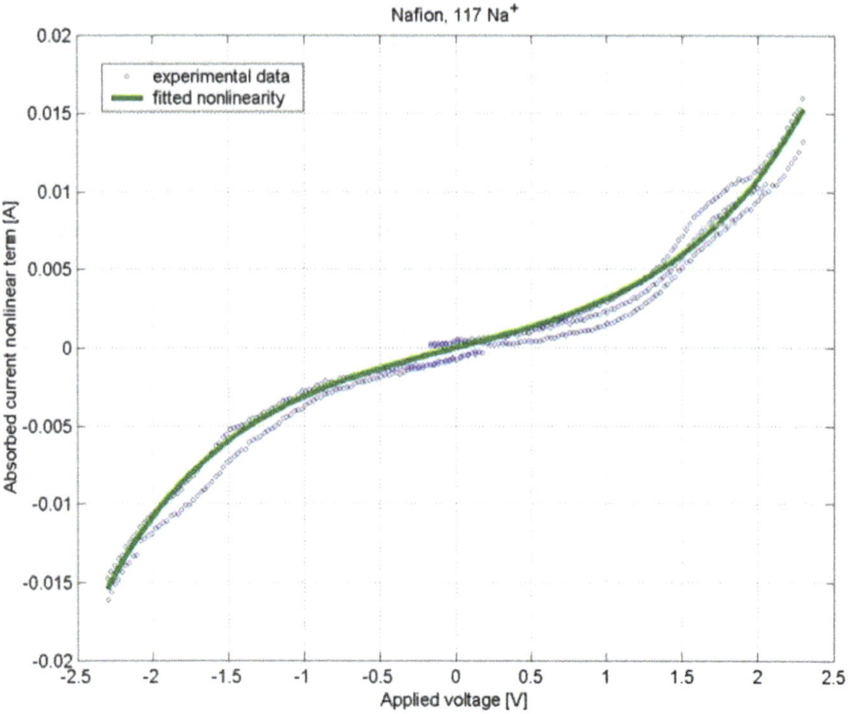

Figure 16.37 One of the identified nonlinearities. Reported graph refers to the case of an IPMC realised by using Nafion® 117 and Na^+ as the counterion.

The optimisation of the values of d and Y was performed considering both the blocked force and the free deflection. Figure 16.39 shows the results obtained by again using a membrane based on Nafion® 117, with Na^+ as the counterion and of size $L_{clamp} = 5$ mm, $w = 6$ mm, and $L_t = 20$ mm. The deflection was measured at a distance of $L_s = 15$ mm from the fixed end.

The model was then validated by using a number of different signals. Moreover, a cross-validation between the time and frequency domains confirmed the good performance of the proposed model. As an example, Figure 16.40(a) shows the modulus of the quotient of the cross-power spectral density (PSD) between the free deformation and the applied voltage obtained by the elaboration of experimental data and by using the proposed model, while in Figure 16.40(b), the corresponding phase is reported. It is worth noting that though the model is a nonlinear one and was able to estimate accurately the resonant-like behaviour of the IPMC.

Finally, as widely indicated, the aim of the described model was to obtain a tool that is useful for people working with IPMCs as candidate actuators for real applications. The model was therefore required to be capable of estimating the effects of IPMC cooling in different ways.

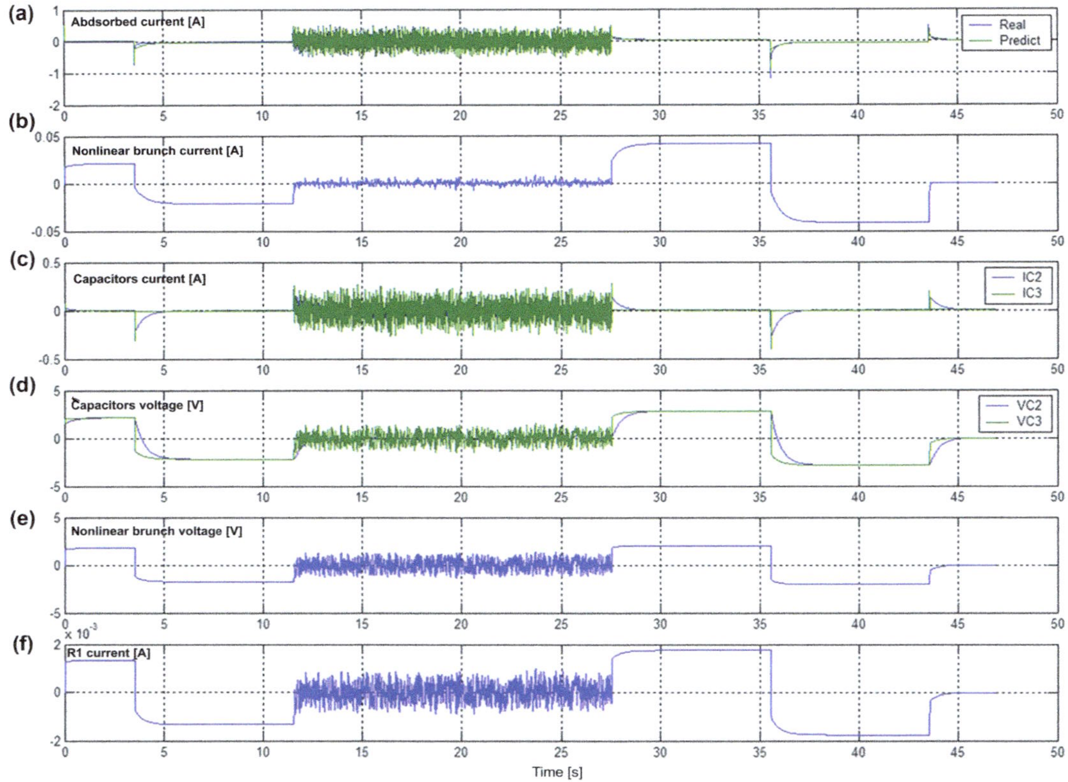

Figure 16.38 The current absorbed by an IPMC actuator and the corresponding estimation (a). The signals estimated by the model for the electric components in the model are also shown in (b) through (f).
Reprinted from C. Bonomo, L. Fortuna, P. Giannone, S. Graziani and S. Strazzeri, A nonlinear model for ionic polymer metal composites as actuators, *IOP Smart Mater. Struct.*, 2007, **16**(1), 12.

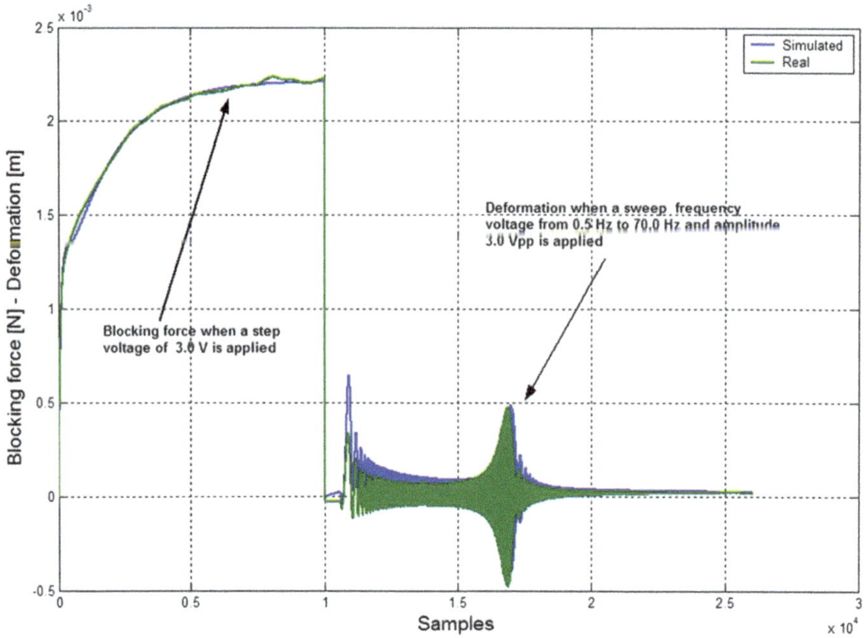

Figure 16.39 Signals used to estimate the IPMC values of d and Y and corresponding model estimates obtained after the parameter identification. Reprinted from C. Bonomo, L. Fortuna, P. Giannone, S. Graziani and S. Strazzeri, A nonlinear model for ionic polymer metal composites as actuators, *IOP Smart Mater. Struct.*, 2007, **16**(1), 1.

As an example, in Figure 16.41, the effects of length scaling on the resonance frequency are shown. More specifically, Figure 16.41 reports the simulation of an IPMC when the length of the membrane was decreased from 25 mm to 10 mm while the width was held constant at 6 mm. The shown graph was obtained by using the parameters characterising the model of a Nafion® 117-based actuator with Li^+ as the counterion.

The experimentally observed values of the resonance frequency are seen, for the same sample, in Figure 16.42.

As is widely mentioned, IPMC applications are envisaged for use in a number of fields where they can be used as actuating and sensing components. Their response to modifying quantities is therefore of interest: any undesired influence can, as a matter of fact, have a negative effect on device reproducibility and therefore on performance. The lack of reproducibility of course limits the possibility of using any transducer, since it introduces a limit to its accuracy. For the case of IPMC transducers, this corresponds to a limit on the actuator position or developed force for the case of the actuator, or on the measured position (and derived quantities) for the case of IPMCs used as motion sensors.

The lack of reproducibility, which has been widely reported in the literature, can be considered the resulting effect of two contributions: the

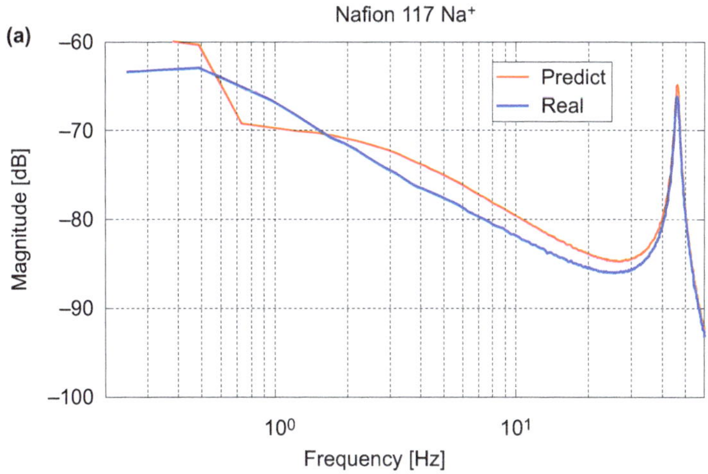

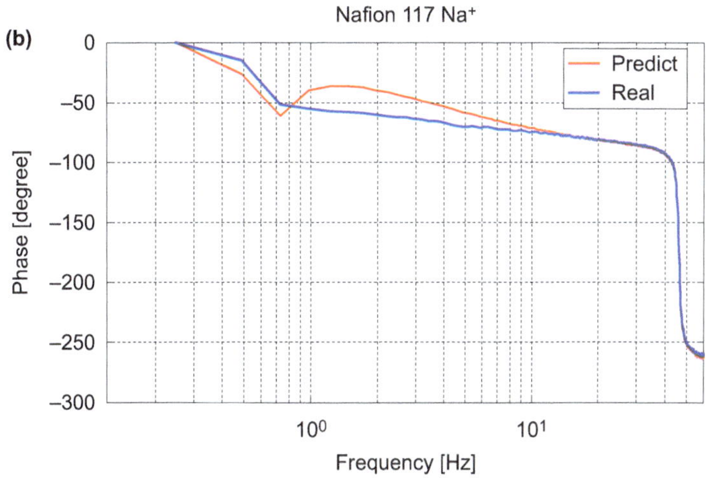

Figure 16.40 The modulus (a) and the phase (b) of the quotient of the cross-power spectral density between the free deformation and the applied voltage obtained by the elaboration of experimental data and by using the proposed model.

dispersion of the transducer response with equal operating conditions; and the dependence of the transducer response on the influencing quantities.

As an interesting example of a lack or reproducibility, consider the case of the models proposed by Newbury and Leo[19,20] and compare them with the model described here.[40] Both models used IPMCs in a bender configuration and identified an effective electromechanical coupling term *d*. Notwithstanding this analogy, Newbury and Leo reported a significant presence of the phenomenon of the back-relaxation and this reflected in the choice of a non-minimum phase structure for the frequency domain

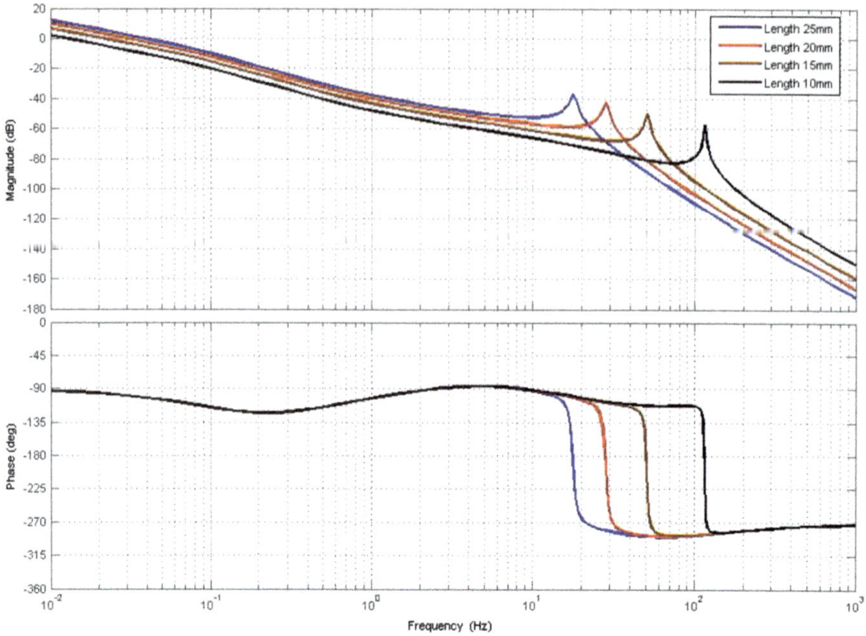

Figure 16.41 The effects of IPMC scaling on the resonant frequency value, as estimated by the described model.

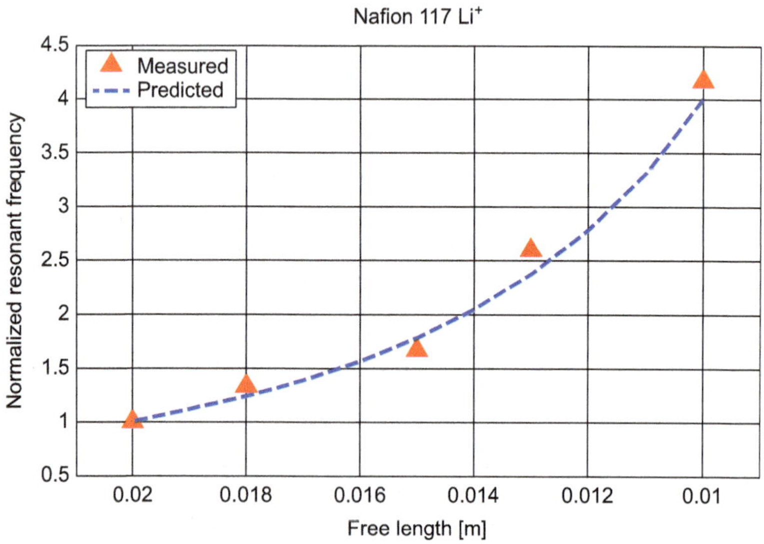

Figure 16.42 The effects of IPMC scaling on the resonant frequency value of a Nafion® 117-based actuator, with Li^+ as the counterion. The comparison of the model estimation end experimentally observed resonance values is reported as a function of the device free length.

representation of *d*. In experimental observations we obtained from tens of IPMCs, we never observed such a phenomenon (*e.g.* see Figure 16.39), and this of course reflected on the structure of *d* that we identified as the basis of our experimental observations.

Such an unresolved problem could be minimised by properly designing the base materials, but in any case, a residual lack of reproducibility cannot be avoided and needs, at least, to be estimated. For this reason, our research activity on IPMC modelling[40] was complemented[41] by a corresponding investigation of the characterisation of IPMC actuators.

Of course, there are infinite candidate modifying quantities for any transducer. Nevertheless, according to the nature of the used materials, some variables can dominate others.

More specifically, for the case of IPMC-based devices, it is reasonable to argue that environmental humidity and temperature represent important modifying parameters. In fact, humidity changes the behaviour of the IPMC transducers working as both sensors and actuators, since it changes the Young's modulus of the device and hence its mechanical response. The influence of temperature was suspected because of the general influence that this quantity has on the characteristics of polymers. When we addressed this problem, a complete characterisation of IPMC transducers as a function of modifying quantities was not accomplished and few analyses were available in the literature.[46,47] More specifically, Nemat-Nasser and Li[46] experimentally investigated the effects of hydration on IPMC mechanical characteristics. Moreover, water content and temperature were suggested as modifying quantities for the case of bare Nafion.[47]

In our work, since the characterisation was intended for devices to be used in biomedical applications, the ranges of 25.0–40.0 °C for temperature and 40.0–100% for relative humidity were investigated. In order to obtain the characterisation of the IPMC actuator, the electromechanical transduction model was identified for a set of working conditions in the ranges presented above. The dispersion of device behaviour with respect to model predictions was therefore estimated.

The investigation of IPMC reproducibility is an experimental task and required both an adequate methodology and a suitable experimental setup to be developed. It was in fact necessary to realise an apparatus that allowed varying of the significant inputs (temperature and relative humidity in the considered case) over the ranges fixed by the operating condition of the envisaged applications.

The elaboration of experimentally observed DUTs allowed the development of the desired input–output relations, along with the corresponding uncertainty estimations.

The devices chosen for the experiments were four IPMCs with the characteristics reported in Table 16.3.

The ranges reported in Table 16.4 were investigated for the modifying inputs.

As described above, the identification of the IPMC actuator parameters requires acquiring data on the applied voltage, the absorbed current, the

Table 16.3 The characteristics of the actuators used for the characterisation process (adapted from Brunetto *et al.*[41]).

Nafion type	Nafion® 115
Counterion	Na^+
Free length	20 mm
Width	6 mm

Table 16.4 The modifying inputs along with their variation ranges (adapted from Brunetto *et al.*[41]).

Temperature (°C)	{25, 30, 35, 40}
Humidity	{40%, 60%, 80%, 100%}

developed force, and the produced deflection. Moreover, since the IPMC characterisation with respect to modifying quantities was of interest, such quantities needed to be recorded for each combination of the values of temperature and humidity reported in Table 16.4.

The characterisation of IPMC actuators was realised by using a measuring system developed *ad hoc* for this task. More specifically, the measuring system was required to:

- Impose the working value for the environmental temperature;
- Impose the working value for the environmental relative humidity;
- Impose the input (voltage) signal;
- Measure the absorbed current and the mechanical output reaction (*i.e.* either the free deflection or the blocked force).

A scheme of the instrumentation devoted to the characterisation of the actuator is seen in Figure 16.43.

The system was composed of:

- A chamber to perform the surveys. It was used to reproduce a controlled environment in which both relative humidity and temperature were varied in a controllable way.
- Control devices to impose the input voltage, the temperature, and the humidity level in the ranges of interest.
- Sensors to acquire data on the actual values of the modifying quantities (*i.e.* a temperature and a humidity sensor, respectively).
- A subsystem devoted to measuring the electrical quantities (*i.e.* the applied voltage and the corresponding absorbed current).
- A subsystem devoted to the measurement of the mechanical IPMC reaction.

More specifically, a load cell of GS0-10 Transducers Techniques® was used for measuring the developed blocking force. A laser sensor

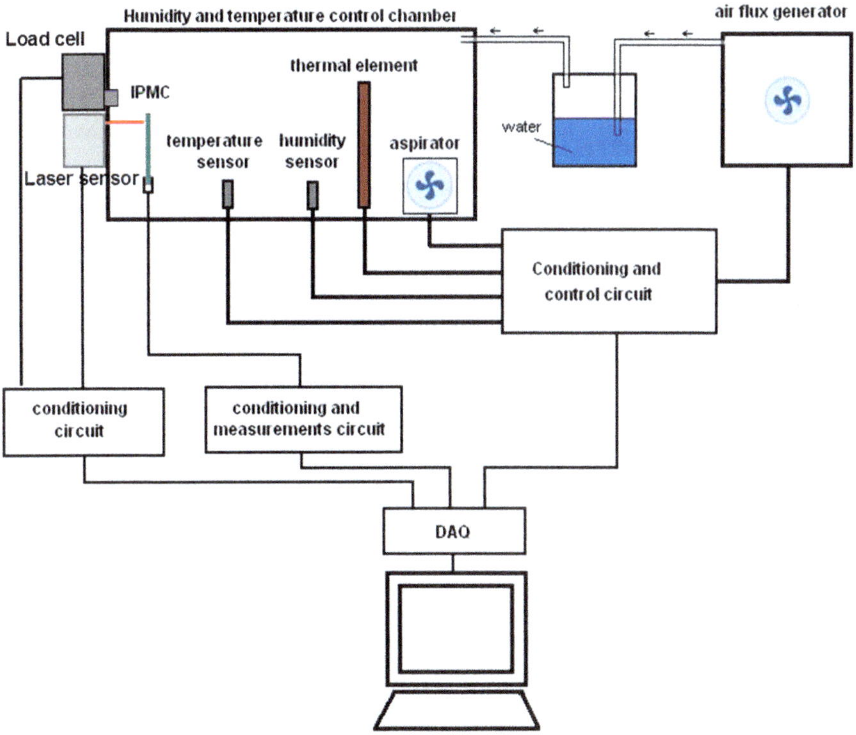

Figure 16.43 A scheme of the setup used to characterise the IPMC actuators. © 2010 IEEE. Reprinted, with permission, from P. Brunetto, L. Fortuna, P. Giannone, S. Graziani and S. Strazzeri, Static and dynamic characterization of the temperature and humidity influence on IPMC actuators, *IEEE Trans. Instrum. Meas.*, 2010, **59**(4), 893–908.

OADM 12U6460/S35A from Baumer was used for the measurement of the IPMC displacement. A temperature sensor LM35 from National Semiconductor and a humidity sensor HIH 3610 from Honeywell were chosen for the monitoring of both temperature and humidity inside the characterisation chamber. The absorbed current was transduced into a corresponding voltage by using a low-value shunt resistor, while the applied voltage was acquired directly by using a data acquisition device DAQ PCI 6052E.

The setup was under the control of LabVIEW VI and the data of all considered quantities were acquired by using the acquisition card mentioned above. A real view of the realised setup is shown in Figure 16.44.

In order to estimate the dispersion of the measured quantities, 20 measurement surveys were executed for each working condition. The samples were stored in deionised water and were casually picked up to be tested. All of the experiments were started after the DUT was maintained inside the chamber for 20 min, in such a way as to allow it to reach the equilibrium conditions with the surrounding environment.

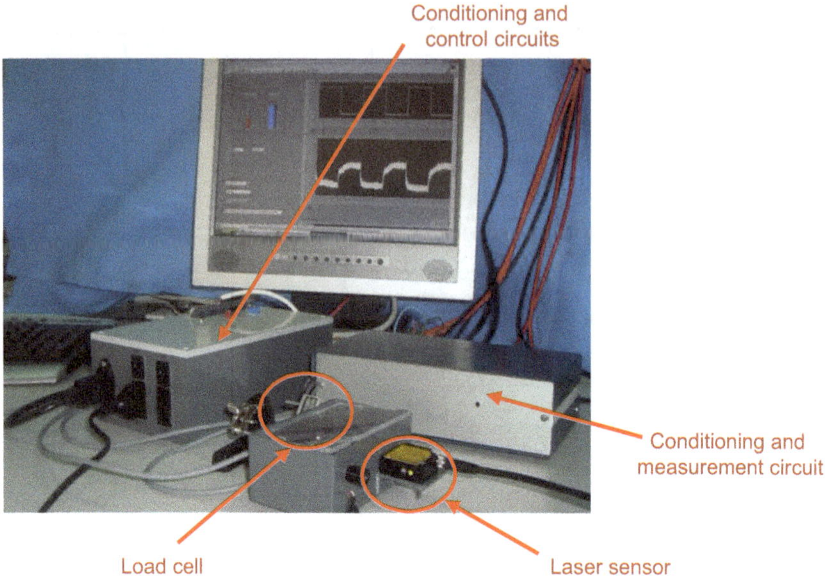

Figure 16.44 A picture of the setup used for the IPMC actuator characterisation. © 2010 IEEE. Reprinted, with permission, from P. Brunetto, L. Fortuna, P. Giannone, S. Graziani and S. Strazzeri, Static and dynamic characterization of the temperature and humidity influence on IPMC actuators, *IEEE Trans. Instrum. Meas.*, 2010, **59**(4), 893–908.

A total number of 20×4×16 valid surveys were therefore considered. This approach allowed both the obtaining of the model-relevant parameters as a function of the working conditions and the estimating of the dispersion of the measured data with respect to the model estimations.

The model identification was performed in three steps devoted respectively to:

- Estimation of the voltage-to-current transduction nonlinearity;
- Estimation of the dynamic components of the voltage-to-current transduction;
- Estimation of the parameters ruling the electromechanical transduction.

More specifically, the static nonlinearity was identified by using a low-frequency triangular signal, with a peak-to-peak amplitude sufficient to outline the system nonlinearity. An example of the applied voltage and of the corresponding recorded currents is reported in Figure 16.45. Observe that the system is nominally in constant working conditions and the dispersion observed in the recorded curves gives evidence of the system non-repeatability. More specifically, all of the following figures refer to a Nafion® 115-based IPMCs working at 30 °C and with relative humidity equal to 80%.

Once the parameters of the nonlinearity were determined, it was possible to estimate the values of the parameters ruling the linear part of the

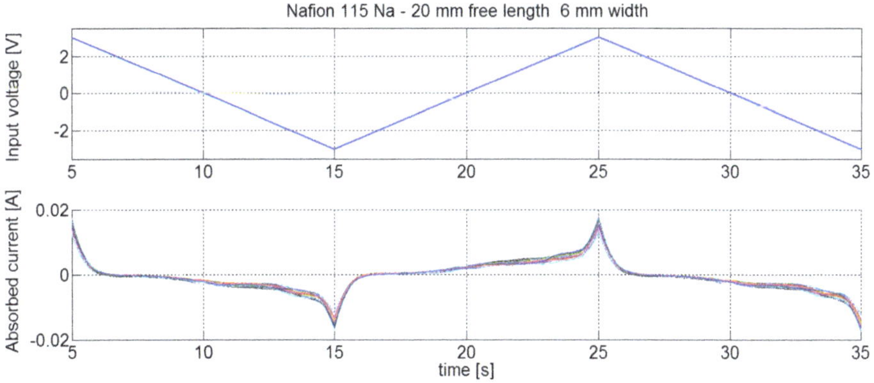

Figure 16.45 The voltage input signal used to estimate the parameters ruling the IPMC actuator nonlinearity and an example of the corresponding absorbed currents.

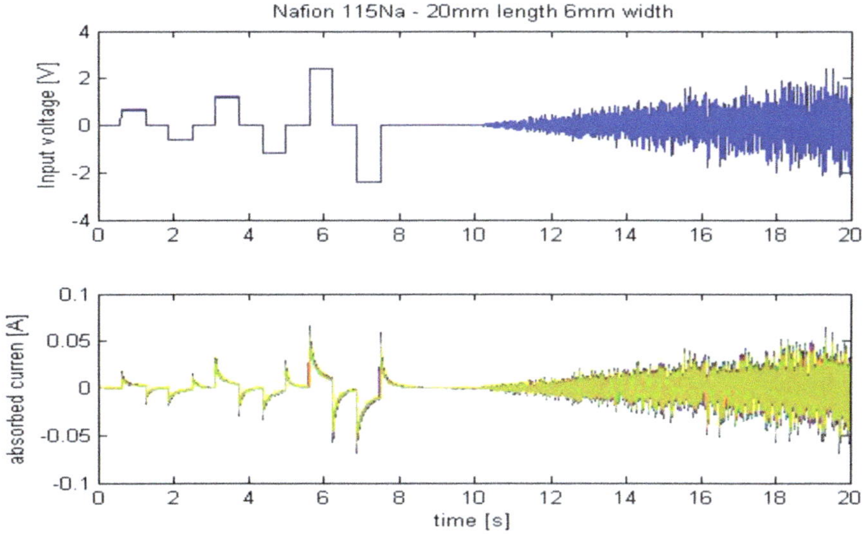

Figure 16.46 The voltage input signal used to estimate the parameters ruling the IPMC actuator linear dynamic components and an example of the corresponding absorbed currents.

electrical realisation. To this purpose, a voltage signal consisting of a train of square waves with alternating signs and of a noise signal with increasing amplitude was forced onto the DUTs. The signal was designed to excite the dynamics of the IPMCs and to investigate any dependence of the system response on the input level. In Figure 16.46, the used input signal and a set of recorded absorbed currents is shown.

Finally, examples of signals used to estimate the electromechanical coupling (*i.e.* the developed blocked force produced by a voltage step and the free deflection when a swept sinusoidal signal is applied) are reported in Figure 16.47(a) and (b), respectively.

The collected data were used to identify the parameters of the IPMC actuator model, along with the corresponding model uncertainty. More specifically, for each couple of values of the modifying inputs (temperature and relative humidity), the model of the actuator was identified, along with the uncertainty estimation.

The identified models were used for the estimation of the actuator output according to the working conditions and the estimated values were compared with the recorded values to obtain the model residuals.

A statistical analysis of the model residuals was performed to obtain an estimation of the actuator uncertainty by using the experimental values of the residual standard deviation.[48] Once the standard deviation was estimated, the coverage factor $k=3$ was used to compute the 3σ confidence level as a function of the modifying inputs.

In the following the results obtained for the case study described so far will be given. However, such an investigation was performed for all working conditions indicated in Table 16.4. Such results could be of interest for researchers involved with IPMCs because we obtained a systematic estimation of IPMC repeatability.

As a first case, the absorbed current was compared with the estimation obtained by using the IPMC actuator model. The results are reported in Figure 16.48.

The same analysis was performed for both the free deflection and the blocked force as quantities relevant to the description of IPMC actuators performance, and the obtained results are shown in Figure 16.49(a) and (b), respectively.

The activity described so far concluded with the investigation of the influence of environmental relative humidity and temperature on the resonance-like behaviour of the IPMC in a cantilever bender configuration as a parameters widely used by competing technologies to realise vibrating sensors.[49] Two examples of the obtained results are shown in Figure 16.50(a) and (b), respectively. More specifically, in Figure 16.50(a), an example is shown of the dependence of both the resonance frequency value and of the corresponding resonant peak value on the relative humidity. Figure 16.50(b) shows the effect produced by the environmental temperature.

The investigation of the reported results allowed confirmation that, at least for the considered ranges of the influencing parameters, larger changes were observed for the case of relative humidity with respect to the effect produced by environmental humidity.

Interest has recently grown on the effects of modifying quantities as a result of multiphysics modelling, as will be further described in a following section.

The sensing properties of IPMCs have been known of since their discovery. The first paper proposing an application of IPMCs as sensors dates back to 1992,[50] when it was shown that IPMCs can sense pressure, though no solvent

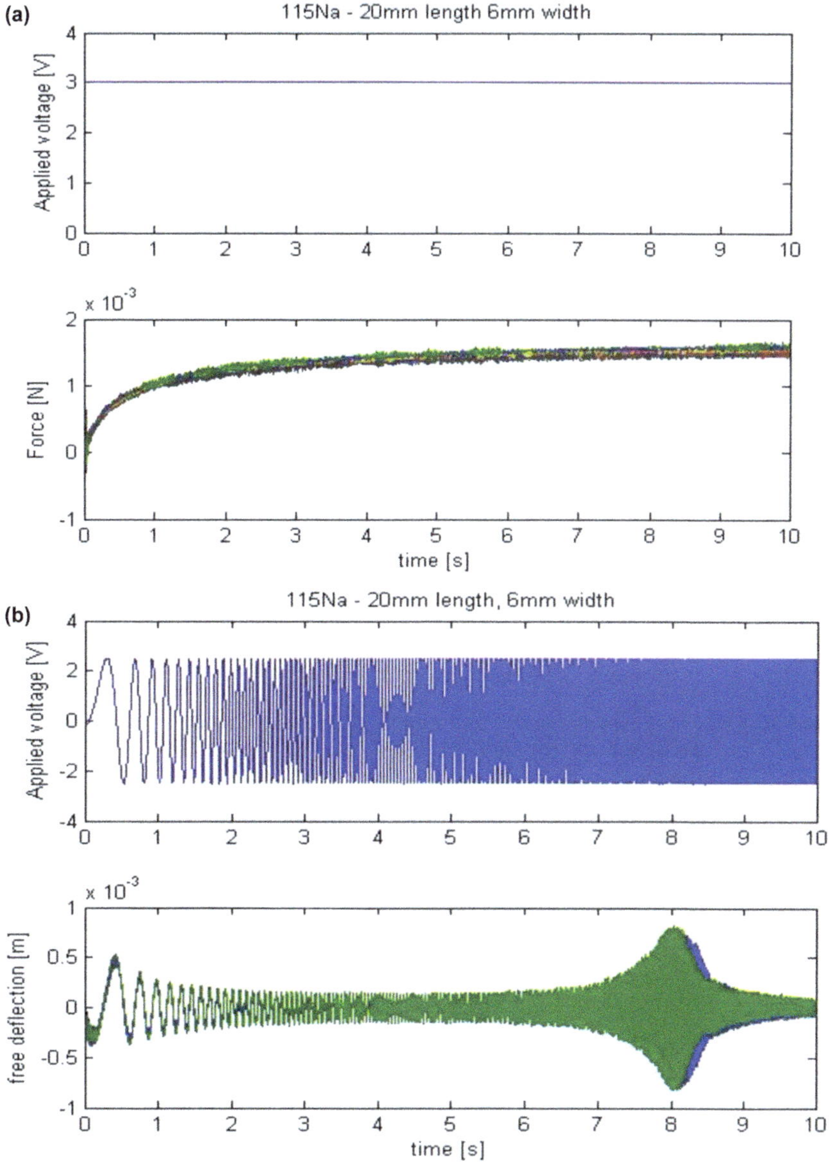

Figure 16.47 The voltage input signals used to estimate the parameters ruling the IPMC actuator electromechanical coupling and the corresponding mechanical reaction. The blocked force (a) and free deflection (b) were considered together in the model parameter estimation.
© 2010 IEEE. Reprinted, with permission, from P. Brunetto, L. Fortuna, P. Giannone, S. Graziani and S. Strazzeri, Static and dynamic characterization of the temperature and humidity influence on IPMC actuators, *IEEE Trans. Instrum. Meas.*, 2010, **59**(4), 893–908.

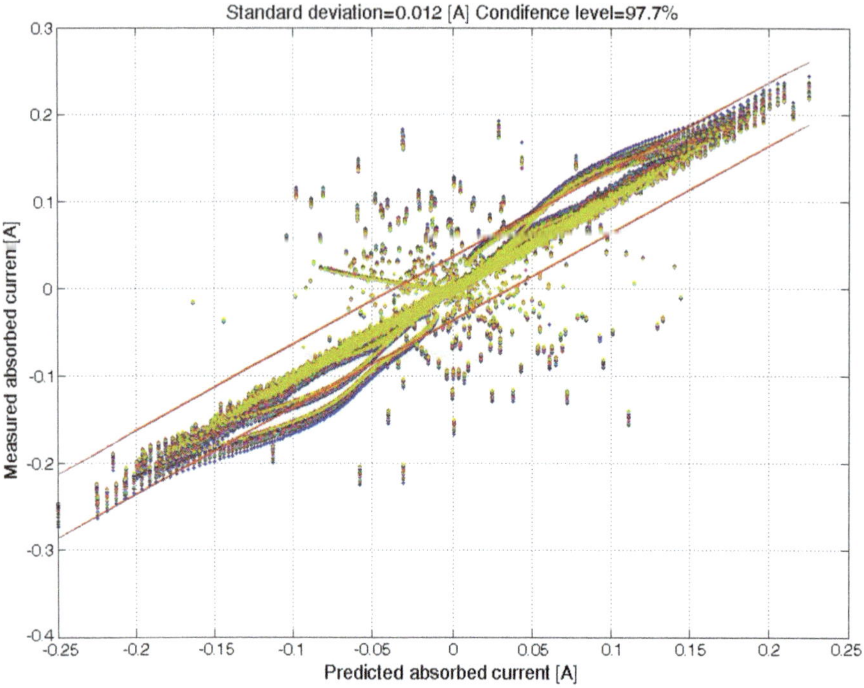

Figure 16.48 Comparison of the measured absorbed current versus the values predicted by the IPMC model. The 3σ values are also shown (red lines).
© 2010 IEEE. Reprinted, with permission, from P. Brunetto, L. Fortuna, P. Giannone, S. Graziani and S. Strazzeri, Static and dynamic characterization of the temperature and humidity influence on IPMC actuators, *IEEE Trans. Instrum. Meas.*, 2010, **59**(4), 893–908.

was used and the device was saturated with hydrogen. Nevertheless, the literature has been biased towards the investigation of the actuation performances of IPMCs for a while, with few exceptions,[51,52] though applications have been proposed in fields recognised as typical for IPMCs, such as robotic and biomedics,[53,54] or general-purpose applications.[55]

In 2005, Paquette *et al.*[56] wrote: "There is potential here for large motion sensors or dampers applications, however, the transduction properties of IPMCs have not been investigated nearly as much as the actuator phenomenon, hence progress has been limited". Such a delay is reflected in the fact that even though surveys on IPMCs have been published almost since the beginning of the studies on this novel technology, the first systematic review of the mechanoelectrical transduction of IPMCs was published only in 2010.[57]

The model proposed by Newbury and Leo[19] is an example of a grey-box model that takes into account the mechanoelectrical transduction properties of IPMCs. This was obtained by introducing a symmetrical model as a consequence of the assumption that the same mechanism is responsible for

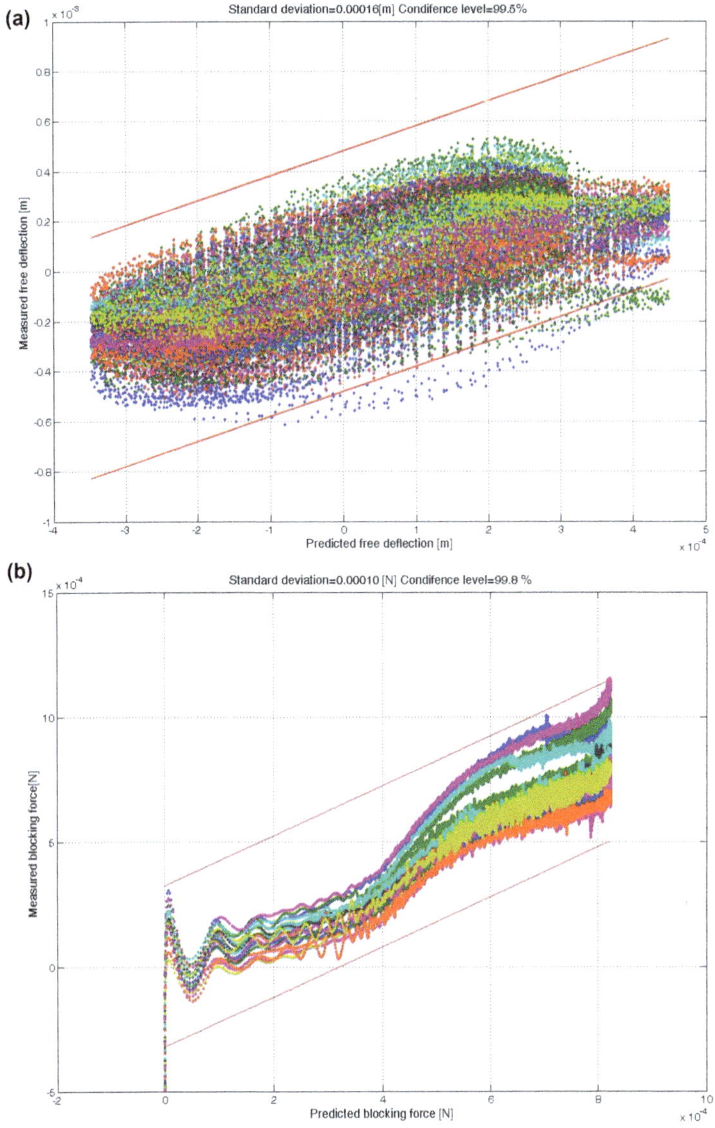

Figure 16.49 The dispersion of the free deflection estimation (a) and of the blocked force (b). The corresponding 3σ values are also shown (red lines).

(a)

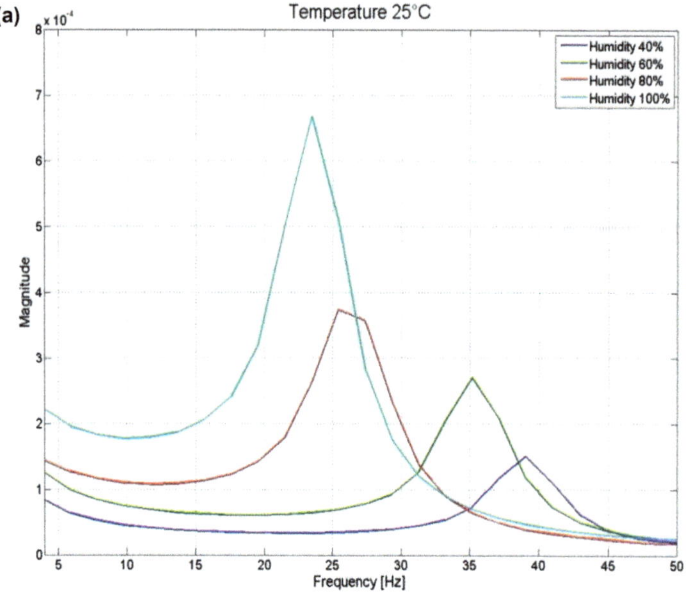

(b)

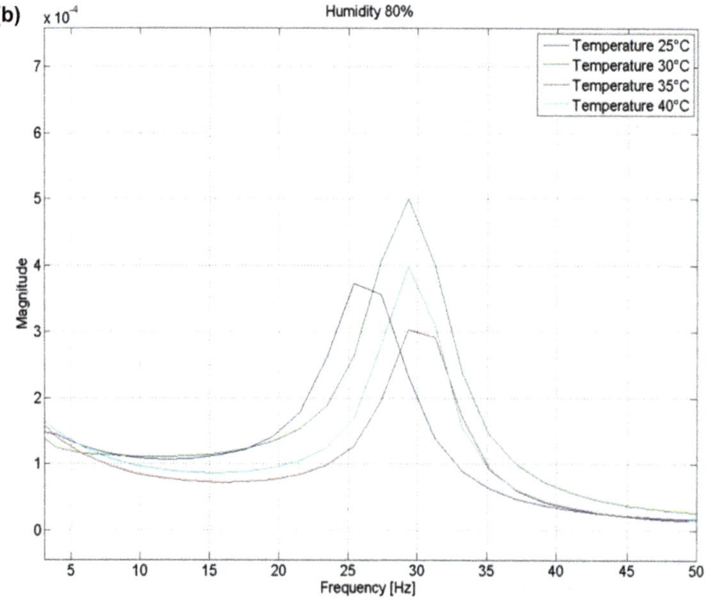

Figure 16.50 Examples of the effect of relative humidity (a) and environmental temperature (b) on the resonance characteristics for an IPMC actuator in a bending configuration.
© 2010 IEEE. Reprinted, with permission, from P. Brunetto, L. Fortuna, P. Giannone, S. Graziani and S. Strazzeri, Static and dynamic characterization of the temperature and humidity influence on IPMC actuators, *IEEE Trans. Instrum. Meas.*, 2010, **59**(4), 893–908.

both the electromechanical and of the mechanoelectrical transduction conversions, regardless of its direction. Nevertheless, operating conditions can greatly influence the IPMC transduction behaviour, and a separate investigation could make sense of the corresponding differences. We investigated[58] the sensing behaviour of IPMCs, focusing on the differences in the sensing of IPMC transduction with respect to the corresponding actuating behaviour that was justified for a different model with respect to the actuator.

One of the first differences between IPMC working modes as actuators and sensors is that, in the latter case, the nonlinearity looked to us to be much less evident, so that the use of a nonlinear model was not justified anymore.

In Figure 16.51, an example of a sinusoidal displacement applied to an IPMC produced by using Nafion® 117 and with Na^+ as the counterion in a cantilever configuration and of the corresponding produced short-circuit current is shown. It is quite evident that the produced current does not represent the significant nonlinearities that characterise the corresponding actuator working mode.

Further evidence of the absence of any significant distortion is obtained if the PSDs of the input and output signals are considered, as is seen in Figure 16.52, referring to the same IPMC sensor described above. More specifically, the PSDs reported in the figure refer to the case when the input

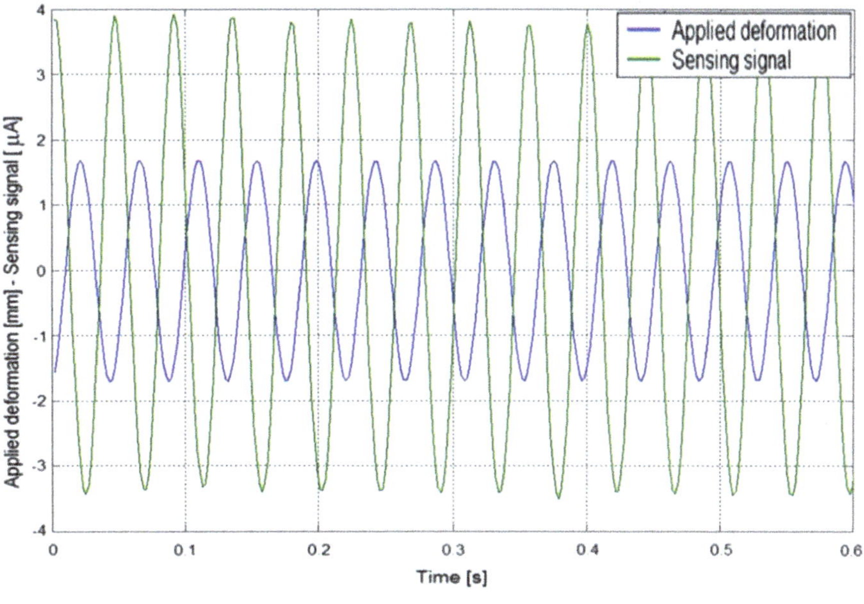

Figure 16.51 An example of the short-circuit current produced by an IPMC when a sinusoidal bending motion is imposed to an IPMC membrane. Reprinted from C. Bonomo, L. Fortuna, P. Giannone, S. Graziani and S. Strazzeri, A model of ionic polymer metal composites as sensors, *IOP Smart Mater. Struct.*, 2006, 15(3), 749–758.

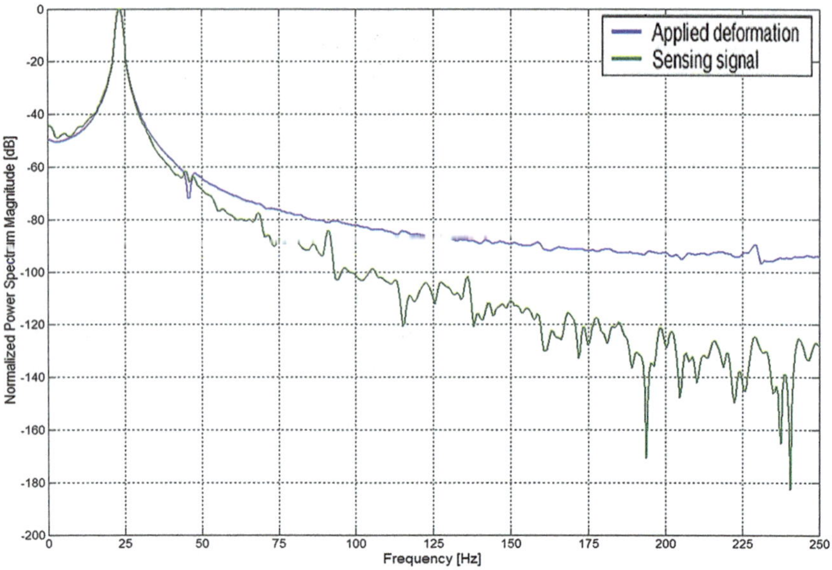

Figure 16.52 The normalised PSDs of the imposed IPMC motion and of the corresponding short-circuit current collected at the IPMC electrodes. Reprinted from C. Bonomo, L. Fortuna, P. Giannone, S. Graziani and S. Strazzeri, A model of ionic polymer metal composites as sensors, *IOP Smart Mater. Struct.*, 2006, **15**(3), 749–758.

frequency was 24.0 Hz. Nevertheless, similar results were obtained when the IPMC was tested by using input sinusoidal signals with frequencies in the range 0.5–100.0 Hz. This was considered a consequence of the low level of electrical signals (voltage and/or current) produced by the IPMCs when they are used as sensors. Another significant difference when an IPMC is used as a sensor with respect to the corresponding working mode as an actuator is the role of the solvent content (generally water).

In the literature, the role of the hydration level of IPMC membranes is thoroughly described for actuators. On the other hand, very few investigations had been reported for sensors when we were investigating the described activity. As an example, Newbury and Leo[19,20] limited their investigations to the use of transducers in the same operative conditions and care was paid to perform experiments on wet samples. This choice was used as a justification to develop a reversible model. We, too, had to keep the IPMC wet during the experiments devoted to the development and characterisation of IPMC actuators.

We observed that the hydration conditions play quite a different role for IPMC sensors. In fact, the sensing behaviour of IPMCs looks better when the membrane is in equilibrium with the environment and the excess water is lost. Figure 16.53 shows two frames of a long-lasting recording of the short-circuit current produced by an IPMC when a sinusoidal deformation is applied. More specifically, in Figure 16.53(a), the signal recorded a few

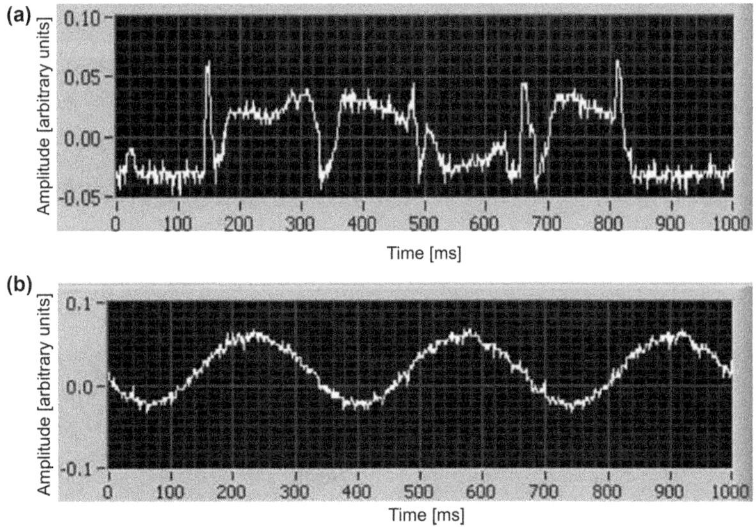

Figure 16.53 Short-circuit current produced by an IPMC sensor when a sinusoidal signal is applied soon after the IPMC is mounted (a) and after it has reached an equilibrium condition with the external environment (b). Reprinted from C. Bonomo, L. Fortuna, P. Giannone, S. Graziani and S. Strazzeri, A model of ionic polymer metal composites as sensors, *IOP Smart Mater. Struct.*, 2006, **15**(3), 749–758.

minutes after the experiment was started is seen, while Figure 16.53(b) shows the signal recorded after about 1.0 h of continuous work.

It is quite evident that the produced signal is initially very noisy and unstable. By contrast, the signal produced after the sensor reached an equilibrium condition was revealed to be stable and much less noisy. As a further example, Figure 16.54 shows two short recordings of the sensor output signal after it was continuously working for about 24 h. The two zoomed images in Figure 16.54 show windows lasting about 2 s of the recorded output signal produced by a mechanical input of frequency equal to 1.5 Hz. The two recordings are separated by more than 1000 full cycles.

Based on the considerations reported above, the opportunity to investigate the IPMC as a sensor in "dry" conditions seemed evident to us. Since the characteristics of IPMCs greatly depend of the hydration level, we decided to model the sensor in this condition. More specifically, the deflection to a short current was rewritten as it follows:[19]

$$\frac{i}{\delta} = s\frac{3dtwY}{4L_s} \tag{16.35}$$

and the focus was on the estimation of the material parameters d and Y in the novel working conditions.

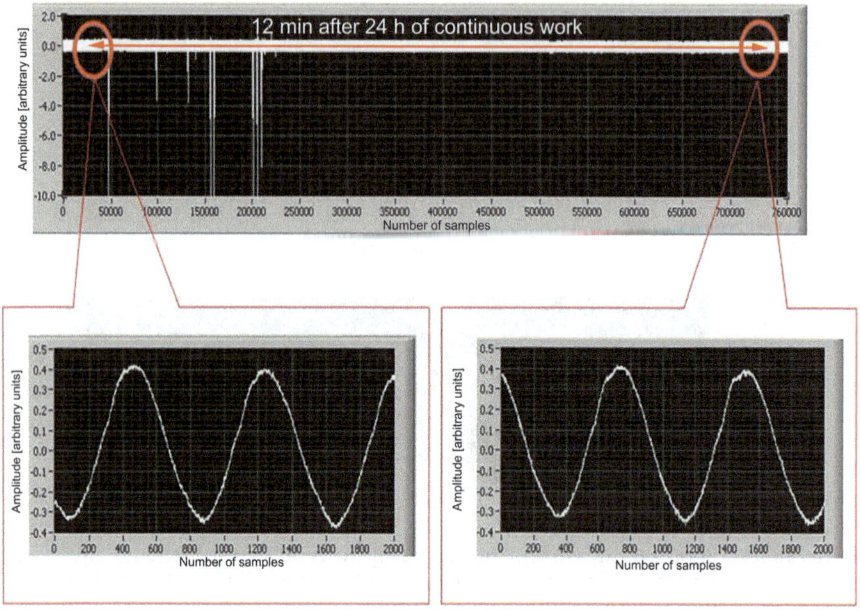

Figure 16.54 Two widows extracted from a long-lasting recording of the sensing current produced by an IPMC sensor after 24 h of continuous working (adapted from Bonomo *et al.*[58]).

The Young's modulus was determined by using the relationship that links the beam deformation to the applied force according to the Euler–Bernoulli theory:

$$\delta = f\frac{4L_S^3}{Ywt^3} \tag{16.36}$$

The adopted identification procedure was organised in two steps. The first one was devoted to the estimation of the complex Young's modulus Y by using eqn (16.36). More specifically, this required the realisation of a mechanical system able to apply both the force f and the corresponding deformation δ and to acquire data to be used during the optimisation step. The following cost function J_Y was used:

$$J_Y = 100 \cdot \frac{\sqrt{\sum_i (\delta^{\text{real}} - \delta^{\text{estimated}})_i^2}}{\sqrt{\sum_i (\delta^{\text{real}})_i^2}} \tag{16.37}$$

where δ^{real} was the deformation experimentally obtained by the measurement system and $\delta^{\text{estimated}}$ was the deformation as predicted by eqn (16.36).

After Y had been determined, eqn (16.35) was used as the coordination function to optimise the coupling term d. In this second step, the input quantities for the optimisation algorithm were the applied deformation δ

and the corresponding short current i. The following cost function J_d was used:

$$J_d = 100 \cdot \frac{\sqrt{\sum_i \left(i^{\text{real}} - i^{\text{estimated}}\right)_i^2}}{\sqrt{\sum_i \left(i^{\text{real}}\right)_i^2}} \tag{16.38}$$

where i^{real} was the sensing current experimentally obtained by the measurement system and $i^{\text{estimated}}$ was the corresponding sensing current as predicted by eqn (16.35).

After a number of trials, we decided to use a composite signal consisting of a sequence of a train of bipolar pulses, a sweep signal, and a noise signal. An example of the recorded signals is given in Figure 16.55. More specifically, the applied deformation, the blocking force, and the recorded short-circuit sensing current are shown.

As broadly outlined previously, the modelling of IPMCs requires a continuous development of new experimental setups. A schematic of the system developed for the described application and a real view of the realised system are shown in Figure 16.56(a) and (b), respectively.

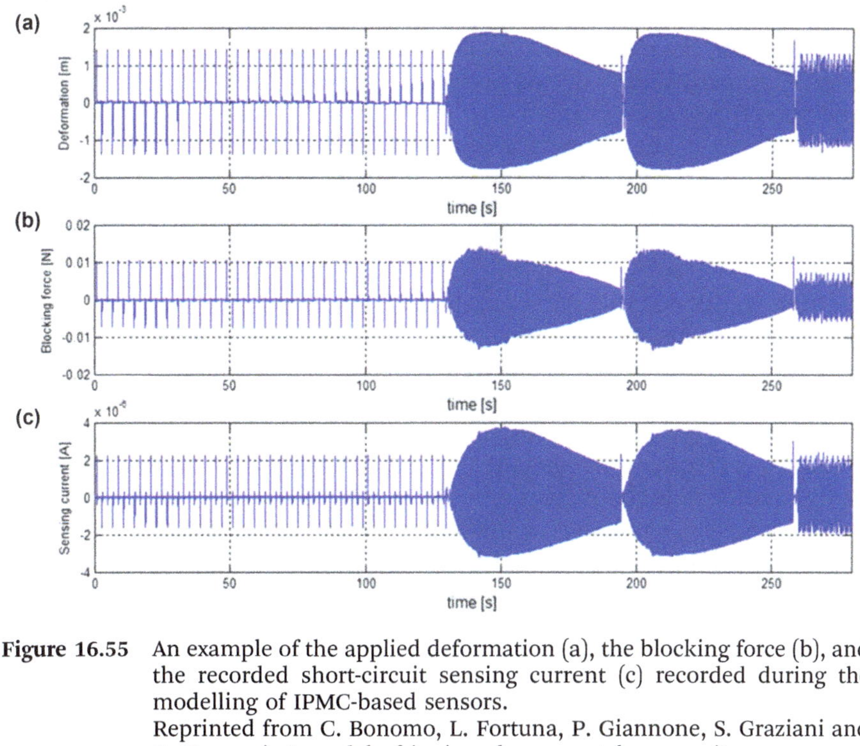

Figure 16.55 An example of the applied deformation (a), the blocking force (b), and the recorded short-circuit sensing current (c) recorded during the modelling of IPMC-based sensors.
Reprinted from C. Bonomo, L. Fortuna, P. Giannone, S. Graziani and S. Strazzeri, A model of ionic polymer metal composites as sensors, *IOP Smart Mater. Struct.*, 2006, 15(3), 749–758.

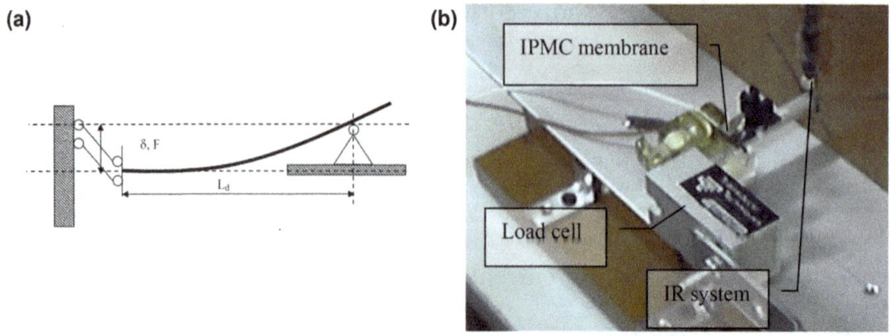

Figure 16.56 A scheme of the system used to estimate the material values for IPMCs working as sensors in air (a) and a real view of the realised setup (b). Reprinted from C. Bonomo, L. Fortuna, P. Giannone, S. Graziani and S. Strazzeri, A model of ionic polymer metal composites as sensors, *IOP Smart Mater. Struct.*, 2006, **15**(3), 749–758.

The deformation δ was measured by using an IR-based system, while a load cell (GS0-10 from Transducer Techniques®) was used to measure the blocking force.

The most relevant results of the research activity described are the differences in the structures that we determined for d and Y with respect to the corresponding parameters that were identified for the case of the wet actuators. As broadly outlined, this was a direct result of our choice to use and model the IPMC sensors in air.

An example of the Young's modulus of an IPMC strip realised by using Nafion® 117 and with Na^+ as the counterion is shown in Figure 16.57. The IPMC modulus was determined by the elaboration as explained before. For comparison, the Young's modulus for the IPMC sample when highly hydrated (reported in Figure 16.57(a)) is compared with the one estimated after reaching equilibrium working conditions (reported in Figure 16.57(b)). In both cases, the structure of the Young's modulus was modelled according to the GHM theory for viscoelastic materials.

Figure 16.57 shows that the hydration level changes the value of the complex Young's modulus. In particular, at the lowest values of the investigated frequency range (10^{-2} Hz), the Young's modulus of the sample in dry conditions is about eight times larger than the corresponding value for the wet sample. Eventually, the GHM model used for the sensor reduced to a constant value in the considered frequency interval. For the dry IPMC, the optimisation procedure gave the value $Y = 2.03$ GPa. The reported results are in accordance with measurements of the IPMC complex Young's modulus that we obtained more recently and that will be introduced later in this chapter when the topic of white-box modelling will be addressed.

The same comparison was performed for the case of the coupling term d. Results obtained for the Nafion® 117-based IPMC with Na^+ as the counterion are given in Figure 16.58(a) and (b) for the wet and dry working

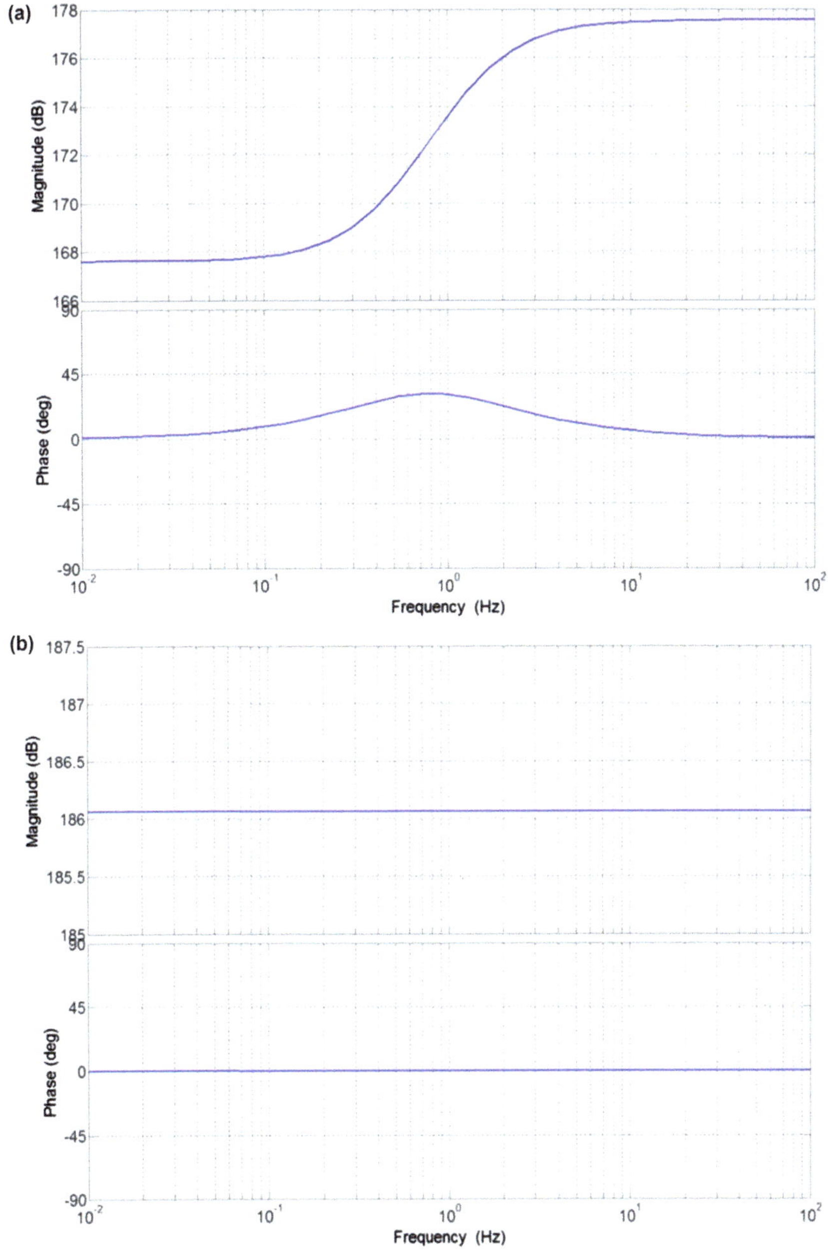

Figure 16.57 Young's modulus estimates according to the GHM model for an IPMC strip that was highly hydrated (a) and in equilibrium with the working environent in air (b).
Reprinted from C. Bonomo, L. Fortuna, P. Giannone, S. Graziani and S. Strazzeri, A model of ionic polymer metal composites as sensors, *IOP Smart Mater. Struct.*, 2006, **15**(3), 749–758.

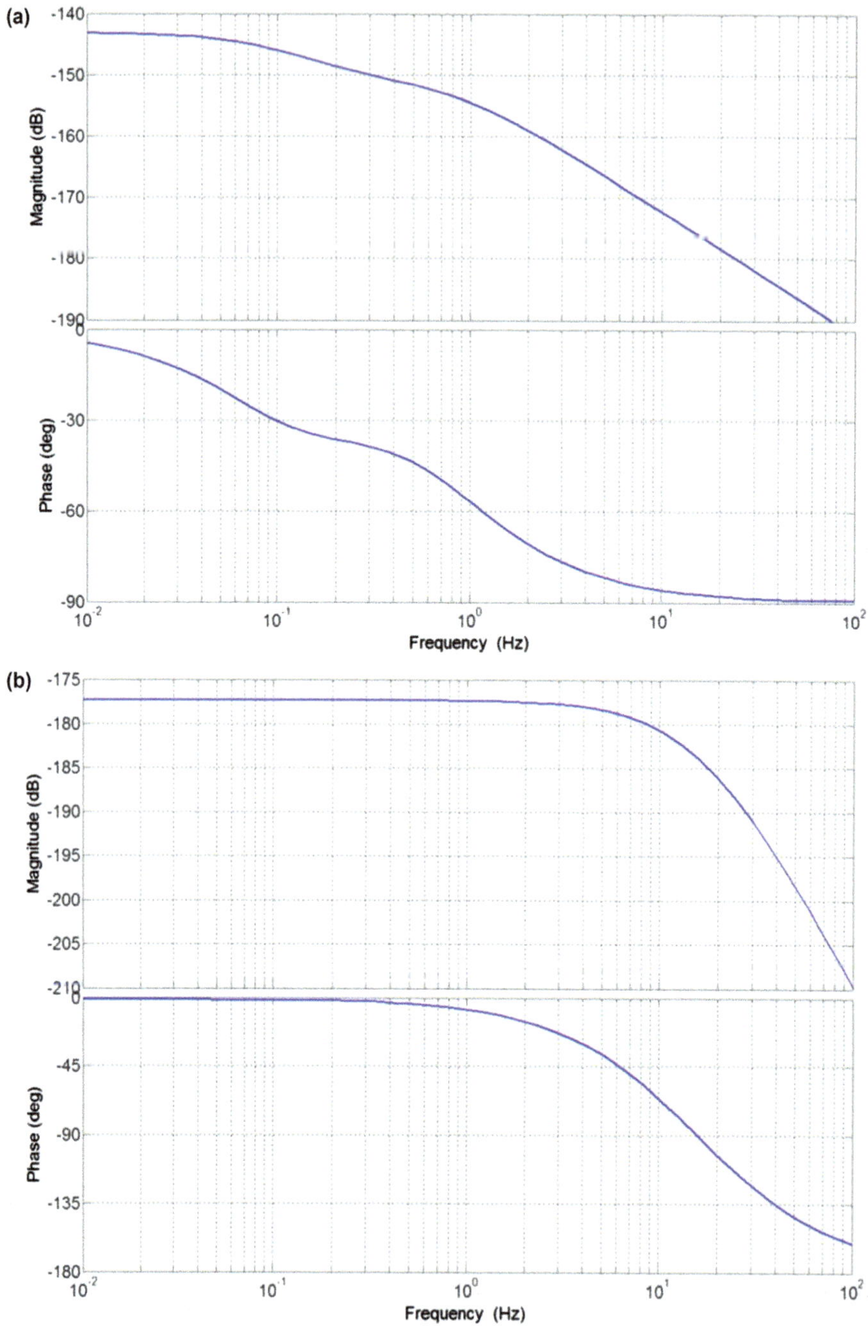

Figure 16.58 Coupling term *d* estimates for an IPMC strip that was highly hydrated (a) and in equilibrium with the working environent in air (b). Reprinted from C. Bonomo, L. Fortuna, P. Giannone, S. Graziani and S. Strazzeri, A model of ionic polymer metal composites as sensors, *IOP Smart Mater. Struct.*, 2006, 15(3), 749–758.

conditions, respectively. Note that in this case a different structure was adopted for the transfer function used to model the coupling term in wet and in dry conditions.

The grey-box model proposed for IPMC sensors was further verified against real data. More specifically, we investigated the performances of the model in both the time and frequency domains.

An example of the performance obtained in the time domain is seen in Figure 16.59(a), while in Figure 16.59(b), the performance of the model in the frequency domain is shown. More specifically, Figure 16.59(a) shows the capability of the model to estimate the short-circuit current when the IPMC is inputted with a train of impulsive deformations, while Figure 16.59(b) shows the model Bode diagram and the corresponding estimations obtained from experimental data.

As with the case of the actuator model, the IPMC sensor model is scalable as a function of the device's geometry. This again was considered a key advantage, since it provided the possibility of estimating the consequences of changes in the device's geometry. As an example, the effects of changes in the IPMC sensor width are seen in Figure 16.60, where the modulus of the Bode diagram as predicted by the model is compared with the corresponding estimation obtained on the basis of acquired data.

Also, for the case of IPMC sensors, we felt that it was necessary to proceed to their characterisation with respect to the influence of the modifying inputs, and again we investigated the environmental temperature and relative humidity as relevant modifying quantities for IPMCs.[59] In a sense, the sensor characterisation completes the analogous activity we performed on IPMC actuators, and it was intended to give more confidence on the results that can be achieved by using IPMCs both as actuators and sensors. The obtained results should act as stimuli for the use of this technology in the development of real applications.

In this case, the investigation was done in a range that was compatible with biological applications, and again the relative humidity revealed a larger influence with respect to temperature.

The devices chosen for the experiments were IPMCs with the characteristics reported in Table 16.5.

The ranges reported in Table 16.6 were investigated for the modifying inputs.

In order to investigate the dispersion of the IPMC sensor behaviour, 20 acquisitions were collected for each combination of temperature and environmental relative humidity values in Table 16.6. This resulted in the acquisition of 20×20 valid surveys, with the first 20 being the number of repeated observations and the second 20 being the number of combinations of the entries in Table 16.6. Moreover, for each of the investigated working conditions, the applied deformation, the blocking force, and the produced current were measured.

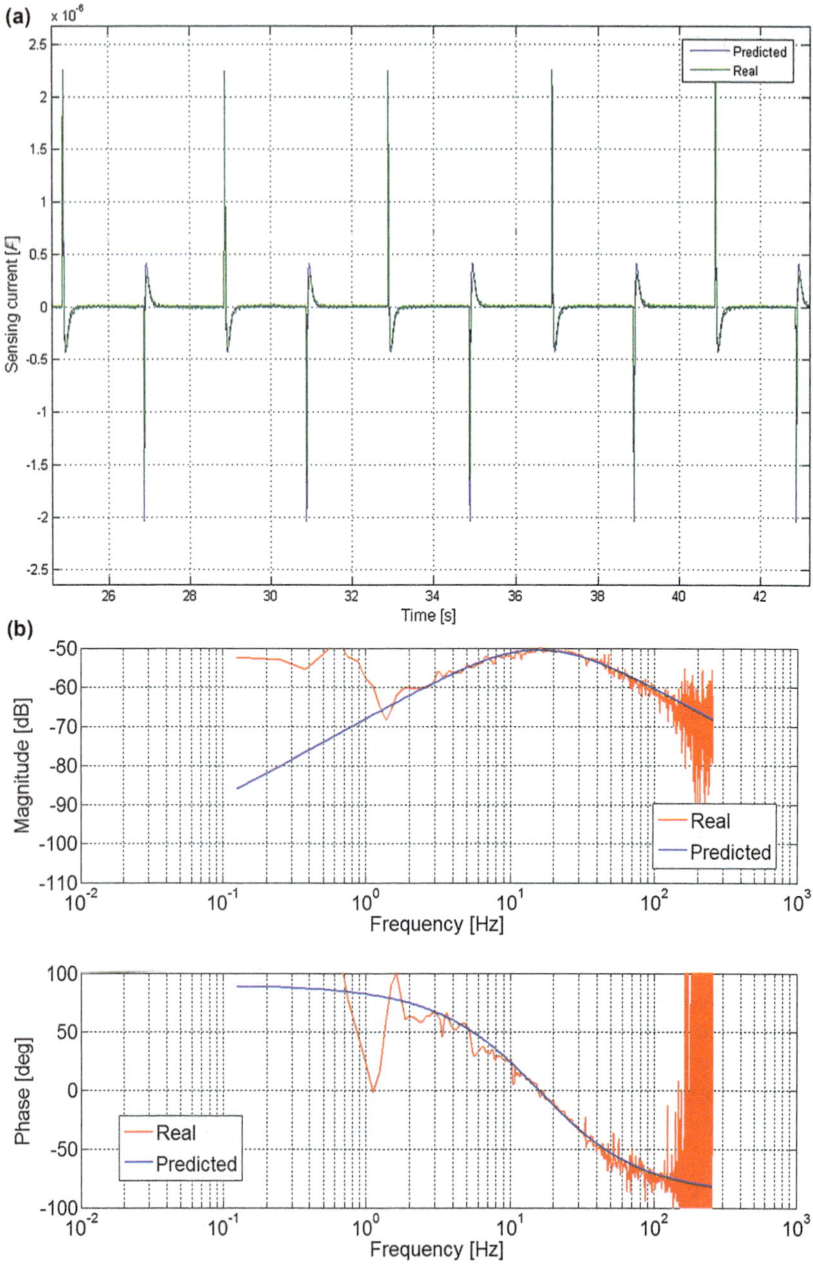

Figure 16.59 The time (a) and frequency domain (b) performances of the model developed for IPMC sensors working in air.
Reprinted from C. Bonomo, L. Fortuna, P. Giannone, S. Graziani and S. Strazzeri, A model of ionic polymer metal composites as sensors, *IOP Smart Mater. Struct.*, 2006, **15**(3), 749–758.

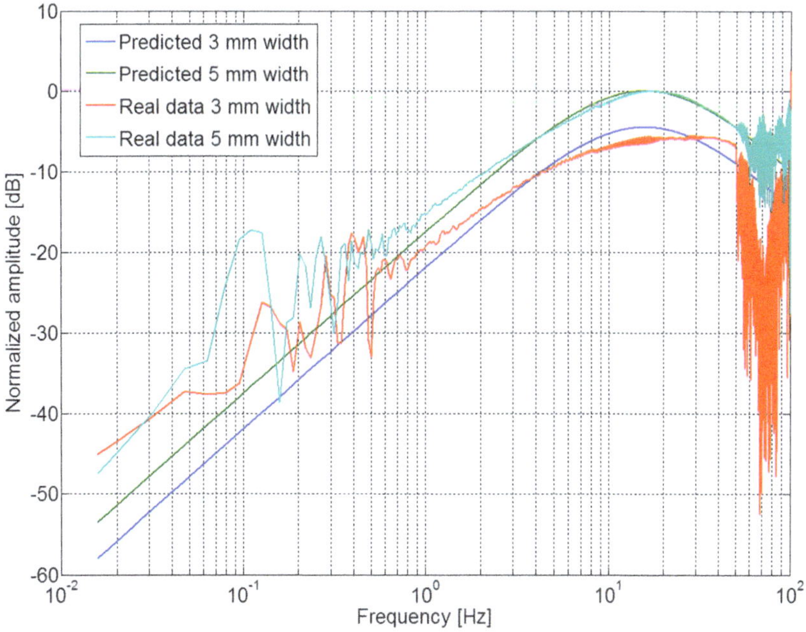

Figure 16.60 Comparison of the scaling effect as predicted by the model and as obtained from the experimental data. The influence of the IPMC sensor width is shown.
Reprinted from C. Bonomo, L. Fortuna, P. Giannone, S. Graziani and S. Strazzeri, A model of ionic polymer metal composites as sensors, *IOP Smart Mater. Struct.*, 2006, **15**(3), 749–758.

Table 16.5 The characteristics of the sensors used for the characterisation process (adapted from Brunetto *et al.*[59]).

Nafion type	Nafion® 117
Counterion	Na^+
Free length	20 mm
Width	6 mm

Table 16.6 The modifying inputs, along with their variation ranges (adapted from Brunetto *et al.*[59]).

Temperature (°C)	{28, 30, 32, 34}
Humidity	{40%, 60%, 80%, 90%, 95%}

The characterisation of the IPMC-based sensors was realised by using a measuring system that was developed *ad hoc* for this task. More specifically, the measuring system was able to:

- Impose the working value for the environmental temperature;
- Impose the working value for the environmental relative humidity;

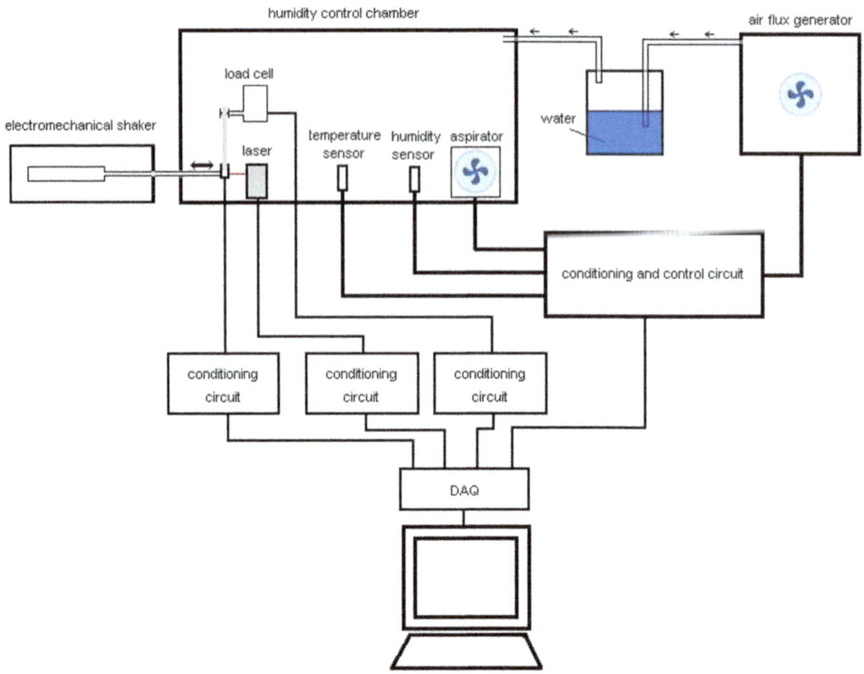

Figure 16.61 Scheme of the setup used to characterise the IPMC sensors. © 2011 IEEE. Reprinted, with permission, from P. Brunetto, L. Fortuna, P. Giannone, S. Graziani and S. Strazzeri, Characterization of the temperature and humidity influence on ionic polymer-metal composites as sensors, *IEEE Trans. Instrum. Meas.*, 2011, **60**(8), 2951–2959.

- Impose the desired input (displacement) signal;
- Measure the output reaction (*i.e.* the developed short-circuit current).

Actually, the developed system was much more complex, since it was used both to acquire data for the estimation of the material parameters required for the model identification and to acquire data for the IPMC characterisation. A scheme of the setup devoted to the modelling and characterisation of the IPMC sensors is shown in Figure 16.61.

The system was composed of:

1. A electromechanical shaker able to impose the desired deformations;
2. A chamber to perform the surveys, where both relative humidity and temperature were varied in a controllable way;
3. A thermal element to control the temperature, a system to control the humidity level, and measurement devices to sense the values of the modifying inputs (*i.e.* one humidity and one temperature sensor, respectively);

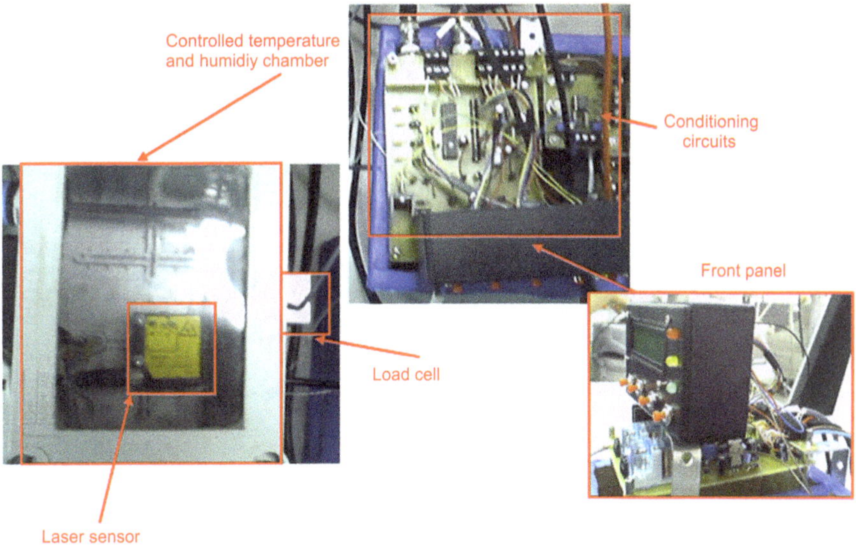

Figure 16.62 View of the setup used to characterise the IPMC sensors with respect to environmental temperature and relative humidity.

4. Conditioning circuitry;
5. A laser sensor and a load cell to reveal IPMC-imposed deformation and force, respectively.

Data were acquired by means of a DAQ card PCI 6052E by using a dedicated software tool developed with LabVIEW. A view of the real system is seen in Figure 16.62.

Swept voltage signals were applied to the shaker and the corresponding signals were acquired by using the experimental set up shown in Figure 16.62.

Acquired data were used for the IPMC sensor model identification and for the corresponding uncertainty estimation as a function of the modifying input values. During the characterisation of the sensor, the model was used for the estimation of the produced signal and therefore of the model residuals.

A statistical analysis of the model residual was performed to obtain an estimation of the model uncertainty[48] by using the experimental values of the standard deviation of the residuals. Once the standard deviation was estimated, the coverage factor $k=3$ was used to compute the 3σ confidence level as a function of the modifying input values.

Since, for each couple of the modifying input (temperature and humidity) values, the model of the sensor was identified and the uncertainty estimation was performed, on the basis of the performed surveys, the uncertainty estimation did not depend on the considered modifying inputs. It was the effect of unconsidered and/or not modelled phenomena.

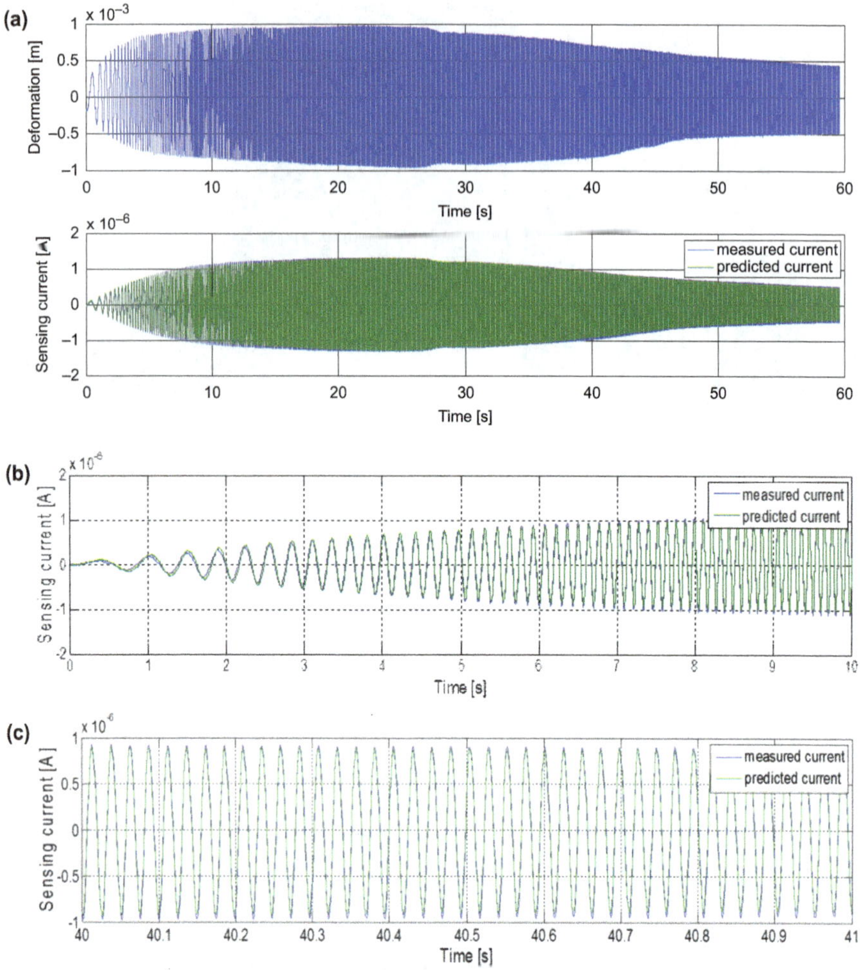

Figure 16.63 An example of the deformation applied to the IPMC sensor and of the corresponding absorbed current (a), and two zoomed images of the estimated currents (b) and (c). Both the measured and predicted current values are shown.

A typical example of the data used to characterise the IPMC sensors is shown in Figure 16.63(a), while in Figure 16.63(b) and (c), two zoomed images of the signals seen in Figure 16.63(a) are shown to highlight the estimation capabilities of the model at different values of the input signal frequency. Reported signals refer to an IPMC sensor working with an environmental temperature equal to 30 °C and a relative humidity equal to 90%, while the deformation was applied 20 mm away from the fixed end.

Also, as a further investigation, the analysis of model residuals along with the uncertainty estimation was performed. As an example, in Figure 16.64, some statistical plots derived by the elaboration of the available data are

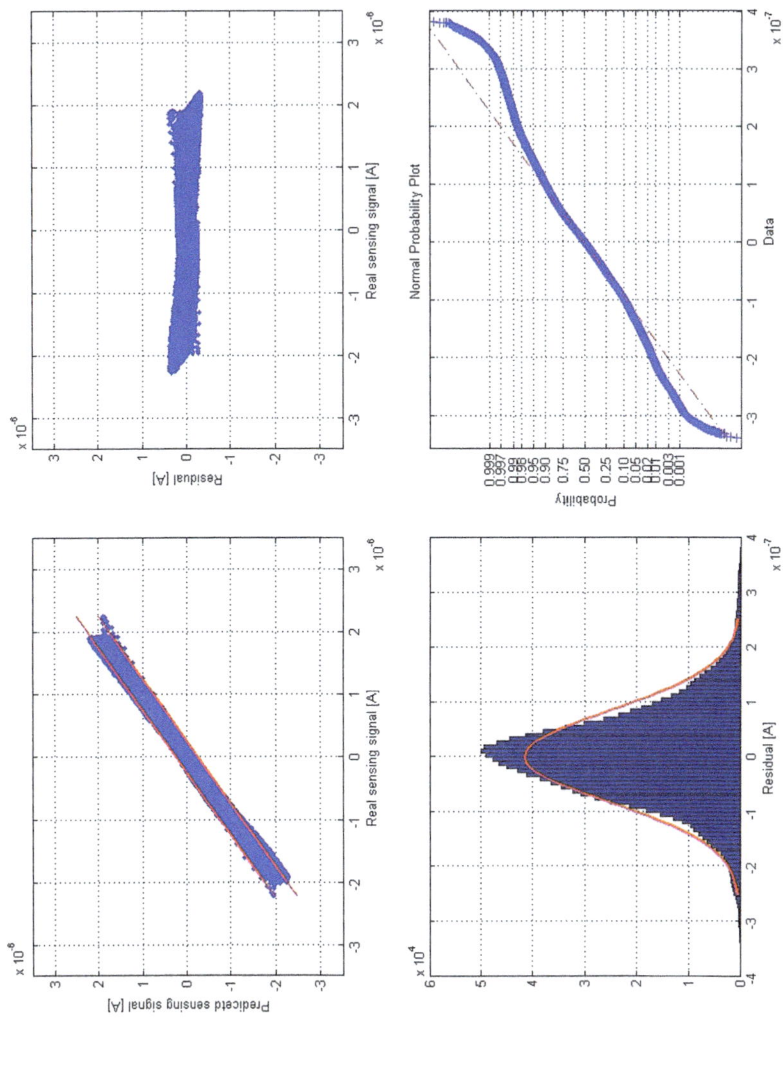

Figure 16.64 The residual analysis of the sensing signal model estimation for a sensor working with a temperature equal to 30 °C and a relative humidity equal to 70%.

shown. More specifically, in the top-left graph in Figure 16.64, the real sensing signal obtained from different measurements and the model estimation along with the uncertainty are drawn on an *X–Y* graph, while in the top-right graph in Figure 16.64, the real sensing signal obtained from different measurements and the corresponding residuals are drawn on an *X–Y* graph. In the bottom-left graph in Figure 16.64, the distribution of the residuals is shown. Finally, the bottom-right graph in Figure 16.64 represents the normal probability plot. The data reported Figure 16.64 refer to an IPMC sensor working with an environmental temperature equal to 30 °C and a relative humidity equal to 70%, while the distance of the application point from the fixed end was again 20 mm.

The frequency behaviour of the IPMC sensor was also investigated as a type of information relevant to sensor users. An example of the dispersion of the experimental frequency response estimated for the IPMC sensing current is seen in Figure 16.65.

As an example of the performed investigation in the frequency domain, in Figure 16.66, the effect of one modifying input while the other is maintained constant is reported. The product d*Y was considered.

The analysis in the frequency domain was repeated for each of the working conditions in Table 16.6. This analysis produced a set of surfaces, such as the ones shown in Figure 16.66.

It was generally observed that, in the investigated frequency range, the amplitude of the signals increased with the humidity level in the working environment, while no relevant effect was observed for the influence of temperature. Moreover, the noise of the sensing signal increased with the humidity level.

The effects of temperature on IPMC sensing characteristics were investigated by another group of researchers.[60] This convergence on the same topic at same time does not surprise me and it is a consequence of the growing interest in this topic and in the direct influences that new results have on the raising of new questions. Nevertheless, the scenario investigated by Ganley *et al.*[60] was quite different, since IPMC sensors in water were taken into account and the range of investigated temperatures (23–65 °C) was wider than in our analysis.[59]

Ganley *et al.*[60] used a black-box approach to identify the experimental IPMC short-circuit frequency response produced by the IPMC deformation. An example of experimental frequency responses is shown in Figure 16.67.

The frequency response was described in terms of coefficients that depend on the working bath temperature. Suitable polynomial functions were used to describe the dependence of such parameters on the bath temperature and this allowed for the estimating of the sensor behaviour for temperatures different from those that were experimentally investigated.

It is worth noting that although the authors observed a significant dependence of the IPMC sensing behaviour on the temperature value, such a dependence in quite evident for the higher-frequency values in the considered range (compare the curves given in Figure 16.67 with surfaces shown

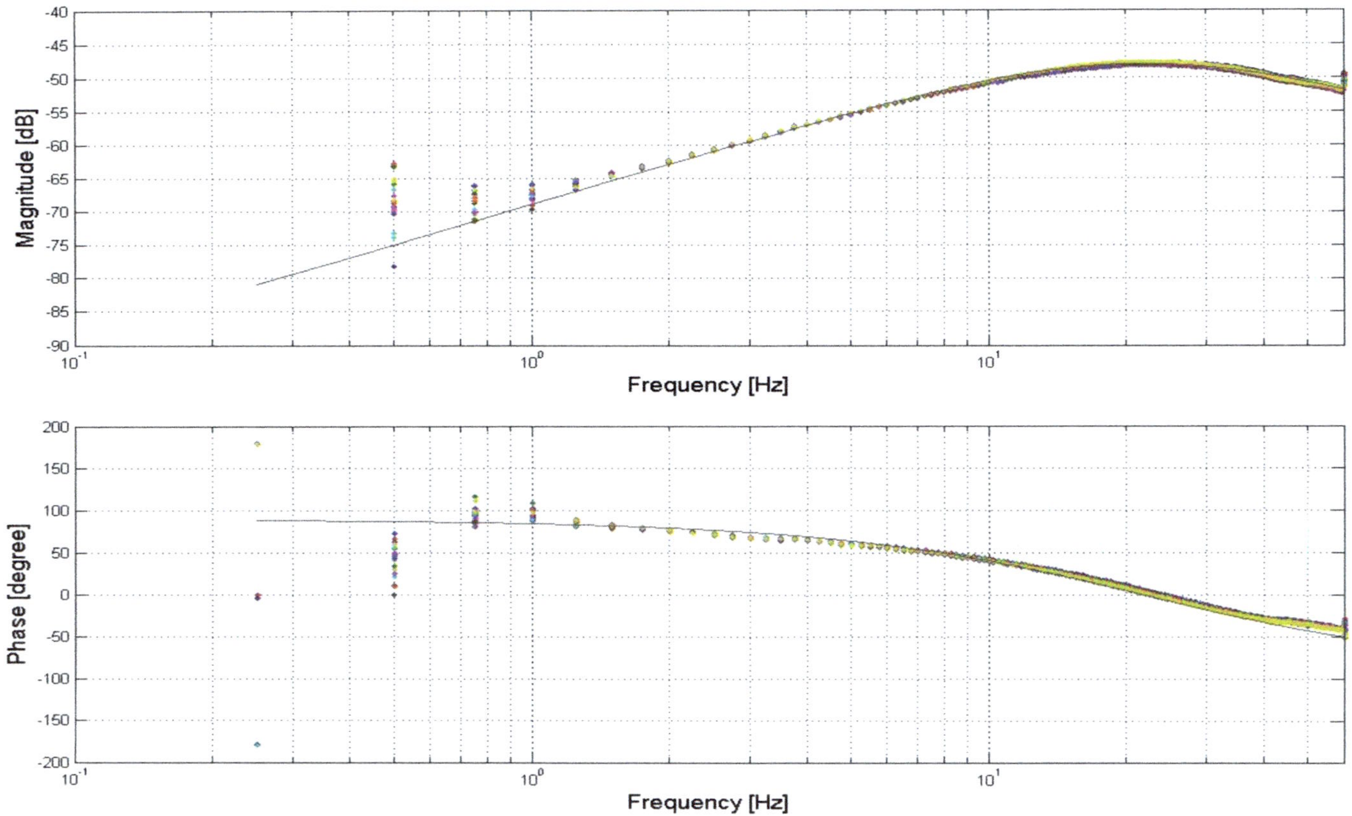

Figure 16.65 Bode plots of the sensor transfer function estimated from the acquired time signals.

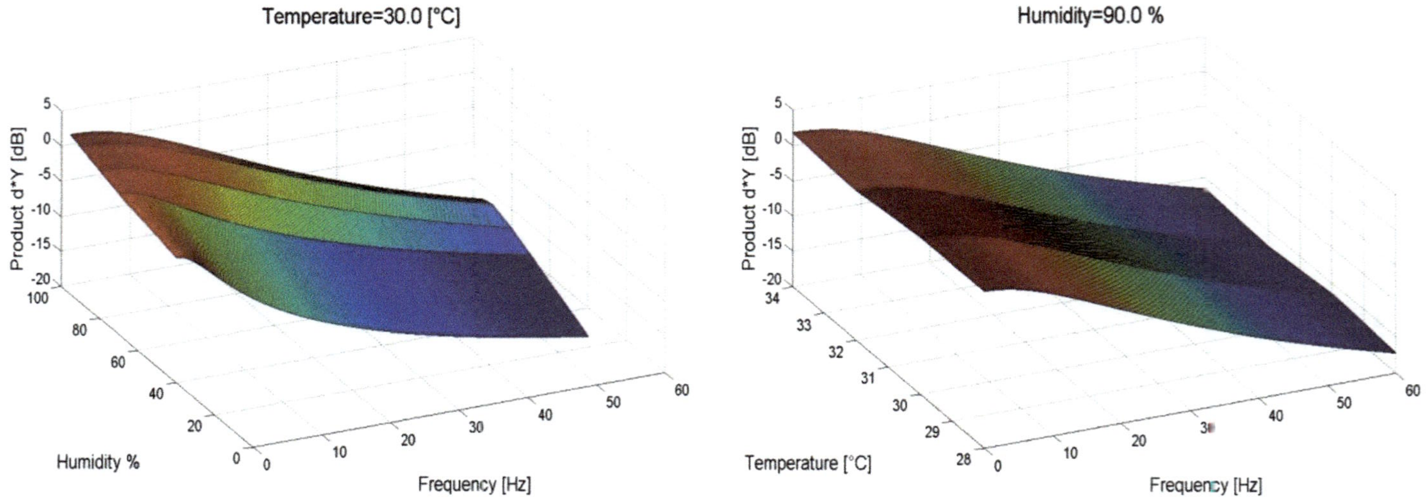

Figure 16.66 Example of the investigation in the frequency domain of the effect of environmental temperature (a) and relative humidity (b). © 2011 IEEE. Reprinted, with permission, from P. Brunetto, L. Fortuna, P. Giannone, S. Graziani and S. Strazzeri, Characterization of the temperature and humidity influence on ionic polymer-metal composites as sensors, *IEEE Trans. Instrum. Meas.*, 2011, **60**(8), 2951–2959.

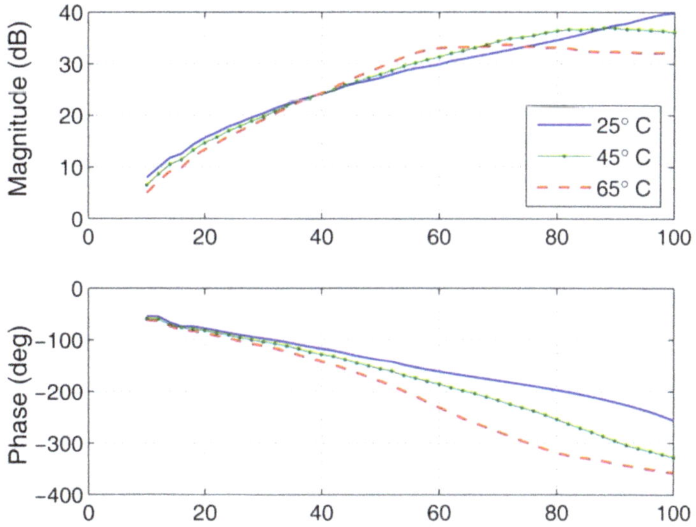

Figure 16.67 Experimental frequency responses of the IPMC sensor in water obtained in ref. 60.

in Figure 16.66). By contrast, if intervals of frequency and temperature comparable with the ones that we considered[59] are taken into account, the temperature dependence becomes much less evident, so that it could be possible to conclude that the reported results are somewhat in agreement.

Finally, Ganley *et al.*[60] completed their study of the IPMC-based sensor with the estimation, by using the corresponding sensor short-circuit current, of the sensor input (*i.e.* the deformation), which is the quantity we are interested in when using a motion sensor.

Part of the IPMCs' success is represented by their capacity to work in wet environments or even immersed in water. This paved the way to underwater applications of IPMCs,[61,62] including autonomous underwater vehicles or biomimetic robotic fish[63–65] and sensors,[66–68] which have been constantly proposed in the literature. The importance of describing the behaviour of IPMC transducers in water is therefore sound. We approached this problem[69] by modelling an IPMC vibrating beam immersed in a viscous incompressible fluid by again using a grey-box model. More specifically, the classical Euler–Bernoulli beam theory and the concept of hydrodynamic function were exploited to model the interaction between the vibrating beam and the surrounding fluid.[70] Also, the coupling term d was used and it was modelled again as a complex function of the frequency. The model was developed referring to the usual pinned IPMC configuration. In the following, L will be the free length of the beam and $A = bw$ its cross-section area. The beam deflection will be indicated with $w(x)$ and the applied mechanical moment with $M(t)$.

As a first step, we used the Castigliano theorem to link the produced moment to the applied voltage:

$$M = \frac{d\,h\,b\,Y}{2}V \tag{16.39}$$

Then, the beam was modelled by using the classical fourth-order differential equation, valid for the slender beam:

$$YI\frac{\partial^4 w(x,t)}{\partial x^4} + C\frac{\partial w(x,t)}{\partial t} + \rho A\frac{\partial^2 w(x,t)}{\partial t^2} = f(x,t) \tag{16.40}$$

where I is the area moment of inertia with respect to the x-axis, ρ is the specific density of the actuator, and C is a damping coefficient.

The general solution of the motion in eqn (16.40) was obtained by using the classical mode summation method.

According to this method and using eqn (16.39), the deflection of an IPMC produced by the applied voltage can be written as:

$$\begin{aligned} w(x,t) &= \frac{1}{\rho AL}\sum_i \frac{H_i(\omega)}{\omega_i^2}\varphi_i'(L)\varphi_i(x)M_0 e^{j\omega t} \\ &= \frac{1}{\rho AL}\sum_i \frac{H_i(\omega)}{\omega_i^2}\varphi_i'(L)\varphi_i(x)\frac{d(\omega)h\,w\,Y}{2}Ve^{j\omega t} \end{aligned} \tag{16.41}$$

$$H_i(\omega) = \frac{1}{1-(\omega/\omega_i)^2 + j2\xi_i\omega/\omega_i}$$

where the summation is extended to infinitely many terms, ϕ_i is the i-th mode, and ω_i and ξ_i are the corresponding beam natural frequency and damping factor, respectively.

The proposed model was based on an adaptation of eqn (16.41) to explicitly take into account the interaction between the active beam and the surrounding viscous fluid. In fact, when an active IPMC oscillates in a viscous fluid, this exerts forces on the moving beam. While for the case of the air such an interaction can be neglected, this is not true when a viscous fluid such as water surrounds the beam, and a further force g needs to be added to the equation ruling the IPMC active beam. Such a force will result from two contributions: an additive inertial term, linked to the fluid dragged by beam, and a damping term.

An expression exists for such a force for the case of small oscillations. In fact, for steady-state oscillations:

$$w = w_0\ sin(\omega t) \tag{16.42}$$

this force assumes the form:[71]

$$g = m_d \omega^2 w_0 [\mathrm{Re}(\Gamma)\sin(\omega t) + \mathrm{Im}(\Gamma)\cos(\omega t)]$$
$$m_d = \rho_f (\pi / 4) b^2 \tag{16.43}$$

where $\Gamma(\omega)$ is the hydrodynamic function and ρ_f is the fluid density.

Though an exact form exists for the case of a beam with a circular section, no such solution has been so far proposed for the case of a beam with a rectangular section, as is the case for the IPMC. Nevertheless, an approximate expression has been proposed by Sader for the case of the rectangular section:

$$\Gamma(\mathrm{Re}) = \Omega(\mathrm{Re})\left[1 + \frac{4iK_1\left(-i\sqrt{i\mathrm{Re}}\right)}{\sqrt{i\mathrm{Re}}K_0\left(-i\sqrt{i\mathrm{Re}}\right)}\right] \tag{16.44}$$

where K_0 and K_1 are the modified Bessel functions of the third type, Re is the Reynolds number, and $\Omega(\mathrm{Re})$ is the correction function to be applied to the corresponding hydrodynamic function, valid for the circular cross-section.

For the range of frequencies relevant to IPMC vibrations, a simplified form of the hydrodynamic function can be used:

$$\Gamma' = c_1 + c_2 \frac{\delta}{b}$$
$$\Gamma'' = c_3 \frac{\delta}{b} + c_4 \left(\frac{\delta}{b}\right)^2 \tag{16.45}$$
$$\delta = \sqrt{\frac{2\eta}{\rho_{\mathrm{fluid}}\omega}}$$

where Γ' and Γ'' are the real and imaginary parts of the hydrodynamic functions, respectively.

An example of the approximation capabilities of eqn (16.45) in the frequency range of interest for the case of IPMCs can be seen in Figure 16.68.

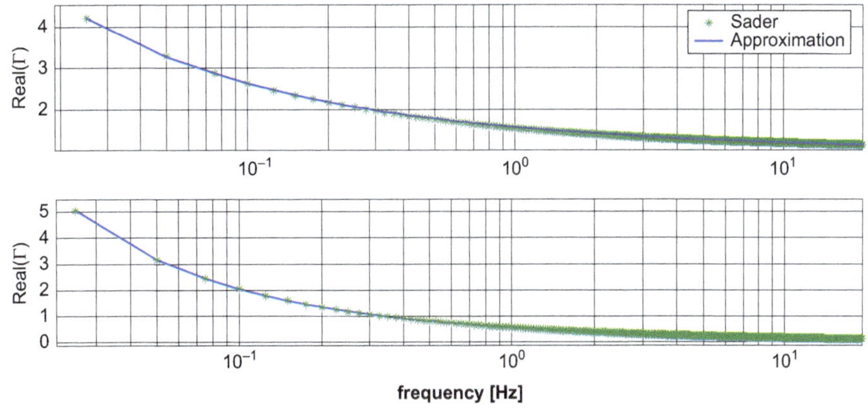

Figure 16.68 Real and Imaginary parts of the hydrodynamic function for a beam of width $b = 2$ mm.

By using the simplified form of the hydrodynamic function, the motion equation of the IPMC immersed in a fluid was rewritten as:

$$YI\frac{\partial^4 w(x,t)}{\partial x^4} + C_\mathrm{v}\frac{\partial w(x,t)}{\partial t} + \left(\rho A + C_\mathrm{m}\rho_\mathrm{f}\frac{\pi}{4}b^2\right)\frac{\partial^2 w(x,t)}{\partial t^2} = f(x,t) \tag{16.46}$$

where:

$$\begin{aligned} C_\mathrm{m} &= a_1 + a_2\frac{\delta}{b} \\ C_\mathrm{V} &= -m_\mathrm{d}\omega\left(b_1\frac{\delta}{b} + b_2\left(\frac{\delta}{b}\right)^2\right) \end{aligned} \tag{16.47}$$

By using eqn (16.39) and the mode summation method to the solution of eqn (16.46), the following expression was finally obtained:

$$\begin{aligned} w^\mathrm{f}(x,t) &= \frac{1}{\rho A + m_\mathrm{d}}\sum_i \frac{H_i{}^\mathrm{f}(\omega)}{(\omega_i{}^\mathrm{f})^2}\varphi_i'(L)\varphi_i(x)M_0 e^{j\omega t} \\ &= \frac{1}{\rho A + m_\mathrm{d}}\sum_i \frac{H_i{}^\mathrm{f}(\omega)}{(\omega_i{}^\mathrm{f})^2}\varphi_i'(L)\varphi_i(x)\frac{d(\omega)hwY}{2}Ve^{j\omega t} \end{aligned} \tag{16.48}$$

where:

$$H_i^\mathrm{f}(\omega) = \frac{1}{1-(\omega/\omega_i^\mathrm{f})^2 + j2\xi_i^\mathrm{f}\omega/\omega_i^\mathrm{f}} \tag{16.49}$$

$$\begin{aligned} \omega_i^\mathrm{f} &= \omega_i\left(1 + \frac{\pi\rho b}{\pi\rho_\mathrm{c}h}C_\mathrm{m}(\omega_i^\mathrm{f})\right)^{-1/2} \\ \xi_i^\mathrm{f} &= \frac{C_\mathrm{v}(\omega_i^\mathrm{f})}{2\rho A\omega_i^\mathrm{f}} \end{aligned} \tag{16.50}$$

As usual, the proposed model was verified against experimental data. As typical examples in Figures 16.69 and 16.70, it is possible to observe the simulated and experimental frequency response of the IPMC cantilever in air and immersed in deionised water. The experimental frequency responses were obtained by imposing a swept voltage stimulus on the IPMC actuator and recording the deformation produced. A Matlab® script was used to estimate the modulus and phase of the frequency responses. Reported results refer to an IPMC of size 30 mm×4 mm×0.180 mm.

As part of our interest in the realisation of novel sensing systems, we further investigated the changes induced in the frequency response of the vibrating IPMC by the rheological characteristics of the fluid. An example of the obtained results is shown in Figure 16.71, where the behaviour of IPMC actuators immersed in air, a synthetic fluid, deionised water, sea water, and human blood is shown, while the rheological characteristics of the considered fluids are given in Table 16.7. The figure shows the amplitude of the frequency responses of the IPMC considered before, immersed in the five

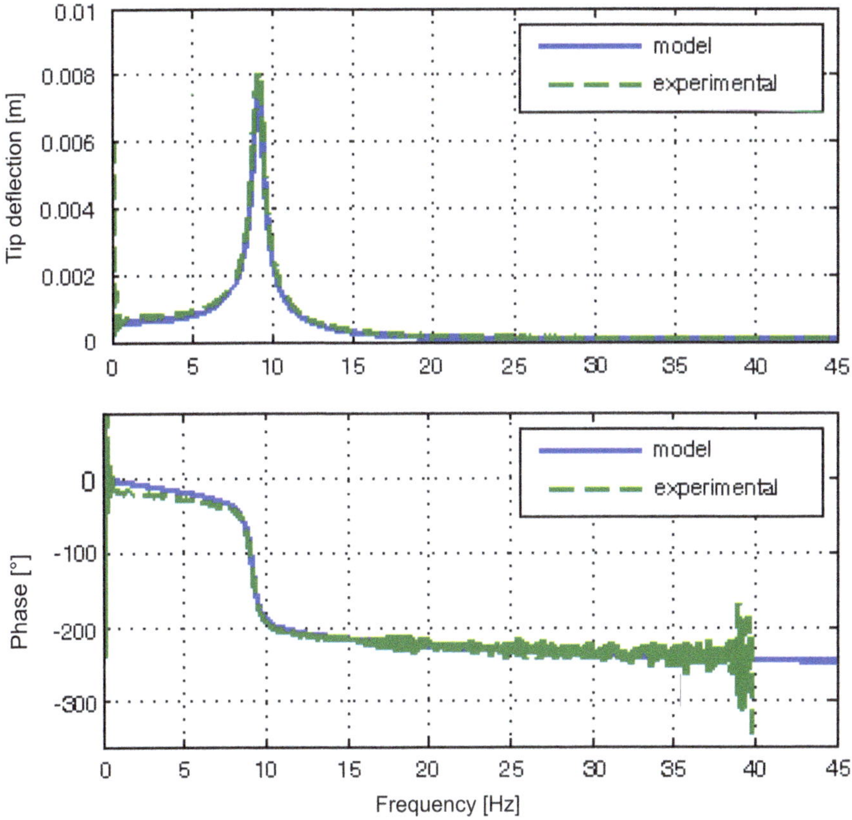

Figure 16.69 Tip vibration amplitude and phase for an IPMC in air. Reprinted with permission from P. Brunetto, L. Fortuna, S. Graziani and S. Strazzeri, A model of ionic polymer-meal composite actuators in underwater operations, *IOP Smart Mater. Struct.*, 2008, **17**, 025029.

fluids mentioned above. It is possible to observe that the proposed model allowed the prediction of the changes in the IPMC behaviour produced by the fluid characteristics. The graphs shown in Figure 16.71 were considered as a proof of concept of the possibility of using IPMCs as active vibrating beams to realise the prototype of a viscometer that we proposed subsequently.[66,72]

A number of authors have been interested in the interaction between IPMC actuators and the surrounding fluid to realise IPMC-based underwater robots. Part of this research was aimed at studying the thrust attainable by a vibrating IPMC as relevant to the realisation of propulsion systems.

As an example, in Peterson *et al.*,[73] the authors used the particle image velocimetry (PIV) technique to analyse the flow field produced by a vibrating IPMC. The estimation of the mean flow field was further used to estimate the produced thrust. The authors concluded that the mean thrust grows with the Reynolds number, while is marginally influenced by the relative tip displacement. The same research group reconsidered the problem in

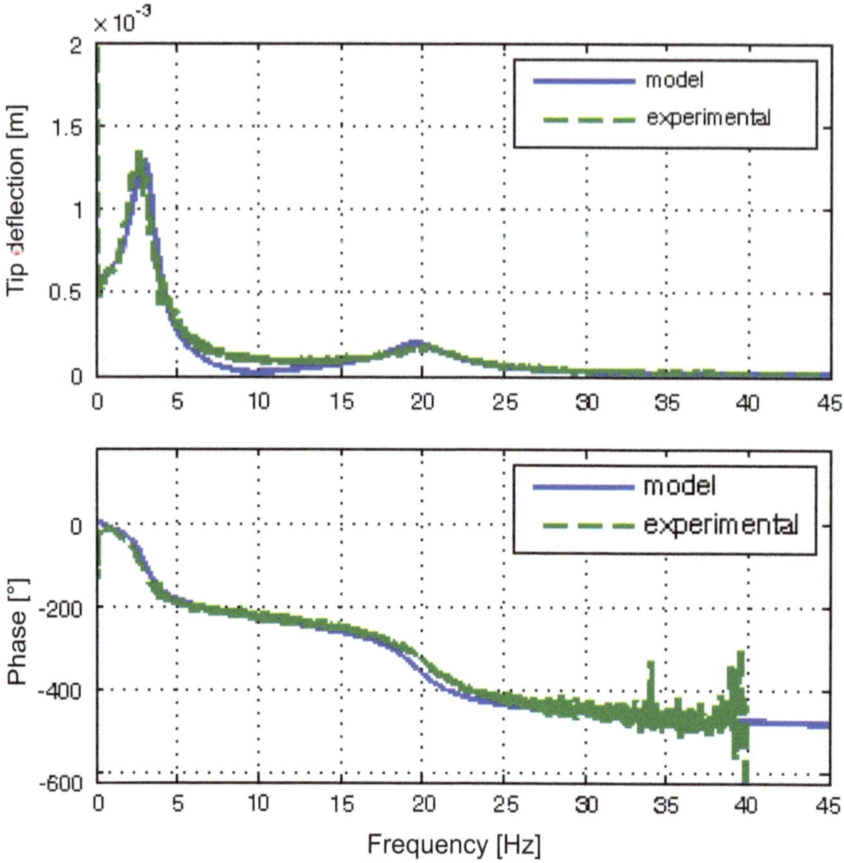

Figure 16.70 Tip vibration amplitude and phase of an IPMC in deionised water. Reprinted with permission from P. Brunetto, L. Fortuna, S. Graziani and S. Strazzeri, A model of ionic polymer-meal composite actuators in underwater operations, *IOP Smart Mater. Struct.*, 2008, **17**, 025029.

Abdelnour *et al.*,[74] where, by using the Navier–Stokes equations, both the thrust and the lateral forces per unit actuator width were estimated.

The same framework was further used to estimate the moment resultant of the forces exerted by the IPMC on the encompassing fluid and the delivered power. The authors validated their results against estimations obtained by using a package for fluid dynamics simulation. The investigation of the numerical results allowed the description of the system vorticity as the cause of the thrust generation capability of the IPMC. Finally, the numerical estimations were compared with experimental data obtained by the same PIV described previously.[73]

The concept of hydrodynamic function as a tool for estimating the free locomotion of underwater vehicles was later proposed in the literature. Aureli *et al.*[75] developed a modelling framework for the free locomotion of underwater vehicles propelled by vibrating IPMCs in quiet water. More

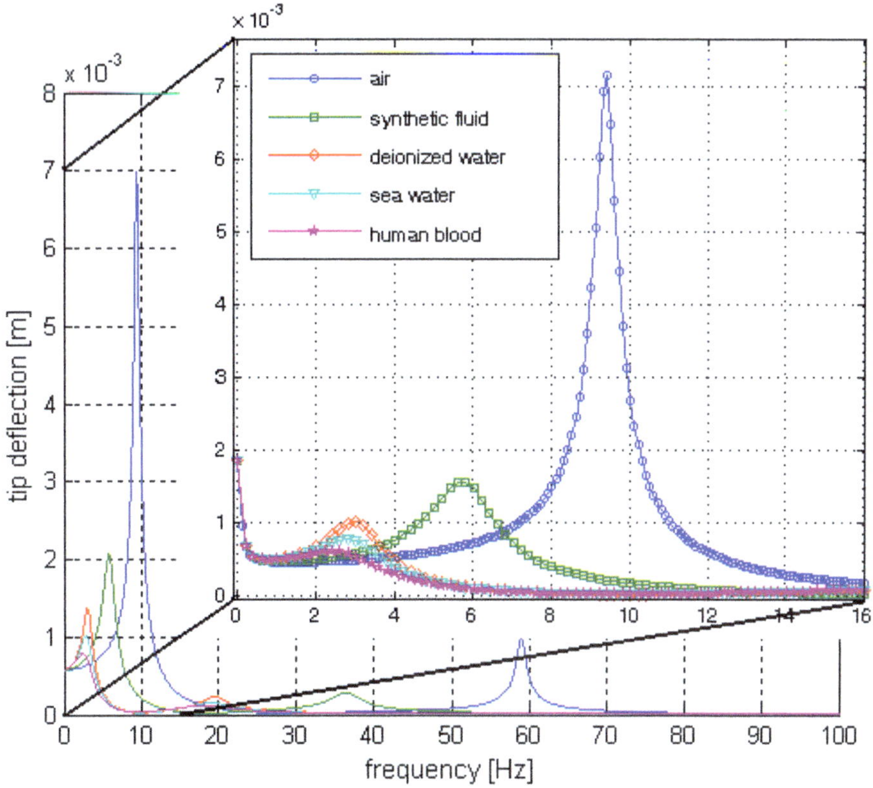

Figure 16.71 Frequency responses of IPMC actuators immersed in air, a synthetic fluid, deionised water, sea water, and human blood.
Reprinted with permission from P. Brunetto, L. Fortuna, S. Graziani and S. Strazzeri, A model of ionic polymer-meal composite actuators in underwater operations, *IOP Smart Mater. Struct.*, 2008, **17**, 025029.

Table 16.7 Density and viscosity values of the fluids analysed in Figure 16.71 (adapted from Brunetto *et al.*[69]).

	ρ (kg m^{-3})	η (Pa s)
Air	1	0.0001
Synthetic fluid	200	0.0002
Deionised water	997	0.0008
Sea water	1030	0.0018
Human blood	1125	0.004

specifically, the modelling restricted to the case of a vehicle's planar motion (3 degrees of freedom – 3 DOFs) and the hydrodynamic function was used to determine fish-like propulsion along with the lift and the moment, which act on the system as motion disturbances. The system's tail consists of an active IPMC and a trapezoidal-shaped passive fin and was used in an experimental setup to validate the proposed model.

Almost contemporaneously, another group of researchers used the Lighthill theory of the elongated body and the hydrodynamic function to estimate the steady-state speed produced by a passive fin constituted by a vibrating IPMC.[64]

More specifically, the authors used the elongated body hypothesis to link the mean thrust and the mean speed of the robot. Then, by balancing the mean thrust to the experienced drag force, the cruising speed was estimated, showing that the speed depended on the lateral velocity and on the slope of the trailing edge. The further step was then to use the Euler–Bernoulli beam theory and the concept of the hydrodynamic function to model the deflection of an IPMC immersed in a viscous fluid. Finally, a form of the transduction phenomenon proposed by Nemat-Nasser and Li[46] was used to model the bending moment produced by the applied voltage (details about the required computations and formulae can be found in the referenced work). Here, it suffices to say that the proposed model was experimentally investigated by the authors. As a significant example of the model's estimations, in Figure 16.72, a comparison between the estimated cruising velocity and the frequency of a square wave applied voltage is seen for various forms of the attached passive fin.

Through using the hydrodynamic function concept, the proposed model estimates the acting moment with a microscopic model, so that it is even closer to the class of white-box models that are the topic of the next subsection. This is evidence of the artificial nature of the classification and outlines the flourishing activity in IPMCs, where researchers share a lot of new findings and spontaneously import or propose new tools for the study of the IPMC transduction, with a general tendency to move towards white-box models.

16.2.3 White-box Modelling

White-box models, also known as first-principle models, have been proposed in the literature for a long time with regards to EAPs.[10] As an example of activities that are similar to those for IPMCs, back in 1990s, Shahinpoor proposed a microscopic-scale model for large electrically induced deformations in a number of ionic polymeric gels (*i.e.* for the case of cross-linked polymer networks swollen in a liquid medium, but in the absence of any electrode).[76,77]

White-box models describe the behaviour of IPMC transducers *via* the comprehension of the physical and chemical phenomena involved in the transduction process. The complexity of the proposed models has greatly increased due to both the better understanding that the research community has gained on this topic and the more powerful numerical simulation tools that have become available and that allow for the simulation of very sophisticated models that generally come under the form of partial differential equations (PDEs).

White-box models can play a fundamental role in the understanding of the microscopic phenomena that cause the IPMC transduction capabilities,

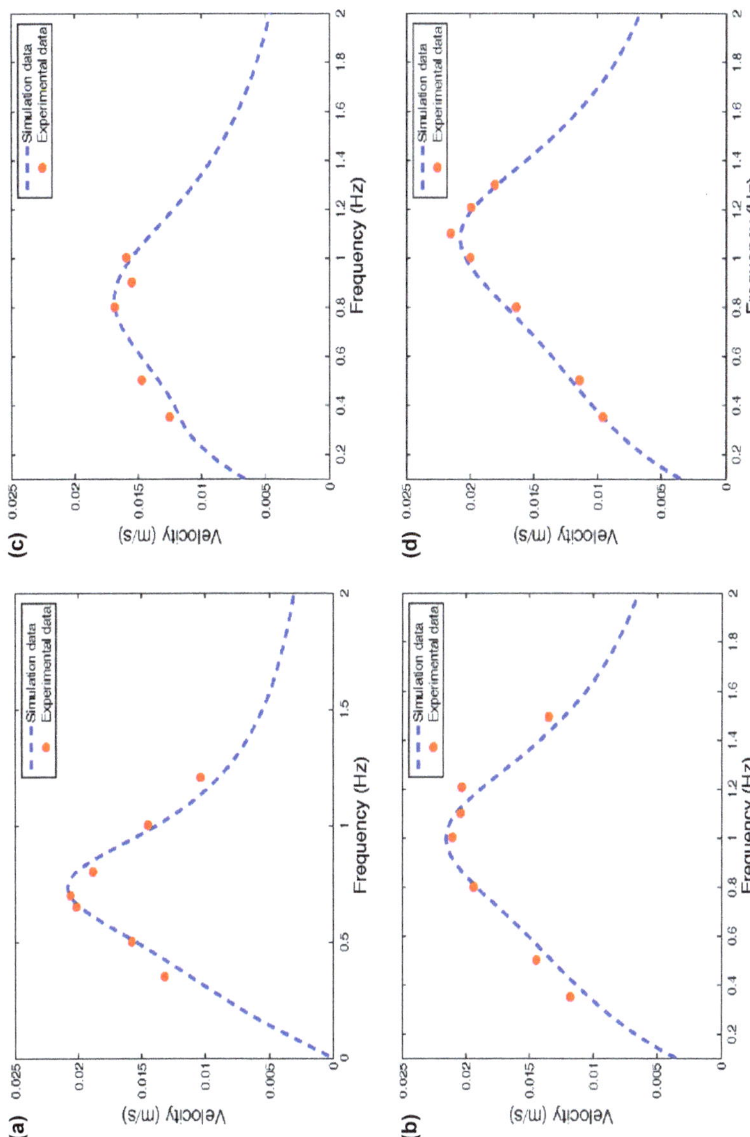

Figure 16.72 Dependence of the robotic fish cruising velocity as a function of the input frequency and of the passive tail shape. © 2010 IEEE. Reprinted, with permission, from Z. Chen, S. Shatara and X. Tan, Modeling of biomimetic robotic fish propelled by an ionic polymer-metal composite caudal fin, *IEEE/ASME Transactions on Mechatronics*, 2010, **15**(3), 448–459.

so they could allow for predicting the consequences of modifications in IPMC production steps. This aspect could provide a formidable contribution to the optimisation of IPMC macro-scale performances. This is very much true since when the numerical solution of PDE systems becomes possible. This makes possible the validation of the proposed models against experimental observations.

From the year 2000, a number of white-box models have been proposed for the IPMC actuation mechanism. A taxonomy of physics-based models, useful for lending order to the large mass of produced contributions, has been proposed by Zhu *et al.*,[78] who classified models as thermodynamics of irreversible process models, frictional models, and Nernst–Planck (NP) equation models.

Following the classification introduced above, after a brief discussion of the white-box models proposed in the literature, the discussion will focus on the more recent contributions and on the activity we have taken in this field of research.

In the first category, approached for the first time by de Gennes *et al.* in 2000,[79] mass transport, which is a non-equilibrium thermodynamic process, was modelled based on the assumption of local equilibrium. This model is suitable for describing both the direct and inverse electroactive effect under static fields. The authors proposed that the transduction is caused by the mobile cations (Na^+ in the described case) that drift under the effect of the electric field. Eventually, the cations carry a certain number of solvent molecules. The molecules pile up near the cathode and create an overpressure that deforms the membrane. A schematic view of the system is seen in Figure 16.73.

The transduction phenomenon was modelled by using two transport equations, one for the charges and one for the solvent transport. The charge

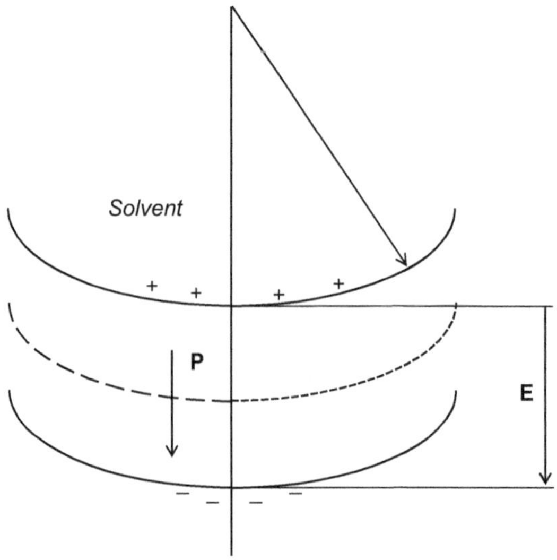

Figure 16.73 Scheme of the transduction phenomenon proposed in ref. 79.

transport occurs because of a current density J normal to the polymer surface, while the solvent transport is represented by the corresponding flux Q. Relevant forces are the electric field E and pressure p. The quantities introduced so far were linked by a set of two equations, derived from thermodynamics as:

$$\begin{aligned} J &= \sigma E - L_{12} \nabla p \\ Q &= L_{21} E - K \nabla p \end{aligned} \tag{16.51}$$

where σ is the membrane conductance, K is the Darcy permeability, and $L_{12} = L_{21} = L$ is a cross-coefficient.

The equations (16.51) were used to derive a direct effect for the case of the actuator produced by the applied electric voltage, and an inverse effect for the case of the sensor produced by the applied torque Γ.

This model has become very popular and accepted as a standard model for IPMC transduction,[80] to the point that it was used to estimate the IPMC transducer's behaviour as a function of the working conditions. As an example, the model in eqn (16.51) was used to estimate the IPMC conductivity as a function of the environmental temperature.[56] More specifically, the investigators used the model in an attempt to explain the phenomena that allowed IPMC actuators to work at temperatures below 0 °C.

The model, even in modified forms, was used and further developed by other authors. Asaka and Oguro[81] used the direct part of the model (*i.e.* the electromechanical transduction term) to model the bending of an IPMC. The authors found that the curvature of an IPMC bender has first-order dynamics. Also, the authors modelled a further action generated at the polymer–electrode interface, which was found to be ruled by the acidity of the water solution in which the IPMC was immersed.[82] They demonstrated that this second contribution, at least for IPMCs with platinum-based electrodes, rules the IPMC bending after the contribution of the first-order dynamics is over, which they called the electrokinetic effect.

A further, yet different, application of Onsager equations to the modelling of IPMCs was proposed by McGee,[83] who described the mass transport inside the IPMC as being governed by diffusion phenomena.

In the category of frictional models, the transport process is described based on the hypothesis that, at steady state, the driving forces are balanced by frictional forces.

As for the case of the thermodynamics of irreversible process models, the first frictional model to describe IPMCs was introduced in 2000 by Tadokoro *et al.*,[84] who proposed the so-called Yamagami–Tadokoro model. They considered the case of a Nafion-based IPMC with Na^+ as the counterion. In the proposed model, the cations, driven by the electrostatic force, move from the anode to the cathode and carry with them some hydrating water molecules.

The charge motion is modelled in an ionic migration model that rules the motion of both the sodium ions and of the water.

The electrostatic force acting on the sodium ions is balanced by the frictional force linked to the ion velocity and the diffusion forces induced by the concentration gradients of both sodium and water. A similar model is proposed for the motion of water molecules, though in this second case, the equilibrium is imposed only between the diffusion and viscous forces.

Finally, a stress-generation model was proposed that takes into account both the effect of swelling and contraction produced by water, the effect of electrostatic forces produced by the fixed negative charges (sulphonic acid groups in the considered polymer matrix), and the effect of moment conservation.

A number of simulations are shown that depict the concentration of the involved species. Also, the authors show that by taking into account the effects of both water migration and repulsive electrostatic forces produced by the fixed charges, the model is capable of predicting the time dynamics observed for IPMC benders.

The Yamagami–Tadokoro model was further developed 2 years later into the Fukuhara–Tadokoro model[85] by considering the rise of lateral strain produced by the migration of sodium ions and consequently by the attractive electrostatic forces acting between the fixed sulphonic groups and the electrode. The authors proposed that the combined effect of water back-diffusion and lateral strain can explain the back-diffusion phenomenon.

Toi and Kang[86] adapted the Yamagami–Tadokoro model to the case of two-dimensional IPMC beams and simulated the model by using a finite element approach. Eventually, the authors linked the back-relaxation phenomenon to the action of water diffusion towards the anode.

In 2006, Costa Branco and Dente proposed a further friction model,[87] but in their model, the mechanical action was produced by electrostatic forces acting on the fixed negative charges, while the contribution of water was considered negligible.

The NP equation-based approach to IMPC modelling is at the basis of our contributions to the IPMC white-box modelling approach and will be dealt with below. Actually, we took care to enrich IPMC physics-based modelling with our interest in experimental validation and we used information gathered from measurements to tune our NP white-box-based models.

NP models are based on the NP equations describing the ion transport in an electrochemical system. Again, research on NP-based IPMC models started in 2000, when Nemat-Nasser proposed a physical model describing the rise of stress in Nafion because of fixed charge imbalance,[46] while a marginal role was given to the effect of water distribution. A further contribution was published in 2002.[22]

The model proposed by Nemat-Nasser and Li is an example of a reversible model. The authors compared experimental data with values estimated by the model for the case of IPMC sensors. The model was capable of predicting that the voltage induced by a sudden imposition of curvature (*i.e.* by using the IPMC as a sensor) is two orders of magnitude less than the voltage required to produce that curvature (*i.e.* by using the IPMC as an actuator).

In 2004, Farinholt and Leo[88] used the model by Nemat-Nasser and Li to develop a physics-based model of IPMCs working as sensors. In their model, the imposed deformation produces charge accumulation at the electrode surface, and then a sensing current. Based on these assumptions and on the hypothesis that a constant links the charge density to the mechanical stress,[46] the authors derived equations for the charge density, the electric field, and the electric potential when short-circuit boundary conditions are considered. By considering a cantilevered beam, the current was estimated to be:

$$I(t) = -\lambda w(L)\frac{Y}{\beta\Psi_s}\frac{3}{2}\frac{bh}{L}\left(\frac{\cosh\beta h}{\sinh\beta h} - \frac{1}{\beta h}\right)e^{-\lambda t} \tag{16.52}$$

The interested reader can find the full list of the used symbols in the referenced paper. A further example of the use of NP equations to model sensing behaviour of IPMCs can be found in Chen *et al.*[89]

NP equations have been used in a number of different forms. More specifically, various simplifications have been considered acceptable by Johnson and Amirouche,[90] Pugal *et al.*,[91] Nemat-Nasser and Zamani,[92] Porfiri,[93] Aureli and Porfiri,[94] Wallmersperger *et al.*,[95] Nardinocchi and Pezzulla,[96] Galante *et al.*,[97] and Zhang and Yang.[98]

Further examples of physics-based approaches exist in the literature. Del Bufalo *et al.*[99] exploited the mixture theory to model IPMCs as the superposition of three species: a charged solid (the backbone polymer matrix), an uncharged fluid (the solvent, *e.g.* water), and a charged gas (the mobile ionic species). That model was improved by Porfiri in 2009,[100] who proposed a white-box model accounting for fundamental microscopic phenomena, such as the electric dipole formation and electrostatic stress generation studied from a micromechanics standpoint.

The efforts in IPMC white-box modelling have been devoted to the chemo-electro-mechanical processes that result in multiphysics modelling of IPMC transduction. The models are in the form of partial differential equations. A new aspect of white-box modelling of IPMCs is the availability of powerful tools that allow for approximating the relevant equations in the form of finite differential equations.

Nardinocchi *et al.*[101] modelled the IPMC actuator as a thin, three-dimensional body resembling the characteristics of hydrated Nafion, sandwiched between two metallic layers. The authors modelled the chemically induced deformations in terms of a volumetric distortion field induced by the redistribution of ions and solvent molecules, while the electric and chemical physics were described through standard equations of the electrostatics and mass conservation. Thermodynamic issues were finally used to derive the stress T, the electric field e, and the chemical flux j. The commercially available finite element code COMSOL Multiphysics® was used to perform numerical simulations for the cases of both IPMC-based actuators and sensors.

We proposed[102] a multiphysics model of IPMC actuators implemented by using COMSOL Multiphysics®. More specifically, a multiphysics model was investigated with the aim of integrating the electric, mechanical, chemical, and thermal phenomena involved in IPMC dynamics.

Moreover, some of the parameters of the model were estimated by fitting the model simulations to experimental data of observable quantities. This required the inclusion of the COMSOL Multiphysics®-based model into a minimisation procedure developed *ad hoc* by using the Matlab® environment.

Investigated IPMCs use ethylene glycol as the solvent, which, like water, consists of polar molecules,[103] thereby avoiding solvent evaporation and allowing for larger values of the input voltage.

The analysis was carried out in the frequency domain. More specifically, the frequency domain investigation was performed by using the experimentally determined Young's modulus of the device.

The geometry of the investigated IPMC actuator is shown in Figure 16.74. The geometrical parameters were: copper electrode length $L_c = 5$ mm; total IPMC strip length $L_t = 26$ mm; width $w = 5$ mm; and thickness of copper and platinum electrodes and Nafion $t_{Cu} = 10$ μm, $t_{Pt} = 10$ μm, and $t_{Naf} = 180$ μm, respectively.

Three regions, characterised by different physical and chemical properties, can be identified in the structure:

- The central zone, consisting of an ionic polymer, Nafion;
- The surface electrodes, made of platinum;
- The metal contacts, made of copper, required for the IPMC to work.

In the literature, the roughness of the electrodes is considered fundamental to the enhancement of the IPMC behaviour[104] and a number of approaches are followed to take into account such a characteristic.[105–109]

Nardinocchi and Pezzulla[96] proposed a varying-along-the-thickness relative permittivity ε_R. More specifically, ε_R increases on the composite boundaries with respect to the hydrated bulk. The parameters characterising the spatial dependency were identified *via* experimental data.

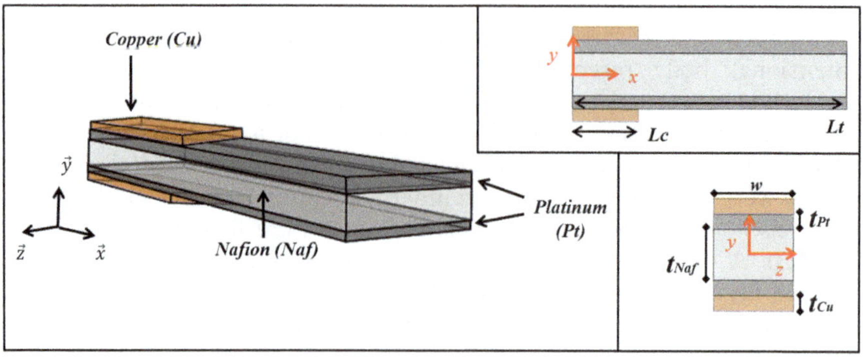

Figure 16.74 Geometry of the IPMC actuator in the cantilever configuration. Reprinted with permission from R. Caponetto, V. De Luca, S. Graziani and F. Sapuppo, An optimized frequency-dependent multiphysics model for an ionic polymer–metal composite actuator with ethylene glycol as the solvent, *IOP Smart Mater. Struct.*, 2013, **22**(12), 125016.

The effects of the electrode roughness can be approximated with a high effective electric permittivity (ε_{Reff}) that can be considered as the spatial average of the space-dependent ε_R.

We fixed the electrical permittivity in the Nafion $\varepsilon_R = \varepsilon_{EG}$ as the minimum permittivity in the bulk. The identification process was chosen to introduce a corrective effect on two electromechanical parameters kf_{eff} and βf_{eff} used in the model.

The complex behaviour of the IPMC actuator was modelled as described in the diagram in Figure 16.75. It includes four coupled physic models:

- *Electrical model* to study the electric potential V;
- *Chemical model* to define the ionic current and to study the phenomenon linked to the transport of the solvent;
- *Mechanical model* to determine the deformation of the structure;
- *Thermal model* to model the temperature distribution on the membrane and its impact on the ionic transport properties.

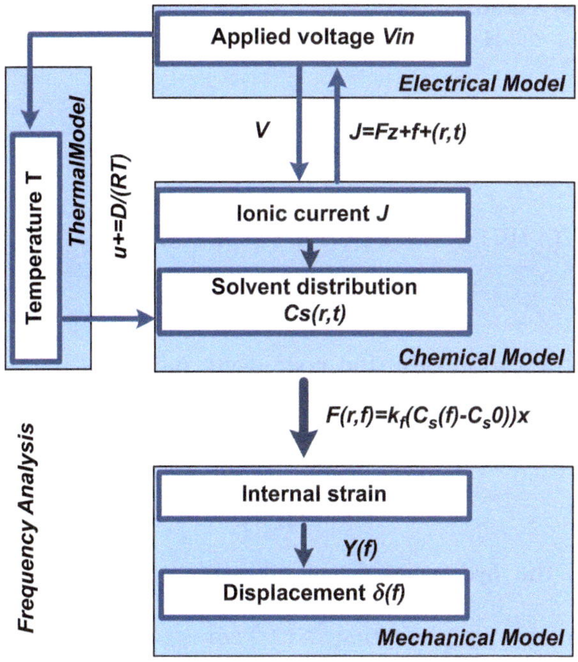

Figure 16.75 The diagram of the multiphysics model proposed in ref. 102. Reprinted with permission from R. Caponetto, V. De Luca, S. Graziani and F. Sapuppo, An optimized frequency-dependent multiphysics model for an ionic polymer–metal composite actuator with ethylene glycol as the solvent, *IOP Smart Mater. Struct.*, 2013, **22**(12), 125016

The models were linked thanks to a number of coupling effects. More specifically, in the proposed model, the H^+ ionic current density (J) and the electric potential (V) cause the interaction between the electrical and the chemical model; the temperature field affects the ionic mobility (μ) and the solvent concentration C_s is the coupling quantity between the chemical and the mechanical model.

The equations required to describe the IPMC electromechanic transduction were developed in the framework of the simplified NP equations. More specifically, the plane electric current is:

$$-\nabla \cdot \frac{\partial(\varepsilon_0 \varepsilon_r \nabla V(r,t))}{\partial t} - \nabla \cdot (\sigma \nabla V(r,t) - J) = 0 \tag{16.53}$$

where ε_0 is the absolute dielectric constant, ε_r is the relative dielectric constant, σ is the conductivity, and $V(r,t)$ is the scalar voltage field. J is the cation current:

$$J = Fz_+ f_+(r,t) \tag{16.54}$$

where F is the Faraday constant, z_+ is the charge number and $f_+(r,t)$ is the total flux of H^+ cations, which was modelled by using the NP equations:[22,78]

$$\frac{\partial C_+(r,t)}{\partial t} = \nabla \cdot \left(Fz_+ u_+ C_+(r,t) \nabla V(r,t) + D_+ \nabla C_+(r,t) + u_+ C_+ M_+ \left(\frac{V_+}{M_+} + \frac{V_S}{M_S} \right) \nabla p - C_+ \bar{v} \right) \tag{16.55}$$

where D_+ is the cation diffusion coefficient and u_+ is the corresponding mobility, M_+ and M_S are the molar weights of the cations and of the solvent, respectively, p is the pressure exerted by the polymer, and the convection velocity $\bar{v}$ is described by Darcy's law:

$$\bar{v} = K_{EG} \nabla p \tag{16.56}$$

where K_{EG} is the hydraulic permeability of solvent (ethylene glycol) in Nafion® 117.

By using suitable simplification,[90–96,98] eqn (16.55) was rewritten as:

$$f_+(r,t) = -(Fz_+ u_+ C_+(r,t) \nabla V(r,t) + D_+ \nabla C_+(r,t)) \tag{16.57}$$

The coefficient u_+ was modelled as a function of the absolute temperature T:

$$u_+ = \frac{D_+}{RT} \tag{16.58}$$

The solvent motion was modelled as:[110]

$$\frac{\partial C_s(r,t)}{\partial t} = -\nabla\cdot(n_+f_+(r,t) - D_s\nabla C_s(r,t)) \tag{16.59}$$

where C_s is the solvent concentration and D_s is the solvent diffusion coefficient. A reduced two-dimensional model has been considered and the variations of the solvent concentrations at the electrodes are:

$$\begin{aligned} C_{s_{up}}(t) &= \tfrac{1}{L_t}\int_0^{x_{max}}(C_s(r,t) - C_{s_0})\mathrm{d}x \quad \text{for } y = \frac{t_{Naf}}{2} \\ C_{s_{low}}(t) &= \tfrac{1}{L_t}\int_0^{x_{max}}(C_s(r,t) - C_{s_0})\mathrm{d}x \quad \text{for } y = -\frac{t_{Naf}}{2} \end{aligned} \tag{16.60}$$

The mechanical load F applied on the boundaries between the platinum and the Nafion subdomains has been considered proportional to the variation of the solvent concentration C_s with respect to the initial concentration C_{s_0}:[111,112]

$$\bar{F} = k_f(C_s - C_{s_0})\hat{x} \tag{16.61}$$

where k_f is a constant to be identified. The stress tensor was computed as:

$$-\nabla\cdot\sigma = \bar{\bar{F}} \tag{16.62}$$

The beam displacement was implemented by using the layer plane strain with a Rayleigh damping model, characterised by the damping parameter α_f set to zero in the considered case, and a stiffness damping parameter β_f that was the second parameter to be determined.

As mentioned above, we used the experimentally determined Young's modulus of the composite. The modulus was determined by using the dynamic mechanical analysis (DMA) at a fixed temperature equal to 25 °C. More specifically, the storage and loss moduli were experimentally determined and results are reported in Figure 16.76.

Finally, the temperature distribution was computed according to the equation:

$$\rho C_p\frac{\partial T}{\partial t} - \nabla\cdot(k\nabla T) = Q + h_{trans}(T_{ext} - T) \tag{16.63}$$

where ρ is the density, C_p is the specific heat capacity, k is the thermal coefficient, T_{ext} is the external temperature, and h_{trans} is the convective heat transfer coefficient. Q is the heat source and takes into account the currents, according to the Joule effect.

After suitable boundary conditions were fixed for all subdomains in the model, this was solved by using COMSOL Multiphysics®. Moreover, since some of the parameters in the model were unknown and it was not possible to measure them directly, they were identified by using experimental data.

To this aim, a suitable setup was realised to measure the IPMC applied voltage (V_{in}) and the corresponding tip deflection (δ). The time domain signals were further elaborated to obtain the corresponding frequency representations.

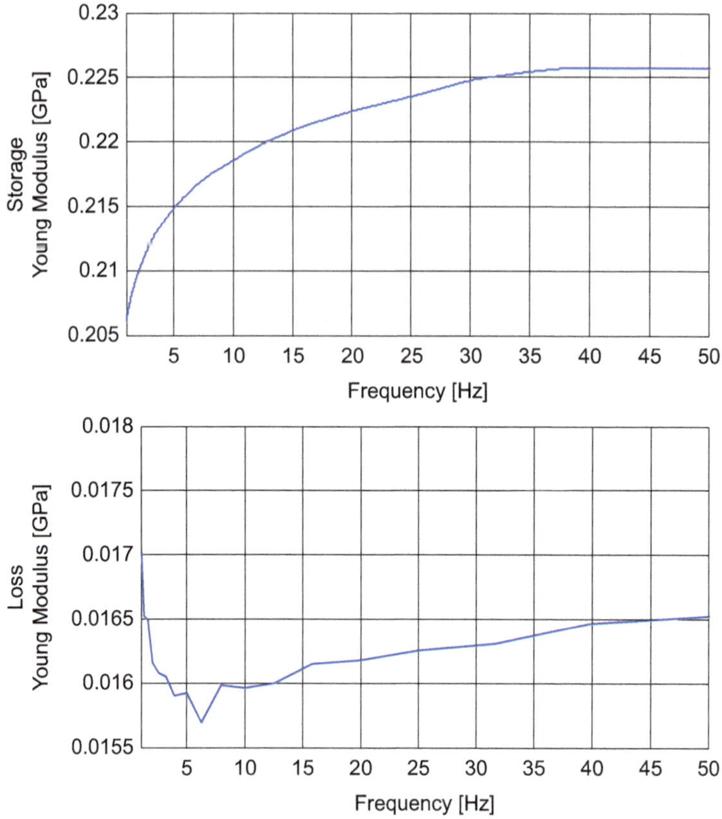

Figure 16.76 Storage Young's modulus (top image) and loss Young's modulus (bottom image) as a function of the frequency.
Reprinted with permission from R. Caponetto, V. De Luca, S. Graziani and F. Sapuppo, An optimized frequency-dependent multiphysics model for an ionic polymer–metal composite actuator with ethylene glycol as the solvent, *IOP Smart Mater. Struct.*, 2013, **22**(12), 125016.

The optimisation of the multiphysics model produced the values of k_f and β_f, respectively. Moreover, because of the simplified assumptions, we indicated the estimations as k_{feff} and β_{feff}. Such parameters are strictly related to the technology and to the working conditions and they are not available as material specifics.

The model was therefore used for a number of investigations. Some examples of the results obtained by using the multiphysics model are reported in the following. In Figure 16.77, an example of the potential field is reported. It is possible to observe the variation of the potential in the Nafion section between the cathode and the anode. The variation in the x-axis direction can also be noticed. The potential is higher in proximity to the copper electrodes and decreases in the direction of the IPMC beam tip as a result of the finite values of the copper and platinum conductivities.

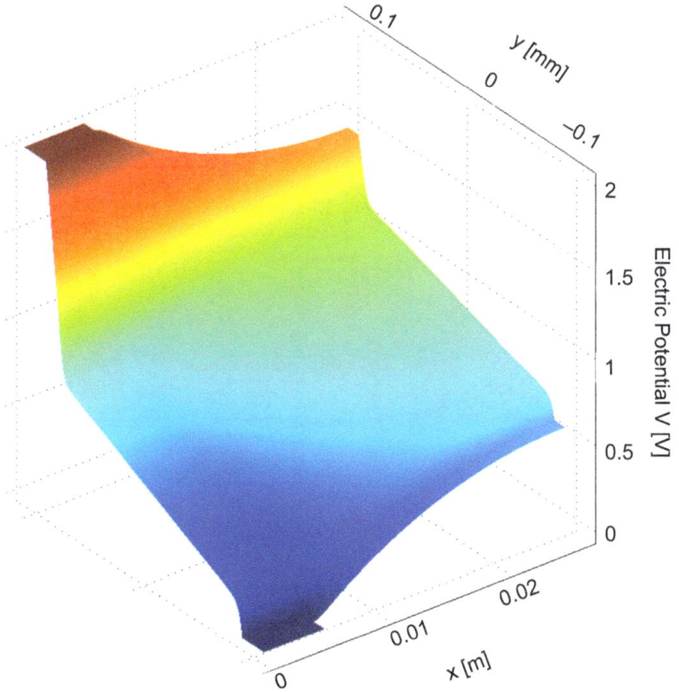

Figure 16.77 Electric potential along an IPMC transversal section at simulation time $t = 1.25$ s (not to scale).
Reprinted with permission from R. Caponetto, V. De Luca, S. Graziani and F. Sapuppo, An optimized frequency-dependent multiphysics model for an ionic polymer–metal composite actuator with ethylene glycol as the solvent, *IOP Smart Mater. Struct.*, 2013, **22**(12), 125016.

A further interesting investigation was performed on the charge and solvent concentrations for both a sinusoidal input and a transient signal. Figure 16.78 shows the solvent distribution when the system is subjected to a square voltage input. The concentration is computed during both the application of the step input (Figure 16.78(a)) and after it has been removed (Figure 16.78(b)).

The accumulation and depletion of the solvent molecules does not occur instantaneously. Instead, it is worth noting that even after tens of seconds following the input signal being removed, a significant charge accumulation was estimated. This is in accordance with the real behaviour observed for IPMC benders with ethylene glycol as the solvent.

With regards to the thermal simulations, an example of obtained results is shown in Figure 16.79, in which one example of estimation along an IPMC transversal section at simulation time $t = 2.5$ s is seen. The solution of the thermal model highlights a higher temperature in the central region of the device near the supply electrodes, and this is in accordance experimental results reported in the literature.[5,90]

Finally, two examples of the performance of the proposed model for estimating the IPMC tip deflection for sinusoidal inputs are seen in

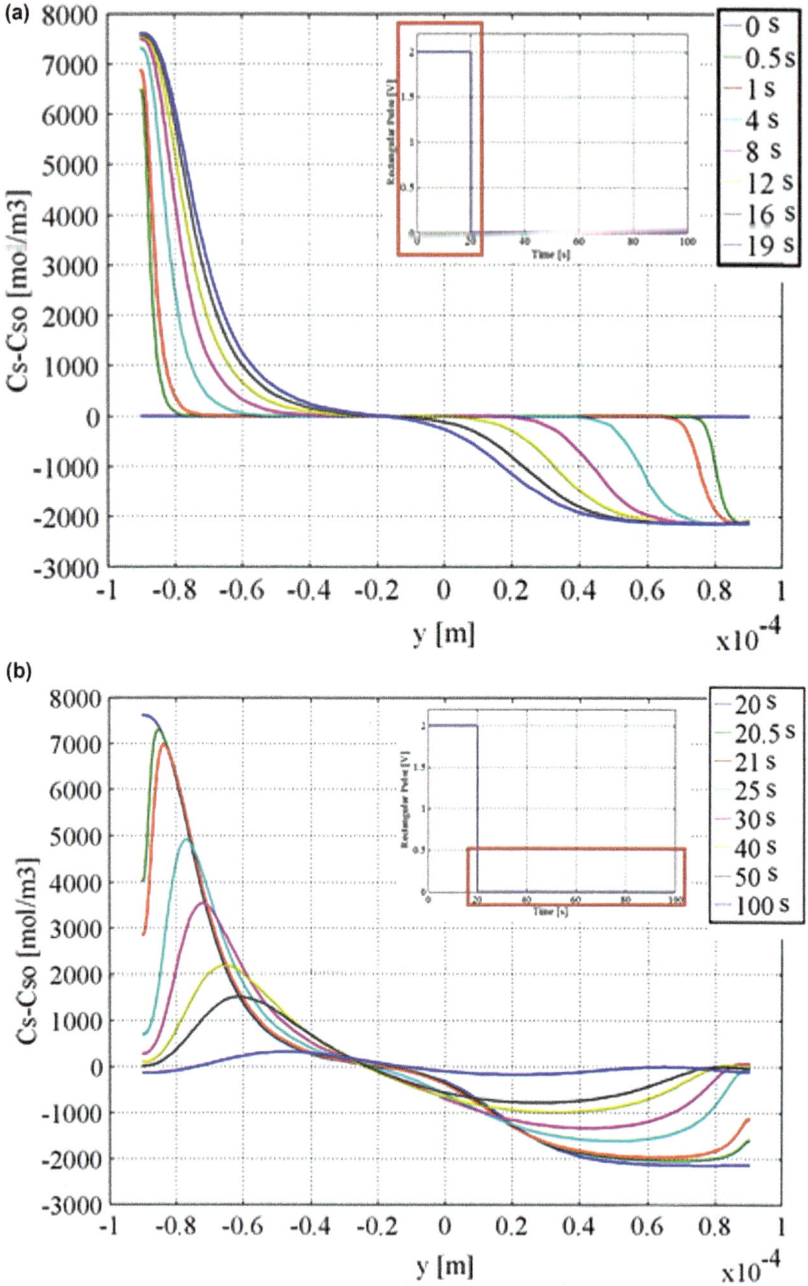

Figure 16.78 Solvent concentration in Nafion in correspondence to a square voltage during T_{on} (a) and the T_{off} (b).
Reprinted with permission from R. Caponetto, V. De Luca, S. Graziani and F. Sapuppo, An optimized frequency-dependent multiphysics model for an ionic polymer–metal composite actuator with ethylene glycol as the solvent, *IOP Smart Mater. Struct.*, 2013, 22(12), 125016.

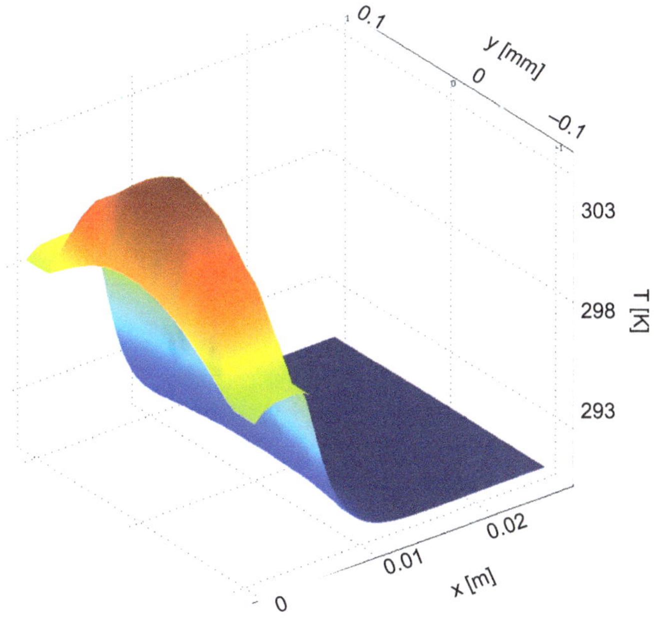

Figure 16.79 Electric potential along an IPMC transversal section at simulation time $t = 1.25$ s (not to scale).
Reprinted with permission from R. Caponetto, V. De Luca, S. Graziani and F. Sapuppo, An optimized frequency-dependent multiphysics model for an ionic polymer–metal composite actuator with ethylene glycol as the solvent, *IOP Smart Mater. Struct.*, 2013, **22**(12), 125016.

Figure 16.80. The reported examples refer to sinusoidal input signals with frequencies equal to 10 Hz (Figure 16.80(a)) and 26 Hz (Figure 16.80(b)).

As I have already mentioned, research on IPMCs is a multidisciplinary task that focuses on a number of interacting topics. Among these, the production of new materials is linked to the necessity of finding new models. This occurred in our research with IP^{2}Cs[38,39,113] and with their modelling. Actually, we proposed different models of IP^{2}Cs, including black-box models[114] and grey-box models.[39,115,116] Then, we focused on frequency-dependent multiphysics white-box models of IP^{2}C actuators.[117,118] The general structure of the model reflects the multiphysics model proposed for the IPMC actuator; nonetheless, significant differences exist between the two models. The schematic of the modelled device is seen in Figure 16.81. The central zone consists of an ionic polymer, Nafion® 117; the surface electrodes are made by PEDOT:PSS, and the metal contacts are made of copper. In this study, water was used as the solvent.

As with the case of the IPMC model, experimental measurements of the Young's modulus were used in this model. The storage and loss Young's moduli were determined by using DMA analysis and the results are seen in Figure 16.82. Measurements were performed at 25 °C. Note that although the DMA analysis was performed in the same laboratory conditions, the Young's moduli reported in Figure 16.82 look quite different from the corresponding moduli seen in Figure 16.76. This should not be a surprise since the

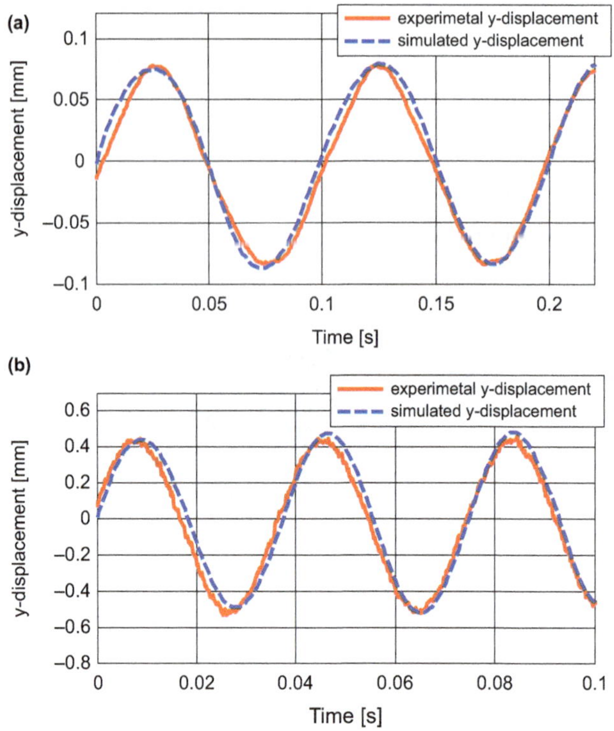

Figure 16.80 Tip displacement simulation for a sinusoidal input signal with frequency equal to 10 Hz (a) and 26 Hz (b).
Reprinted with permission from R. Caponetto, V. De Luca, S. Graziani and F. Sapuppo, An optimized frequency-dependent multiphysics model for an ionic polymer–metal composite actuator with ethylene glycol as the solvent, *IOP Smart Mater. Struct.*, 2013, **22**(12), 125016.

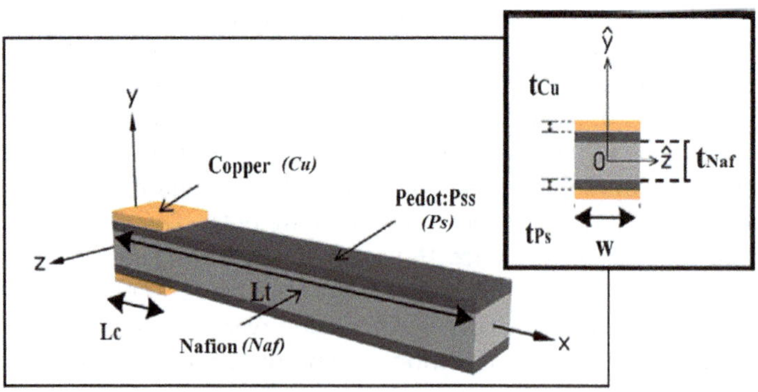

Figure 16.81 A scheme of the modelled IP^{2}C.
© 2014 IEEE. Reprinted, with permission, from R. Caponetto, V. De Luca, G. Di Pasquale, S. Graziani, F. Sapuppo and E. Umana, A multiphysics frequency-dependent model of an IP^{2}C Actuator, *IEEE Trans. Instrum. Meas.*, 2014, **63**(5), 1247–1255.

investigated membranes differ both in the nature of the electrodes and in the solvent used.

Also in this case, the parameters to be identified were the stiffness damping parameter β_f and the coupling term k_f. Nonetheless, in this case, it was not possible to obtain significant approximation capabilities by using two constant terms. Instead, the model was able to fit experimental data when a fractional-order transfer function of the form:

$$k_f(f) = \frac{k_{f0}}{(j2\pi f)^a (j2\pi f + p)^b} \tag{16.64}$$

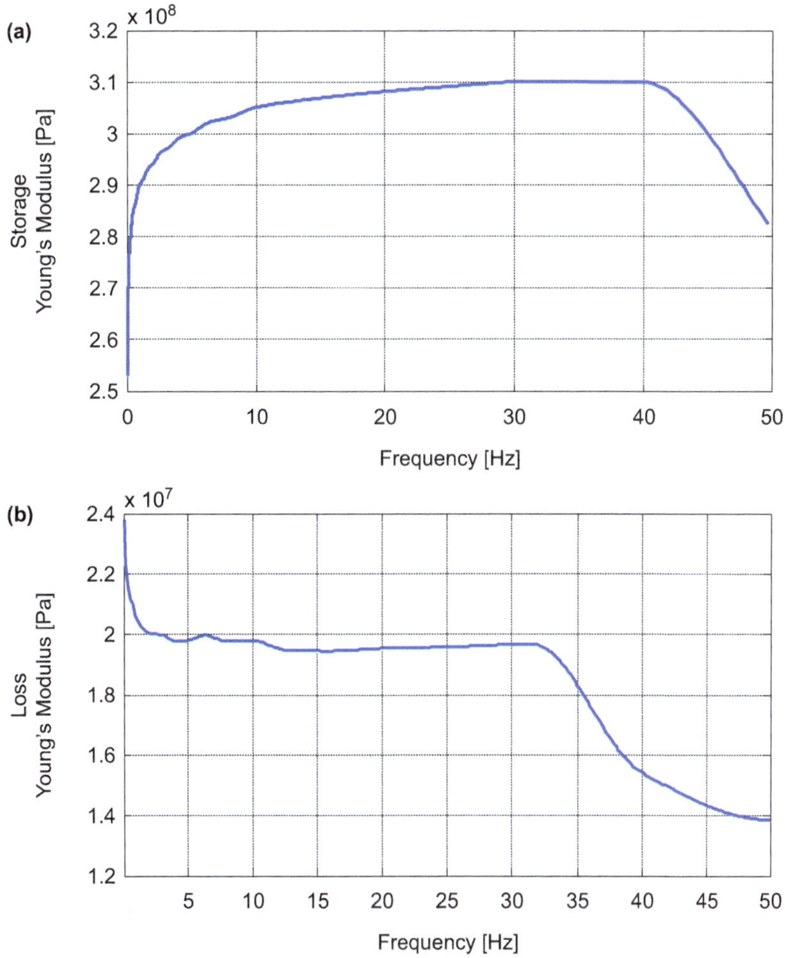

Figure 16.82 Storage (a) and loss (a) Young's moduli as a function of the frequency. © 2014 IEEE. Reprinted, with permission, from R. Caponetto, V. De Luca, G. Di Pasquale, S. Graziani, F. Sapuppo and E. Umana, A multiphysics frequency-dependent model of an IP^{2}C Actuator, *IEEE Trans. Instrum. Meas.*, 2014, **63**(5), 1347–1355.

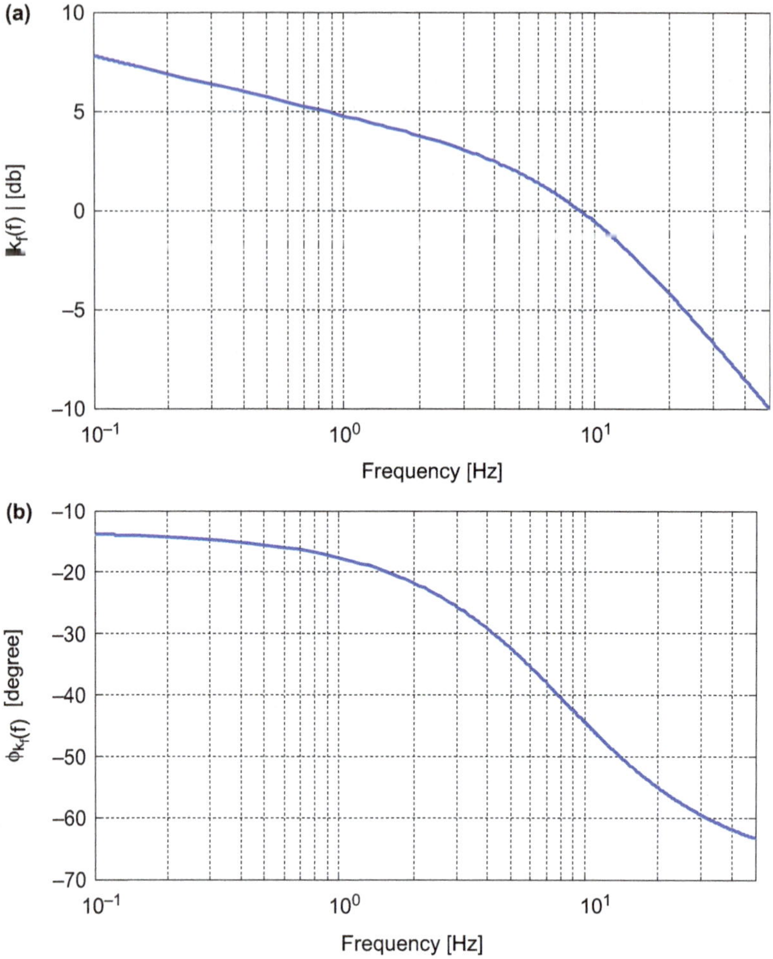

Figure 16.83 Modulus (a) and phase (b) of the coupling term in eqn (16.65). R. Caponetto, V. De Luca, G. Di Pasquale, S. Graziani, F. Sapuppo and E. Umana, A multiphysics frequency-dependent model of an IP^{2}C Actuator, *IEEE Trans. Instrum. Meas.*, 2014, **63**(5), 1347–1355.

was supposed for k_f, which was identified as:

$$k_f(f) = \frac{27.08}{(j2\pi f)^{0.62}(j2\pi f + 53)^{0.15}} \tag{16.65}$$

The plots of the modulus and phase of k_f in eqn (16.65) are shown in Figure 16.83.

An example of the simulated potential field is shown Figure 16.84. The variation in the x-axis direction can be noticed. The potential decreases in the direction of the IP^{2}C beam tip as a result of the organic conductor resistivity.

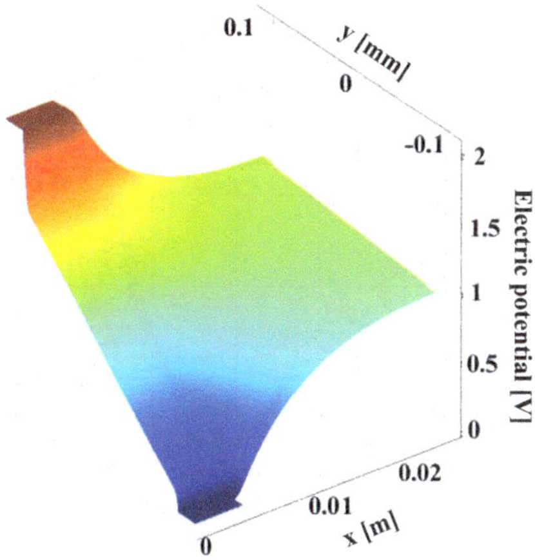

Figure 16.84 Voltage field on the Nafion transversal section. © 2014 IEEE. Reprinted, with permission, from R. Caponetto, V. De Luca, G. Di Pasquale, S. Graziani, F. Sapuppo and E. Umana, A multiphysics frequency-dependent model of an IP^{2}C Actuator, *IEEE Trans. Instrum. Meas.*, 2014, **63**(5), 1347–1355.

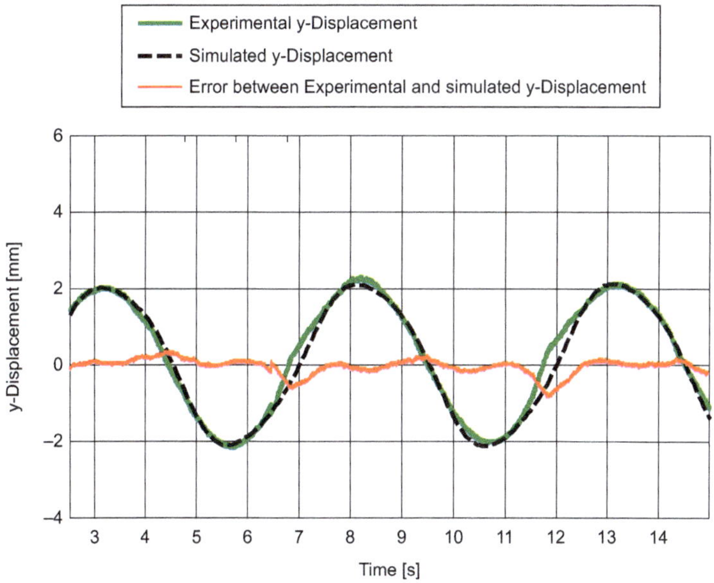

Figure 16.85 Estimation capabilities of the proposed IP^{2}C white-box model. © 2014 IEEE. Reprinted, with permission, from R. Caponetto, V. De Luca, G. Di Pasquale, S. Graziani, F. Sapuppo and E. Umana, A multiphysics frequency-dependent model of an IP^{2}C Actuator, *IEEE Trans. Instrum. Meas.*, 2014, **63**(5), 1347–1355.

An example of the estimation capabilities of the obtained model is reported in Figure 16.85 for a sinusoidal input $f = 0.2$ Hz. The model estimation error is seen in the same figure.

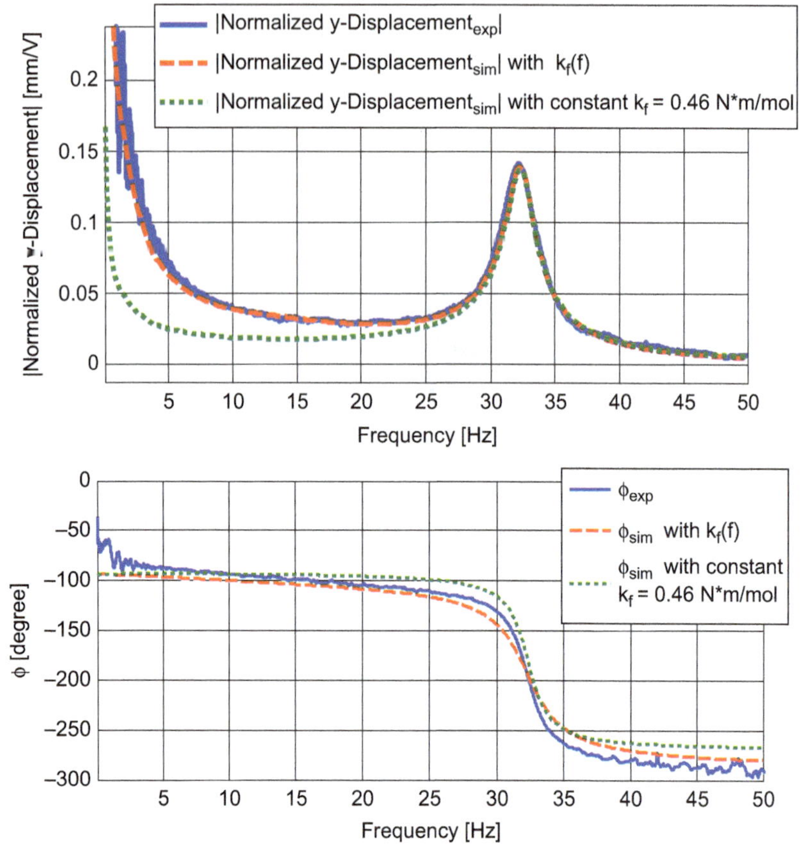

Figure 16.86 Estimation capabilities of the proposed model in the frequency domain.
© 2014 IEEE. Reprinted, with permission, from R. Caponetto, V. De Luca, G. Di Pasquale, S. Graziani, F. Sapuppo and E. Umana, A multiphysics frequency-dependent model of an IP^{2}C Actuator, *IEEE Trans. Instrum. Meas.*, 2014, **63**(5), 1347–1355.

Finally, the modelling performances were investigated and the results are reported in Figure 16.86, in which the modulus of the experimental frequency response is compared with the results obtained by using both a constant value of the coupling coefficient k_f and the result reported from eqn (16.65).

References

1. I. Chopra, *AIAA J.*, 2002, **40**, 2145.
2. *Electroactive Polymer (EAP) Actuators and Artificial Muscles – Reality, Potential, and Challenges*, ed. Y. Bar-Cohen, SPIE Press, Bellingham, WA, USA, 2nd edn, 2004

3. S. Katzir, *The Beginnings of Piezoelectricity: A Study in Mundane Physics*, Springer, Dordrecht, The Netherlands, 2006.
4. M. Shahinpoor, *Proc. of IEEE Int. Conf. Robotics and Automation*, IEEE, New York, NY, 1993, 380–385.
5. M. Shahinpoor and K. J. Kim, *Smart Mater. Struct.*, 2001, **10**, 819.
6. C. Bonomo, L. Fortuna, P. Giannone and S. Graziani, *Proc. of IEEE Sensors – 3rd Conference on Sensors*, IEEE, New York, NY, 2004, 489–492.
7. L. Ljung, *System Identification: Theory for the User*, ed. T. Kailath, Prentice Hall, Englewood Cliffs, 2nd edn, 1999.
8. R. Kanno, A. Kurata, M. Hattori, S. Tadokoro and T. Takamori, *Proc. of Japan–USA Symp. on Flexible Automation*, 1994, 691–698.
9. V. De Luca, P. Di Giamberardino, G. Di Pasquale, S. Graziani, A. Pollicino, E. Umana and M. G. Xibilia, *J. Polym. Sci., Part B: Polym. Phys.*, 2013, **51**, 699.
10. M. Shahinpoor and K. J. Kim, *Smart Mater. Struct.*, 2004, **13**, 1362.
11. K. Oguro, Y. Kawami and K. Takenaka, *Proc. of 1992 Japan Micromachine Symposium* (in Japanese), 1992.
12. C. Bonomo, L. Fortuna, P. Giannone and S. Graziani, *IEEE Trans. Circuits Syst.*, 2006, **53**, 338.
13. M. Shahinpoor, *Proc. SPIE's 6th Annual International Symposium on Smart Structures and Materials*, SPIE, Bellingham, Washington, 1999, 121–139.
14. R. Kanno, S. Tadokoro, T. Takamori, M. Hattori and K. Oguro, *Proceedings of the 1995 IEEE IECON 21st International Conference on Industrial Electronics, Control, and Instrumentation*, IEEE, New York, NY, 1995, 913–918.
15. X. Bao, Y. Bar-Cohen and S.-S. Lih, *Proc. of the SPIE Smart Structures and Materials Symposium, EAPAD Conference*, SPIE, Bellingham, Washington, 2002, 4695-27.
16. J. Paquette, K. J. Kim, J.-D. Nam and Y. S. Tak, *J. Intell. Mater. Syst. Struct.*, 2003, **14**, 633.
17. Y. Cha, M. Aureli and M. Porfiri, *J. Appl. Phys.*, 2012, **111**, 124901-1.
18. R. Kanno, S. Tadokoro, T. Takamori, M. Hattori and K. Oguro, *Proc. of IEEE Intern. Conf. on Robotics and Automation*, IEEE, New York, NY, 1996, 219–225.
19. K. M. Newbury and D. J. Leo, *J. Intell. Mater. Syst. Struct.*, 2003, **14**, 333.
20. K. M. Newbury and D. J. Leo, *J. Intell. Mater. Syst. Struct.*, 2003, **14**, 343.
21. M. Shahinpoor and K. J. Kim, *Smart Mater. Struct.*, 2000, **9**, 543.
22. S. Nemat-Nasser, *J. Appl. Phys.*, 2002, **92**, 2899.
23. R. Caponetto, G. Dongola, L. Fortuna, A. Gallo and S. Graziani, *New Trends in Nanotechnology and Fractional Calculus Applications*, ed. D. Baleanu, Z. B. Guvenc and J. A. T. Machado, Springer, Berlin, Heidelberg, 2010, pp. 263–272.
24. P. Arena, R. Caponetto, L. Fortuna and D. Porto, *Nonlinear Noninteger Order Circuits and Systems – An Introduction*, ed. L. O. Chua, World Scientific, Singapore, 2000.

25. D. W. Marquardt, *J. Soc. Ind. Appl. Math.*, 1963, **11**, 431.
26. R. Caponetto, S. Graziani, F. L. Pappalardo, E. Umana, M. G. Xibilia and P. Di Giamberardino, *IFAC 7th Vienna International Conference on Mathematical Modelling (MATHMOD 2012)*, IFAC, Elsevier Ltd., Oxford, UK, 2012, 593–596.
27. K. Hornik, *Neural Networks*, 1989, **2**, 359.
28. B. Kosko, *IEEE Trans. Comput.*, 1994, **43**, 1329.
29. K. S. Narendra and K. Parthasarathy, *IEEE Trans. Neural Networks*, 1990, **1**, 4.
30. T. Takagi and M. Sugeno, *IEEE Trans. Sysy. Man Cybern.*, 1985, **15**, 116.
31. D. Q. Truong and K. K. Ahn, *J. Intell. Mater. Syst. Struct.*, 2011, **22**, 253.
32. M. Annabestani and N. Naghavi, *Sens. Actuators, A*, 2014, **209**, 140.
33. D. N. C. Nam and K. K. Ahn, *Sens. Actuators, A*, 2012, **183**, 105.
34. A. D. Lantada, P. L. Morgado, J. L. M. Sanz, J. M. M. Guijosa and J. E. Otero, *J. Signal Information Process.*, 2012, **3**, 137.
35. L. Fortuna, S. Graziani, A. Rizzo and M. G. Xibilia, *Soft Sensors for Monitoring and Control of Industrial Processes*, ed. M. G. Grimble and M. A. Jhonson, Springer, Dordrecht, The Netherlands, 2006.
36. V. De Luca, E. Hosseini-Asl, S. Graziani and J. M. Zurada, *Proceedings of Eurosensors 2014*, Elsevier Ltd., Oxford, UK, 2014, pp. 424–427.
37. S. J. Higgins, V. K. Lovell, R. M. G. Rajapakse and N. M. Walsby, *J. Mater. Chem.*, 2003, **13**, 2485.
38. G. Di Pasquale, L. Fortuna, S. Graziani, M. La Rosa, D. Nicolosi, G. Sicurella and E. Umana, *Proc. of IEEE IMTC*, IEEE, New York, NY, 2008, 771–776.
39. L. Fortuna, S. Graziani, M. La Rosa, D. Nicolosi, G. Sicurella and E. Umana, *Eur. Phys. J.: Appl. Phys.*, 2009, **46**, 12513.
40. C. Bonomo, L. Fortuna, P. Giannone, S. Graziani and S. Strazzeri, *Smart Mater. Struct.*, 2007, **16**, 1.
41. P. Brunetto, L. Fortuna, P. Giannone, S. Graziani and S. Strazzeri, *IEEE Trans. Instrum. Meas.*, 2010, **59**, 893.
42. O. Brand, *Proc. of IEEE Sensors*, IEEE, New York, NY, 2005, 129-132.
43. R. Kornbluh, R. Pelrine, H. Prahlad and R. Heydt, *Proceedings of SPIE 2004*, ed. S. W. Janson and A. K. Henning, SPIE, Bellingham, WA, USA, 2004, pp. 13–27.
44. A. P. Gerratt, M. Tellers and S. Bergbreiter, *Proc. of MEMS 2011*, IEEE, New York, NY, 2011, 332–335.
45. K. M. Newbury and D. J. Leo, *J. Intell. Mater. Syst. Struct.*, 2002, **13**, 51.
46. S. Nemat Nasser and J. Y. Li, *J. Appl. Phys.*, 2000, **87**, 3321.
47. F. Bauer, S. Denneler and M. Willert-Porada, *J. Polym. Sci., Part B: Polym. Phys.*, 2005, **43**, 786.
48. JCGM 100:2008, *GUM 1995 with minor correction, Evaluation of measurement data – Guide to the expression of uncertainty in measurement*, 2008.
49. E. Benes, M. Gröschl, W. Burger and Schmid, *Sens. Actuators, A*, 1995, **48**, 1.

50. K. Sadeghipour, R. Salomon and S. Neogi, *Smart Mater. Struct.*, 1992, **1**, 172.
51. M. Shahinpoor, Y. Bar-Cohen, T. Xue, J. O. Simpson and J. Smith, *Proceedings of SPIE-EAPAD 1998 Conference*, ed. Y. Bar-Cohen, SPIE Press, Bellingham, WA, USA, 1998, pp. 3324–3327.
52. M. Shahinpoor and K. J. Kim, *Smart Mater. Struct.*, 2004, **13**, 1362.
53. L. Ferrara, M. Shahinpoor, K. J. Kim, H. B. Schreyer, A. Keshavarzi, E. Benzel and J. W. Lantz, *Proceedings of SPIE-EAPAD 1999*, ed. Y. Bar-Cohen, SPIE Press, Bellingham, WA, USA, 1999, pp. 394–401.
54. A. Keshavarzi, M. Shahinpoor, K. J. Kim and J. W. Lantz, *Proceedings of SPIE-EAPAD 1999*, ed. Y. Bar-Cohen, SPIE Press, Bellingham, WA, USA, 1999, pp. 369–376.
55. M. Konyo, Y. Konishi, S. Tadokoro and T. Kishima, *Proceedings of SPIE-EAPAD 2004*, ed. Y. Bar-Cohen, SPIE Press, Bellingham, WA, USA, 2004, pp. 307–318.
56. J. W. Paquette, K. J. Kim and D. Kim, *Sens. Actuators, A*, 2005, **118**, 135.
57. D. Pugal, K. Jung, A. Aabloo and K. J. Kim, *Polym. Int.*, 2010, **59**, 279.
58. C. Bonomo, L. Fortuna, P. Giannone, S. Graziani and S. Strazzeri, *Smart Mater. Struct.*, 2006, **15**, 749.
59. P. Brunetto, L. Fortuna, P. Giannone, S. Graziani and S. Strazzeri, *IEEE Trans. Instrum. Meas.*, 2011, **60**, 2951.
60. T. Ganley, D. L. S. Hung, G. Zhu and X. Tan, *IEEE/ASME Transactions on Mechatronics*, 2011, **16**, 80.
61. J. W. Paquette and K. J. Kim, *IEEE J. Oceanic Eng.*, 2004, **29**, 729.
62. K. J. Kim, W. Yim, J. W. Paquette and D. Kim, *J. Intell. Mater. Syst. Struct*, 2007, **18**, 123.
63. S.-W. Yeom and I.-K. Oh, *Smart Mater. Struct.*, 2009, **18**, 085002.
64. Z. Chen, S. Shatara and X. Tan, *IEEE/ASME Transactions on Mechatronics*, 2010, **15**, 448.
65. Q. Shen, T. Wang, L. Wen and J. Liang, *Int. J. Adv. Robot. Syst.*, 2013, **10**, 13.
66. P. Brunetto, S. Graziani, S. Strazzeri and M. G. Xibilia, *Proceedings of 2nd IFAC Intelligent Control Systems and Signal Processing*, IFAC, Elsevier Ltd., Oxford, UK, 2009, pp. 153–157.
67. A. T. Abdulsadda and X. Tan, *Int. J. Smart Nano Mater.*, 2012, **3**, 226.
68. A. T. Abdulsadda and X. Tan, *Smart Mater. Struct.*, 2013, **22**, 045010.
69. P. Brunetto, L. Fortuna, S. Graziani and S. Strazzeri, *Smart Mater. Struct.*, 2008, **17**, 025029.
70. J. E. Sader, *J. Appl. Phys.*, 1998, **84**, 64.
71. S. Kirstein, M. Mertesdorf and M. Schönhoff, *J. Appl. Phys.*, 1998, **84**, 1782.
72. P. Brunetto, L. Fortuna, P. Giannone, S. Graziani and F. Pagano, *Proc. of IEEE 2010 Inst. and Meas. Tech. Conf. I2MTC*, IEEE, New York, NY, 2010, 585–589.

73. S. D. Peterson, M. Porfiri and A. Rovaldi, *IEEE/ASME Transactions on Mechatronics*, 2009, **14**, 474.
74. K. Abdelnour, E. Mancia, S. D. Peterson and M. Porfiri, *Smart Mater. Struct.*, 2009, **18**, 085006.
75. M. Aureli, V. Kopman and M. Porfiri, *IEEE/ASME Transactions on Mechatronics*, 2010, **15**, 603.
76. M. Shainpoor, *Proc. of IEEE International Conference on Robotics and Automation*, IEEE, New York, NY, 1993, 380–385.
77. M. Shahinpoor, *J. Intell. Mater. Syst. Struct.*, 1995, **6**, 307.
78. Z. Zhu, H. Chen, L. Chang and B. Li, *AIP Adv.*, 2011, **1**, 040702.
79. P. G. de Gennes, K. Okumura, M. Shahinpoor and K. J. Kim, *Europhys. Lett.*, 2000, **50**, 513.
80. M. Shahinpoor, *Electrochim. Acta*, 2003, **48**, 2343.
81. K. Asaka and K. Oguro, *J. Electroanal. Chem.*, 2000, **480**, 186.
82. K. Asaka, K. Oguro, Y. Nishimura, M. Mizuhata and H. Takenaka, *Polym. J.*, 1995, **27**, 436.
83. J. McGee, *Mechano-electrochemical Response of Ionic Polymer-metal Composites*, PhD Dissertation, University of California, 2002.
84. S. Tadokoro, S. Yamagami, T. Takamori and K. Oguro, *Proceedings of SPIE-EAPAD 2000*, ed. Y. Bar-Cohen, SPIE Press, Bellingham, WA, USA, 2000, pp. 92–102.
85. S. Tadokoro, M. Fukuhara, Y. Maeba, M. Konyo, T. Takamori and K. Oguro, *Proc. of 2002 IEEE/RSJ*, IEEE, New York, NY, 2002, 2010–2017.
86. Y. Toi and S.-S. Kang, *Comput. Struct.*, 2005, **83**, 2573.
87. P. J. Costa Branco and J. A. Dente, *Smart Mater. Struct.*, 2006, **15**, 378.
88. K. Farinholt and D. J. Leo, *Mech. Mater.*, 2004, **36**, 421.
89. Z. Chen, X. Tan, A. Will and C. Ziel, *Smart Mater. Struct.*, 2007, **16**, 1477.
90. T. Johnson and F. Amirouche, *Microsyst. Technol.*, 2008, **14**, 871.
91. D. Pugal, K. J. Kim and A. Aabloo, *J. Appl. Phys.*, 2011, **110**, 084904.
92. S. Nemat-Nasser and S. Zamani, *J. Appl. Phys.*, 2006, **100**, 064310.
93. M. Porfiri, *J. Appl. Phys.*, 2008, **104**, 104915.
94. M. Aureli and M. Porfiri, *Continuum Mech. Thermodyn.*, 2013, **25**, 273.
95. T. Wallmersperger, D. J. Leo and C. S. Kothera, *J. Appl. Phys.*, 2007, **101**, 024912.
96. P. Nardinocchi and M. Pezzulla, *J. Appl. Phys.*, 2013, **113**, 224906.
97. S. Galante, A. Lucantonio and P. Nardinocchi, *Int. J. Non-Linear Mech.*, 2013, **51**, 112.
98. L. Zhang and Y. W. Yang, *Smart Mater. Struct.*, 2007, **16**, S197.
99. G. Del Bufalo, L. Placidi and M. Porfiri, *Smart Mater. Struct.*, 2008, **17**, 045010.
100. M. Porfiri, *Smart Mater. Struct.*, 2009, **18**, 015016.
101. P. Nardinocchi, M. Pezzulla and L. Placidi, Thermodinamically based multiphysics modeling of ionic polymer metal composites, *J. Intell. Mater. Syst. Struct.*, 2011, **22**, 1887.
102. R. Caponetto, V. De Luca, S. Graziani and F. Sapuppo, *Smart Mater. Struct.*, 2013, **22**, 125016.

103. S. Nemat-Nasser and S. Zamani, *Proceedings of SPIE-EAPAD 2003*, ed. Y. Bar-Cohen, SPIE Press, Bellingham, WA, USA, 2003, pp. 233–244.
104. R. Tiwari and K. J. Kim, *Appl. Phys. Lett.*, 2010, **97**, 244104.
105. M. Porfiri, *Phys. Rev. E: Stat., Nonlinear, Soft Matter Phys.*, 2009, **79**, 041503.
106. M. Aureli and M. Porfiri, *Smart Mater. Struct.*, 2012, **21**, 105030.
107. Y. Cha, M. Aureli and M. Porfiri, *J. Appl. Phys.*, 2012, **111**, 124901.
108. Y. Cha and M. Porfiri, *Phys. Rev. E: Stat., Nonlinear, Soft Matter Phys.*, 2013, **87**, 022403.
109. B. J. Akle, W. Habchi, T. Wallmersperger, E. J. Akle and D. J. Leo, *J. Appl. Phys.*, 2011, **109**, 074509.
110. W. Jong Yoon, P. G. Reinhall and E. J. Seibel, *Sens. Actuators, A*, 2007, **133**, 506.
111. C. Jo, D. Pugal, I. K. Oh, K. J. Kim and K. Asaka, *Prog. Polym. Sci.*, 2013, **38**, 1037.
112. D. Pugal, H. Kasemägi, K. J. Kim, M. Kruusmaa and A. Aablo *Proceedings of SPIE-EAPAD 2007*, ed. Y. Bar-Cohen, SPIE Press, Bellingham, WA, USA, 2007, 65240B.
113. G. Di Pasquale, L. Fortuna, S. Graziani, M. La Rosa, A. Pollicino and E. Umana, *Smart Mater. Struct.*, 2011, **20**, 045014.
114. S. Graziani, E. Umana and M. G. Xibilia, *Proc. MED 2012*, IEEE, New York, NY, 2012, 216–221.
115. G. Di Pasquale, L. Fortuna, S. Graziani, M. La Rosa, D. Nicolosi, G. Sicurella and E. Umana, *IEEE Trans. Instrum. Meas.*, 2009, **58**, 3731.
116. G. Di Pasquale, S. Graziani, M. La Rosa, G. Sicurella and E. Umana, *IEEE Trans. Instrum. Meas.*, 2013, **62**, 1284.
117. R. Caponetto, V. De Luca, G. Di Pasquale, S. Graziani, F. Sapuppo and E. Umana, *Proc. of 2013 I2MTC*, IEEE, New York, NY, 2013, 971–975.
118. R. Caponetto, V. De Luca, G. Di Pasquale, S. Graziani, F. Sapuppo and E. Umana, *IEEE Trans. Instrum. Meas.*, 2014, **63**, 1347.

CHAPTER 17

Ionic Polymer Metal Composites as Post-silicon Transducers for the Realisation of Smart Systems

SALVATORE GRAZIANI

Dipartimento di Ingegneria Elettrica Elettronica e Informatica, Università degli Studi di Catania, Viale Andrea Doria 6, Catania, Italy
Email: salvatore.graziani@dieei.unict.it

17.1 Introduction

Stimulus-responsive materials are key elements for the realisation of smart systems and, thanks to their contributions, applications that once belonged to the science fiction domain are becoming available at an impressive rate. In fact, the envisaged systems are the result of continuous advances in material performances and functions, modelling, and technology. As a matter of fact, new materials have become available that respond to a number of stimuli such as heat (thermo-responsive materials), stress/pressure (mechano-responsive materials), electric current/voltage (electro-responsive materials), magnetic fields (magneto-responsive materials), pH change/solvents/moisture (chemo-responsive materials), and light (photo-responsive materials).[1]

For the case of interest of this chapter, it suffices here to mention that thanks to their peculiar characteristics, electroactive polymers (EAPs) have greatly stimulated researchers to imagine applications in a large variety of

RSC Smart Materials No. 18
Ionic Polymer Metal Composites (IPMCs): Smart Multi-Functional Materials and Artificial Muscles, Volume 2
Edited by Mohsen Shahinpoor

Published by the Royal Society of Chemistry, www.rsc.org

fields. It is possible to envisage a trend towards the development of EAP-based smart systems able to solve even the most complex problems with little or no human intervention in strategic sectors such as bio-inspired robotics, aerospace, and medicine, to name just a few.[2]

For example, we are talking about bio-inspired robots that will take care of repetitive or dangerous tasks, active prostheses to help with the rehabilitation of patients, smart textiles to distribute drugs on the basis of well-established protocols, or even artificial organs and tissues that can both eliminate the need for donors and solve the problem of rejection, with a dramatic improvement in the quality of life of people undergoing transplantation.

The number and the quality of the prototypes realised have clearly increased with time, according to the improvements in the knowledge of EAP behaviour and in the technologies for their fabrication.

In fact, on the one hand, new production technologies, along with models capable of better describing transducer capabilities, are of chief importance in the development of meaningful applications. On the other hand, newly envisaged applications demand better materials and efficient yet accurate models.[3]

A number of functions are relevant to the realisation of smart systems. Such systems need, in fact, to be capable of power generation and storage,[4] signal detection and processing, and actuation capabilities.[5]

Stimulus-responsive polymeric materials are a hot research topic since they have been shown to have the functionalities mentioned above and therefore will play a main role in the development of smart systems.[6–8]

Studies have been reported on the capability of this class of materials to scavenge energy from the surrounding environment,[9–12] to realise all organic electronic devices,[13] and to obtain reversible energy transduction.[6,7]

Of course, such capabilities can be obtained by using more traditional inorganic materials. Nonetheless, these suffer for a number of shortcomings that do not plague polymers: to give an example, energy scavenging can be obtained by using piezoelectric ceramics, but they are very rigid and brittle, or to give another example, signal processing can be obtained by using very fast silicon-based integrated circuits, but these need to be mounted on rigid cards, while flexible electronics are of great interest.[14]

A lot of scientific contributions have been published on ionic polymer metal composite (IPMC)-based applications since their introduction in the early 1990s. It is worth noting that the possibility of using IPMCs in engineering applications that pose hard working conditions to competing technologies or with performances that those technologies cannot fulfil (*e.g.* think of the capacity of IPMCs to act in water without any needing to encapsulate them or to the large deformations that an IPMC can produce under the effect of a quite small applied difference voltage) has been, in the last two decades, a tantalising prospect that raised interest in this novel technology. A perusal of the extensive literature on IPMC applications published so far would reveal in fact that there has been a constant contribution to IPMC-based applications that, along with contributions to IPMC production technology and modelling, represent three research fields that influence each other.

Such an interest also characterised my research activity and attention that was paid both to production processes, modelling, and applications. Also, it is hard for me to define a clear border between these fields. New IPMC-based applications (*e.g.* demanded for novel models[10]) and IPMCs with different geometries were developed based on some desired characteristics.[15]

Though a complete review of all of the IPMC-based applications is quite impossible and beyond the objective of the present chapter, I will describe, in the following, our contribution to the topic of IPMC-based applications, along with an overview of related applications proposed in the literature. More specifically, some applications will be described that we developed during our research activity, together with production improvement and modelling, which were necessary to enabling IPMCs to become a mature technology.

The focus will be on the capability of IPMCs to be used in applications as components relevant to the realisation of smart devices. More specifically, the capability of an IPMC membrane to work in a reversible way, both as an actuator and as a sensor, has had the effect of producing two main directions in the development of applications and it is possible to classify them according to their exploitation of IPMC sensing capabilities or IPMCs used as actuators. Finally, applications have been proposed in which IPMC transducers, used as both sensors and actuators, cooperate to realise smart devices.[16]

A nice review of the state of IPMC applications was published in a paper by Shahinpoor and Kim,[17] which represents a good starting point for understanding the envisaged fields of applications relevant to IPMC technology. More recently, tutorial contributions on IPMC applications have been proposed by Bhandari *et al.*,[18] De Luca *et al.*,[16] or, finally, Pugal *et al.*,[19] who focused on the applications of IPMCs as mechanoelectrical transducers.

17.2 IPMC-based Actuators

It has been widely noted that the research activity on IPMC transducers has been focused mainly on IPMCs as bending actuators and, as a consequence, the main part of the available literature on IPMC applications deals with the use of bending IPMCs. It is worth saying here that such a simple actuation structure (*i.e.* an IPMC, pinned at one end, bending in a beam configuration) has enough potential to allow for the introduction, in time, of even more sophisticated structures, ranging from microgrippers, with both industrial and bio-manipulation relevant applications,[20] to bio-inspired worms,[21] jellyfish,[22] or swimming fish,[23,24] as will be described in the following.

The realisation of grippers was one of the first proposed applications for IPMCs: a three-finger gripper for robotic applications was described by Shahinpoor *et al.* back in 1998.[25] Nevertheless, even such a simple application requires a number of problems to be solved to improve the performances of the realised prototypes. The micromanipulator proposed by Lumia and Shahinpoor[20] is a good example of the complex interplay between production, modelling, and application development that rules the research on IPMCs. The referred paper proposes the realisation of an IPMC-based micromanipulator suitable for use in both dry and wet environments and with relevant industrial and bio-applications, in which the manipulation of micro-sized objects is required. The manipulator was realised as a micro-gripper with two fingers. Though it is a typical applicative contribution, the authors faced the modelling of the interaction between the fingers and the manipulated objects and addressed the optimisation of the load carrying capacity as a function of the finger dimensions and shape. An assembled computer-aided design (CAD) view of the microgripper is seen in Figure 17.1(a), while in Figure 17.1(b), the realised microgripper lifting a flexible hydrogel object is shown.

We addressed the realisation of a three-finger gripper[15] and, interestingly, in our case the intended application also forced us to solve problems beyond the pure prototype realisation.

One of the main drawbacks that affects IPMC applications as actuators is the small value of the force that can be produced. It has been predicted that the value of the blocked force increases with IPMC actuator thickness,[26] such that it makes sense to try to realise thicker IPMC samples.

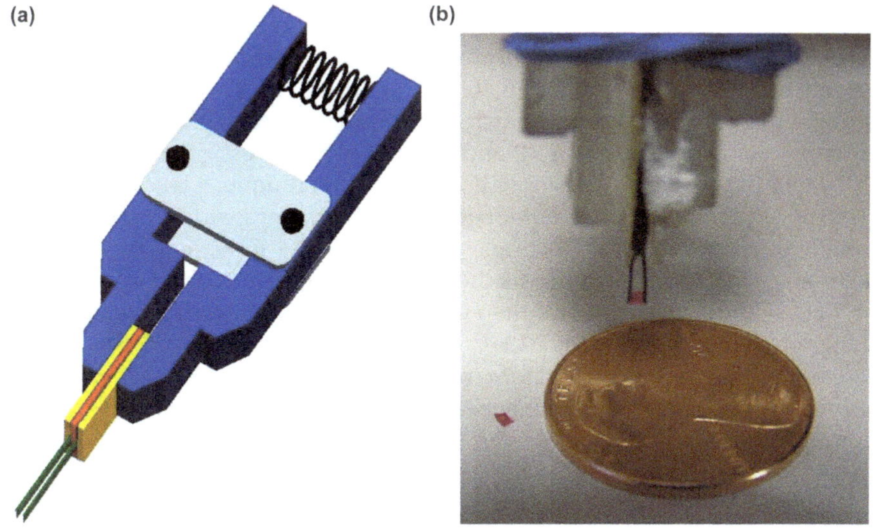

Figure 17.1 The CAD model of a microgripper proposed in ref. 20 (a), and the realised microgripper lifting a flexible hydrogel object (b).
Reprinted with permission from R. Lumia and M. Shahinpoor, IPMC Microgripper Research and Development, 4th World Congress on Biomimetics, Artificial Muscles and Nano-Bio, *J. Phys.: Conf. Ser.*, 2008, **127**, 012002.

Unfortunately, commercially available Nafion® membranes, the ionic exchange polymer generally used to realise IPMCs, come in sheets and the most widely used is only about 180 μm thick, and even thicker Nafion® 1110 membranes are only about 250 μm thick.

In order to improve the IPMC actuating force, we decided to produce three-dimensional IPMCs (3D-IPMCs).[27–31] The concept of 3D-IPMCs was not new at that time, and the most commonly used approach was to obtain 3D-IPMCs by dissolving received ion exchange membranes in appropriate solvents and evaporating the solvents out of the solution,[27–30] so that the ion exchange membranes were obtained. This method had the problem of low reproducibility; in fact, it was necessary to control accurately the process variables (temperature, nature of solvent, concentration of solvent, *etc.*). A further method for fabrication of 3D-IPMCs using Nafion® 117 film was proposed by Lee *at al.* in 2006.[31] This uses a hot-pressing system to make several thin Nafion films adhere together. We adopted this method because of the better control of the thickness of the IPMC and its greater reproducibility. Moreover, to improve the performance of the membranes, IPMCs were fabricated utilising additional post-processes. We produced IPMC membranes by performing more cycles of platinum electroless-plating, as widely suggested in the literature.[32]

The effects of dispersing agents were also investigated.[7,33,34] More specifically, polyvinylpirrolydone (PVP) and dodecyl sulphate sodium (DSS) salt were investigated as additives. For the case of PVP, which was already proposed in the literature as a dispersing agent,[7,33,34] the effects of the molecular weight of the utilised PVP were investigated. The surfactants[35,36] were known to play an important role as size controllers. DSS salt gave good results on controlling the dimensions of silver particles,[37] so we decided to study its effects on platinum particles. More specifically, the influences of different concentrations of PVP and DSS salt were investigated.

To obtain the 3D-IPMCs with a hot-pressing technique, a laboratory press (Graseby Specac P/N15800, equipped with spacer rings with nominal thickness 500 μm) was used and samples of Nafion® 117 films in groups of four were stacked. The membranes were held (20 min) in the preheated press (170 °C) without pressure and then were pressed (20 min) to make the films adhere together (170 °C, 100 MPa).

The 3D-IPMCs were therefore fabricated by applying three cycles of primary plating and using dispersing agents. The used additives and the concentrations to be used were determined during a preliminary experimental optimisation phase.

An experimental setup was realised to test the actuators in the bender configuration. The system was designed to measure both the developed blocked force and the free deflection. Moreover, the relationship between the blocked force and the IPMC tip deflection was investigated. The system was equipped with a laser sensor (OADM 12I6430/S35A from Baumer) for the free deflection detection, while the force exerted by the polymer actuators was measured using the load cell GS0 of Transducers Techniques®.

To enable precise positioning of the load cell relative to the actuator, it was mounted on a micro positioning system, as shown in Figure 17.2.

An example of the force versus deflection curves is reported in Figure 17.3. From the figure, it can be seen that the ability of the IPMC actuator to generate force decreases linearly with the deflection. Also, the maximum blocked force is of order of 20 mN, while for the corresponding thin IPMCs, values one order of magnitude smaller were observed.

As mentioned above, the 3D-IPMC was used to produce the prototype of a gripper, which is shown in Figure 17.4. It was able to sustain grip on objects with weights of about 10 mN.

The reported example, along with the results reported by Lumia and Shahinpoor,[20] are typical examples of the multidisciplinary nature of

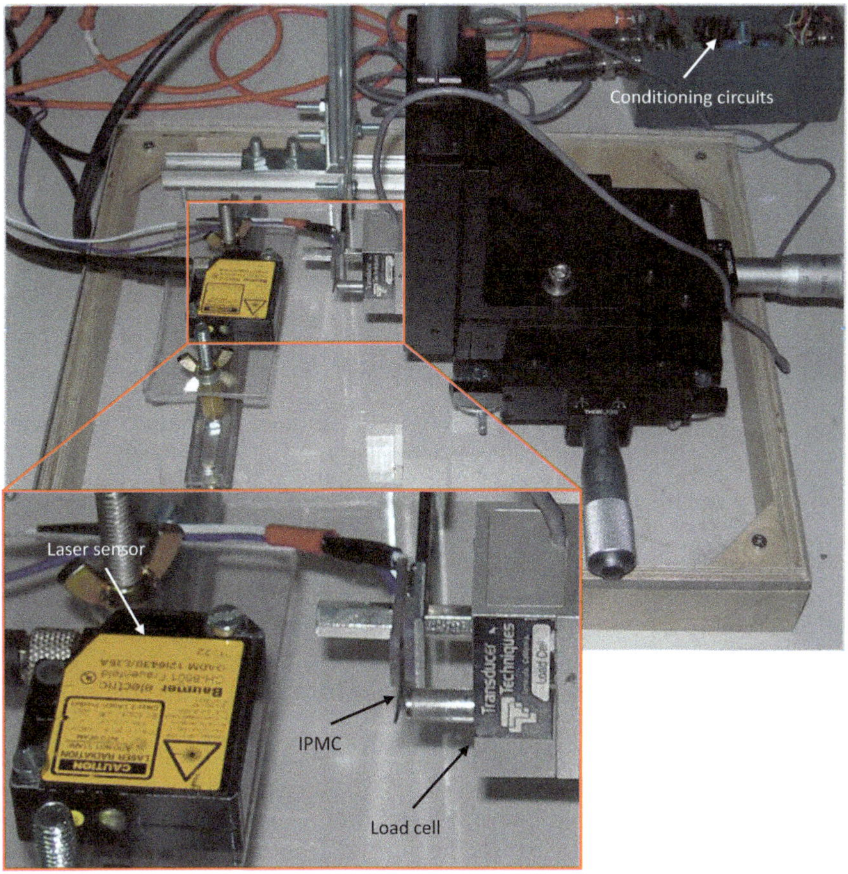

Figure 17.2 The experimental setup used to investigate the performance of the 3D-IPMCs.
Reprinted with permission from C. Bonomo, M. Bottino, P. Brunetto, G. Di Pasquale, L. Fortuna, S. Graziani, and A. Pollicino, Tridimensional ionic polymer metal composites: optimization of the manufacturing techniques, *IOP Smart Mater. Struct.*, 2010, **19**, 055002.

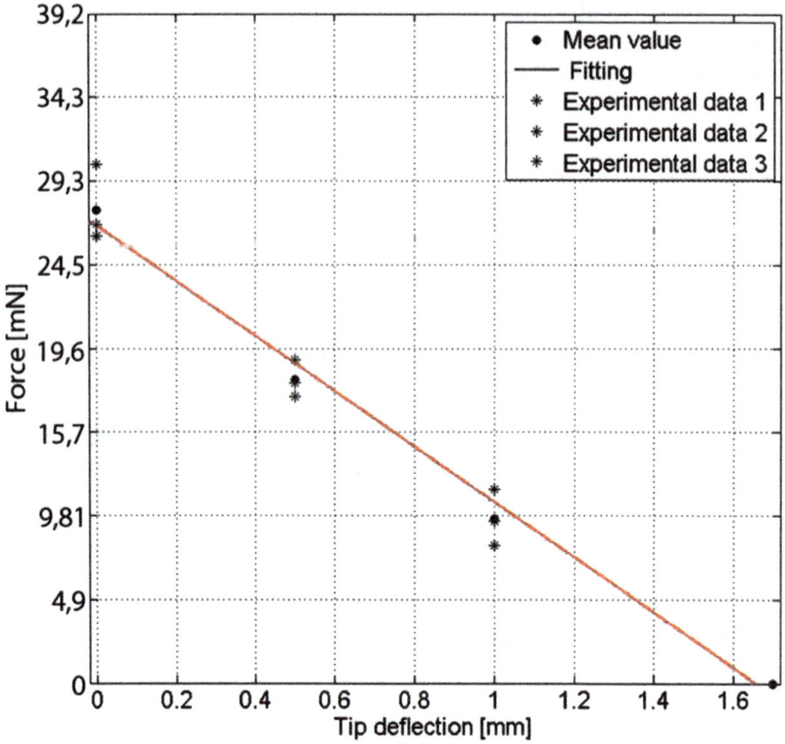

Figure 17.3 An example of the force versus deflection curves obtained for a 3D-IPMC.
Reprinted with permission from C. Bonomo, M. Bottino, P. Brunetto, G. Di Pasquale, L. Fortuna, S. Graziani, and A. Pollicino, Tridimensional ionic polymer metal composites: optimization of the manufacturing techniques, *IOP Smart Mater. Struct.*, 2010, **19**, 055002.

IPMC research: in both cases, problems other than the simple prototype design and realisation needed to be solved. More specifically, the authors had to develop a model of the prototype in order to control the system design. In the case of our gripper, we had to face the problem of thick IPMC production as a way of producing IPMCs capable of developing larger forces.

A further example of the application of IPMCs in a pinned beam configuration was described by McDaid *et al.*,[38] who proposed a stepper motor and realised it by using a couple of IPMC actuators, forced by using adequate voltage signals. The proposed application has the main advantage of realising a system that exploits the bending motion of IPMC actuators to obtain a system capable of a continuous rotational motion. The CAD model of the proposed motor and its rapid prototyping realisation are seen in Figure 17.5(a) and (b), respectively.

We developed a different stepper motor in our laboratories, in which a couple of IPMCs are mounted on the rotor. This allowed us to realise a

Figure 17.4 A gripper realised by using a 3D-IPMC is used to grasp a foam object. Reprinted with permission from C. Bonomo, M. Bottino, P. Brunetto, G. Di Pasquale, L. Fortuna, S. Graziani and A. Pollicino, Tridimensional ionic polymer metal composites: optimization of the manufacturing techniques, *IOP Smart Mater. Struct.*, 2010, **19**, 055002.

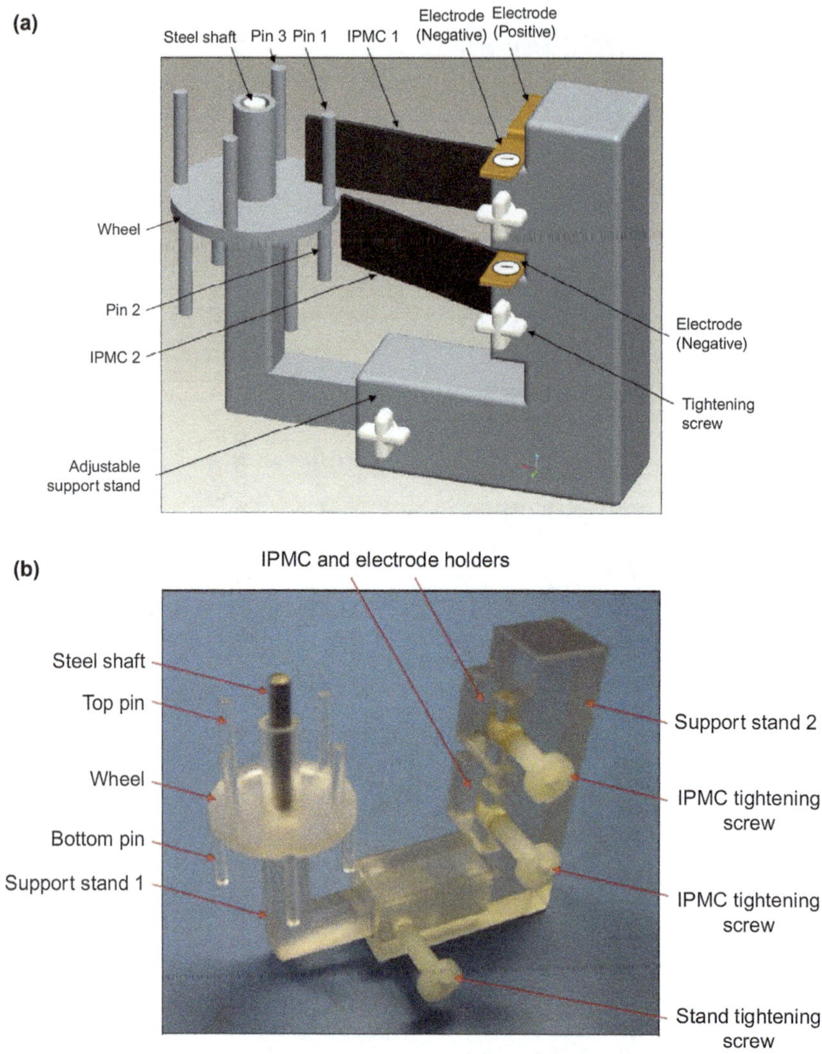

Figure 17.5 A CAD model of the stepper motor proposed in ref. 38 (a) and its rapid prototype realisation (b).
Reprinted with permission from A. J. McDaid, K. C. Aw, K. Patel, S. Q. Xie and E. Haemmerle, Development of an ionic polymer-metal composite stepper motor using a novel actuator model, *Int. J. Smart Nano Mat.*, 2010, **1**(4), 261–277.

system consisting only of a central shaft and two moving elements. A schematic view of our stepper motor and two snapshots acquired during the rotation of the device are seen in Figure 17.6.

IPMCs in pinned configurations have also been widely used in robotic applications as in the contributions by Ryu *et al.*[39] and Kim *et al.*,[40] where a ciliary-based eight-legged micro-robot was proposed, suitable for moving in

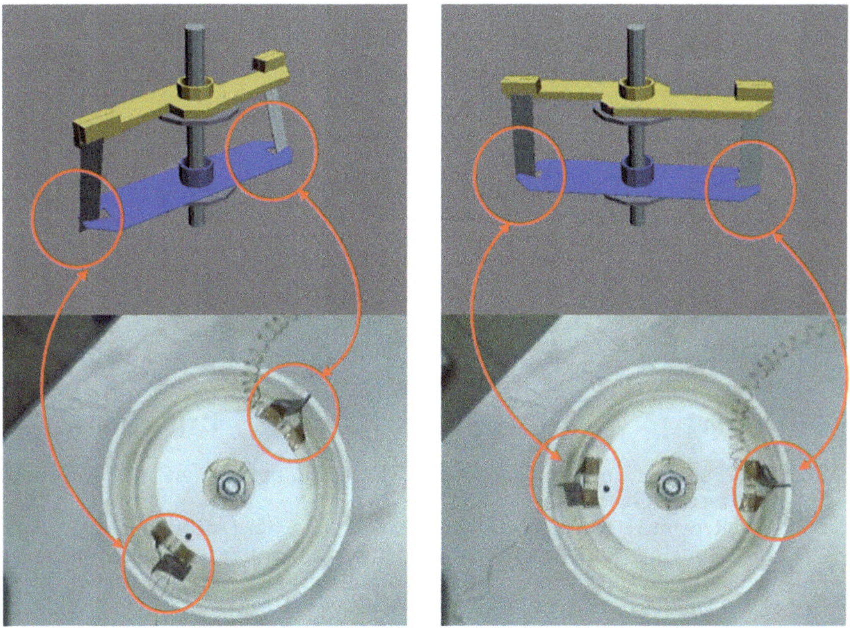

Figure 17.6 A stepper motor designed and realised at our laboratories.

aqueous environments. The legs of the robot were realised by using thick-cast Nafion as a solution for obtaining larger generative forces.

Nevertheless, significant applications have been proposed for IPMCs that differ from the pinned configuration. The realisation of linear actuators based on IPMCs has been widely reported in the literature,[17,25] along with possible robotic applications.[41,42] A possible linear actuator was modelled and optimised by Lee *et al.*,[43] who addressed the difference between biological muscles, which are capable of linear actuation, and IPMCs, which produce a bending deformation. An elementary rectangular structure composed of adequately connected IPMC strips and bare Nafion was therefore introduced.

A schematic view of the linear muscle-like actuator, composed of the proposed rectangular elementary units, is seen in Figure 17.7. From the analysis of the reported scheme, it emerges that by connecting a number of elementary units in series, large linear deformations can be obtained, while if the units are in a parallel connection, large forces can be produced. The authors concluded that the proposed structure is capable of a 25% free linear strain under an applied voltage of 2 V, comparable with the strain observed for biological muscles, while somewhat smaller blocked forces were measured.

In an effort to increase the diversity of IPMC-based actuator possibilities, in a more recent contribution,[44] a two degree of freedom (DOF) actuator, obtained by the realisation of a cylindrical IPMC with four inter-digitate

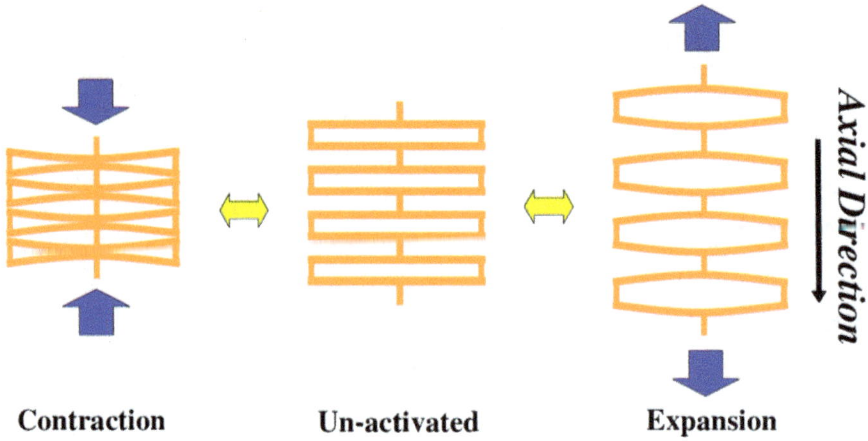

Figure 17.7 The linear muscle-like actuator, based on the elementary IPMC-based rectangular unit.[43]
Reprinted with permission from S. Lee, K. J. Kim and I.-S. Park, Modeling and experiment of a muscle-like linear actuator using an ionic polymer metal composite and its actuation characteristics, *Smart Mater. Struct.*, 2007, **16**(3), 583–588.

electrodes, was introduced. The fabrication technique was described with particular attention to the realisation of the electrodes. Experimental results, recorded on a prototype that was 20 mm long and with a diameter of 1 mm, showed the effectiveness of the actuator, being able to bend up to 50° in both the vertical/horizontal and diagonal directions. A view of the tip deflections recorded from the realised actuator is shown in Figure 17.8.

The same robotic structure of a biaxial bending actuator, but with a different manufacture procedure and with a square cross-section, has been also proposed.[45] A prototype was demonstrated to be capable of circular motion.

The applications reported so far are good examples of the efforts devoted to mimicking biological muscles as relevant to bio-inspired robotics. Both ground robots and underwater swimming vehicles have been proposed in the literature. Starting with ground robots, a biped robot, actuated by the linear IPMC described previously,[41,42] has been realised and tested.[46] A suitable parallel combination of elementary IPMC cells was mounted on a prototype robot and experiments were conducted showing the capability of the robot to walk for a short time.

A number of bio-inspired robots use legged structures. A bio-inspired four-legged micro-robot was presented by Tomita *et al.*[47] Unlike the described biped robot, the prototype of the quadruped robot was fully realised by IPMC material with a suitable shape. The legs of the robot were further patterned in such a way as to allow for their independent control. Both a lizard-like and a turtle-like gait were attempted, but the system was able to move only in the second case.

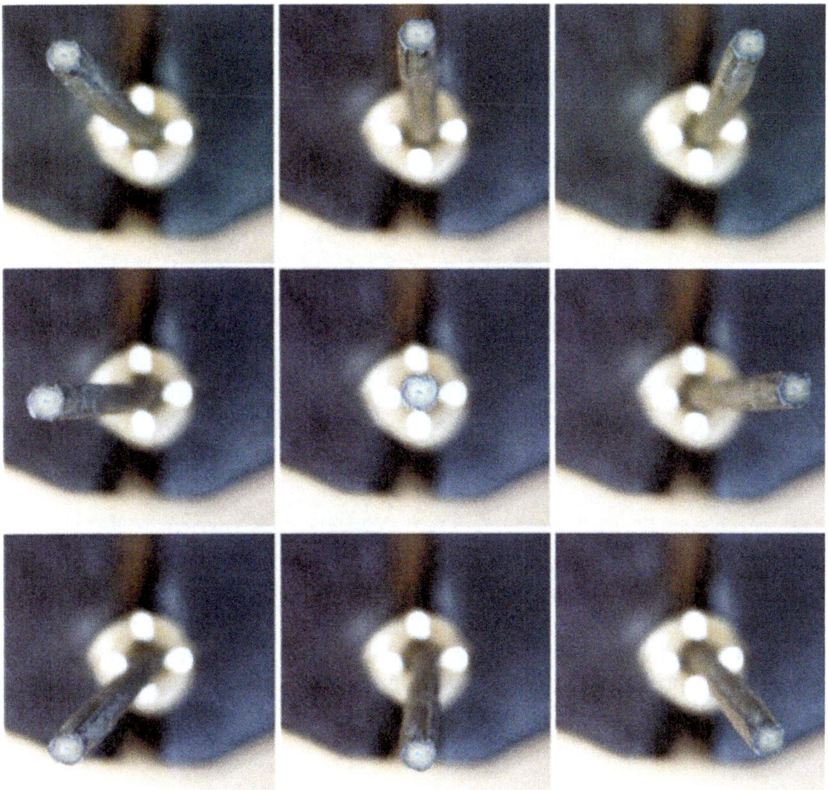

Figure 17.8 A top view of the tip deflections recorded for the cylindrical IPMC. Reprinted with permission from S. J. Kim, D. Pugal, J. Wong, K. J. Kim and W. Yim, A bio-inspired multi degree of freedom actuator based on a novel cylindrical ionic polymer-metal composite material, *Rob. Aut. Syst.*, 2014, **62**(1), 53–60.

A further example of bio-inspired legged robots has been proposed by Guo *et al.*[48] They introduced an eight-legged micro-robot designed to walk and rotate when submerged in a fluid environment (water) as an attempt to realise a micro-robot with multiple DOF, capable of efficient locomotion and great maneuverability in limited spaces. Eventually, the robot ingeniously used the electrolysis of the water surrounding the IPMC actuators to let the micro-robot float. The number of IPMC actuators was increased to ten in an inchworm-inspired robot proposed more recently.[49] This increased number was instrumental in obtaining the micro-robot multi-functionality. Four IPMCs were used for the robot locomotion, while the remaining six were introduced for object grasping. A picture of the prototype realised in shown in Figure 17.9.

The concept of legged robots was exploited by Vahabi *et al.*,[50] who proposed a four-legged robot, with two legs mounted on each side of a "main body", for in-pipe applications.

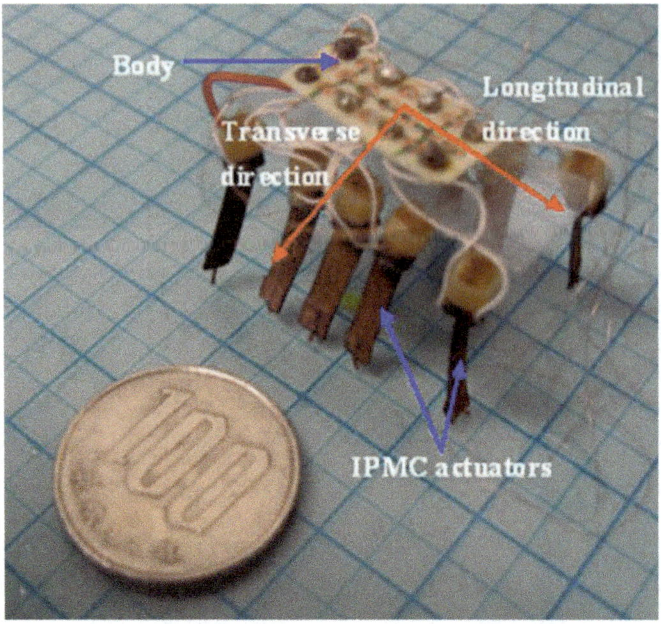

Figure 17.9 The ten-legged prototype realised to obtain robot multi-functionality. Reprinted with permission from S. Guo, L. Shi, N. Xiao and K. Asaka, A biomimetic underwater microrobot with multifunctional locomotion, *Rob. Aut. Syst.*, 2012, **60**(12), 1472–1483.

A number of bio-inspired robots actually mimic the motion capabilities of un-legged creatures. A typical un-legged locomotion is the undulatory mechanism.[51] We proposed a prototype mimicking the undulatory locomotion of a worm,[21] in which a fully IPMC-based structure, composed of a chain of IPMC actuators, was presented. The use of IPMCs allowed for realising an active structure with a beneficial effect on the system dimensions when compared with bio-inspired robots based on traditional actuators. The structure consisted of four connected segments and its total length was 10.5 cm. Each segment was powered at its middle point, in such a way as to obtain symmetrical bending. A schematic of the structure, along with the segment sizes, is shown in Figure 17.10.

At that time, our interest was to complete the bio-inspired structure with bio-inspired locomotion generation to make the resulting system closer to a real biological system. The actuation and then the motion were obtained by propagating a periodic signal from the tail to the head, imitating the travelling waves observed in the real-world undulatory locomotion adopted by some worms.[52,53] More specifically, the locomotion mechanism was implemented according to the case of transversal undulatory locomotion adopted from polychaete worms. The activation signals were generated by a cellular neural network (CNN) that generated autowaves, which can be considered as the electrical transposition of neural waves observed during

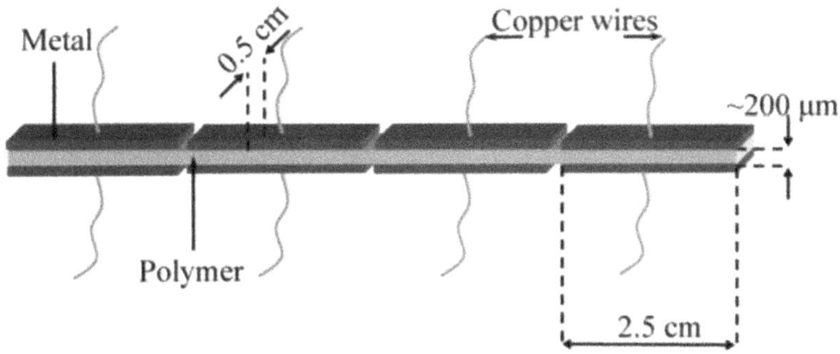

Figure 17.10 A scheme of the IPMC-based worm.
© IEEE 2006. Reprinted, with permission, from P. Arena, C. Bonomo, L. Fortuna, M. Frasca and S. Graziani, Design and control of an IPMC wormlike robot, *IEEE Trans. Syst., Man Cyb. B.*, 2006, **36**(5), 1044–1052.

the locomotion process in the nervous systems of animals. The produced analogue autowaves propagated in the CNN and the signals could be directly fed to the IPMC actuators. The control scheme was based on the central pattern generator (CPG) to control the robot locomotion. In this control scheme, each CNN cell is a nonlinear oscillator, producing cycle-limit oscillations, while the locomotion pattern depends on the connection between the CPG cells. The CPG was eventually implemented by using a CNN-reconfigurable digital architecture, in connection with a powering system capable of converting the digital signals into the corresponding analogue actions required by the IPMC segments.

In Figure 17.11(a) and (b), a schematic of the locomotion mechanism (the configurations in time are indicated in Figure 17.11(a) as A to G) and a view of the realised prototype are shown, respectively.

An example of the control signals used to actuate the four IPMC segments, lasting for 20 s, is seen in Figure 17.12.

Interestingly, in the very same year, a further worm-like robot was proposed,[54] based on the polychaete locomotion strategy, and also in this case, it consisted of four segments. This is the result of the common feeling that undulatory locomotion schemes, mimicking un-legged animal locomotion strategies, represent an efficient solution to realising robots capable of exploring unstructured environments. In this case, IPMC strips were connected by using suitable realised clamps. The clamps were moreover designed in such a way as to realise parapodia-like structures, with the aim to reduce the whole system complexity. The modules were activated by using square waves, with a constant lag between one module and the successive one.

As a further example, more recently in 2009, a similar but somewhat smaller worm-like robot was described,[55] again mimicking the undulatory locomotion scheme of a worm, *Caenorhabditis elegans*. The proposed worm-like robot consisted of five segments and was 25 mm long. The dimensions

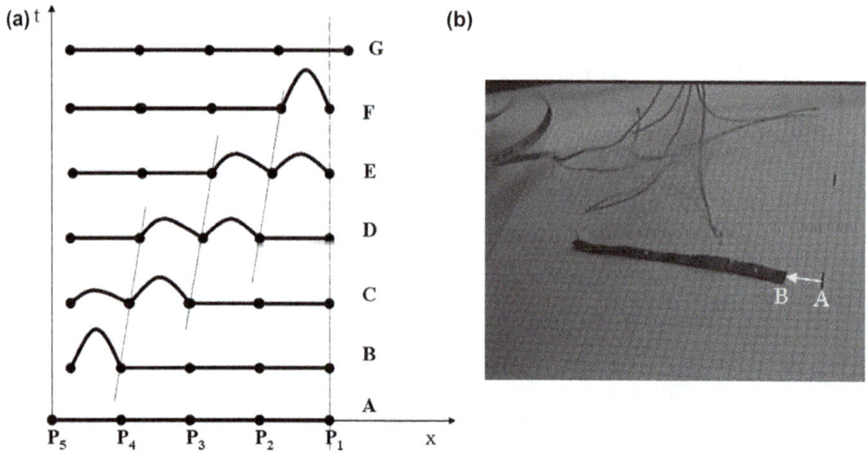

Figure 17.11 A schematic of the locomotion mechanism (a) and a view of the realised prototype (b) of a bio-inspired worm moving from A to B. © IEEE 2006. Reprinted, with permission, from P. Arena, C. Bonomo, L. Fortuna, M. Frasca and S. Graziani, Design and control of an IPMC wormlike robot, *IEEE Trans. Syst., Man Cyb. B.*, 2006, **36**(5), 1044–1052.

of the wormbot were considered key issues for the authors. The robot was, in fact, intended as a tool to complement experiments on the biological worm and worm's size greatly influences the nature of the interaction with the fluid it is immersed in, because of the dependence of the Reynolds number on both system size and oscillation frequency. The five segments were separately actuated by sinusoidal signals and the wave propagation was obtained by applying the sinusoidal waves to the segments with equal phase lags.

While both of the robots proposed in our contribution[21] and in the paper by Pak *et al.*[54] were tested in air, the worm-like robot proposed by Nguyen *et al.*[55] was tested in water. The possibility of realising robots capable of autonomous motion with un-legged structures is even more relevant for the case of underwater applications. As a matter of fact, the capability of IPMC actuators to work in a fluid environment has stimulated a great number of researchers to propose robotic underwater applications, with particular attention devoted to fish-like mobile structures.

Some preliminary considerations on the possibility of using IPMC actuators for robotic motion in fluid environments were presented by Paquette and Kim.[56] Then, a number of contributions were proposed that focused on the interaction between moving objects (including artificial fins) and the surrounding fluid, in order to estimate the thrust attainable by a vibrating IPMC as part of the realisation of propulsion systems.

As an example, Peterson *et al.*[57] used the particle image velocimetry (PIV) technique to analyse the flow field produced by a vibrating IPMC. The estimation of the mean flow field was further used to estimate the produced thrust. The same research group reconsidered the problem[58] by using the

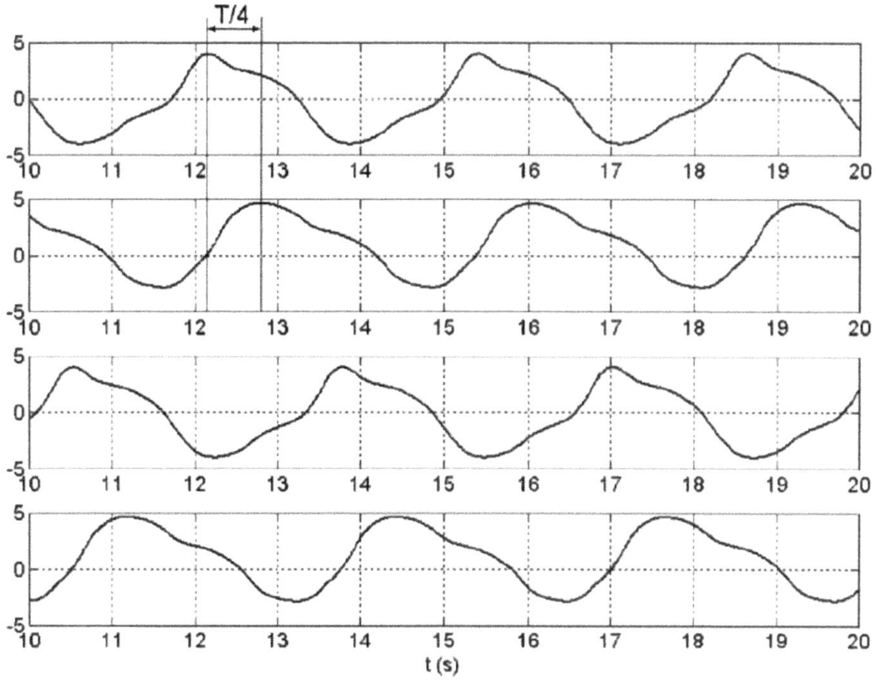

Figure 17.12 An example of the signals generated by the CNN-based CPG used to control the worm-like robot.
© IEEE 2006. Reprinted, with permission, from P. Arena, C. Bonomo, L. Fortuna, M. Frasca and S. Graziani, Design and control of an IPMC wormlike robot, *IEEE Trans. Syst., Man Cyb. B.*, 2006, **36**(5), 1044–1052.

Navier–Stokes equations to estimate both the thrust and the lateral forces per unit actuator width. The same framework was further used to estimate the moment resultant of the forces exerted by the IPMC on the encompassing fluid and the delivered power. The authors validated their results against estimations obtained by using a package for fluid dynamics simulation. The investigation of the numerical results enabled the description of the system vorticity as the cause of the thrust generation capacity of the vibrating IPMC. Finally, the numerical estimations were compared with experimental data obtained by the same PIV described in Peterson *et al.*[57]

The concept of hydrodynamic function[59] was adopted as a suitable theoretical tool in a number of works to estimate the locomotion of underwater vehicles. Aureli *et al.*[60] developed a modelling framework for the free locomotion of underwater vehicles propelled by vibrating IPMCs in quiet water. More specifically, modelling restricted to the case of a vehicle's planar motion (three DOFs) and the hydrodynamic function was used to determine a fish-like propulsion along with the lift and the moment, which act on the system as motion disturbances. The system's tail, consisting of an active IPMC and a trapezoidal-shaped passive fin, was used in an experimental setup to validate the proposed model.

Almost contemporaneously, another group of researchers used the Lighthill theory of the elongated body and the hydrodynamic function to estimate the steady-state speed produced by a passive fin acted on by a vibrating IPMC.[61] The authors used the elongated body hypothesis to link the mean thrust to the mean speed of the robot. Then, by balancing the mean thrust to the experienced drag force, the cruising speed was estimated, showing that this depends on the lateral velocity and on the slope of the trailing edge.

The contributions described so far on the study of the thrust produced by fin-like structures are part of a more general interest in the realisation of fish-like swimming structures and in the need to provide such structures with a propulsion system. In fact, a huge number of applications have been proposed for bio-inspired fish-like robots capable of motion in fluid environments.

A prototype realisation of a bio-inspired fish actuated by IPMCs has been proposed by Guo *et al.*,[62] who proposed a robot that was realised by using two IPMC actuators to mimic the body of a fish, while the swimming speed was regulated by changing the input voltage frequency. A more sophisticated system was proposed by Lee *et al.*,[63] who described a system capable of undulatory underwater motion, obtained by using a segmented IPMC. The systems consisted of three segments that were driven by using the output of bio-inspired nonlinear neural oscillators. These were further synchronised in such a way as to obtain the output lags required for wave propagation.

In the same year,[64] a centimetre-scale robotic fish was proposed that was propelled by means of a polyvinyl chloride (PVC) caudal fin actuated by an IPMC membrane, to mimic the caudal fin of a centimetre-scale goldfish. Two out of three prototypes were tested to assess their maneuverability and obstacle avoidance capabilities. A view of a refined version of the fish-like robot[65] is shown in Figure 17.13.

Though not a fish-like robot, it is worth mentioning here the case of a tadpole-like robot proposed by Jung *et al.*[66] and by Kim *et al.*.[67] The structure had a caudal fin actuated by a thick-cast IPMC. The fin was realised by using both a polymer film and a polyethylene terephthalate (PET) film in order to obtain undulatory and oscillatory motion, respectively. A schematic view of the tadpole-like robot and of the corresponding prototype are shown in Figure 17.14(a) and (b), respectively.

The prototype was tested, verifying that the undulatory motion is more effective than oscillating fin motion in propelling the tadpole robot.

Of course, there is an enormous number of living creatures that live in fluid environments and that, as a result of evolution, are optimised to their surrounding environment and can provide outstanding solutions to engineering problems. As a result, underwater IPMC-based robots implementing solutions that are different from fish-like robots have been proposed, taking inspiration from the diversity observed in nature.

A snake-like swimming robot was investigated by Yamakita *et al.*[68] The robot consisted of three elements realised by using styrene foam, connected

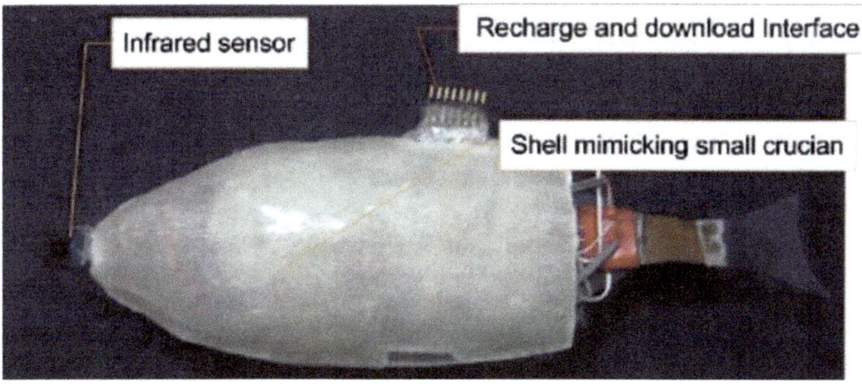

Figure 17.13 A view of the fish-like prototype mimicking a crucian fish.
© 2008 IEEE. Reprinted with permission, from X. Ye, Y. Su, S. Guo and L. Wang, Design and realization of a remote control centimeter-scale robotic fish, Proc. of 2008 *IEEE/ASME Int. Conf. Adv. Int. Mech.*, Xian, China, 2008, 25–30.

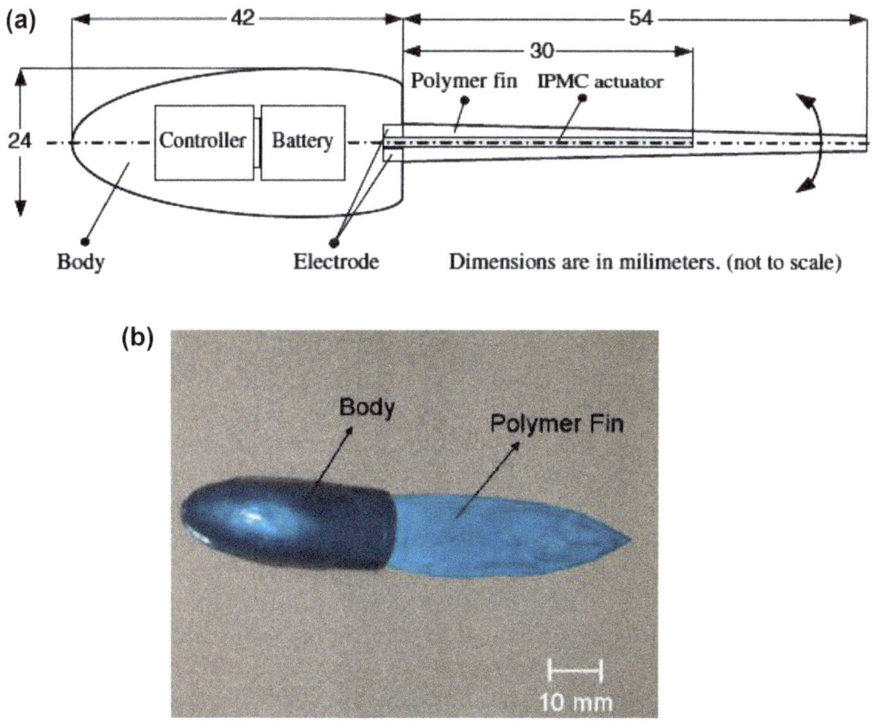

Figure 17.14 A scheme of the tadpole-like robot (a) and the corresponding prototype (b).
Reprinted with permission from B. Kim, D.-H.Kim, J. Jung, and J.-O. Park, A biomimetic undulatory tadpole robot using ionic polymer metal composite actuators, *IOP Smart Mat. and Struct.*, 2005, **14**(6), 1579–1585.

by using two IPMC actuators that were 20 mm long. The robot was 120 mm long and its total mass was 0.6 g. Experimental results showed that the robot was characterised by an undulatory motion and was capable of moving forward. Interestingly, the influence of the counterion used in the IPMC actuators on the robot's performance was investigated, and the authors showed that the counterion nature affects both the robot speed and the corresponding consumed power.

As a further example of bio-inspired robots, the case of jellyfish structures is worth mentioning. A micro-robot has been proposed[69] consisting of a main body mimicking the jellyfish body and four appendages. A shape memory alloy (SMA) actuator is used both to extrude the water out of the robot body in such a way as to produce its floating, and to let the water in during the sinking phase of the robot. The IPMC-based legs are used to allow the robot to walk.

A jellyfish-inspired robot was also proposed by Yeom and Oh.[70] The robot is still based on IPMC actuators, but with a different configuration. In this case, thermally pre-shaped IPMCs were used to realise eight curved jellyfish legs. These were further covered with cellophane paper to realise the bell-shaped body of the jellyfish robot. The actuators were excited with a sequence of bio-inspired fast pulses and slow recovery phases to generate the robot's vertical motion. An interesting comparison of the performances obtained by using a pure sinusoidal actuation signal and the proposed bio-inspired one is given in the paper. The authors show that the second signal greatly outperforms the pure signal input. Real views of a jellyfish during swimming and of the prototype realised are shown in Figure 17.15(a) and (b), respectively.

As further examples of the enormous range of animals that can provide useful suggestions for bio-inspired robots, structures moving like a ray fish were described[71,72] to implement rajiform motion, in which pectoral fins are used to produce undulatory locomotion. Each pectoral fin of the ray-like robot proposed by Punning *et al.*[71] was 110 mm long and realised by using eight bottle-like IPMC strips that were 40 mm long. A latex foil connected the actuators to the flexible fin structure. The authors were able to coordinate the action of each actuator in the fins in such a way as to generate an undulatory motion and propel the robot forward.

Takagi *et al.*[72] proposed a further ray fish-like structure with each fin actuated by using eight IPMCs. The IPMC actuators (each 5 mm×50 mm) were connected by using a thin polyethylene film so that each fin was 45 mm tall by 75 mm long. The fins were mounted on the robot body realised by using a polystyrene foam. The prototype of the robot is shown in Figure 17.16. The authors concluded by experimentally investigating the performance of the fins.

More recently, a manta ray-inspired robot was proposed in 2012.[73] The bio-inspired manta ray robot used two artificial pectoral fins, realised by using two large IPMC actuators, to generate undulatory flapping motions. The IPMCs were mounted on the leading edges of the pectoral fins, while the trailing edge was realised by using a passive polydimethylsiloxane (PDMS) membrane. When the IPMC was actuated, the passive trailing edge followed

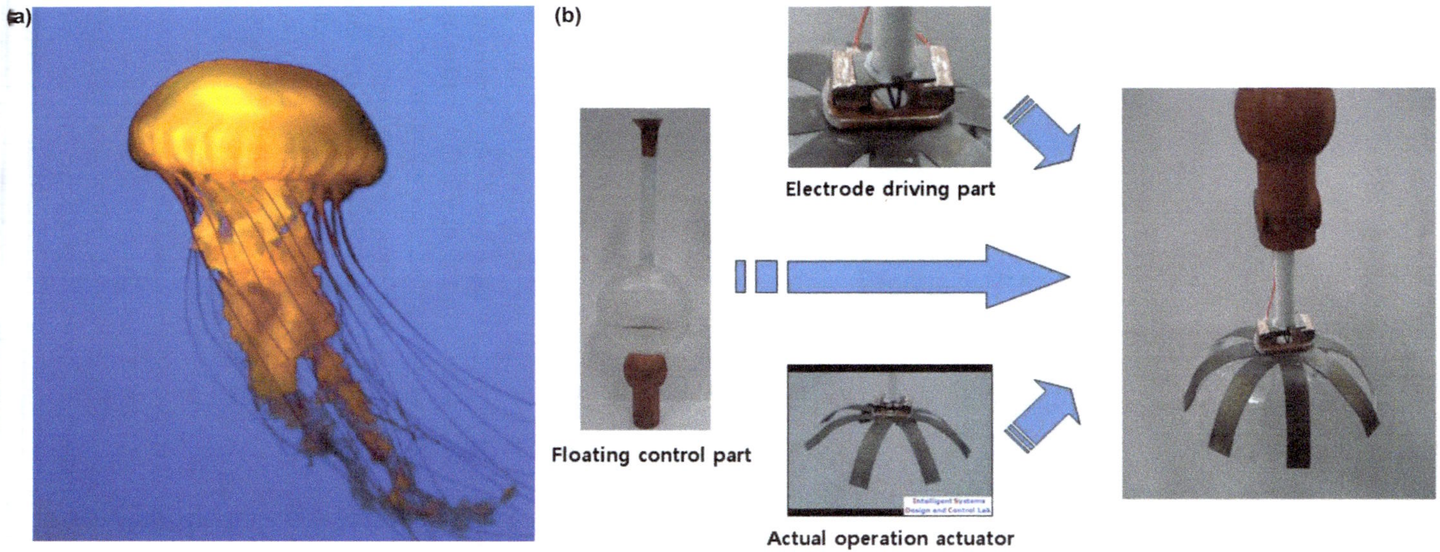

Figure 17.15 Jellyfish during swimming (a) and components of the jellyfish robot that was realised (b). Reprinted with permission from S.-W. Yeom and K. Oh, A biomimetic jellyfish robot based on ionic polymer metal composite actuators, *IOP Smart Mat. Struct.*, 2009, **18**(8), 085002.

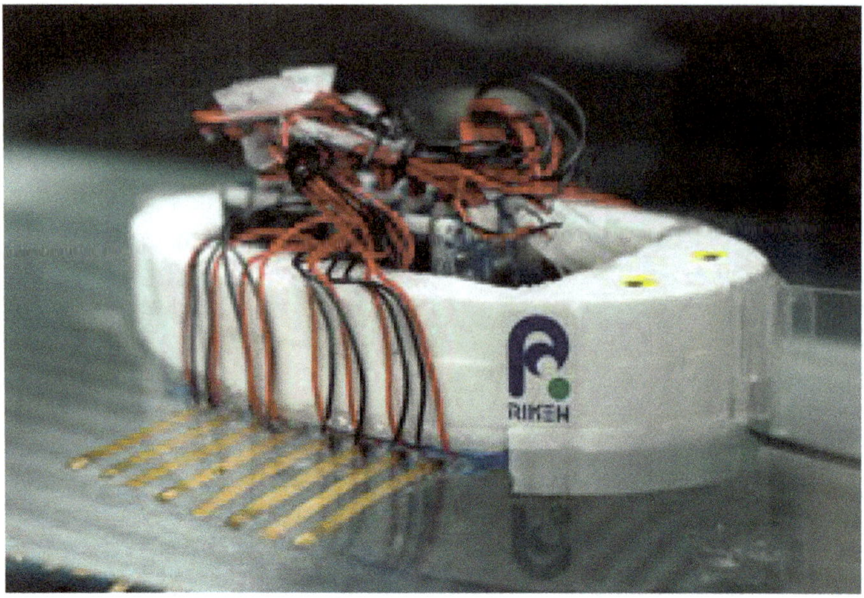

Figure 17.16 The prototype of the ray fish-like robot.
© 2006 IEEE. Reprinted with permission from K. Takagi, M. Yamamura, Z.-W.Luo, M. Onishi, S. Hirano, K. Asaka and Y. Hayawaka, Development of a rajiform swimming robot using ionic polymer artificial muscles, *Proc. IEEE/JRS Int. Conf. Intell. Rob. Syst*, 2006, 1861–1866.

the IPMC bending with the phase delay necessary to implement the undulatory fin motion. The robot was 11 cm long, 21 cm wide, and 2.5 cm thick, with a total mass of 55 g. A photograph of the robot is shown in Figure 17.17.

In the applications described so far, especially for mini- and micro-realisations, the powering of the system has been obtained either by using batteries or by flexible wires. On the one hand, batteries represent a significant payload, especially when compared with the mechanical power that can be produced by using IPMCs. Also, the significant current required by IPMCs during fast motion can greatly limit the autonomy of the envisaged systems. On the other hand, flexible wires can only be considered as a laboratory prototype solution for robotic systems, since they greatly impair the robot's mobility and can represent a challenge, especially when complex and/or segmented robots are considered. The opportunity to solve the problem of IPMC robotic application powering is therefore sound and contributions have been proposed on this topic.

A first solution to this problem was given in 2012 by Lee *et al.*,[74] who suggested the possibility of realising a power link between the IPMC actuator and its remote powering system. The envisaged powering system is based on the realisation of microstrip patches on the IPMC actuator surface that act as receivers. The receiver converts a microwave signal into DC power that is

Figure 17.17 A picture of the manta ray-like robot.
Reprinted with permission Z. Chen, T. I. Um and H. Bart-Smith, Bio-inspired robotic manta ray powered by ionic polymer-metal composite artificial muscles, *Int. J. Smart and Nano Mat.*, 2012, **3**(4), 296–308.

used to actuate the IPMCs. The receivers can eventually be selectively powered according to the used microwave frequency. The authors show that the system is feasible at least for receivers realised by using standard rigid supports, while poor results were obtained when the receivers were realised directly on the IPMC surface.

The problem of wireless powering of IPMCs was addressed, again in 2012,[75] with the aim of replicating hovering flight and swimming in biological systems. A conceptual scheme of the robotic microswimmer to be powered is shown in Figure 17.18.

17.3 IPMC-based Sensors

Since the first sensing application of IPMCs was proposed in 1992 by Sadeghipur,[76] the fields of application of IPMC-based sensor devices have been the same as for IPMC-based actuators (*i.e.* those fields in which their mechanical, chemical, and electrical characteristics guarantee them a privileged role with performances that cannot be obtained by using competing technologies). Also, most of the sensing applications use the common cantilever configuration widely adopted for actuators. Nonetheless, IPMC-based sensing systems share most of the corresponding actuator limitations.

When we started our research activity on IPMC sensors, one of the reported limitations was the impossibility of transducing constant time

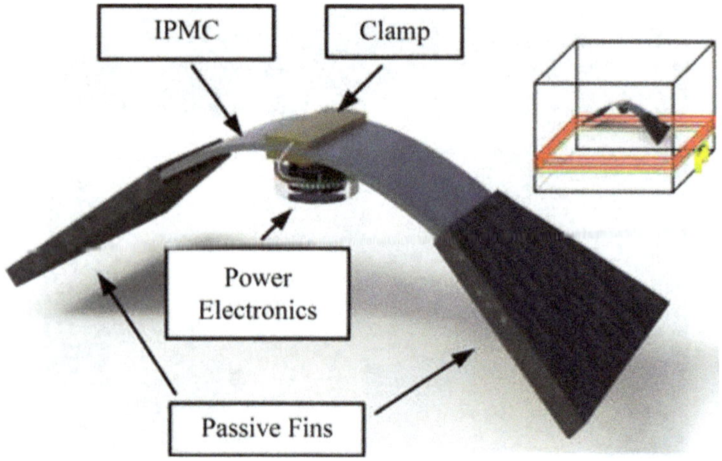

Figure 17.18 A schematic of a robotic microswimmer.

mechanical inputs,[77] and the corresponding investigations ranged down to quasi-static inputs (sub-hertz frequencies).[78] Moreover, IPMCs in cantilevered configurations were investigated as near-DC accelerometers capable of working down to 0.03 Hz.[79] We were therefore interested in overcoming such a limitation, expanding the possibility of measuring DC mechanical loads. Eventually, due to the appealing possibility of downscaling IPMC transducers, we were interested in measuring techniques suitable for polymer microelectromechanical system (MEMS) realisation.[80] Based on these considerations, we developed a resonant force sensor capable of DC measurement by using a vibrating cantilevered IPMC.[81] As a matter of fact, sensor-resonant cantilever beams were already widely proposed in the literature to realise MEMS based on piezoelectric materials,[82,83] and we extended that technology to the case of IPMC benders. An IPMC mounted in a cantilever configuration was activated while a force was applied to the second end. This changed the natural resonant frequencies of the beam and we showed that the value of the first resonance frequency of the beam could be used to estimate the tensile force applied in the axial direction. A model was developed for a vibrating IPMC with the applied force and an example of the simulated first resonance mode deformation is reported in Figure 17.19 for an IPMC that was 45 mm long. It is possible to observe that the application of the external force produces both the increment of the value of the resonance frequency corresponding to the first mode and the decrease in the maximum deformation.

The proposed model was experimentally validated by using an IPMC bender, based on Nafion® 117 with Li^+ as the counterion, with a length of

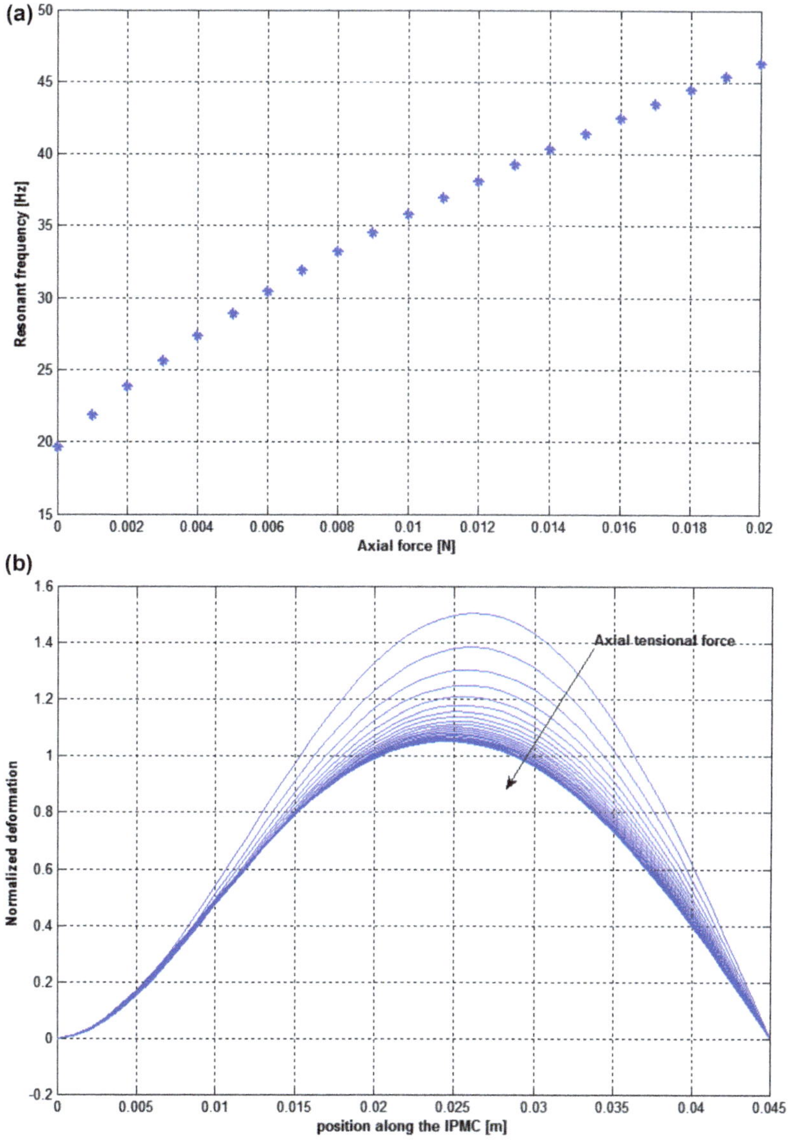

Figure 17.19 An example of the changes in the first resonance mode frequency (a) and in the corresponding amplitude (b).
Adapted from ref. 81.

$L = 45.0$ mm and width of $b = 4.0$ mm. The second end of the beam was connected to a mass lying on a slanting plane, whose inclination was varied to change the force applied to the IPMC. A sinusoidal swept voltage with a peak-to-peak amplitude equal to 8.0 V and a frequency ranging from 20.0 Hz up to 60.0 Hz at a time interval of 40.0 s was used as input to the IPMC.

Input–output data were acquired by means of a PCI-6052E card from National Instruments™. Data were processed offline using a Matlab® code in order to estimate information relevant for the characterisation process. An example of the experimental IPMC frequency response obtained when a load equal to 11.3 mN was applied is shown in Figure 17.20, where the typical second-order system-like behaviour can be observed.

The system was further investigated by changing the applied load in a suitable range to obtain a family of experimental values of the resonance frequency as a function of the applied load. An image of the developed experimental setup is shown in Figure 17.21.

The setup also consisted of a subsystem to read the IPMC deformation and of an electronic circuit capable of supplying the IPMC actuator with a voltage signal whose frequency followed the value of the beam resonance frequency as it changed because of the applied axial external load.

A circuit able to lock the phase displacement between the applied voltage and deformation was developed. The sharp phase change occurring near the resonance frequency allowed for large selectivity with respect to the frequency working conditions. More specifically, a phase-locked loop (PLL) was used and an *ad hoc* designed circuitry was realised to adapt the PLL to the described application. The voltage input into the PLL voltage controlled oscillator (VCO) was used to estimate the axial load, while its output was used to force the IPMC. The obtained results are shown in Figure 17.22.

Notwithstanding the huge number of proposed underwater applications of IPMCs working as actuators, few contributions have been proposed for IPMC sensors working in water or immersed in other fluids.

We again used a vibrating IPMC immersed in a fluid to realise a small-scale viscometer.[84] More specifically, a vibrating IPMC was immersed in a still fluid while an infrared (IR) sensing system recorded its tip deflection. The system was intended to determine the rheological properties of fluids. A model of the system vibrating in a fluid was proposed based on the concept of the hydrodynamic function for the case of an active IPMC beam with a rectangular cross-section,[59] adapted to the case of the range of frequencies of interest for an IPMC.[85] The model was used to describe both the resonant frequency and the quality factor of the vibrating system. Since these parameters depend both on the fluid density and viscosity, observations of the system frequency response were used to estimate the fluid rheological properties. The system adopted the well-known approximation of the frequency response of a cantilever beam in the neighbourhood of each resonant peak to a simple harmonic oscillator, provided that the quality factor Q greatly exceeds unity. This means that the amplitude frequency response can be approximated as:

$$w_i^f(\omega) = \frac{W_0 \omega_r^2}{\sqrt{\left(\omega^2 - \omega_r^2\right)^2 + \frac{\omega^2 \omega_r^2}{Q^2}}} \tag{17.1}$$

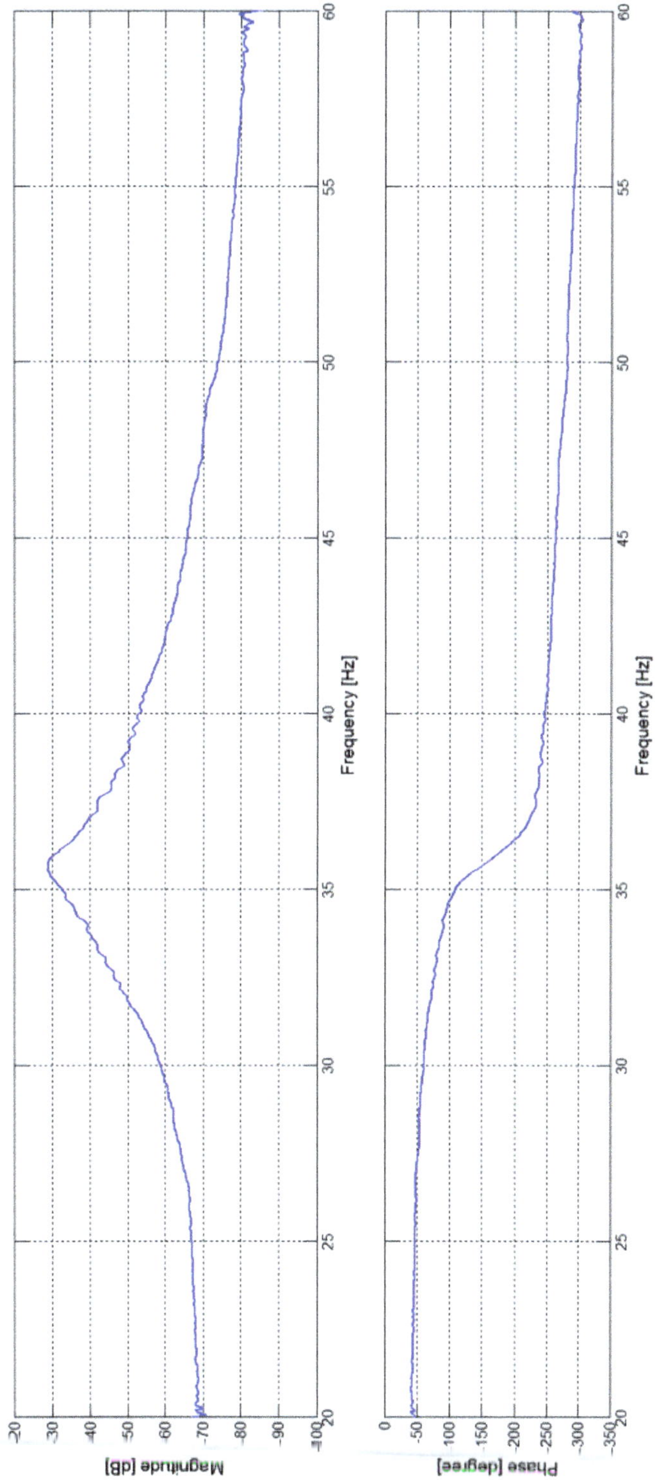

Figure 17.20 An example of the experimental frequency response of the vibrating IPMC. Adapted from ref. 81.

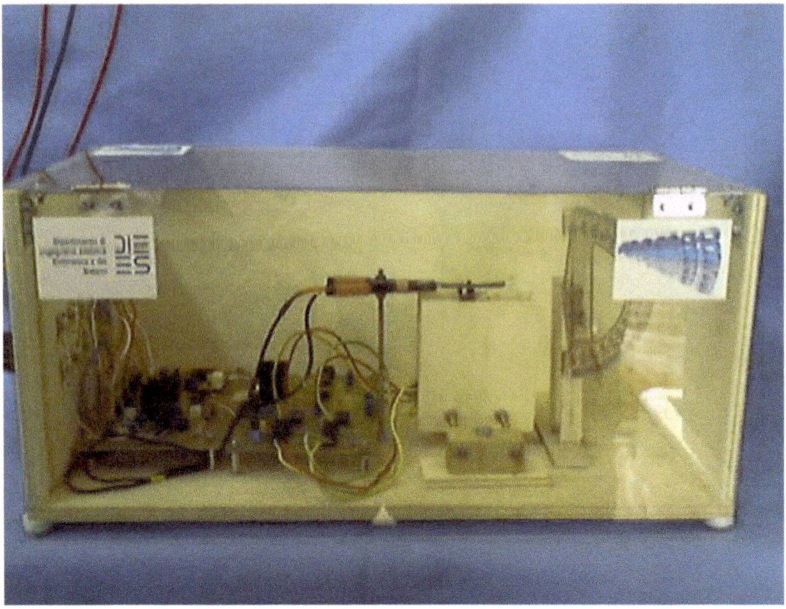

Figure 17.21 The experimental setup used for the force sensor characterisation.

where W_0 denotes the amplitude at the frequency $\omega = 0$ and Q is the quality factor, defined as:

$$Q = \frac{\frac{4\mu}{\pi\rho b^2} + \Gamma_r(\omega_r)}{\Gamma_i(\omega_r)}. \tag{17.2}$$

ω and ω_r are the radial frequency and resonant frequency, respectively (the interested reader can find the meaning of the other quantities in eqn (17.1) and (17.2) in the referenced paper). Eqn (17.1) and (17.2) were both used for the viscometer calibration and for the estimation of the rheological parameters of the investigated fluids.

In order to validate the theoretical analysis, an IPMC vibrating system was tested on sugar solutions, which are valuable test systems for such a device because many biological liquids contain a remarkable amount of sugar. The IPMC cantilever immersed in the fluids was actuated with a sinusoidal swept input voltage. The frequency response was measured and estimations of the viscosity and density of the fluid were derived. A photograph of the developed system is shown in Figure 17.23(a), while examples of the comparison between the obtained experimental frequency responses with the corresponding estimations for two typical solutions, obtained by using the model that describes the viscometer, are shown in Figure 17.23(b). A further version of the vibrating sensor has been developed and will be described, along with the obtained results, in Section 17.4.

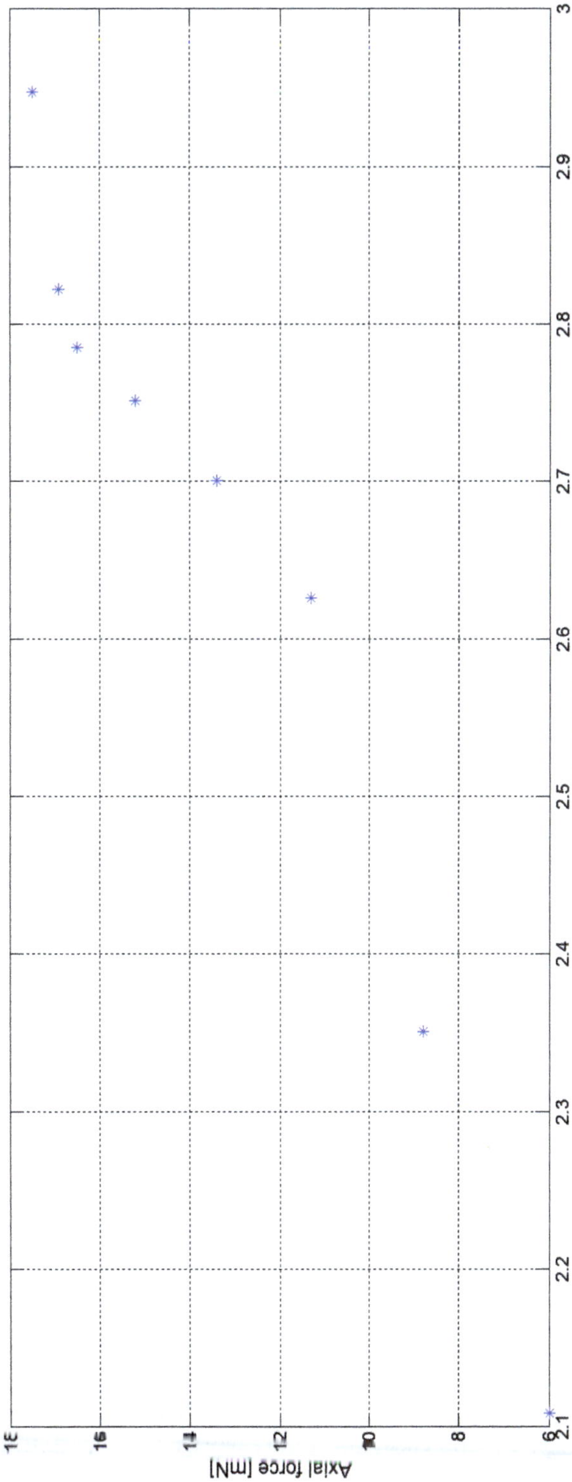

Figure 17.22 The experimental dependence between the applied force and the VCO output voltage. Adapted from ref. 81.

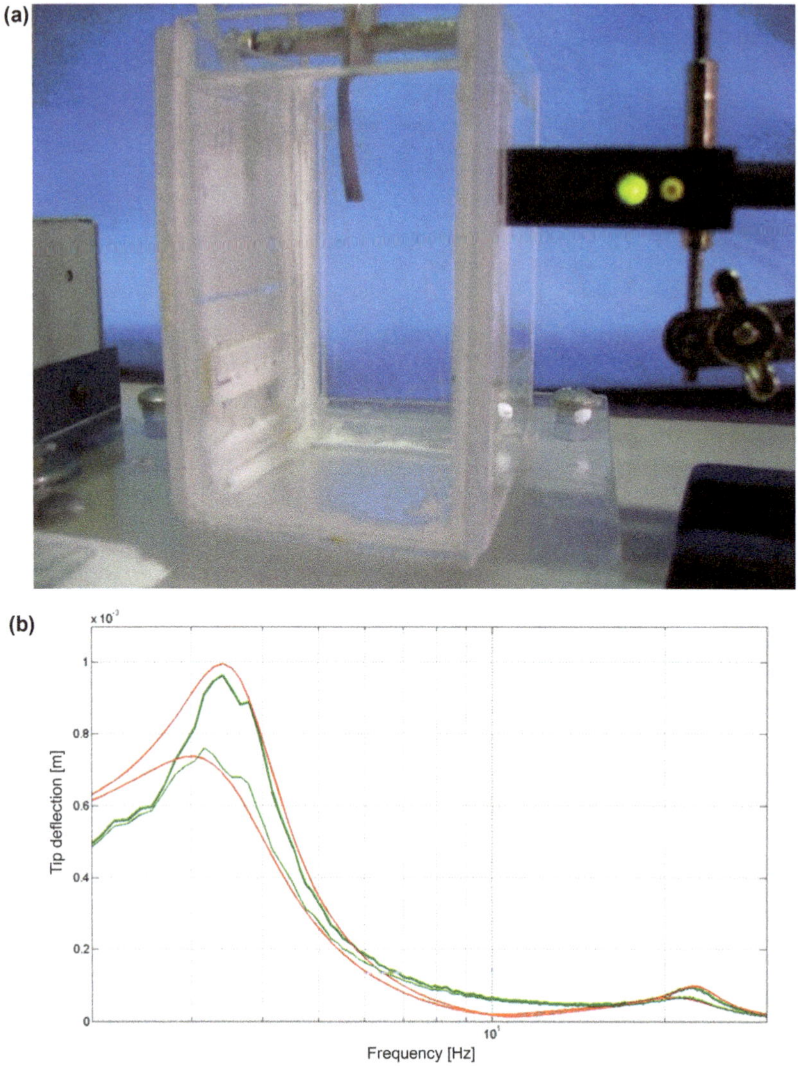

Figure 17.23 A photograph of the IPMC-based viscometer and an example of the obtained curves as a function of the fluid rheological properties.

A different approach for determining the rheological properties of fluids has been proposed by Chen *et al.*,[86] who used an IPMC membrane to sense the rheological characteristics of non-aqueous fluids. In the paper, a multi-segmented modelling approach was adopted as a tool to model the large deformations of the system while maintaining low computational efforts. A model describing the interaction of an IPMC vibrating beam with the surrounding fluid was tuned and therefore proposed for estimating the fluid rheological characteristics in the form of the product of the drag coefficient by the fluid density. In this case, the IPMC works as a generating system under the effect of the mechanical action in the typical cantilever configuration. The proposed system works provided that a pulsating flow is produced, with relevant applications in automotive industry.

Biological systems have also been relevant sources of inspiration for the case of IPMC-based actuators. Moreover, the possibility of using IPMC devices in water or in fluids more generally has partially polarised the research activity on bio-inspired IPMC-based sensing systems. This is the case for the contributions by Abdulsadda and Tan,[87–89] dealing with IPMCs in cantilever configurations that vibrate under the effect of external mechanical inputs.

More specifically, in the contributions by Abdulsadda and Tan,[87,88] a bio-inspired lateral line (an array of hair cell sensors), used by underwater organisms to obtain information on flow conditions, obstacles, prey/predators, and moving objects, was proposed, replacing the hair cells with IPMC strips and using an artificial neural network for processing the sensor signals. The developed sensor array was 8 cm long and the lateral line consisted of five IPMC sensors that were 8 mm long, with a sensor-to-sensor separation of 2 cm. The prototype and the experimental setup are shown in Figure 17.24. Experimental results showed the prototype's capability to localise a dipole up to four to five body lengths away.

The same authors[89] extended their research on the localisation of a dipole vibrating in a fluid by using an IPMC-based artificial lateral line for the localisation a dipole moving in a plane. More specifically, the artificial lateral line in this second case consisted of six IPMC sensors, with a sensor-to-sensor distance equal to 2 cm, and the problem was solved when the dipole moved in the same direction as the lateral line axis.

Biomedical applications have been a prosperous field of applications for IPMC transducers, and IPMC-based sensing systems have been considered in the literature because of a number of interesting properties that make them suitable for medical applications, such as their required low voltage levels, capability to work in wet environments, softness, and absence of any biological activity, along with their further possibility of being coated with compatible materials, to mention just a few.[90,91]

As a matter of fact, biomedical applications of IPMC sensors have been proposed in the past.[90,91] More specifically, Ferrara *et al.*[90] proposed the use of IPMCs to sense the pressure within the spine, as a technology suitable for allowing the realisation of small sensors that can fit within the facet joints in the spine, whose thickness is in the range of 1–3 mm. Such a size is generally too small for traditional transducers. By contrast, the authors proposed the use of IPMCs as pressure sensors, with the mechanical compressive load applied on the IPMC thickness (*i.e.* about 200 μm) as usual. To prove the suitability of IPMCs as pressure sensors in the envisaged application, the authors realised devices 2 cm×2 cm in size and tested them against compressive loads of 200 N and 350 N on a surface of 4 cm^2. Experiments were performed by using loading and unloading cycles.

Figure 17.24 An IPMC-based lateral line prototype (a) and the tank for the experimental investigation of the system (b).
© IEEE 2011. Reprinted with permission from A. T. Abdulsadda and X. Tan, Underwater source localization using an IPMC-based artificial lateral line, proc. of 2011 *IEEE Int. Conf. Rob. Aut.*, 2011, 2719–2724.

The authors recorded a maximum output voltage of 80 mV and 108 mV in the two cases, along with a quite long relaxation phenomenon in the sensor response, which they attributed to the IPMC viscoelastic nature.

Keshavarzi *et al.*[91] proposed the use of IPMCs as systolic and diastolic blood pressure, pulse rate, and rhythm sensors, which could eventually be mounted on the inner surface of a cuff. The IPMC sensor was intended for the realisation of portable blood pressure monitors. Also in this case, the IPMC response to compressive loads was exploited and experiments were performed on a prototype whose size was 1 cm×1 cm, with an active area equal to 0.5 cm×0.5 cm. Experiments were performed with loads in the range

15–165 mm Hg, a range that is relevant to the envisaged application. The corresponding observed maximum value of the sensor output was in the range 5–240 mV. The authors concluded that the sensor showed a roughly linear relationship between the applied load and the corresponding output, though a kind of saturation occurs as the load approaches the upper limit of the considered range. Also, though the system was capable of providing a time-dependent output linked to the load variations, discrepancies were observed between the sensor input and output, which the authors attributed, again, to the IPMC viscoelastic nature.

We were also interested in the biomedical applications of IPMC sensor capabilities and proposed[92] a system consisting of a set of IPMC sensors, used to mimic ciliate cells of the vestibular labyrinth. This is a biological component of living beings, situated in the inner ear and devoted to the perception of the angular acceleration to which the head is subjected.

Ferrofluids,[93] which are biphasic systems made of small solid ferromagnetic particles melted in suitable, non-magnetic carrier liquids, were used in this application to stimulate the IPMC sensors that produce a coded signal used to detect system motion. Unlike the systems previously described,[90,91] we used the IPMC in the bending configuration and the sensor output was produced by the sensor deflection. In the proposed system, the motion of a ferrofluid sphere stimulates different IPMC membranes working as sensors, placed along the sphere's trajectory in a circular tube. The position of the sphere can be therefore estimated as a function of the IPMC array output signals. More specifically, the system was inspired by the human equilibrium control system. This consists of:

a) Ciliate cells. These are able to detect variations in the inclination of the body, in particular sensing the direction, intensity, and duration of the stimulus, which work continuously and act as the sensorial receivers.
b) The vestibular system consisting of the utricle, the sacculus, and three semicircular channels, which are placed on three roughly perpendicular planes. Each channel expands near the common outlet (the utricle), forming an ampulla. The ampulla is totally closed by the ampullary crest, on which the sensorial cells are arranged, and the cupola, which is a gelatinous mass in which the cilia are partially contained.

If the head is subjected to a rotational acceleration, the endolymph inside the channels preserves its stationary position and this causes the bending of both the cupola and the cilia of the sensorial cells inside it. The semicircular channels therefore behave like a three-axes accelerometer. The vestibular system is able to perceive extremely weak accelerations, even lower than $1^\circ\ s^{-2}$.

The proposed IPMC-based system was intended to reproduce the vestibular labyrinth. More specifically, a semicircular channel has been realised according to the scheme shown in Figure 17.25, in which the pipette substitutes for the bony structure of the channel, the endolymph is replaced by a

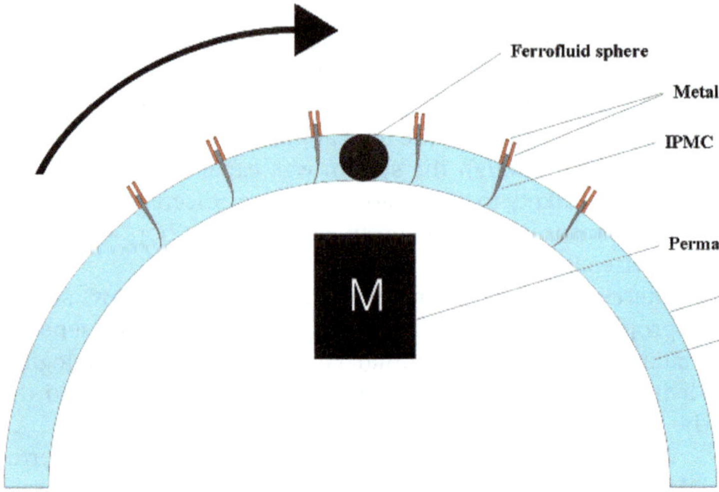

Figure 17.25 A scheme of the human ear-inspired artificial circular channel. Reprinted with permission from B. Andò, C. Bonomo, L. Fortuna, P. Giannone, S. Graziani, L. Sparti and S. Strazzeri, A bio-inspired device to detect equilibrium variations using IPMCs and ferrofluids, *Sens. Actuators A: Phys.*, 2008, **144**, 242–250.

sphere of ferrofluid immersed in deionised water, and the ciliate cell function is performed by the IPMC sensors.

It is worth noting that the proposed system actually works as an incremental encoder capable of reading both the discrete position and its direction (as IPMC sensors change the sign of the produced output according to their deflection direction). By contrast, the human ear circular channels are angular accelerometers. Nonetheless, once the output signals are acquired from the array of IPMCs, both the angular speed and the corresponding acceleration can be estimated by using their discrete approximations.

A prototype of the proposed system was realised in order to validate the bio-inspired circular channel. More specifically, a Petri dish with diameter 8×10^{-2} m was used. The dish was filled with deionised water, and a quantity of ferrofluid capable of realising a sphere of diameter 5 mm was injected. The ferrofluid was aggregated by using a permanent magnet. Once it reached its spherical form, the magnet was removed while the ferrofluid maintained its spherical form, without dissolving in the water. As a further difference with respect to the scheme shown in Figure 17.25, in the actual prototype the Petri dish was fixed, while a set of eight electromagnets, forced by using suitable currents, were used to move the ferrofluid sphere. This allowed for the production of relative motion of the ferrofluid with respect to the IPMC array. Photographs of the realised prototype are shown in Figure 17.26.

In a first trial, one IPMC sensing element was immersed in water and the signal was acquired, while the ferromagnetic sphere was forced to rotate in the channel. The currents forcing the electromagnets were delayed with

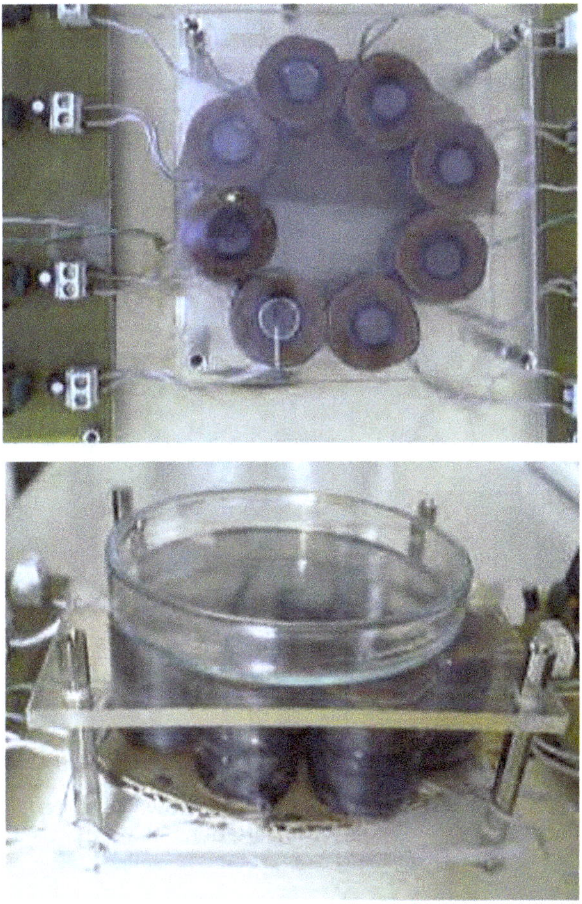

Figure 17.26 A view of the prototype used to realise the artificial circular channel. Reprinted with permission from B. Andò, C. Bonomo, L. Fortuna, P. Giannone, S. Graziani, L. Sparti, L. and S. Strazzeri, A bio-inspired device to detect equilibrium variations using IPMCs and ferrofluids, *Sens. Actuators, A: Phys.*, 2008, **144**, 242–250.

respect to each other, in such a way that the sphere rotated in the channel with a frequency equal to 178.5 mHz. The corresponding signal acquired by the IPMC is shown in Figure 17.27. It is possible to observe that the sensor output is characterised by a train of very large pulses, whose frequency corresponds to the turning frequency of the ferrofluid sphere.

The produced signal was actually very rich and the interested reader can look to the referenced paper for further details. Here, it suffices to say that a set of eight IPMC sensing elements was mounted in the prototype and a virtual instrument (VI) was realised by using LabVIEW. A view of the user interface showing the coded position of the ferrofluid sphere is shown in Figure 17.28. It shows the coding used, in which LEDs are turned on by the impinging sphere.

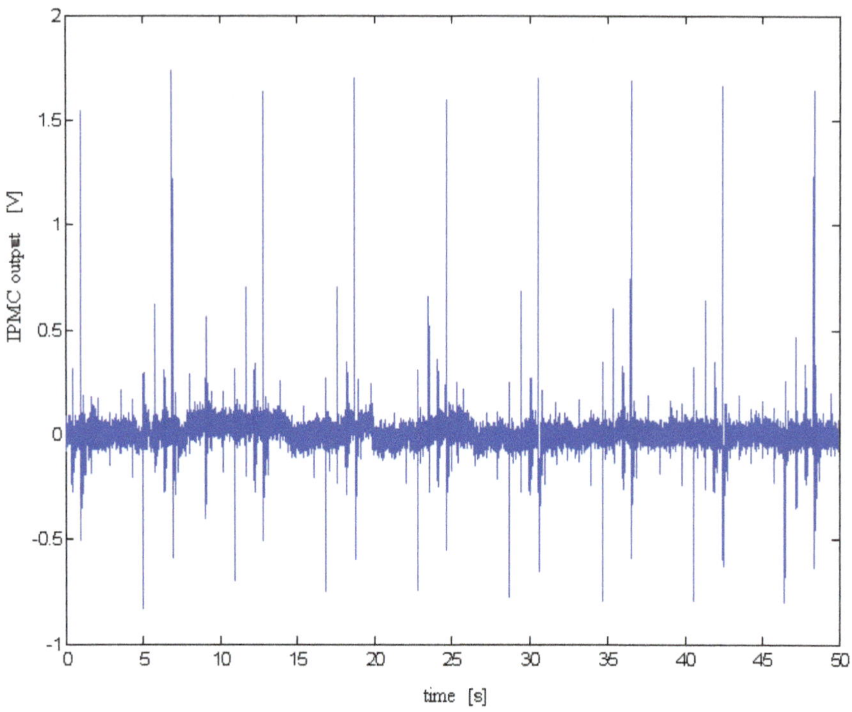

Figure 17.27 An example of the signal acquired by using the IPMC sensor. Reprinted with permission from B. Andò, C. Bonomo, L. Fortuna, P. Giannone, S. Graziani, L. Sparti and S. Strazzeri, A bio-inspired device to detect equilibrium variations using IPMCs and ferrofluids, *Sens. Actuators, A: Phys.*, 2008, **144**, 242–250.

One of the functionalities enabling the realisation of autonomous robotic systems is their efficient powering. Nonetheless, as mentioned above, a lot of the realised laboratory-scale robots use either heavy, large batteries or are directly wired to a kind of powering system. Though this can be considered a suitable solution for a proof of concept, this system of powering could be a weakness when systems capable of operation in real life are of interest. The necessity of adopting adequate powering strategies for the envisaged smart systems is therefore sound. A solution to the power needs of autonomous systems is of course the harvesting of environmental energy, which could improve the system's performances in terms of augmented maneuverability and lifetime and reduced maintenance, weight, and size.[94] The mechano-electric transduction property, already exploited for IPMC-based sensors, suggests the possibility of also using them as power harvesters, capable of scavenging energy from the wasted mechanical vibrations that characterise many systems both in air and underwater. Suitable sources could, in fact, be either human-produced processes, such as automobiles, ships, aircraft, or bridges, or natural sources, such as wind or water waves.

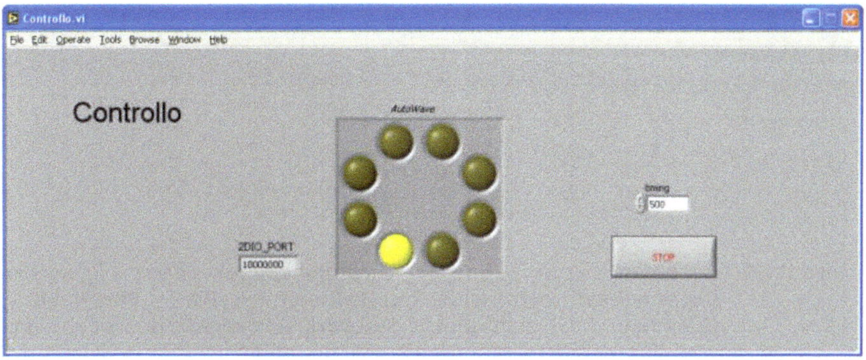

Figure 17.28 The user interface of the VI realised to encode the ferrofluid sphere position in the artificial circular channel.
Reprinted with permission from B. Andò, C.Bonomo, L. Fortuna, P. Giannone, S. Graziani, L. Sparti and S. Strazzeri, A bio-inspired device to detect equilibrium variations using IPMCs and ferrofluids, *Sens. Actuators, A: Phys.*, 2008, **144**, 242–250.

Since the realisation of a power harvester by using IPMCs involves their mechanoelectric transduction, a harvester can be considered as a case of a sensor, and this is why IPMC harvesters are addressed in this section. Nonetheless, it is worth considering here that for a power harvester, the focus is on its capability to efficiently collect energy from the environment, without too much attention paid to the information degradation, as is the case for a sensor. Aspects such as load adaptation are relevant in power harvesting and require the development of harvesting systems that deal with such topics.[95]

The research on IPMC-based power harvesters started quite late with respect to other IPMC applications. This topic has garnered rapidly growing interest as a technology that is relevant to the powering of smart devices required to work in the absence of traditional power sources, to the point that it is now one of the most active fields of research. The first contribution in fact only arose in 2005,[96] when B.R. Martin addressed this aspect in his Master's thesis in mechanical engineering. Further contributions were proposed in 2008[9,10] when the power harvesting capabilities of IPMCs in air were investigated.

More specifically, we exploited the available knowledge on the IPMC grey-box modelling[26,97–99] to model the power generation capability of IPMC transducers in air under the effect of base excitation vibrations.[10] At that time, we were interested in the investigation of the harvesting capabilities of a novel technology in the low frequency range, where the competing technologies were not able to provide significant contributions, and we searched for both the optimal working region and the optimal harvester dimensions for maximum power generation. As a matter of fact, we showed that the investigated system was able to reach its best performance when a vibrating

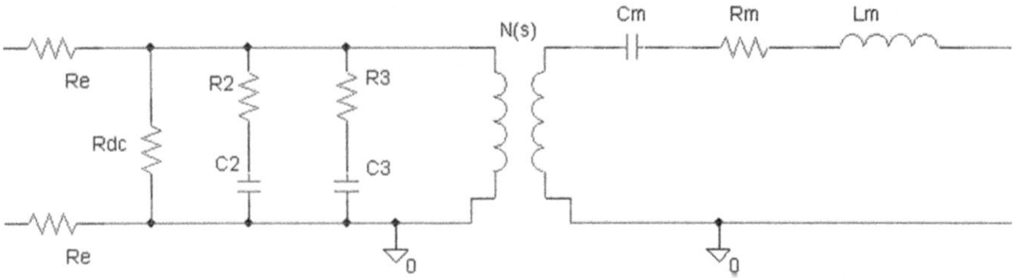

Figure 17.29 Equivalent circuit for an IPMC membrane in a bender configuration. Reprinted with permission from J. Brufau-Penella, M. Puig-Vidal, P. Giannone, S. Graziani and S. Strazzeri, Characterization of the harvesting capabilities of an ionic polymer metal composite device, *IOP Smart Mater. Struct.*, 2008, **17**(1), 015009.

frequency of 7.09 rad s^{-1} was imposed, though a power generation capability of a few nanowatts was estimated.

As a first step, we modelled the IPMC in a bender configuration by using an equivalent lumped circuit model, as shown in Figure 17.29 (for further details about the meaning of the elements introduced in the model, the interested reader can look to the referenced paper). Here, it suffices to say that the left-hand side of the circuit represents the electrical domain, while on the right-hand side, the mechanical domain is represented. The coupling between both domains is modelled by means of a linear transformer.

The unknown parameters ruling the proposed model were therefore experimentally determined by using a suitable setup, and the model was validated under a number of working conditions.

The IPMC bender base was mechanically excited by using a shaker, according to the scheme reported in Figure 17.30, and both the produced short-circuit current and the open circuit voltage were measured in different surveys.

The sample used for the measurements was based on a Nafion® 117 Na^+ membrane that was 32 mm long (where $L_t = 25$ mm was the free portion of the membrane while $L_c = 7$ mm was the clamped part of the IPMC); the width was $w = 5$ mm. The frequency range from 0.1 to 100.0 Hz was investigated.

Assuming an electrical power storage circuit matched with the electrical impedance of the IPMC sample, the short-circuit current and the open circuit voltage can be used to estimate the maximum power that can be extracted from the membrane. In Figure 17.31(a) and (b), the experimental transfer functions of the short-circuit current and of the open-circuit voltage with respect to the base deformation are shown, respectively.

The reported data show that, in the investigated frequency range, the harvester behaves like a high pass filter with a resonance frequency equal to 24 Hz for both the short-circuit current and the open-circuit voltage. At this frequency value, the current transfer function presents a maximum of about −40 dB, corresponding to 0.01 A m^{-1} or 0.439 μA $(m\ s^{-2})^{-1}$, while the voltage

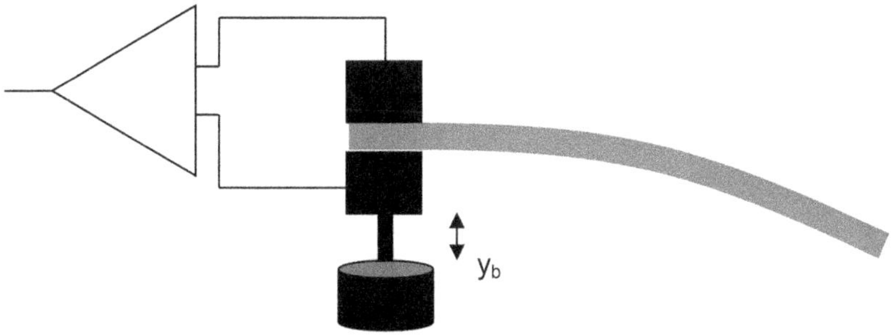

Figure 17.30 Mechanical configuration of the power harvester.
Reprinted with permission from J. Brufau-Penella, M. Puig-Vidal, P. Giannone, S. Graziani and S. Strazzeri, Characterization of the harvesting capabilities of an ionic polymer metal composite device, *IOP Smart Mater. Struct.*, 2008, **17**(1), 015009.

transfer function is about −12 dB, corresponding to 0.25 V m^{-1} or 10.99 V (m s^{-2})$^{-1}$. Supposing that the system is working in matched load conditions, at the resonant frequency, the maximum power that can be generated can be computed by using the equation:[100]

$$P_{av} = \frac{|V_{OC}(s)I_{SC}(s)|}{8} \tag{17.3}$$

which, at the resonance frequency, gives an estimated value of 0.61 pW (m s^{-2})$^{-2}$. It is worth observing that such a value corresponds to an acceleration of 1.0 m s^{-2}, while IPMCs can face much higher accelerations. If, for example, the reasonable value 22 m s^{-2} is considered, the available power rises to 0.29 nW.

A number of investigations were performed to estimate the effects of the system design parameters on the available power, and a generation capability in the order of a few nanowatts was estimated for a number of configurations.

Also in 2008, Tiwari *et al.*[9] proposed a power harvester based on IPMCs that were soaked with ionic liquids. Moreover, various deformations (*i.e.* bending, tension, and shear) were considered in order to evaluate the IPMC power generation capabilities.

Of course, research activity on IPMCs as power harvesters has also been carried out for the case of submerged systems, as in the contribution by Aureli *et al.*,[101] where the authors investigated the power harvesting capabilities of an IPMC immersed in a fluid and connected to a suboptimal resistive load because of mechanical base excitation. The mechanical vibration of the IPMC was modelled trough the Kirchhoff–Love plate theory, while its interaction with the encompassing fluid was described by using the linearised solution of the Navier–Stokes equations and the concept of hydrodynamic function. The corresponding chemo-electric response was described by using the Poisson Nernst–Planck model, widely used to

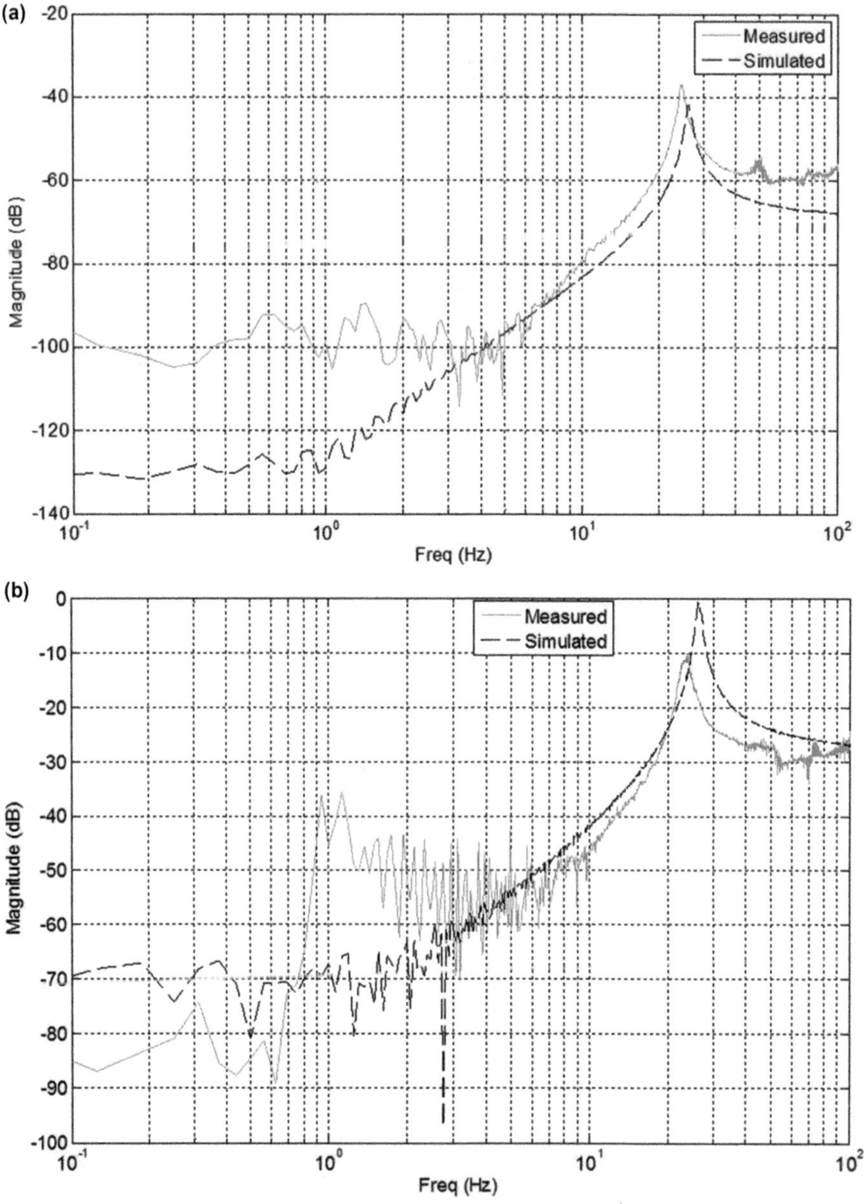

Figure 17.31 Experimental and simulated short-circuit current (a) and open-circuit voltage (b) for the IPMC harvester.
Reprinted with permission from J. Brufau-Penella, M. Puig-Vidal, P. Giannone S. Graziani and S. Strazzeri , Characterization of the harvesting capabilities of an ionic polymer metal composite device, *IOP Smart Mater. Struct.*, 2008, **17**(1), 015009.

determine white-box models of IPMCs. Also in this case, a developed power in the order of nanowatts was predicted.

The same research group has addressed a number of different configurations for scavenging energy from a fluid by using IPMCs. Giacomello and Porfiri[102] investigated the possibility of harvesting power from the flutter instability of a highly compliant system caused by fluid flow. The system experiences flutter motion at moderately low flow speeds and an IPMC transducer is mechanically mounted on a flag-shaped system. A schematic view of the proposed system is shown in Figure 17.32.

Finally, the possibility of scavenging energy from the torsional deformation of an IPMC beam was investigated[103] and an experimental setup was realised to confirm the suitability of the proposed approach. More specifically, a patterned IPMC consisting of two sensors divided by a gap of 2 mm was considered, while the total dimensions of the IPMC were 64 mm×35 mm×200 μm. The authors proposed a model that takes into account both of the interactions of the twisted device with the surrounding fluid by using the hydrodynamic function concept, and of the electrical reaction of the sensor to the twisting deformation by using a black-box approach. Power harvesting densities in the order of a few picowatts per millimetre cubed were obtained.

Though the power harvested by using IPMCs reported so far is very small and does not allow for powering any real device, this field of research is quite promising. It can be argued, in fact, that a significant improvement in the collected power could be obtained, for example, by the connection of a huge number of energy scavenging units that could be feasible thanks to the low

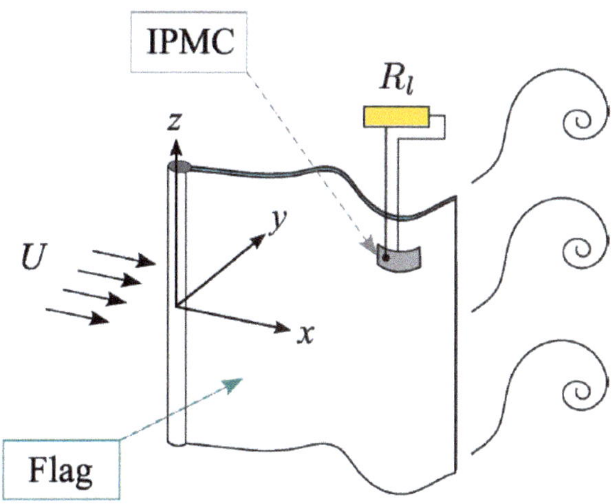

Figure 17.32 A schematic view of the flexible heavy flag and of the IPMC harvesting element.
Reprinted with permission from A. Giacomello and M. Porfiri, Underwater energy harvesting from a heavy flag hosting ionic polymer metal composites, *J. Appl. Phys.*, 2011, **109**(8), 084903.

cost of polymeric devices. Such a solution could also have the beneficial effect of enlarging the usable bandwidth of the scavenging system, if the units are purposefully chosen with unmatched mechanical characteristics. IPMCs collect energy in the very low frequency range (from 1 to 10 Hz), where a lot of environmental mechanical power is available.

17.4 Smart IPMC-based Devices

Smart systems need to adequately react to environmental stimuli, and as such, they need to be capable of both sensing and acting. Applications have been proposed in which such sensing and actuation functionalities both can be fulfilled by using IPMC transducers. The cases in which IPMCs are used in a combined action as sensors and as actuators are not rare, especially in micro-scale or in bio-inspired realisations. In this section, such systems are described as relevant contributions towards the realisation of post-silicon smart systems.

We were attracted to this area from the very beginning of our research activity on IPMCs due to this fascinating possibility of using one technology to realise both sensing and acting functions. One of my first conference contributions on IPMCs[104] was, as a matter of fact, the realisation of a very simple prototype in which an integrated motion sensor–actuator system was proposed. The system consisted of a couple of IPMC strips working respectively as a sensor and as an actuator and was meant to show the capability of IPMC transducers to realise smart devices through the realisation of a controlled system in which both of the working modes of IPMCs were exploited. Though very simple, it can be considered a kind of study, which has been addressed a few times in the literature,[105,106] in which actuator–sensor couples patterned on one IPMC strip were proposed.

We developed a number of smart sensing systems based on the vibrating beam approach. In Brunetto *et al.*,[107] we reconsidered the realisation of an IPMC-based viscometer. Though the working principle is the same as described for the previously proposed version,[84] a coupled actuator–sensor IPMC-based system was used in this second version to measure the rheological properties of still fluids. The model of the vibrating system was used again to estimate the rheological properties of the fluid surrounding the vibrating beam. A schematic view of the viscometer realised in this second case is shown in Figure 17.33.

Both the actuator and the sensor were realised by using Nafion® 117 with platinum electrodes. The sizes of the devices are given in Table 17.1 (adapted from Brunetto *et al.*[107]).

It is worth noting that, in this case, the IPMC sensor output (short-circuit current) needed to be inverted in order to determine the corresponding deflection. Such a correction is shown in Figure 17.34. More specifically, in Figure 17.34(a), the modulus of the system frequency response as sensed by the IPMC sensor is shown, while in Figure 17.34(b), this modulus, after the

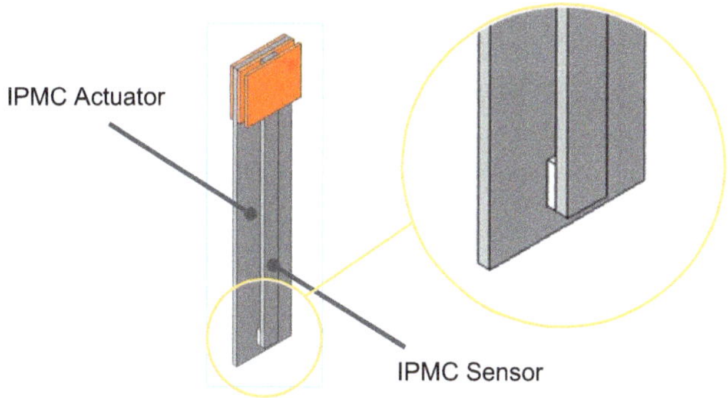

Figure 17.33 A scheme of the IPMC-based smart viscometer.

Table 17.1 Geometrical parameters of the IPMC actuator and sensor used in the IPMC-based viscometer (adapted from Brunetto *et al.*[107]).

	Length (mm)	Width (mm)	Thickness (mm)
Actuator	23	6.0	0.180
Sensor	22	1.5	0.180

correction of the sensor dynamics, is compared with its estimation obtained by a laser sensor, used as a reference.

As a final result, in Figure 17.35, the magnitude of the theoretical system frequency responses and the corresponding experimental data are reported for the case of the vibrating viscometer immersed in deionised water and in a 30% mass sucrose solution.

Also, a comparison between the real values of some rheological fluids and the corresponding estimations, obtained by using the described system, are reported in Table 17.2 (adapted from Brunetto *et al.*[107]).

Biomedical applications have been proposed based on IPMCs. Yoon *et al.*[108] proposed a scanning endoscope that used IPMC actuators to move the distal end, while an optical sensor was proposed for image capturing. We further exploited the use of IPMC actuator–sensor-coupled systems as useful tools for realising smart sensing probes for biomedical applications. We proposed the combined action of a coupled IPMC actuator and IPMC sensor for designing a tactile system.[109] In the system, both the actuator and the sensor are mounted in a cantilever configuration. The actuator brings the sensor into contact with the investigated material that deforms in accordance with its stiffness. The sensor electrical reaction can be therefore used to estimate the material stiffness.

The proposed smart system was intended for possible applications in catheters, laparoscopy, and surgical resection of tumors, in which the knowledge of the investigated tissue's mechanical properties can be a discriminating element. Such a system could eventually be used in assisted navigation, where the tactile sensor can help with recognising the relative position of the probe with respect to surrounding objects.

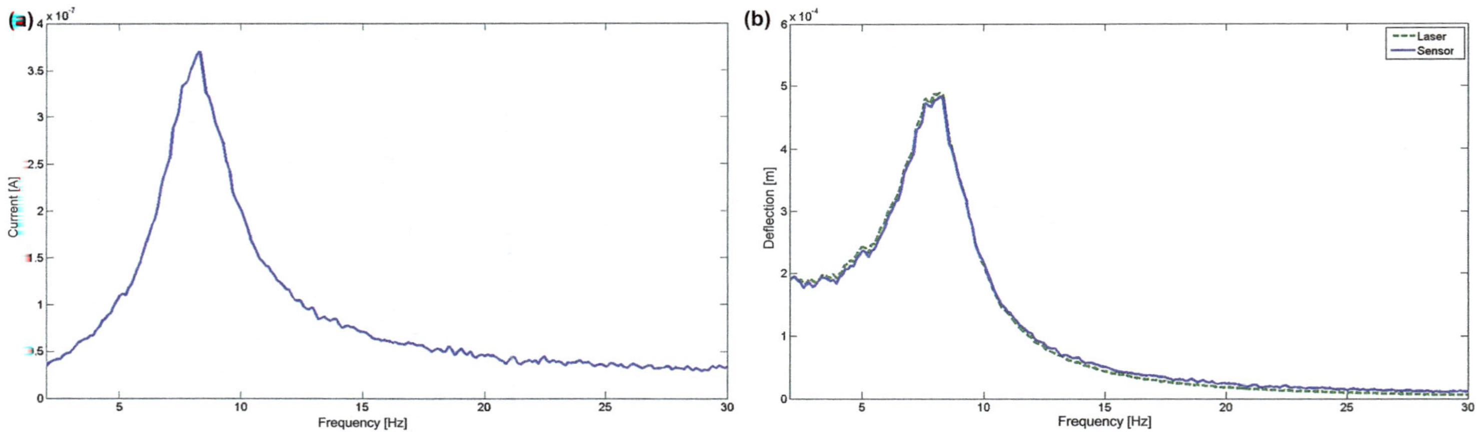

Figure 17.34 Modulus of the vibrating system frequency response (a) and the same modulus after the correction of the IPMC sensor dynamics (b).

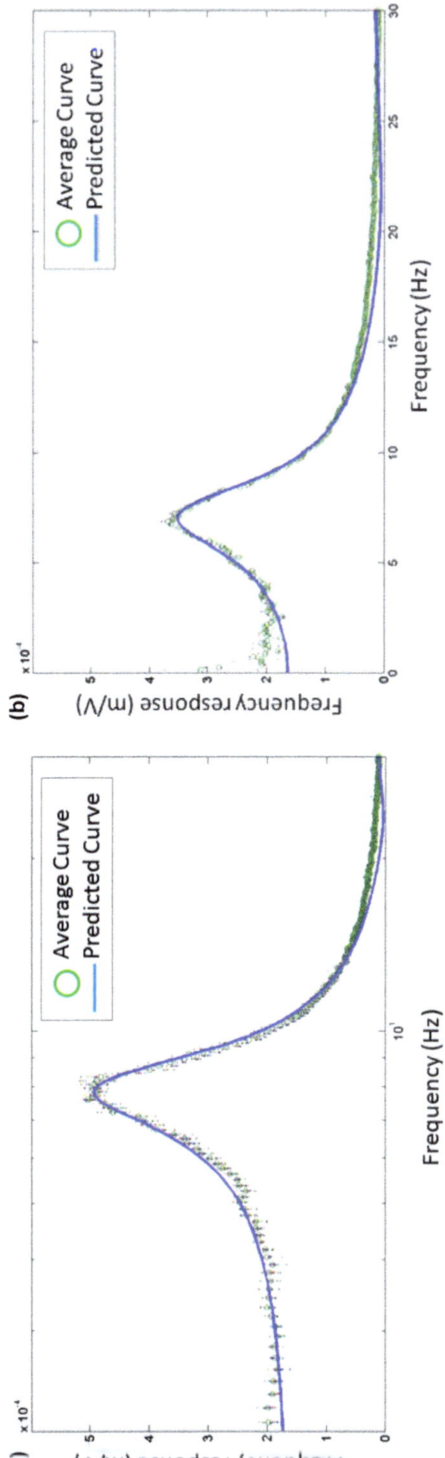

Figure 17.35 Frequency response of the IPMC-based viscometer immersed in deionised water (a) and a 30% sucrose solution (b).

Table 17.2 Experimental values and estimations of sucrose solution densities and viscosities obtained by using the IPMC-based viscometer (adapted from Brunetto *et al.*[107]).

	Density ρ_f (kg m^{-3})		Viscosity η (Pa s)	
	Experimental	Theoretical	Experimental	Theoretical
Deionised water	895.76	998	0.00091	0.001
0.3 sucrose solution	1031.7	1028	0.0022	0.0019
0.6 sucrose solution	1151.4	1134	0.0033	0.0031
Saturated sucrose solution	1260.4	1282	0.0041	0.0044

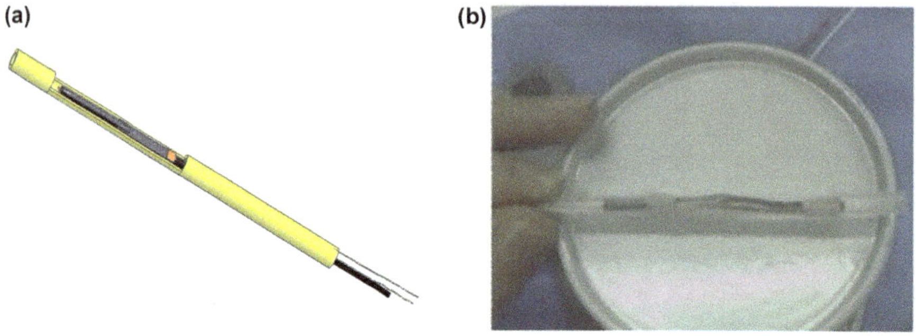

Figure 17.36 Scheme of the biomedical tactile probe (a) and a photograph of the realised prototype (b).

A scheme of the proposed devise is shown in Figure 17.36(a), while a photograph of the realised prototype is shown in Figure 17.36(b).

The sensing capabilities of the realised prototype were tested by using a specific setup and performing a number of experiments. As an example, the discrimination capabilities of the system, investigated by using samples of materials with Young's moduli equal to 100 Pa (sample 1) and 300 Pa (sample 2), are reported in Figure 17.37 (a third, harder sample was also investigated, but the system was not able to produce any output signal, giving evidence that the upper threshold of its investigation range was reached).

A further smart probe based on an IPMC actuator–sensor combination was proposed,[110] in which the transducers were arranged in a different way with respect to the system previously proposed[109] and it was forced to work in resonant conditions.[111] Such a working mode is reported to be capable of improved accuracy with respect to the corresponding structures that investigate the mechanical characteristics of the material relying on the amplitude of the deformation caused by the touching probe.

In the proposed measuring strategy, the changes in the resonance frequencies measured in correspondence with the interaction with different materials make it possible to distinguish the material itself from its mechanical characteristics.

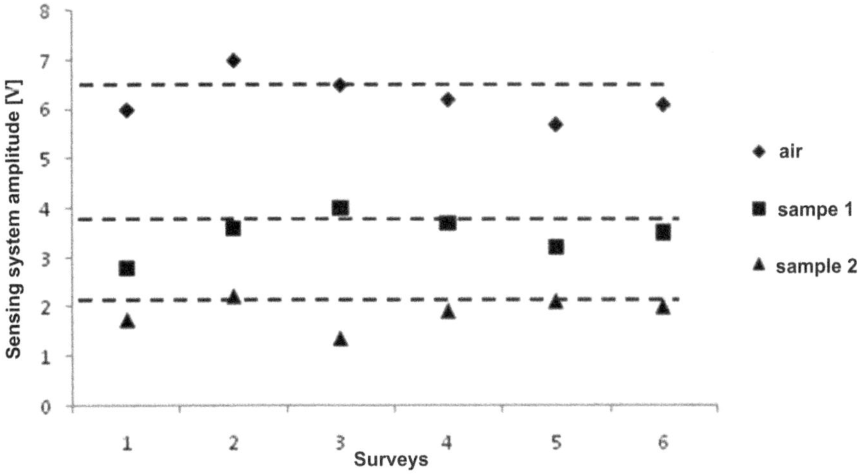

Figure 17.37 Amplitudes of the output of the IPMC sensing system (after signal conditioning) for various materials.

After its modelling, a system was realised and its performances were investigated against gels. More specifically, the CH1077 from Carlo Erba was used. The investigated concentrations were 1%, 3%, 5%, and 7%.

In Figure 17.38, a scheme of the system (Figure 17.38(a)), the experimental setup (Figure 17.38(b)), and the system behaviour for different materials (Figure 17.38(c)) are shown.

Integrated actuator–sensor systems have also been proposed as bio-inspired systems. An interesting smart system in this context is represented by the contribution of Shahinpoor,[112] in which a biomimetic robot Venus flytrap, a carnivorous plant that uses antenna-like sensors to detect the presences of prey and to trigger the rapid motion of a pair of jaw-like lobes that capture the prey, was proposed. Moreover, the author suggested that the acting and sensing mechanisms adopted by the Venus flytrap are similar to our understanding of the actuation/sensing mechanisms that rule the transduction phenomena in IPMCs. In the robot, the sensing elements are realised by IPMC-based bristles, while two IPMC actuators with a common middle electrode realise the robotic flytrap lobes. The hair-like sensor were 3 mm×0.5 mm×0.1 mm, while the lobes were 80 mm× 40 mm×0.2 mm. The prototype was capable of closing its jaw-like lobes in only 0.3 s.

Photographs of a Venus flytrap and of the corresponding robot are shown in Figure 17.39(a) and (b), respectively.

A similar system was proposed by Shi *et al.*,[113] who again suggested a bio-inspired flytrap-like robot with three lobes. The robot was further improved by mounting it on an eight legged IPMC-based robot capable of translational

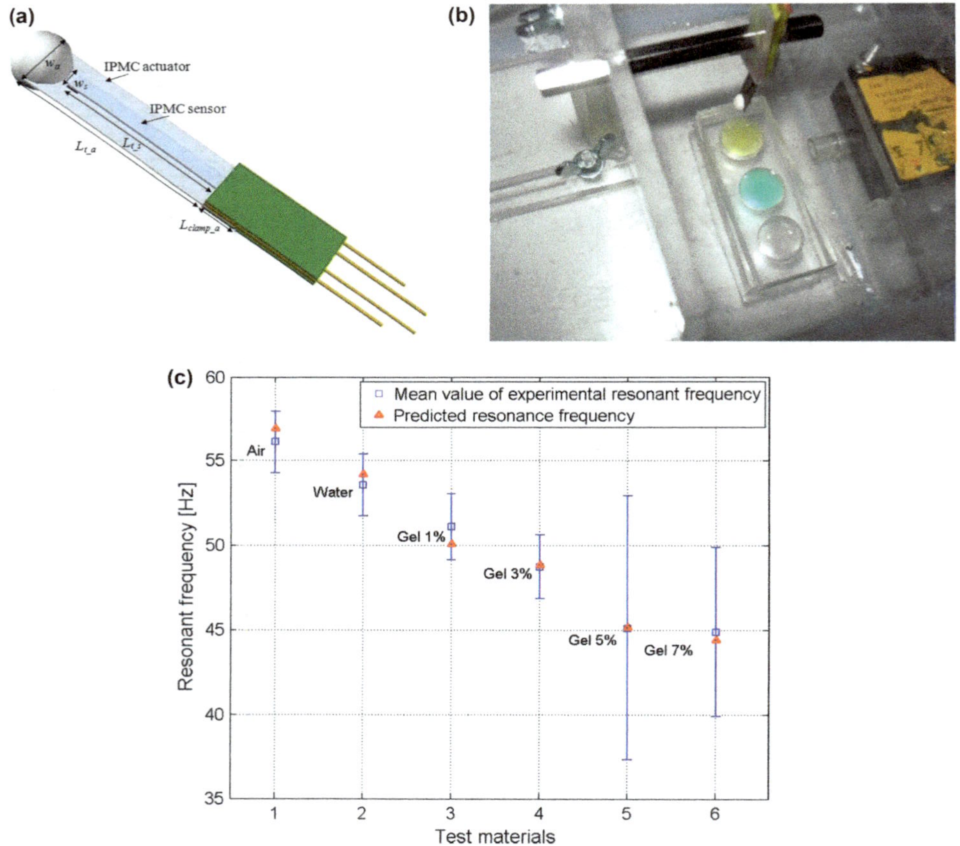

Figure 17.38 Views of the system scheme (a), the experimental setup corresponding prototype (b), and the system behaviour for different materials (c).
© 2010 IEEE. Reprinted, with permission, from P. Brunetto, L. Fortuna, P. Giannone, S. Graziani and F. Pagano, A resonant vibrating tactile probe for biomedical applications based on IPMC, *IEEE Trans. Inst. Meas.*, 2010, **59**(5), 1453–1462.

Figure 17.39 A Venus flytrap (a) and a view of the corresponding robot (b). Reprinted with permission from M. Shahinpoor , Biomimetic robotic Venus flytrap (*Dionaea muscipula Ellis*) made with ionic polymer metal composites, *IOP Bioinsp. Biomim.*, 2011, **6**(4), 046004.

and rotational motion. Also, the bristles were realised by using a proximity sensor. In such a way, a robot with functionalities similar to those of the grasping mobile robot proposed by Guo *et al.*[49] were improved, realising a kind of a "centaur", with sensing capabilities. The authors proposed such a solution to obtain a better working robot, yet this was less close to the original inspiring biological system.

The realisation of smart IPMC-based systems capable of complex deformation or even of multi-functionality on the same substrate is of interest since it allows for realising light and small structures, and this is mandatory for the envisaged applications of IPMCs in fields such as bio-inspired robotics or biomedical devices. A technological solution to this goal can be fulfilled by using a kind of patterning,[114] and as a matter of fact, this has been proposed for realising both the sensing and acting characteristics on the same membrane.[106] Unfortunately, the attempts to use IPMC-based actuators and sensors in such a way were not very successful because of the interferences between the acting and sensing elements.

Other solutions are possible for realising IPMC-based smart systems. Electrodes can be a useful source of information: the sensing element is realised by using the metallic electrodes that deform and change their resistance as a consequence of actuator motion. It is worth considering that the sensor is now a resistive element and that this requires different conditioning circuitry. The dependence of the electrode resistance on the IPMC actuator or sensor deformation was investigated and modelled, with the aim of providing a more accurate description of the transduction phenomenon.[115] More specifically, the authors proposed an improved version of the Kanno *et al.* model[116] in which varying resistances were introduced. The values of such resistances depend on the IPMC curvature. The same authors proposed the exploitation of such a change to obtain a self-sensing actuator with the aim of realising feedback-controlled IPMC actuators, without using bulky sensors.[117] In the referenced paper, a description of how the self-sensing actuator works is given, though no mathematical model is proposed for the sensing phenomenon.

The patterning approach was exploited again by Kruusamäe *et al.*[118] to realise the IPMC actuator and the resistive sensing element. More specifically, it was proposed to measure the changes in conductivity of resistive elements patterned on the surfaces of an IPMC in order to track the deformation of the actuator–sensor system. This patterning was realised with the same geometry on both sides of the IPMC strip, allowing for a differential measurement scheme and, hence, filtering out the effects of common mode noise.

The pattern proposed for the IPMC surface is shown in Figure 17.40. It consists of three separate parts: a section for the actuator (in the inner part); a C-shaped sensor (in the outer part); and a shielding electrode to reduce the cross-talk (between the two previously mentioned elements). The resistance of the two sensors is in correlation with their bending curvature.

The shielding electrodes were revealed to be unable to remove the cross-talk completely. The sensing elements realised on both faces of the IPMC strip were, therefore, connected in a Wheatstone bridge configuration. It was assumed that the cross-talk was a common mode noise and that the bridge configuration could remove such undesired components. Eventually, the bridge configuration increased the sensitivity to the variation of resistance, since the resistive sensors undergo opposite variations during IPMC system

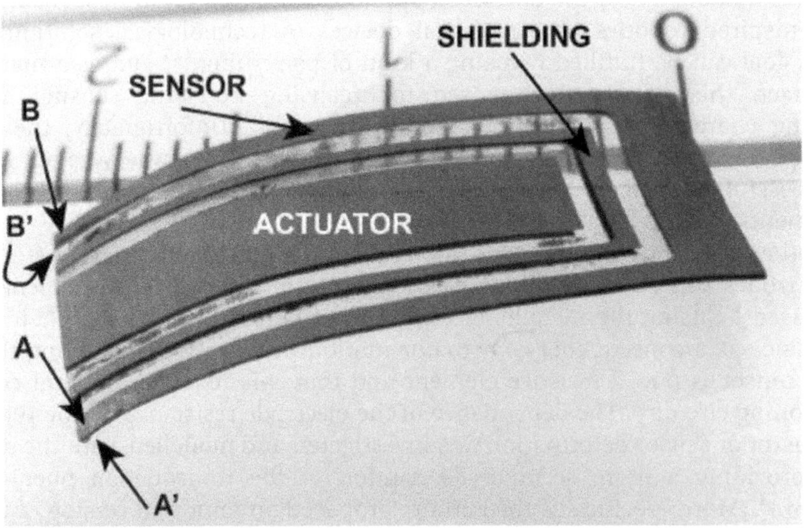

Figure 17.40 The patterned IPMC smart actuator, with the resistive sensor. Reprinted with permission from K. Kruusamäe, P. Brunetto, S. Graziani, A. Punning, G. Di Pasquale and A. Aabloo, Self-sensing ionic polymer-metal composite actuating device with patterned surface electrodes, *Polym. Int.*, 2010, 59(3), [illegible].

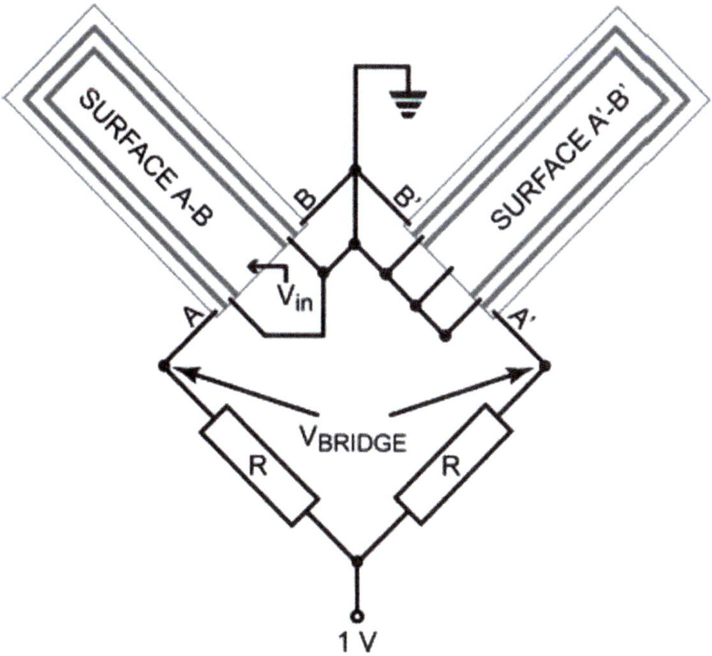

Figure 17.41 A scheme of the IPMC-based self-sensing actuator along with the conditioning circuitry.
Reprinted with permission from K. Kruusamäe, P. Brunetto, S. Graziani, A. Punning, G. Di Pasquale and A. Aabloo, Self-sensing ionic polymer-metal composite actuating device with patterned surface electrodes, *Polym. Int.*, 2010, **59**(3), 300–304.

bending. A scheme of the proposed system and of the conditioning circuitry is shown in Figure 17.41.

The system was experimentally investigated and the results showed a good correlation between the IPMC actuator motion and the sensory system output. An example of the sensing signal produced by the Wheatstone bridge, compared with IPMC deflection recorded by using a laser sensor, is shown in Figure 17.42.

The described structure was further modelled[119] after a further patterning technique was adopted based on laser ablation.

In Leang *et al.*,[120] the motion of IPMC actuators was sensed again by using resistive elements. Those elements were not realised by patterning the IPMC electrodes. Strain gages were used instead. Finally, Chen *et al.*[121] proposed an IPMC actuator equipped with a polyvinylidene fluoride (PVDF) based sensing element to realise a smart acting/sensing system.

The applications described in this section are just a part of what can be found in the literature: I have tried to give an ordered description of the research activity at the University of Catania on IPMCs—of course, also in collaboration with researchers working in other universities—and of the

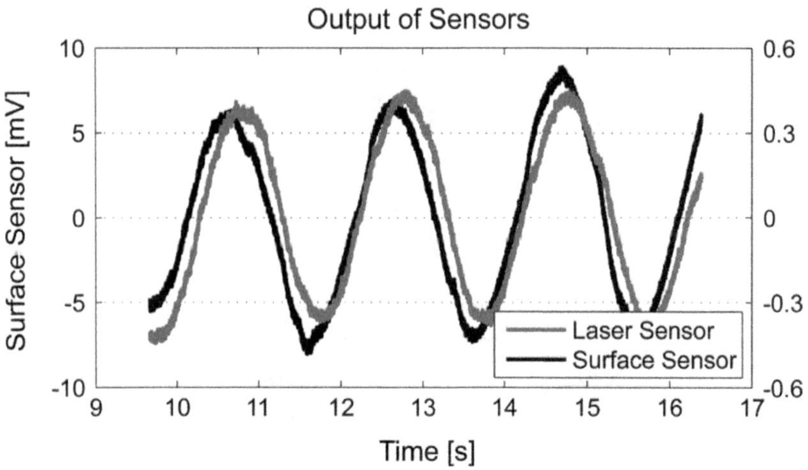

Figure 17.42 An example of the sensing signal produced by the Whetstone bridge compared with IPMC deflection recorded by using a laser sensor. Reprinted with permission from K. Kruusamäe, P. Brunetto, S. Graziani, A. Punning, G. Di Pasquale and A. Aabloo, Self-sensing ionic polymer-metal composite actuating device with patterned surface electrodes, *Polym. Int.*, 2010, **59**(3), 300–304.

influences I experienced from what became available over time in the scientific community. Even if this was not intended to be an exhaustive report of the applications proposed, the large number of referenced contributions and the even larger number of papers that can be found in the literature give an idea of the huge research activity and of the vivid interest in IPMC-based applications.

Among the application fields investigated for IPMCs, continuous interest has been devoted to robotic applications, and in particular, IPMC peculiarities have been exploited to realise bio-inspired systems with locomotion based on tails or fins (for underwater propulsion) and legs. Nonetheless, biomedical applications have been constantly proposed because of the tantalising possibility of realising systems that require low voltage levels, are small, light, and flexible, and can work in wet environments. Last, but no less important, a constant evolution towards systems with increased complexity that grow closer and closer to the realisation of autonomous, multifunctional smart systems can be observed.

Most of described applications are laboratory-scale prototypes and have been developed as a proof of concept more than as systems ready for the market. Of course, IPMC-based applications suffer for the drawbacks that still affect IPMC production and modelling phases.

Real-life applications need more repeatable processes and these can be obtained only if tighter control of the chemo-physical properties of obtained IPMCs becomes available. Efforts are needed to better define the relationships between production steps and their consequences on produced transducers.

Though sophisticated control strategies can greatly alleviate the effects of unmodelled uncertainty and time variance, actions are needed to improve material stability, mainly due to solvent loss.

Finally, in the experiments reported so far, some fatigue effects have been observed for IPMC-based tranducers. Nevertheless, organised experimental campaigns are required before the useful lifespan can be defined for IPMC transducers.

References

1. L. Sun, W. M. Huang, Z. Ding, Y. Zhao, C. C. Wang, H. Purnawali and C. Tang, *Materials Design*, 2012, **33**, 577.
2. I. Chopra, *AIAA J.*, 2002, **40**, 2145.
3. P. J. Costa Branco and J. A. Dente, *Smart Mater. Struct.*, 2006, **15**, 378.
4. P. D. Mitchelson, E. M. Yeatman, G. K. Rao, A. S. Holmes and T. C. Green, *Proc. IEEE*, 2008, **96**, 1457.
5. M. A. Cohen Stuart, W. T. S. Huck, J. Genzer, M. Müller, C. Ober, M. Stamm, G. B. Sukhorukov, I. Szleifer, V. V. Tsukruk, M. Urban, F. Winnik, S. Zauscher, I. Luzinov and S. Minko, *Nat. Mater.*, 2010, **9**, 101.
6. *Electroactive Polymer (EAP) Actuators as Artificial Muscles – Reality, Potential and Challenges*, ed. Y. Bar-Cohen, SPIE Press, Bellingham, WA, USA 2nd edn, 2004.
7. M. Shahinpoor and K. J. Kim, *Smart Mater. Struct.*, 2001, **10**, 819.
8. B. Andò and S. Graziani, *IEEE Instr. Meas. Mag.*, 2009, **12**, 30.
9. R. Tiwari, K. J. Kim and S. M. Kim, *Smart Struct. Syst.*, 2008, **4**, 549.
10. J. Brufau-Penella, M. Puig-Vidal, P. Giannone Graziani and S. Strazzeri, *Smart Mater. Struct.*, 2008, **17**, 015009.
11. C. Jean-Mistral, S. Basrour and J.-J. Chaillout, *Smart Mater. Struct.*, 2010, **19**, 085012.
12. K. M. Farinholt, N. A. Pedrazas, D. M. Schluneker, D. W. Burt and C. R. Farrar, *J. Intell. Mater. Syst. Struct.*, 2009, **20**, 633.
13. M. La Rosa, N. Malagnino, A. Marcellino, D. Nicolosi, L. Occhipinti, F. Porro, G. Sicurella, R. Vecchione, L. Fortuna, M. Frasca, E. Umana, Printed Organic Electronics: From Materials to Circuits, in *Nanoelectronics: Nanowires, Molecular Electronics, and Nanodevices*, ed. K. Iniewski, Mc Graw Hill, New York, NY, 2010.
14. S. R. Forrest, *Nature*, 2004, **428**, 911.
15. C. Bonomo, M. Bottino, P. Brunetto, G. Di Pasquale, L. Fortuna, S. Graziani and A. Pollicino, *Smart Mater. Struct.*, 2010, **19**, 055002.
16. V. De Luca, P. Di Giamberardino, G. Di Pasquale, S. Graziani, A. Pollicino, E. Umana and M. G. Xibilia, *J. Polym. Sci., Part B: Polym. Phys.*, 2013, **51**, 699.
17. M. Shahinpoor and K. J. Kim, *Smart Mater. Struct.*, 2005, **14**, 197.
18. B. Bhandari, G.-Y. Lee and S.-H. Ahn, *Int. J. Precis. Eng. Man.*, 2012, **13**, 141.

19. D. Pugal, K. Jung, A. Aabloo and K. J. Kim, *Polym. Int.*, 2010, **59**, 279.
20. R. Lumia and M. Shahinpoor , *Proc. of 4th World Congress on Biomim., Art. Mus. and Nano-Bio, J. Phys.: Conf. Ser.*, IOP, Bristol, UK, 2008, **127**, 012002.
21. P. Arena, C. Bonomo, L. Fortuna, M. Frasca and S. Graziani, *Systems, Man, Cybernetics, Part B: Cybernetics, IEEE Transactions on*, 2006, **36**, 1044.
22. S.-W. Yeom and I.-K. Oh, *Smart Mater. Struct.*, 2009, **18**, 085002.
23. Z. Chen, S. Shatara and X. B. Tan, *IEEE/ASME Transactions on Mechatronics*, 2010, **15**, 448.
24. Q. Shen, T. M. Wang, L. Wen and J. Liang, *Int. J. Adv. Robot. Syst*, 2013, **10**, 350.
25. M. Shahinpoor, Y. Bar-Cohen, J. O. Simpson and J. Smith, *Smart Mater. Struct.*, 1998, 7, R15.
26. K. M. Newbury and D. J. Leo, *J. Intell. Mater. Syst. Struct.*, 2003, **14**, 333.
27. R. B. Moore, K. M. Cable and T. L. Croley, *J. Membr. Sci.*, 1992, 75, 7.
28. G. Gebel, P. Aldebert and M. Pineri, *Macromolecules*, 1987, **20**, 1425.
29. K. J. Kim, M. Shahinpoor, *Proc. of SPIE-EAPAD 2001*, ed. Y. Bar-Cohen, SPIE Press, Bellingham, WA, USA, 2001, 223–232.
30. K. J. Kim and M. Shahinpoor, *Polymer*, 2002, **43**, 797.
31. S. J. Lee, M. J. Han, S. J. Kim, J. Y. Jho, H. Y. Lee and Y. H. Kim, *Smart Mater. Struct.*, 2006, **15**, 1217.
32. J. J. Pak, S. E. Cha, H. J. Ahn, S. K. Lee, *Proc. of the 32nd ISR (Int. Sym. on Rob.)*, Seoul, Ed. Korea Institute of Science and Technology, Korea, 2001, pp. 1.
33. K. J. Kim and M. Shahinpoor, *Smart Mater. Struct.*, 2003, **12**, 65.
34. K. J. Kim, M. Shahinpoor, Proc. of SPIE-EAPAD 2001, ed. Y. Bar-Cohen, SPIE Press, Bellingham, WA, USA, 2001, 189–198.
35. S. H. Wu and D. H. Chen, *J. Colloid Interface Sci.*, 2004, **273**, 165.
36. X. Song, S. Sun, W. Zhang and Z. Yin, *J. Colloid Interface Sci.*, 2004, **273**, 463.
37. K. D. Kim, D. N. Han and H. T. Kim, *Chem. Eng. J.*, 2004, **104**, 55.
38. A. J. McDaid, K. C. Aw, K. Patel, S. Q. Xie and E. Haemmerle, *Int. J. Smart Nano Mater.*, 2010, **1**, 261.
39. J. Ryu, Y. Jeong, Y. Tak, B. Kim, B. Kim, J. -. O. Park, *Proc. of the 2002 Int. Sym. Micromech. Hum. Sci.*, IEEE, New York, NY, 2002, 85–91.
40. B. Kim, J. Ryu, Y. Jeong, Y. Tak, B. Kim, J.-O. Park, *Proc. of 2003 IEEE Int. Conf. on Rob. & Aut.*, Taipei, Taiwan, IEEE, New York, NY, 2003, 2490–2495.
41. Y. Kaneda, N. Kamamichi, M. Yamakita, K. Asaka, Z. W. Luo, *Proc. of 2003 SICE Annual Conf.*, IEEE, New York, NY, 2003, 1650–1655.
42. N. Kamamichi, Y. Kaneda, M. Yamakita, K. Asaka, Z. W. Luo, *Proc. of 2003 SICE Annual Conf.*, IEEE, New York, NY, 2003, 123–128.
43. S. Lee, K. J. Kim and I.-S. Park, *Smart Mater. Struct.*, 2007, **16**, 583.
44. S. J. Kim, D. Pugal, J. Wong, K. J. Kim and W. Yim, *Robot. Auton. Syst.*, 2014, **62**, 53.

45. G.-Y. Lee, J.-O. Choi, M. Kim and S.-H. Ahn, *Smart Mater. Struct.*, 2011, **20**, 105026.
46. M. Yamakita, N. Kamamichi, T. Kozuki, K. Asaka, Z. W. Luo, *Proc. of 2005 IEEE/ASME Int. Conf. on Adv. Intel. Mech.*, IEEE, New York, NY, 2005, 48–53.
47. N. Tomita, K. Takagi, K. Asaka, *Proc. of 2011 SICE annual conf.*, IEEE, New York, NY, 2011, 1687–1690.
48. S. Guo, L. Shi, K. Asaka, L. Li, *Proc. of the 2009 IEEE Inter. Conf. Mech. Aut.*, IEEE, New York, NY, 2009, 3330–3335.
49. S. Guo, L. Shi, N. Xiao and K. Asaka, *Robot. Auton. Syst.*, 2012, **60**, 1472.
50. M. Vahabi, E. Mehdeizadeh, M. Kabganian and F. Barazadeh, *Proc. of the Institution of Mechanical Engineers, Part I: Journal of Systems and Control Engineering*, 2011, **225**, 63.
51. C. Wilbur, W. Vorus, Y. Cao, S. Currie, *Neurotechnology for Biomimetic Robots*, ed. J. Ayers, J. L. Davis and A. Rudolph, MIT Press, Cambridge, MA, 2002, pp. 285–296.
52. R. M. Alexander, *Principles of Animal Locomotion*, Princeton University Press, Princeton, NJ, 2006.
53. E. Niebur and P. Erdös, *Biophys. J.*, 1991, **60**, 1132.
54. N. N. Pak, S. Scapellato, G. La Spina, G. Pernorio, A. Menciassi, P. Dario, *Proc. of first IEEE/RAS-EMBS Int. Conf. Biom. Rob. Biomech.*, IEEE, New York, NY, 2006, 666–671.
55. B. K. Nguyen, J. H. Boyle, A. Dehghani-Sanij, N. Cohen, *Proc. of 2009 IEEE Int. Conf. Rob. Biomim.*, IEEE, New York, NY, 2009, 765–770.
56. J. W. Paquette and K. J. Kim, *IEEE J. Oceanic Eng.*, 2004, **29**, 729.
57. S. D. Peterson, M. Porfiri and A. Rovaldi, *IEEE/ASME Transactions on Mechatronics*, 2009, **14**, 474.
58. K. Abdelnour, E. Mancia, S. D. Peterson and M. Porfiri, *Smart Mater. Struct.*, 2009, **18**, 085006.
59. J. E. Sader, *J. Appl. Phys.*, 1998, **84**, 64.
60. M. Aureli, V. Kopman and M. Porfiri, *IEEE/ASME Transactions on Mechatronics*, 2010, **15**, 603.
61. Z. Chen, S. Shatara and X. Tan, *IEEE/ASME Transactions on Mechatronics*, 2010, **15**, 448.
62. S. Guo, Y. Ge, L. Li, S. Liu, *Proc. of 2006 IEEE Int. Conf. Mech. Aut.*, IEEE, New York, NY, 2006, 249–254.
63. J. S. Lee, S. Gutta, W. Yim, *Proc. of 2007 IEEE/RJS Int. Conf. Int. Rob. Sys.*, IEEE, New York, NY, 2007, 2132–2137.
64. X. Ye, Y. Su, S. Guo, *Proc. of 2007 IEEE Int. Conf. Rob. Biomim.*, IEEE, New York, NY, 2007, 262–267.
65. X. Ye, Y. Su, S. Guo, L. Wang, *Proc. of 2008 IEEE/ASME Int. Conf. Adv. Int. Mech.*, IEEE, New York, NY, 2008, 25–30.
66. J. Jung, B. Kim, Y. Tak, J.-O. Park, *Proc. of 2003 IEEE/RSJ Intl. Conf. Int. Rob. Sys.*, IEEE, New York, NY, 2003, 2133–2138.
67. B. Kim, D.-H. Kim, J. Jung and J.-O. Park, *Smart Mater. Struct.*, 2005, **14**, 1579.

68. M. Yamakita, N. Kamamichi, T. Kozuki, K. Asaka, S.-W. Luo, *Proc. of 2005 IEEE/RJS Int. Conf. Intell. Robots Syst. (IROS)*, IEEE, New York, NY, 2005, 2035–2040.
69. S. Guo, L. Shi, X. Ye, L. Li, *Proc. of the 2007 IEEE Int. Conf. Mech. Aut.*, IEEE, New York, NY, 2007, 509–514.
70. S.-W. Yeom and I.-K. Oh, *Smart Mater. Struct.*, 2009, **18**, 085002.
71. A. Punning, M. Anton, M. Kruusmaa, A. Aabloo, *Proc. of Int. IEEE Conf. Mech. Rob.*, IEEE, New York, NY, 2004, 241–245.
72. K. Takagi, M. Yamamura, Z.-W. Luo, M. Onishi, S. Hirano, K. Asaka, Y. Hayawaka, *Proc. of IEEE/JRS Int. Conf. Intell. Rob. Sys.*, IEEE, New York, NY, 2006, 1861–1866.
73. Z. Chen, T. I. Um and H. Bart-Smith, *Int. J. Smart Nano Mater*, 2012, **3**, 296.
74. J. S. Lee, W. Yim, C. Bae and K. J. Kim, *Int. J. Smart Nano Mater*, 2012, **3**, 244.
75. K. Abdelnour, A. Stinchcombe, M. Porfiri, J. Zhang and S. Childress, *IEEE/ASME Transactions on Mechatronics*, 2012, **17**, 924.
76. K. Sadeghipour, R. Salomon and S. Neogi, *Smart Mater. Struct.*, 1992, **1**, 172.
77. K. M. Newbury *Characterization, Modeling, and Control of Ionic Polymer Transducers*, PhD Dissertation, Ed. Virginia Tech, Blacksburg, Virginia, 2002, edt-09182002-081047.
78. M. Shahinpoor, Y. Bar-Cohen, T. Xue, J. O. Simpson and J. Smith, *Proc. of SPIE-EAPAD 1998*, ed. Y. Bar-Cohen, SPIE Press, Bellingham, WA, USA 1998, 3324–3327.
79. B. K. Henderson, S. Lane, M. Shahinpoor, K. Kim, D. Leo, *AIAA Space 2001 – Conference and Exposition*, AIAA, Reston, VA, paper 2001–4600.
80. A. P. Gerratt, M. Tellers, S. Bergbreiter, *Proc. of MEMS2011*, IEEE, New York, NY, 2011, 332–335.
81. C. Bonomo, L. Fortuna, P. Giannone, S. Graziani and S. Strazzeri, *Smart Mater. Struct.*, 2008, **17**, 015014.
82. O. Brand, *Proc. of IEEE Sensors 2005*, IEEE, New York, NY, 129–132.
83. N. Sepulveda, D. Aslam and J. P. Sullivan, *Diamond Relat. Mater.*, 2006, **15**, 398.
84. P. Brunetto, S. Graziani, S. Strazzeri, M. G. Xibilia, *Proc. of 2009 IFAC Intel. Cont. Sys. Sign. Proc.*, Elsevier, Oxford, UK, 2009, pp. 153–157.
85. A. Maali, C. Hurth, R. Boisgard, C. Jai, T. Cohen-Bouhacina and J.-P. Aimè, *J. Appl. Phys.*, 2005, **97**, 074907.
86. X. Chen, G. Zhu, X. Yang, D. L. S. Hung and X. Tan, *IEEE/ASME Transactions on. Mechatronics*, 2013, **18**, 932.
87. A. T. Abdulsadda, X. Tan, *Proc. of 2011 IEEE Int. Conf. Rob. Aut.*, IEEE, New York, NY, 2011, 2719–2724.
88. A. T. Abdulsadda and X. Tan, *Int. J. Smart Nano Mater*, 2012, **3**, 226.
89. A. T. Abdulsadda and X. Tan, *Smart Mater. Struct.*, 2013, **22**, 045010.

90. L. Ferrara, S. Shahinpoor, K. J. Kim, B. Schreyer, A. Keshavarzi, E. Benzel, J. Lantz, Proc. of SPIE-EAPAD 1999, ed. Y. Bar-Cohen, SPIE Press, Bellingham, WA, USA, 1999, pp. 394–401.
91. A. Keshavarzi, M. Shahinpoor, K. J. Kim, J. Lantz, *Proc. of SPIE-EAPAD 1999*, ed. Y. Bar-Cohen, SPIE Press, Bellingham, WA, USA, 1999, 369–376.
92. B. Andò, C. Bonomo, L. Fortuna, P. Giannone, S. Graziani, L. Sparti and S. Strazzeri, *Sens. Actuators, A*, 2008, **144**, 242.
93. *Magnetic Fluids and Applications Handbook*, ed. B. Berkovski, V. Bashtovoy, Begell House, New York, NY, 1996.
94. H. A. Sodano, D. J. Inman and G. Park, *Shock Vib. Dig*, 2004, **36**, 197.
95. R. Tiwari and K. J. Kim, *Smart Mater. Struct.*, 2013, **22**, 015017.
96. B. R. Martin, *Energy harvesting applications of ionic polymers*, Mast. Of Sci. Th., Virginia Polytechnic Inst. and State Univ., 2005.
97. K. M. Newbury and D. J. Leo, *J. Intell. Mater. Syst. Struct*, 2003, **14**, 343.
98. C. Bonomo, L. Fortuna, P. Giannone and S. Graziani, *Smart Mater. Struct.*, 2006, **15**, 749.
99. P. Brunetto, L. Fortuna, P. Giannone, S. Graziani and S. Strazzeri, *IEEE Trans. Instrum. Meas.*, 2011, **60**, 2951.
100. J. Brufau-Penella, *Smart Materials for Microrobotics. Motion Control and Power Harvesting*, PhD Thesis in Enginyeria I technologies Electrò-niques, Univesitat de Barcelona, Barcellona, Spain, 2008.
101. M. Aureli, C. Prince, M. Porfiri and S. D. Peterson, *Smart Mater. Struct.*, 2010, **19**, 015003.
102. A. Giacomello and M. Porfiri, *J. Appl. Phys.*, 2011, **109**, 084903.
103. Y. S. Cha, L. Shen and M. Porfiri, *Smart Mater. Struct.*, 2013, **22**, 055027.
104. C. Bonomo, L. Fortuna, P. Giannone, S. Graziani, *Proc. of IEEE Sensors – 3rd Conference on Sensors*, IEEE, New York, NY, 2004, 489–492.
105. M. Yamakita, A. Sera, N. Kamamichi, K. Asaka, Z.-W. Luo, *Proc. of the 2006 IEEE Int. Conf. Rob. Aut.*, 2006, IEEE, New York, NY, 1834–1839.
106. H. Nakadoi, A. Sera, M. Yamakita, K. Asaka, Z.-W. Luo, K. Ito, *Proc. of 4th IEEE Int. Conf. Mech.*, IEEE, New York, NY, 2007, 1–6.
107. P. Brunetto, L. Fortuna, P. Giannone, S. Graziani, F. Pagano, *Proc. of 2010 IEEE Inst. Meas. Tech. Conf.*, IEEE, New York, NY, 2010, 585–589.
108. W. J. Yoon, P. G. Reinhall and E. J. Seibel, *Sens. Actuators, A*, 2007, **133**, 506.
109. C. Bonomo, P. Brunetto, L. Fortuna, P. Giannone, S. Graziani and S. Strazzeri, *IEEE Sens. J.*, 2008, **8**, 1486.
110. P. Brunetto, L. Fortuna, P. Giannone, S. Graziani and F. Pagano, *IEEE Trans. Instrum. Meas.*, 2010, **59**, 1453.
111. T. Hemsel, R. Stroop, D. Oliva Uribe and J. Wallaschek, *J. Sound Vib.*, 2007, **308**, 441.
112. M. Shahinpoor, *Bioinspiration Biomimetics*, 2011, **6**, 046004.
113. L. Shi, Y. He, S. Guo, H. Kudo, M. Li, K. Asaka, *Proc. of 2013 ICME Int. Conf. Comp. Med. Eng.*, IEEE, New York, NY, 2013, 375–378.
114. Z. Chen and X. Tan, *Sens. Actuators, A*, 2010, **157**, 246

115. A. Punning, M. Kruusmaa and A. Aabloo, *Sens. Actuators, A*, 2007, **133**, 200.
116. R. Kanno, S. Tadokoro, T. Takamori, M. Hattori, K. Oguro, *Proc. of the IEEE Int. Conf. Rob. Aut.*, IEEE, New York, NY, 1996, 219–225.
117. A. Punning, M. Kruusmaa and A. Aabloo, *Sens. Actuators, A*, 2007, **136**, 656.
118. K. Kruusamäe, P. Brunetto, S. Graziani, A. Punning, G. Di Pasquale and A. Aabloo, *Polym. Int.*, 2010, **59**, 300.
119. K. Kruusamäe, P. Brunetto, A. Punning, S. Graziani, M. Kodu, R. Jaaniso, S. Graziani, L. Fortuna and A. Aabloo, *Smart Mater. Struct.*, 2011, **20**, 124001.
120. K. K. Leang, Y. Shan, S. Song and K. J. Kim, *IEEE/ASME Transactions on Mechatronics*, 2012, **17**, 345.
121. Z. Chen, Y. Shen, N. Xi and X. Tan, *Smart Mater. Struct.*, 2007, **16**, S262.

CHAPTER 18

Micromachined Ionic Polymer Metal Composite Actuators for Biomedical Applications

GUO-HUA FENG

Department of Mechanical Engineering, National Chung Cheng University, 168 University Road, Min-Hsiung, Chia-Yi 621, Taiwan
Email: imeghf@ccu.edu.tw

Although electroceramic materials that provide compact and effective actuation have been applied in many commercial products, such as ultrasonic motors and precision manipulators, electroactive polymers are now appearing as relatively new actuation materials with displacement abilities that are difficult to match with strain-constrained and rigid ceramics. Whereas a shape memory alloy can provide a larger deformation than an electroactive polymer in actuation, its response time is normally much slower than those of electroactive polymers.

Electroactive polymers applied to microdevices have shown great advantages for fabricating sensor and actuator arrays, thus enabling commercialization. For example, Braille displays that could improve blind people's lives have been developed using conducting polymers, dielectric elastomers, polyvinylidene difluoride, and ionic polymer metal composites (IPMC) in recent years.[1] Manipulating or grasping targets of any size is always a demand in our daily lives, though this can be a big challenge when the targets become very tiny, even down to the micron scale, and when the targets are soft. For these applications, electroactive polymer actuators such

RSC Smart Materials No. 18
Ionic Polymer Metal Composites (IPMCs): Smart Multi-Functional Materials and Artificial Muscles, Volume 2
Edited by Mohsen Shahinpoor

Published by the Royal Society of Chemistry, www.rsc.org

as microgrippers, allowing proper manipulation of such small objects, show great importance.

Employing electroactive polymers' biomimetic flexibility and the capability to engineer their properties to meet the goals of micro-robotic applications has been a booming research issue.[2] In addition, micromachining technology has been successfully applied in many currently available commercial products, such as microelectromechanical system inertial sensors and biochips, providing the potential to handle electroactive polymer materials.[3] Among the versatile electroactive polymers, studies have reported that micromachining technology applied to IPMCs represents a practical technology for fabricating micro-sized sensor and actuator microarrays and disposable microbiosensors for various applications.[4–7] For example, IPMC actuator microarrays demonstrate applicability in micromirror-based photonic optical fiber switches. Furthermore, IPMC microgrippers can be operated at a fast speed with relatively low voltages. Flexible IPMC devices also demonstrate major advantages for biomedical applications. In this chapter, the fabrication of IPMC actuators with micromachining techniques is presented and the operation characteristics and analysis of these low-density and large-actuation devices is discussed. Finally, we will present some representative micromachined IPMC actuators that are designed and fabricated for biomedical applications.

18.1 Fabrication of Micromachined IPMC Actuators

Fabricating IPMC actuators using versatile micromachining technology has unique advantages, and this section presents some representative micromachining methods for producing IPMC devices. These methods can be divided into surface machining, bulk micromachining, and micromolding. Surface micromachining indicates either IPMC devices integrated onto the surface of a silicon wafer or a commercial Nafion membrane as a standalone substrate to be processed. Micromachining schemes are applied on patterned electrodes on selected surfaces of a Nafion membrane to shape the designed IPMC actuators. With bulk micromachining, the IPMC actuators are built by modifying Nafion, which normally uses water as its ionic medium. This category also includes IPMC devices fabricated with micromachining techniques by shaping the planar Nafion film to be designed structures. In addition, a promising technique using micromachining to fabricate molds along with liquid Nafion to structure IPMC actuators has been implemented, and versatile actuators have demonstrated potential applications. The fabrication details are covered in the last subsection.

18.1.1 Fabrication by Surface Micromachining

Zhou and Li reported a surface micromachining technique for fabricating IPMC actuators.[8] They used the DuPont Nafion solution SE-5012 made to fabricate the Nafion membrane as the IPMC active layer. Using the silicon wafer as a substrate, the fabrication process started with thermal oxidization of a SiO_2 layer on the silicon substrate, which is a typical process used in complementary metal–oxide-semiconductor fabrication as an insulating layer. The aluminium layer was deposited and patterned as a sacrificial layer, and gold was selected for the metal electrode. Since the adhesion between SiO_2 and gold is relatively poor, chromium was deposited as a seed layer before the gold deposition. Instead of using a chemical method to grow the metal electrodes, the gold layer was coated on the processed silicon wafer with a physical vapour deposition and patterned by a lift-off process.

This deposition was followed by spin coating the Nafion solution. To form each Nafion layer, the coated Nafion solution was heated to the range of 70–150 °C for several minutes to solidify it. The thickness can be controlled by multiple spinning and baking, although residual thermal stress could cause the Nafion film to crack as the number of Nafion layers increases. Based on experimental results, a <1 μm thick Nafion membrane was implemented and tested in Zhou and Li's study. To make an Au/Nafion/Au sandwiched structure for the IPMC transducer, a gold layer was then sputtered and patterned by chemical etching to form the top electrode. The patterned top electrode also served as a mask to protect the underlying Nafion membrane as oxygen plasma etched away the remaining Nafion membrane in the inactive region. At this point, the entire device was finished but anchored on the silicon wafer, so a release process was implemented.

With this method, the releasing procedure plays a critical role in determining the flatness of the fabricated IPMC actuator. The pH value greatly affects the curvature of the device. For removal, the entire silicon wafer is immersed in undiluted phosphoric acid and heated to approximately 40 °C for 2 h to remove the sacrificial aluminium layer and then taken from the acid solution with a DI water rinse. At this point, the IPMC actuator displays a flat cantilever shape. Nevertheless, if the releasing process is executed with phosphoric acid diluted with DI water, curling of the IPMC beam actuator was observed, and diluting the acid solution with more DI water makes the IPMC beam curl even more. This could be due to the different volume of expansion of Nafion membrane in the acid solution with a different pH level or hydrogen ion concentration.

Chen and Tan presented a lithography-based monolithic fabrication process for building a structure resembling a sunfish pectoral fin.[9] In order to achieve a multiple degree-of-freedom deformation for the designed actuator, multiple electrode pairs were made on a Nafion film, and a plasma etching method was applied to make the Nafion thinner at the passive region. The major fabrication procedure is as follows.

Starting with a 225 μm thick Nafion membrane, two transparencies with predetermined openings were used to cover front and back sides of the Nafion membrane as shadow masks, although the scheme for patterning the transparency was not clearly described. If conventional knife cutting is used, creating micro-sized patterns might be difficult. Covering the membrane is followed by e-beam aluminium deposition through the shadow masks onto both Nafion surfaces. The aluminium-coated region is used for protecting the Nafion beneath during the subsequent plasma etching. According to the report,[9] with only oxygen-based chemical plasma etching, the etching rate is less than that of oxygen–argon mixed plasma etching and the etching process over 2 h could damage the Nafion due to temperature rises. A Nafion removal thickness of approximately 200 μm is feasible through this process. The protected aluminium layer was removed with a hydrochloric acid solution.

A method to stiffen the Nafion membrane was presented by placing the Nafion sample into a platinum salt solution for ion exchange, making a subsequent photolithography process possible.[9] In contrast to the normal photoresist being directly spin-coated on a flat silicon wafer, the curved Nafion membrane needs to be sandwiched between two metal frames due to the non-rigid property of Nafion, even with ion exchanging stiffening. By stretching the four edges of the membrane to provide the necessary tensile stress to flatten it, the following spin-coating and patterning process could be executed.

With selective regions protected by the photoresist a second ion exchange process and chemical reduction by adding 5 wt% $NaBH_4$ solution several times was continued to plate the platinum electrodes on the unprotected Nafion membrane. This study also sputtered 0.1 μm thick gold film on the device to increase the electrical conductivity. The surface resistance could be reduced by half, but the real resistance values were not described. Finally, the researchers used hydrogen ions to replace the platinum complex ions by boiling the processed device in a hydrochloric acid solution. This made the inactive region of the IPMC device regain its flexibility because the previously ion exchanged platinum complex ions did not precipitate on the Nafion surface as electrodes. Then, the finished planar IPMC sheet with patterned electrode patterns was cut to designed dimensions and placed into an alkaline solution (*e.g.* sodium or lithium ions) to enhance the actuation.

One important comment for this fabrication process is that even though certain regions of the Nafion membrane were protected by the photoresist the reducing agent may penetrate the boundary of the photoresist and cause some platinum plating. This could be a critical issue when the protected pattern of the photoresist has micro-sized features.

The process of making multiple electrodes on an electroactive polymer such as a commercial Nafion membrane has been broadly studied since motion could be manipulated through shape deformation of the actuator surface. Several schemes have been investigated based on an electrodeless chemical reduction with a masking technique. For example, applications of masking tape, crepe paper, jigs, and silicon rubber can easily found in the literature. However, these masking methods all have drawbacks. For one, the patterned tapes—usually made of polymers—are relatively easily peeled off from the intended protected Nafion membrane, especially when performing a long period of electrode plating processes. In addition, wrinkling could be an issue while utilizing a jig mask to pattern a Nafion membrane.

Jeon *et al.* experimentally compared three different tapes as the masks for fabricating patterned electrodes on the Nafion-117 film:[10] (1) polyimide film tape 5413; (2) plastic film tape 472; and (3) ultra-high molecular weight polyethylene tape 3M 5423. With the first two tapes, adhesion problems and softening due to sensitivity to temperature in chemical reduction caused fundamental difficulties with further processing. Although 3M 5423 tape could supply proper adhesion and possibly preserve the patterned shapes during chemical reduction, two limitations should be noted according to the researchers' report. One is that the platinum particles infuse into the micro-gaps between the Nafion membrane and the boundaries of the masking tape when the chemical reduction time lasts for days. On the other hand, shrinkage and expansion of the tape causes the edges of the patterned tape to be insufficiently sharp, which induces rough edges due to platinum penetration. This could result in unwanted platinum electrodes appearing on the original inactive region.

18.1.2 Fabrication by Bulk Micromachining

Akle *et al.* indicated two major limitations of ionomeric polymers serving as electromechanical actuators: unstable operation in air and solvent breakdown at lower voltages.[11] In order to overcome these issues, they developed an ionic liquid–ionomeric composite with a modified electrode composite and pursued fabricated actuators with a high strain property. Therefore, highly stable ionic liquids at room temperature were first utilized as solvents for Nafion films, and then a metal powder painting technique was applied for building the electrodes. This increased the capacitance of the processed Nafion film so that maximum strain could be produced during operation.

There are three motivations for employing an ionic liquid as a solvent: (1) the stability of the ionic liquid decreases the evaporation, so it is more suitably manipulated in an air environment; (2) in general, ionic liquids have higher electrochemical stability than water, so the range of the driving voltage could be improved even more; and (3) ionic liquids are ionically conductive and this would facilitate ion movement in the Nafion film. To realize this work, a bar of Nafion film is treated with NaOH to acquire sodium ions and then baked to a dry condition under vacuum. Then, the processed Nafion film is immersed into 1-ethyl-3-methyimidazolium trifluoromethanesulfonate ionic liquid for 4.5 h at 150 °C to absorb a proper amount of ionic liquid. The presented metal powder plating method is used to grow the electrodes, and a polymer/metal solution was prepared with the recipe 47% of 5% Nafion solution, 47% glycerol, and 6% RuO_2 metal powder (with a particle size of 3–5 nm). The prepared solution was directly applied on each side of the Nafion film treated by the ionic liquid, and then heated to 130 °C under vacuum to remove the solvent. After the same processing four times, two thin conductive gold foils were placed on the top and bottom metal powder plated electrodes and bound to the electrodes through a hot-pressing procedure. The experimental results show that the surface-to-volume ratio of the metal particulate is critical for increasing the capacitance of the actuator. The increased capacitance produces a larger strain output for the IPMC actuator at low driving frequencies. Furthermore, a larger conductivity of the metal particulate can improve the response of the IPMC actuator at an operation frequency greater than 10 Hz.

In addition, Feng and Zhan presented a micromachining technique to construct a three-dimensional helical IPMC actuator by shaping the planar Nafion film.[12] Figure 18.1 shows the fabrication process flow, which started with the treatment of the Nafion strip by a nickel sulfate hexahydrate solution. Then, the outer electrode region of the spring was specified using a

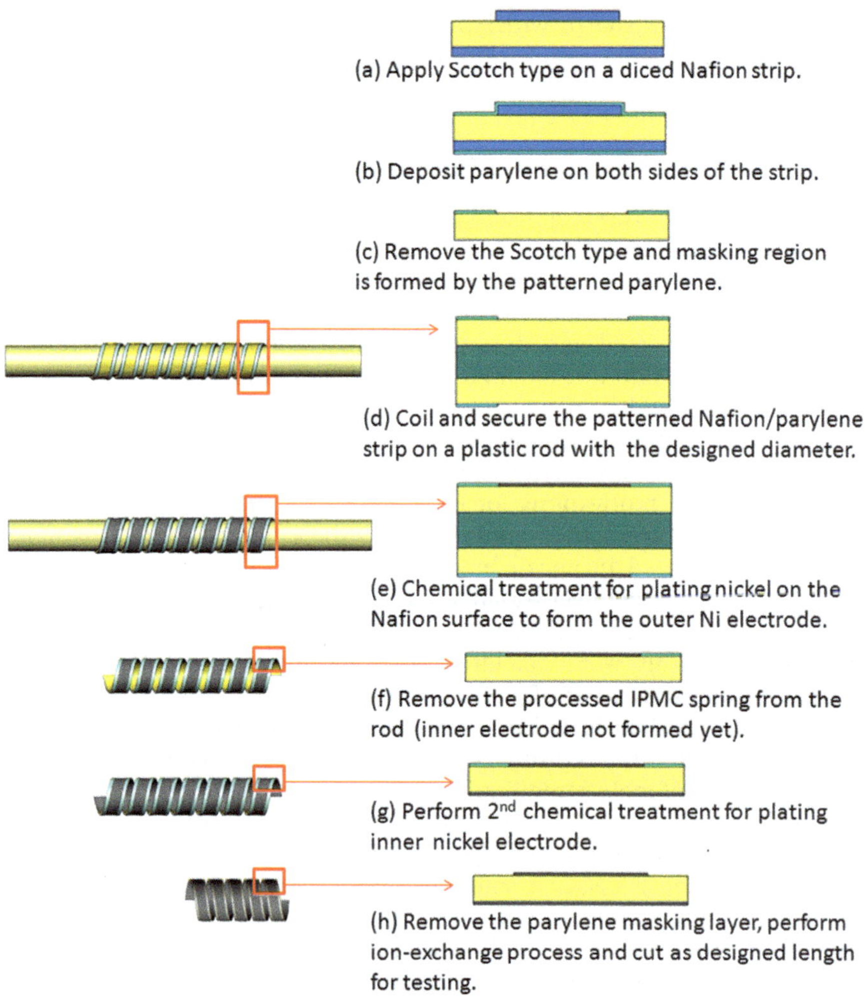

Figure 18.1 Fabrication process flow of the novel room temperature-processed parylene-patterned helical IPMC spring actuator.
Reprinted with permission from ref. 12.

3M Scotch tape glued to the Nafion strip. Meanwhile, the other side of the Nafion strip was fully covered with tape, followed by a 3 μm thick parylene-C coating. The paylene-C film was deposited on the Nafion strip with a vapor deposition polymerization process. The parylene coating layer had two purposes: providing internal stress to shape the Nafion strip to the designed helical shape; and utilization as a mask to selectively delineate the inactive region to prevent the nickel electrode from plating on it.

After parylene deposition, a knife was used to mark parylene along the edges of the tape so the tape could be easily peeled off from the Nafion. The parylene-patterned Nafion strip was then coiled on a rod with the patterned side outwards and its two ends secured to the rod with tape. $LiBH_4$ reduction agent was used to precipitate nickel on the non-parylene protected area. After removing the processed Nafion strip from the rod, the nickel electrode was formed in the central band of the outer surface of the Nafion strip, and no metal was grown on the inner surface of the Nafion strip. In this method, the plated and the coated parylene on the outer surface of the Nafion strip supplied enough stiffness to maintain the helical shape. The second metal plating process was executed by immersing the processed Nafion strip into a reduction agent at room temperature. Finally, the Nafion strip with nickel plated on both surfaces was immersed in a NaOH ion exchange solution after peeling off the protected parylene from the Nafion strip, and the helical IPMC actuator was complete.

18.1.3 Fabrication by Micromolding

Micropumps are a key element of microfluidic devices. Among the many actuation schemes, one is to utilize the IPMC as an oscillating membrane to drive the liquid. Pak *et al.* employed a bulk silicon micromachining method along with a liquid Nafion casting method to create a silicon-based micropump with the following method.[13] Starting with a *p*-type (100) silicon wafer that has thermal oxides grown on both faces, a standard photolithography process was used to pattern two squares aligned on each side of the wafer. This was followed by a wet etching process to remove the unprotected oxide region so that the etching windows could be formed. Then, the penetrations can be created using potassium hydroxide or tetramethylammonium hydroxide (TMAH) solution. At this point, the upper part of the micropump for securing the IPMC membrane actuator is finished. The lower part of the micropump has a similar fabrication process to the upper part up to the removal of unprotected oxide, which is followed by silicon etching until a thin silicon layer is left. This results in shallow cavities forming on both sides of the micromachined silicon wafer. Next, the proper amount of liquid Nafion is injected into the cavities on one side of the wafer to shape the designed thickness. After the Nafion membrane has solidified, the supported silicon layer is etched away by TMAH. The finished lower part of the micropump is aligned and bonded to the upper part of the micropump with epoxy. Finally, the bonded wafers underwent chemical plating to grow the platinum on both sides of the Nafion film to be the IPMC actuator for micropump usage.

Feng and Chen used silicon bulk machining to make a supporting membrane on a silicon wafer and pattern the photoresist coating on the membrane as micromolds for fabricating a Nafion device.[14,15] Figure 18.2 shows the fabrication process flow. After etching the supported membrane, the sidewall of the Nafion device can be secured by the sidewalls of the photoresist. This prevents short-circuits during the chemical electrode plating process, especially when constructing IPMC actuators with micro-sized features. Basically, the low-stress SiN membranes or parylene can be used as the supporting membrane for a whole silicon wafer. JSR photoresist at 80 μm thick is patterned as the micromolds because JSR is a negative-type

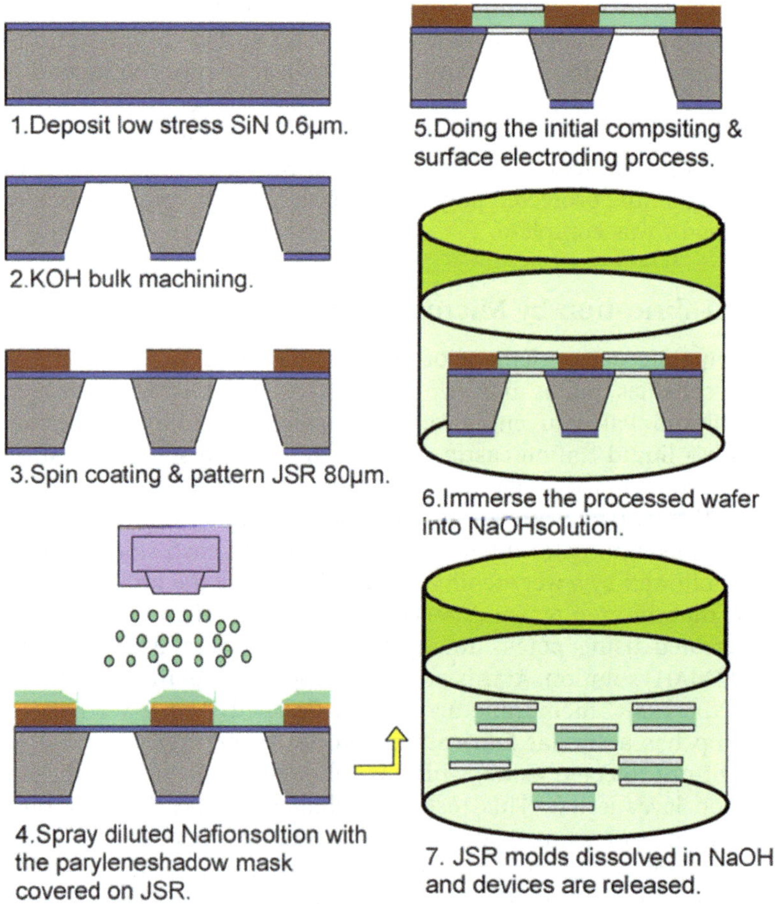

Figure 18.2 Fabricating IPMC actuators with micro-features by micromolding techniques.
Reprinted with permission from ref. 11.

photoresist that is good for making microfeatured structures. In addition, JSR Micro, Inc. photoresist can be dissolved in sodium hydroxide solution, which is important for releasing the IPMC actuator micropump in the final fabrication process. Then, liquid Nafion is sprayed over the whole wafer through a parylene shadow mask and filled into the JSR micromolds. After removing the parylene mask, the processed wafer is heated up so that the solidified Nafion polymer is formed with its sidewalls adhered securely to the sidewalls of the micromold. At this moment, the whole wafer with the shaped IPMC elements is immersed into a small amount of DI water to help prevent the IPMC films from cracking.

Subsequently, reactive ion etching is performed to remove the supporting SiN layer, thereby exposing the top and bottom surfaces of the shaped polymer to air. The initial compositing and surface electroding processes are followed to plate platinum electrodes on both the top and bottom surfaces of the device. Finally, sodium hydroxide solution is employed to remove the JSR micromolds as well as execute ion exchange. The arbitrarily shaped micromachined IPMC actuators with clear contours and sidewalls are then complete.

In addition, Feng and Chu reported an arc-shaped IPMC actuator with curvature change and scanning motion characteristics based on a different micromolding fabrication process as follows.[16] The process began with the patterning of the 2 mm thick negative photoresist SU8 to form predesigned grooves on a glass substrate. The grooves included a structure region for fabricating the IPMC actuator and a refill region to replenish the liquid Nafion. A 3 μm thick parylene layer was then deposited over the grooves, to prevent the succeeding polydimethylsiloxane (PDMS) mold fabrication from sticking to the patterned SU8. After pouring the liquid PDMS (ratio = 10 : 1) and solidification, the PDMS mold could be peeled off easily. Then, the smooth surface of the PDMS mold was anchored to a glass plate. Parylene was coated again onto the entire PDMS/glass unit to release the induced residual stress during the succeeding Nafion solidification. Liquid Nafion was then filled into the PDMS mold. During solvent evaporation, the liquid Nafion had a concave surface in the grooves of the mold, so additional liquid Nafion solution was added to fill these regions of the PDMS mold before the liquid solidified. The surface of the structure became flat after repeating this replenishment process. The solidified Nafion element was removed from the flexible PDMS mold and the refilled region was trimmed to obtain the designed shapes. To produce four electrodes on the Nafion device in a single step, laser machining was applied to pattern a piece of tape as a shadow mask for delineating inactive regions of the device. The patterned tape was glued to the Nafion device and an AZ5214 photoresist was applied to protect the device. The processed Nafion device was then immersed in 1 M $NiSO_4$ solution for 12 h. After rinsing, the processed Nafion device was then immersed in the $LiBH_4$ reduction agent for plating nickel electrodes onto the designated regions. Finally, the ion exchange process was performed by treating the IPMC devices with a 1 M sodium hydroxide solution. This caused the sodium ions to spread inside the IPMC actuator and removed the protected photoresist.

18.2 Analysis and Characterization of Micromachined IPMC Actuators

The actuation mechanism of IPMC actuators has been broadly studied to understand their controllability and to achieve better performance. Shahinpoor *et al.* presented several models for the micro-electro-mechanics of IPMCs as electrically controllable devices.[17] Nemat-Nasser and Li developed a micromechanical model accounting for the coupled ion transport, electric field, and elastic deformation to predict the responses of IPMC.[18] Weiland and Leo used a computational micromechanics model to predict the equilibrium state of a single cluster of solvated ionomeric polymer with the Monte Carlo method.[19] Akle and Leo used an electromechanical coupling model, which relates strain to charge, to study IPMC actuation through the thickness of its membrane.[20] Porfiri employed the Poisson–Nernst–Planck (P–N–P) equations to model the time evolution of the electric potential and the concentrations of mobile counterions of IPMCs.[21] Chen and Tan presented a dynamic model by solving the physics-governing partial differential equation analytically in the Laplace domain.[22]

Further understanding of the dynamic behavior and electromechanical characteristics of the IPMC actuators built with micromachining techniques would help to more accurately manipulate the actuators and fabricate more reliable devices with optimal performance. This section discusses the dynamic behaviour of micromachined IPMC actuators with molecular-scale models and uses an electrical circuit model to characterize the micromachined IPMC actuator.

18.2.1 Investigation of the Dynamic Behavior of Micromachined IPMC Actuators with Molecular-scale Models

The micromachined IPMC actuators using molding technology and liquid Nafion could have different performances than the common IPMC actuators, which plate metal electrodes on a commercial Nafion film. Feng presented a theoretical model to explain the micromachined IPMC devices.[23]

To model the behaviour of the micromachined IPMC actuator at a molecular scale, four conditions are considered: (1) the electrostatic forces cause the cations (sodium ions) inside the Nafion layer to migrate from the anode to the cathode; (2) the water molecules are bonded with the sodium ions due to hydration, so the cations—together with bonded water molecules—travel as particles inside the hydrated Nafion while the external electric field is applied; (3) Stokes' law explains the induced viscous force when the particle is travelling, and the concentration of the moving sodium ions in response to the exerted voltage is considered not necessarily equal to the solution concentration used for Nafion membrane ion exchange; and (4) the fabricated IPMC device is operated in an ambient environment, with the driving signals having different operating frequencies and amplitudes, and with zero bias.

To explain the dynamic response of the micromachined IPMC device, the governing equation can be derived as follows. Figure 18.3 illustrates the mechanism of cat-particles (meaning a moving cation bonded with water molecules as a particle) moving inside the hydrated Nafion film under an external electric field. The backbone matrix of the hydrated Nafion film is composed of the negatively charged long-chain molecules. It is assumed that the long chain remains immobile while the external electric filed is applied, and the degree of hydration of Nafion remains unchanged, so that both cluster size and relative distance between clusters are maintained as invariable during the IPMC actuator operation. Partial cat-particles within the hydrated Nafion are considered to be pushed from their static electrical equilibrium location when there is sufficient external electrostatic force to conquer the bonding force of the anions. The remaining cat-particles remain in the cluster under the applied electric field.

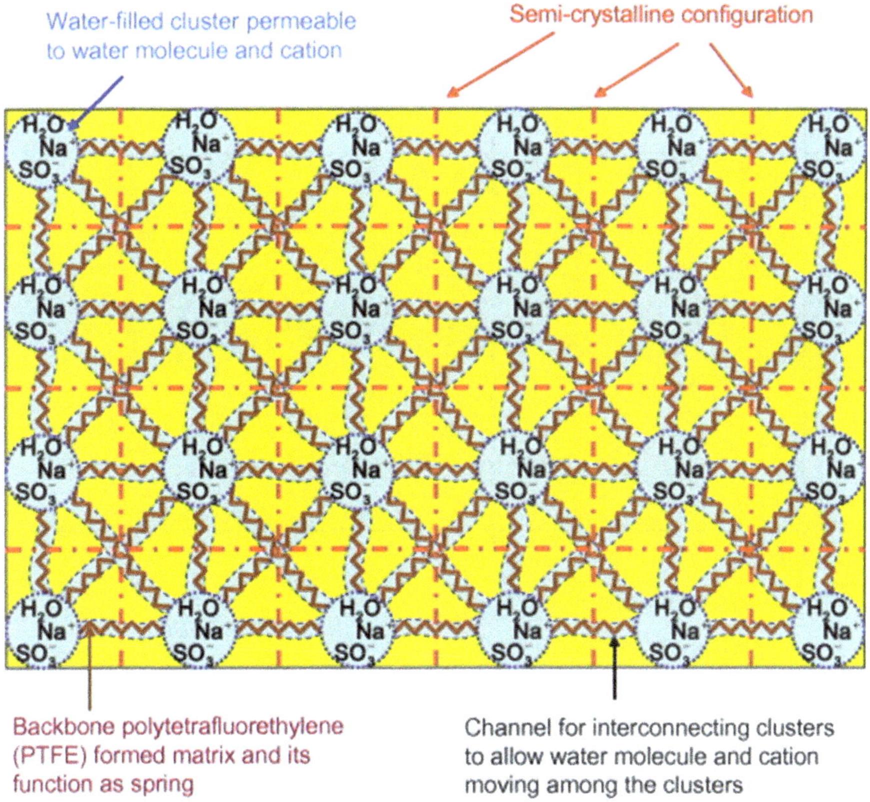

Figure 18.3 Conceptual diagram of the microstructure of hydrated Nafion after the ion exchange process used for modeling.
Reprinted with permission from ref. 23.

$c_{ion}(x, t)$ is defined as the equilibrium cation concentration of the hydrated Nafion after the ion exchange process. The concentration of cations that cannot escape the source bonding link is expressed as $c_{fix}(x, t)$. Thus, $c_{mov}(x, t)$ can be written as the concentration of these moving cations at position x and time t, $c_{mov}(x, t) = c_{ion}(x, t) - c_{fix}(x, t)$, and $c_{mov,0}$ $(= c_{mov}[x, 0])$ is labeled as the initial concentration of these moving cations. For one-dimensional analysis, each moving cat-particle is affected by three kinds of forces: electrostatic force (F_{es}), viscous force (F_{vis}), and diffusion force (F_{dif}) (Figure 18.4). According to Newton's second law of motion, the resultant force can be written as eqn (18.1):

$$F_{es}(x,t) - F_{vis}(x,t) - F_{dif}(x,t) = m_p \frac{\partial v(x,t)}{\partial t} \tag{18.1}$$

in which m_p and $v(x, t)$ are the mass and velocity of the cat-particle, respectively. The electrostatic force $F_{es}(x, t) = eE(x, t)$, where e is the charge of cation and $E(x, t)$ is the entire electric field at location x and time t. The viscous force is $F_{vis} = 6\pi\mu Rv(x, t)$, where μ is the dynamic viscosity of the hydrated Nafion and R is the equivalent radius of the moving cat-particle

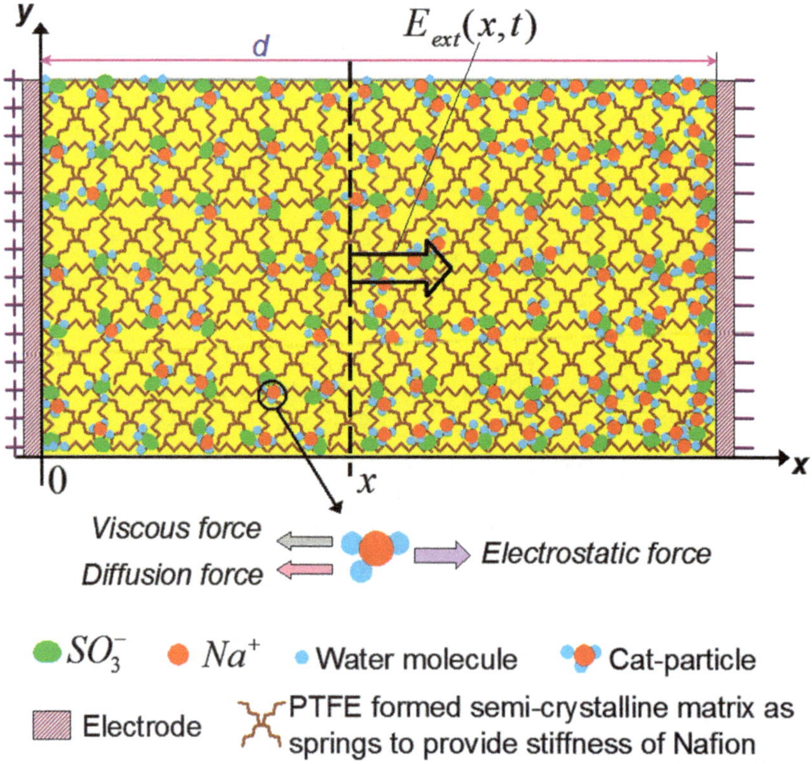

Figure 18.4 The mechanism of cat-particles moving inside the micromachined IPMC device under external electric field.
Reprinted with permission from ref. 23.

and can be represented as $R=(r_i^3+n\cdot r_w^3)^{1/3}$, in which r_i and r_w are the radii of the cation and water molecule, respectively. In addition, n is the number of water molecules bonded to a cation. The diffusion force can be derived as eqn (18.2):

$$F_{\mathrm{dif}}=kT(1+n)\cdot\frac{1}{c_{\mathrm{mov}}(x,t)}\cdot\frac{\partial(c_{\mathrm{mov}}(x,t))}{\partial x} \tag{18.2}$$

along with the principle of mass conservation (continuity equation in eqn (18.3)):

$$\frac{\partial(c_{\mathrm{mov}}(x,t))}{\partial t}+\frac{\partial(c_{\mathrm{mov}}(x,t)\cdot v(x,t))}{\partial x}=0 \tag{18.3}$$

The governing equation can be obtained as eqn (18.4):

$$\begin{aligned}6\pi\mu R\frac{\partial q(x,t)}{\partial t}&=kT(1+n)\frac{\partial^2 q(x,t)}{\partial x^2}\\&\quad-\mathrm{e}\left\{\frac{V(t)}{d}+\frac{2}{\varepsilon_{\mathrm{e}}\cdot S_x}[q(x,t)-q_0(x)]\right\}\frac{\partial q(x,t)}{\partial x}\end{aligned} \tag{18.4}$$

with the initial condition $q_0(x)=\mathrm{Ne}S_x c_{\mathrm{mov},0}\cdot x$ and the boundary conditions $q(0,t)=0$ and $q(d,t)=\mathrm{Ne}S_x\int_0^d c_{\mathrm{mov}}(\lambda,t)\mathrm{d}\lambda$. $q(x,t)$ in the governing equation can then be solved.

18.2.2 Electrical Circuit Model used to Characterize the Micromachined IPMC Actuator

Using the measured voltages and currents to establish a circuit model for describing the behavior of the micromachined IPMC actuator was also studied by Feng and Tsai.[24] They started with the P–N–P equation,[25] which describes the charge movement contributed by drift in the electric field and thermal diffusion as in eqn (18.5):

$$\psi_{\pm}=\pm\mu\cdot n_{\pm}E-D\frac{\partial n_{\pm}}{\partial x} \tag{18.5}$$

where ψ is the ion flux density, μ is the ion mobility, and n is the ion concentration (the subscript $\pm$ indicates positive and negative charges, E is the electric field, and D is the diffusion coefficient). Beunis *et al.* indicated that the solutions of the P–N–P equation with perfect blocking electrode boundary conditions can be characterized as four limiting cases using two key parameters, as in eqn (18.6) and (18.7):[26]

$$\tilde{V}_{\mathrm{A}}=\frac{V_{\mathrm{A}}}{V_{\mathrm{T}}} \tag{18.6}$$

$$\tilde{Q}_{\mathrm{tot}}=\frac{Q_{\mathrm{tot}}}{\left(\frac{\varepsilon\varepsilon_0}{2d}V_{\mathrm{T}}\right)} \tag{18.7}$$

where V_A is the voltage across the electrodes. V_T is the thermal voltage defined as kT/q, in which k is Boltzmann's constant, T is the absolute temperature, and q is the charge of the cation. ε is the relative dielectric permittivity of Nafion, and ε_0 is the dielectric permittivity of the vacuum. d is the Nafion thickness between the two applied electrodes. Q_{tot} is the total charge in the device per unit electrode surface, defined as $q\bar{n}d$, in which $\bar{n}$ is the initial concentration of cations in the Nafion material.

Based on the experimental results,[23] $\tilde{V}_A$ was approximately 100–1000, and the solution can be approximated as regions limited by geometry or space charge, wherein the drift transport mechanism is dominant and the diffusion term in the P–N–P equation is neglected. A geometry-limited region specifies that the effects of diffusion and electric field screening are neglected. A space charge-limited region suggests that the charge content is sufficiently high to almost fully screen the electric field in the bulk region of Nafion by the transient space charge regions.[27] To find the solution in a space charge-limited region, the current must diminish to zero following a power law. However, according to the experimental data, the currents of the operated IPMC actuator exhibited significant non-zero values when reaching a near-steady state. This may be attributed to the reality of imperfect electrodes in fabrication, which causes the Faradic reaction that results in cations due to the electrodes and lowers the concentrations of cations. Therefore, the current solution for the geometry-limiting regime can be applied as follows in eqn (18.8):

$$I = I_0\left(1 - \frac{t}{\tau_{tr}}\right) \tag{18.8}$$

in which $I_0 = \dfrac{2q\bar{n}\mu V_A}{d}$. For the experimental results, τ_{tr} (approximately 30–120 s) was larger than the driving time of interest t by at least ten times, derived from the calculation, which implies that the current can be roughly assumed to be proportional to I_0. Because I_0 was proportional to V_A, the current caused mainly by the geometric effect is defined as eqn (18.9):

$$I_g(t) = \frac{V(t) - V(t=0)}{V(t_f) - V(t=0)} I(t_f) \tag{18.9}$$

where $V(t)$ represents the measured voltage across the electrodes (*i.e.* $V_A(t) = V(t)$) and t_f represents the time required to reach the near-steady state. Thus, considering that I_g was deducted from the measured current I, we defined the space charge current $I_{sc}(t) = I(t) - I_g(t)$. Here, $I_{sc}(t)$ is mainly caused by the space charge effect, and the contribution from the non-dominant diffusion effect was also incorporated.

Hence, the electrical circuit model (Figure 18.5) can be constructed simply as the series connected capacitor C_{sc} and resistor R_{sc}, which are responsible for the $I_{sc}(t)$ contribution, and parallel connected to a resistance R_g, which is

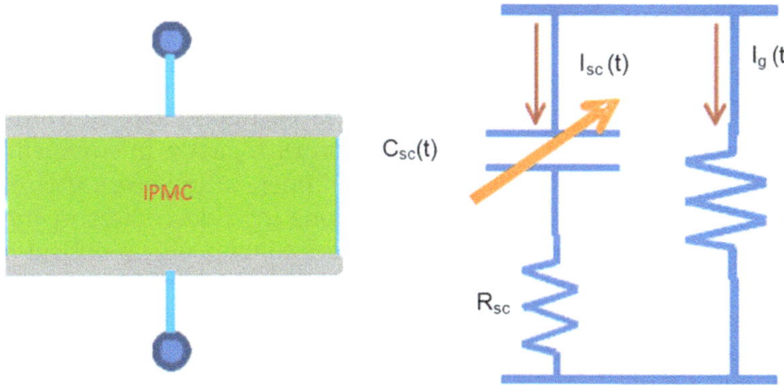

Figure 18.5 The electric circuit model to describe the dynamic current–voltage relation of IPMC actuators.
Reprinted with permission from ref. 24.

responsible for the $I_g(t)$ contribution. R_g can be identified as follows to ensure that its value can be exactly fitted with the measured voltage and geometry current I_g as time varies, as in eqn (18.10):

$$R_g = \frac{V(t_f) - V(t=0)}{I_g(t_f)} \tag{18.10}$$

The values of capacitor C_{sc} and resistor R_{sc} can be determined using the following equations. As indicated by Kirchhoff's voltage law, the voltage across the electrode pair of the IPMC actuator can be expressed as eqn (18.11):

$$V(t) = V_C(t) + I_{sc}(t)R_{sc} \tag{18.11}$$

where $V_C(t)$ stands for the voltage across the capacitor C_{sc}. Based on the principle that capacitor voltages are unable to change instantaneously (which would require an infinite amount of power), the following relation in eqn (18.12) exists (the moment for switching the input voltage polarity is denoted as $t=0^-$ and the next sampling moment denoted as $t=0^+$):

$$V_C(t=0^+) = V_C(t=0^-) \tag{18.12}$$

Thus, R_{sc} can be evaluated as eqn (18.13):

$$R_{sc} = \frac{V(t=0^+) - V(t=0^-)}{I_{sc}(t=0^+)} \tag{18.13}$$

The capacitance $C_{sc}(t)$ can be derived as eqn (18.14):

$$C_{sc}(t) = \frac{\int_0^t I_{sc}(t')\mathrm{d}t'}{V_C(t) - V_C(t=0)} \tag{18.14}$$

18.3 Micromachined IPMC Actuators for Biomedical Applications

This section presents three representative IPMC actuators related to micromachining technology as mentioned in Section 18.1. Firstly, micromachined grippers with the dimensions of several millimeters and feature sizes of hundreds of microns for potential applications in endoscopic surgery are reported. Next, using micromolding techniques to fabricate an optical fiber enclosed by a four-electrode IPMC actuator with multidirectional controllability to manipulate the laser beam is presented. The fabricated device allows emission of the laser light on a mimic eyeball, and moving the laser spot to a specific position on the eyeball is demonstrated. Finally, helical IPMC actuators with two different fabrication processes are described. The helical IPMC actuator possesses mobility in the rotational and longitudinal directions of the actuator, which can be promising for applications as active stents.

18.3.1 Microgrippers for Endoscopic Surgery

Microgripping has many applications, such as micromanipulation of bio-particles and assembly of micro-components to form a system.[28] Usually, bio-particles such as bacteria or cells are compliant and fragile, making them easily damaged with a sharp tip during manipulation. Polymer-based actuators exhibit outstanding potential to handle these micro-objects due to their greater flexibility compared to common metal or glass devices. Furthermore, the manipulation of most bio-molecules must be in a liquid environment. These characteristics make IPMC microgrippers more attractive for handling biological particles. Microgrippers can also be useful for assembling microsystems, where integrating micron-sized feature components is necessary to construct micromachines for diverse functions. Especially at the prototype development stage, microgrippers are useful for grasping tiny objects made from various materials, where improper force control could cause assembly failure, and even broken components.[29] Polymer grippers have a "damper" effect, which allows the polymer micro-grippers to absorb excessive force through shape change during the interaction between the grippers and the targeted object.

Feng and Chen presented micromachined micro-IPMC actuators with arbitrary shapes used as tiny mechanical gripping devices for minimally invasive surgical applications.[14] Minimally invasive surgery avoids large open incisions and instead employs several small local incisions for an endoscope and mechanical or robotic manipulative tools. The goal of developing this device is to replace the end-unit of a conventional laparoscopic gripping device, as illustrated in Figure 18.6.

To improve the traditional IPMC fabrication process, which commonly uses a knife or laser for cutting an IPMC sheet to create single devices, in which ragged edges or short-circuits of the IPMC device could occur, a micromachining technology has been applied. The process starts with fabricating the IPMC microgrippers onto a very thin membrane. Patterned photoresist produces the designed shapes with the membrane as molds. After the liquid Nafion has filled in the molds and solidified as ionic polymer devices, the bottom supporting membrane is removed by dry etching. The processed polymer can be anchored on the sidewalls of photoresist-made molds due to the small size of the devices. Next, the whole wafer along with the shaped ionic polymers is immersed into suitable chemical solutions to perform the initial composing and surface electroding. Because the

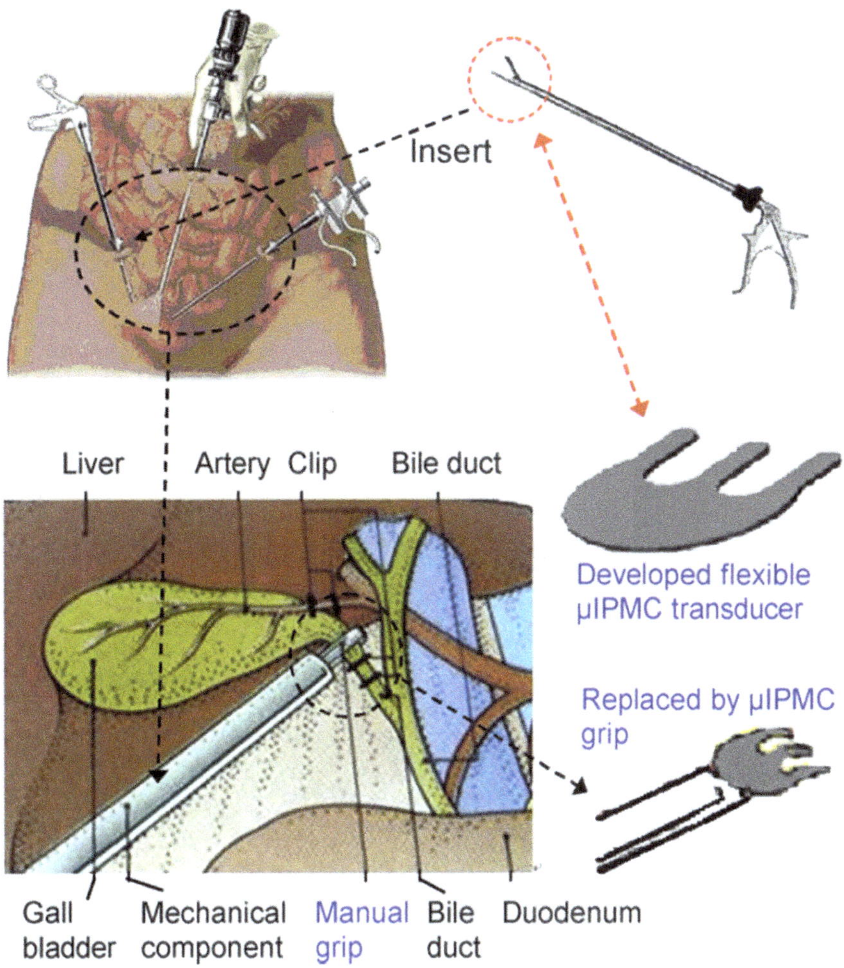

Figure 18.6 Laparoscopic surgery: a surgical technique (left) in which long, narrow instruments are inserted into the body through small incisions. Current laparoscopic surgical tools (*i.e.* gripper in upper right) could be replaced by IPMC flexible grips (middle and lower right).
Reprinted with permission from ref. 12.

chemicals do not attack the patterned photoresist and the interface between the shaped polymers and photoresist molds, the sidewalls of the devices are well protected by photoresist. The photoresist molds are then dissolved by chemicals and the individual IPMC microgrippers are released. Fabrication details are presented in the previous section. Figure 18.7 shows the finished IPMC microgrippers with a featured size in the range of 100 μm and a thickness of approximately 50 μm.

In order to verify the potential of the fabricated devices for applications in laparoscopic surgery, the fabricated device is held in a clamp at one end and the free end is operated to grip a flexible tube with a diameter of 800 μm, which is similar to the size of some ducts in the human body. The maximum displacement and instantaneous applied force of the fabricated IPMC microgrippers are also evaluated. The experimental setup for measuring the load-free displacement of the device involves directing a laser beam at the tip of the transducer and recording the position of the reflected spot by a high-speed camera. Results of the maximum displacement indicate that devices driven with low frequencies have better performance than with high frequencies for either sinusoidal or square waves. The maximum displacement by measuring the point 1 mm away from the anchor achieved is 300 μm at 0.5 Hz for a 12 V (peak-to-peak) square wave of a given amplitude, which is

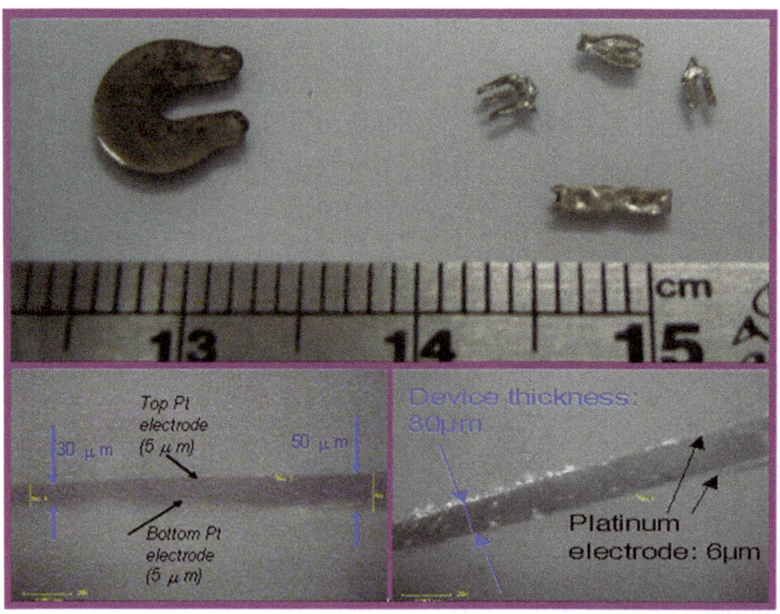

Figure 18.7 (Top) Individual micromachined IPMC grippers with different microfeatures. (Bottom-left) The IPMC actuator with platinum electrodes simultaneously formed on top and bottom surfaces. (Bottom-right) Cross-sectional view of the device with a clear edge in which no short-circuits occur.
Reprinted with permission from ref. 11.

approximately two-times larger than the displacement generated by a sine wave of the same amplitude (peak-to-peak) and frequency, and around six-times larger than the displacement achieved when driven by a same-amplitude square wave of 10 Hz.

The measurement of instantaneous applied force is as follows. Initially, without input voltage, one end of the device is held in an anchoring clamp and the free end of the device remains on a plastic pillar that is fastened to the surface of the measuring plate of a standard electronic balance. Then, a bipolar signal (sinusoidal or square wave) with zero bias and constant frequency is applied to the device. Because of the driving electric signal, the free end of the device moves up and down, pushing down on the pillar during the down cycle and allowing the force to be monitored quantitatively on the balance. The maximum readout from the balance is recorded as instantaneous force output. The results show that the instantaneous maximum force does not increase monotonically as the voltage increases for all tested cases, and lower driving frequencies cause higher instantaneous force output. The maximum value achieved is 5 mN for a 0.5 Hz square wave of amplitude 12 V peak-to-peak.

18.3.2 Optical Fiber Enclosed by Four-electrode IPMC Actuators for Directing Laser Beams

Feng and Tsai present a micro-fabrication process to make a column-structured four-electrode IPMC actuator, which surrounds a portion of optical fiber.[30] The device allows electronic directional control of the conducted laser light, which can be helpful in applications such as ocular surgery (Figure 18.8).

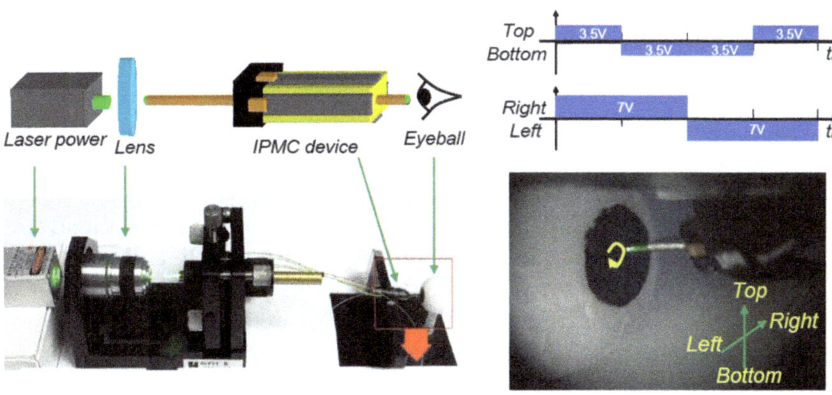

Figure 18.8 Illustration of the micromachined optical fiber-enclosed four-electrode IPMC actuator for biomedical applications such as microendoscopic ocular surgery.
Reprinted with permission from ref. 30.

A 7 mm × 2 mm × 1 mm four-electrode IPMC actuator with an enclosed optical fiber is demonstrated. The process starts by making a thin-wall parylene cuboid mold with one side open for the injection of Nafion solution. The two length-determining walls of the mold are punched with one hole in the center of each wall so that an optical fiber can be inserted through one hole and drawn out from the other hole. Slight tension is applied to the fiber while Nafion solution is put into the mold so that the optical fiber remains straight during solidification of the Nafion. To generate four electrodes on the device in one process, photolithography is employed to delineate the inactive regions of the device. Chemical processing is performed to plate platinum electrodes on the four long sides of the cuboid device. The two fiber-containing ends are free of plating due to the photoresist protection. Finally, the processed IPMC actuator is treated with sodium hydroxide solution to complete both ion exchange processing and photoresist removal.

Experimental data show that square wave actuation generated a larger displacement than sinusoid wave actuation for the same given driving frequency and amplitude. Maximum displacement at the free end of the manufactured IPMC actuator (5 mm away from the anchor) can be 400 μm and 270 μm for a 0.1 Hz driving signal applied to the top–bottom electrode pair and side–side electrode pair, respectively. In general, lower operation frequencies cause larger displacement. One significant characteristic of this IPMC actuator is that it is driven by selected electrode pairs. By applying actuation signals to the top–bottom electrode pair or side–side electrode pair, the generated bending motion is perpendicular to the surface of the top–bottom or side–side electrode pair. Imposing a driving voltage on the top–right electrode pair is also studied. The maximum displacement at the free end of the manufactured IPMC actuator (5 mm away from the anchor) reaches 275 μm for a sinusoidal 0.1 Hz driving signal of 9 V peak-to-peak. The moving direction maintains a near constant angle between the surfaces of the top and bottom electrodes during actuation with different voltage amplitudes and frequencies. These results verify that the selective actuation of various combinations of the electrode pairs displays multi-directional motion abilities.

Since the aim of this work is to demonstrate that the fabricated IPMC actuator could facilitate ocular surgery, the experimental setup of a laser-emitting system with a mimic eyeball is constructed. Figure 18.9 illustrates the preliminary setup for this application, starting with a laser power source that radiates a focused laser beam into the optical fiber through the biconvex lens. The multi-electrode column-structured IPMC actuator encloses a section of the fiber and is located close to the other end of the optical fiber. A fixture with four electrically conductive pads made of copper tapes is used for mechanically securing the IPMC actuator and transmitting the actuation signals to control the motion of the IPMC actuator. A white ball made of Styrofoam selectively painted black

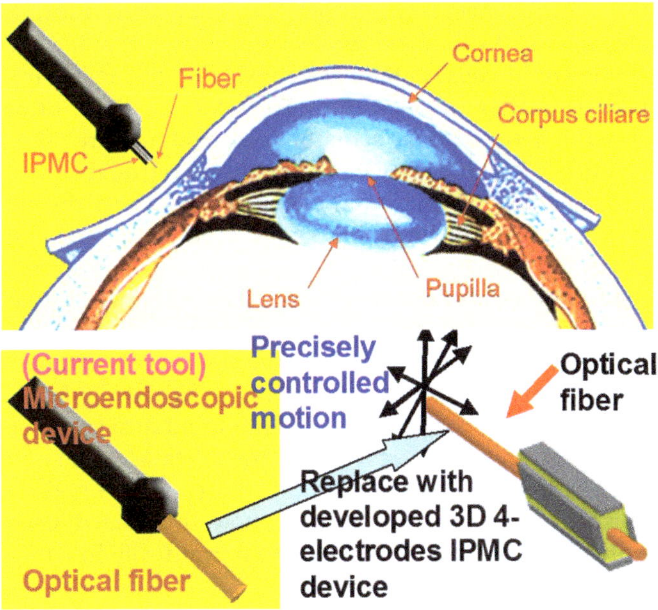

Figure 18.9 Demonstration of the developed optical fiber-enclosed IMPC actuator facilitating ocular surgery.
Reprinted with permission from ref. 30.

represents a human eyeball. A green laser beam is emitted onto the eyeball while the laser power source is turned on. To manipulate the IPMC actuator so that the laser spot can be controlled at a desired location of the eyeball, two specified square wave signals are employed on the top–bottom electrode pair and right–left electrode pair, respectively. The voltage amplitudes for the top–bottom and right–left electrodes are 3.5 V and 7 V, respectively. Initially ($t = t_0$), polarities of the top and right electrodes are both positive. Then, at $t = t_1$, the polarity of the applied signal on the top–bottom electrodes is switched, but the signal on the right–left electrodes remains unchanged. Based on the given driving sequences, the moving direction of the laser spot follows the order of top-right, bottom-right, bottom-left, and top-left. This result demonstrates the promising ability of the fabricated IPMC actuators to be used in ocular surgical applications.

18.3.3 Helical IPMC Actuators with Rotational and Longitudinal Motions for Active Stents

Stents are small, cylindrical devices placed in narrowed coronary arteries to reopen the lumen and restore blood flow, but current stents have certain shortcomings. For instance, stents attack the inner arterial wall because of sudden expansion of the balloon while the stents are being placed. Furthermore, stents could cause lumen restenosis due to the accumulation of intimal hyperplasia after implantation. To minimize these drawbacks, Li *et al.* presented a helical IPMC actuator to actively control the radius of biomedical stents.[31] The advantages of a radius-controllable stent include lessening of the damage to the internal wall of the artery during stent installation and possibly removing the plaque around the stent while operating the actuator.

The fabrication process of the Nafion-based helical IPMC actuators starts with the IPMC film that was obtained by an electroless plating method to grow platinum layers on both surfaces of the Nafion membrane. Then, the fabricated IPMC film is processed by soaking in lithium chloride solution for cation exchange, after which the IPMC film is cut into strips of designed length and width. Subsequently, the strips are helically coiled on glass rods and dipped into distilled water at 90 °C for 1 h to fix the helical shape by thermal treatment. Experimental results show that the diameter of the fabricated helical IPMC actuator plays an important role in controlling the radius of the stent.

In addition, Feng and Zhan present a similar helical IPMC actuator with a selectable active region based on a micromachining fabrication process.[12] Instead of raising the temperature to shape the flat strip as a coil spring and plating platinum metal as electrodes, as previously described, the presented method uses room temperature fabrication for the entire process. The advantage to this fabrication process is that the high processing temperature of the polymer material might increase the Young's modulus and residual stress, leading to dissimilar internal structures of Nafion compared to room temperature-processed devices for the structural mechanism of ion transportation. Meanwhile, the parylene-patterned method selectively deposits the metal electrode on the specific area that allows the active region of the spring actuator to be designable. Thus the process guarantees that the inner electrode never short-circuits with the outer electrode. More detailed fabrication information has been described in Section 18.3.

The motion of the fabricated helical actuator has been experimentally tested (Figure 18.10), and the maximal displacement reaches 1 mm at the endpoint of the spring under a 0.1 Hz, 6 V square wave actuation for the developed actuator with a diameter of approximately 4 mm and a length of 1 cm. A microtensile experiment is executed to characterize the stress and strain relation. The resulting Young's moduli of the Nafion and fabricated IPMC devices are 183 and 227 MPa, respectively. The produced moment, force output, strain energy, and displacement of an arbitrarily point on the fabricated helical actuator have been analyzed using Castigliano's theorem. The results show the moment, force, and strain energy of the operated helical actuator are at the level of 1.5 μN m, 300 mN, and 20–100 μJ, respectively.

Figure 18.10 Images captured from the viewpoints of the front side (a) and bottom side (b) of the spring, with a time interval of 1 s between each sub-image when the device is driven with a 0.1 Hz, 6 V square wave in air. Reprinted with permission from ref. 12.

18.4 Conclusion

IPMCs have attracted wide research attention because of their large dynamic deformation driven by relatively low electric fields compared with other smart materials such as piezoelectric ceramics, due to their larger dynamic frequency response than shape memory alloys. Currently, IPMCs display considerable potential as biomimetic and flexible actuators for soft robotic or artificial muscle applications.

Micromachining technology can provide important advantages for fabricating versatile IPMC actuators. Surface micromachining allows the micro-sized IPMC actuator to be integrated onto a silicon wafer through a process of spin-coating, photolithography, thin film metal deposition, and sacrificial layer releasing. A commercial Nafion film with uniform thickness can be etched by reactive ion etching on the selected surface to a desired thickness through the shadow masking method. Bulk micromachining modifies the ionic liquid used inside the Nafion film and allows metal electrode fabrication with different metal power composition. To chemically plate metal electrodes on a structured non-planar Nafion strip, parylene thin film deposition can provide internal stress to shape the Nafion strip. It can also be used as a mask to selectively delineate inactive regions to prevent the metal electrode from plating there. Micromolding technology offers a scheme for fabricating arbitrarily shaped micro-featured IPMC actuators with clear shape edges, while avoiding laser or knife cutting issues. Moreover, a column-structured IPMC actuator with surface electrodes on different planes can be fabricated. This allows the structure of IPMC actuators to be more versatile and have better actuation performance through fabrication improvement.

Due to the capabilities of constructing the device with precision and miniaturization, yielding more complicated shapes and structures by micromachining technology, numerous micromachined IPMC actuators have been successfully fabricated, demonstrating potential biomedical applications. Microfabricated grippers intended for laparoscopic surgery have been demonstrated to grasp a flexible tube with a diameter of 800 μm, which mimics ducts of the human body. The optical fiber enclosed by a four-electrode IPMC actuator made by micromolding technology has demonstrated its maneuverability to multi-directionally drive an IPMC actuator, and experimentally demonstrated the ability to facilitate ocular surgery. Helical IPMC actuators provide advantages as active stents for treating cardiac disease due to their radial and longitudinal motion. Investigation of the dynamic behavior and electromechanical characteristics of these various micromachined IPMC actuators could be helpful for more accurately manipulating the actuators and fabricating more reliable devices. Thus, more advanced micromachined IPMC actuators for clinical usage can be expected in the future.

References

1. Y. Bar-Cohen, *Proc. SPIE Electroactive Polymer Actuators and Devices (EAPAD)*, 2010, 764206.

2. F. Carpi, R. Kornbluh, P. Sommer-Larsen and G. Alici, *Bioinspiration Biomimetics*, 2011, **6**(4), 045006.
3. G.-H. Feng and K.-M. Liu, *Sensors*, 2014, **14**(5), 8380.
4. M. Shahinpoor and K. J. Kim, *Smart Mater. Struct.*, 2005, **14**, 197.
5. Y. Bahramzadeh and M. Shahinpoor, *Smart Mater. Struct.*, 2011, **20**(9), 094011.
6. M. Shahinpoor, *Bioinspiration Biomimetics*, 2011, **6**(4), 046004.
7. Y. Bahramzadeh and M. Shahinpoor, *Soft Robotics*, 2014, **1**(1), 38–52.
8. W. Zhou and W. J. Li, *Sens. Actuators, A*, 2004, **114**(2–3), 406.
9. Z. Chen and X. Tan, *Sens. Actuators, A*, 2010, **157**(2), 246.
10. J.-H. Jeon, S.-W. Yeom and I.-K. Oh, *Composites, Part A*, 2008, **39**(4), 588.
11. B. J. Akle, M. D. Bennett and D. J. Leo, *Sens. Actuators, A*, 2006, **126**, 173.
12. G.-H. Feng and Z.-H. Zhan, *Smart Mater. Struct.*, 2014, **23**(4), 045002.
13. J. J. Pak, J. Kim, S. W. Oh, J. H. Son, S. H. Cho, S. K. Lee, J. Y. Park and B. Kim, International Society for Optics and Photonics, *Smart Struct. Mater.*, 2004, 272.
14. G.-H. Feng and R.-H. Chen, *Sens. Actuators, A*, 2008, **143**(1), 34.
15. G.-H. Feng and R.-H. Chen, *J. Micromech. Microeng.*, 2008, **18**(1), 015016.
16. G.-H. Feng and G.-Y. Chu, *Sens. Actuators, A*, 2014, **208**, 130.
17. M. Shahinpoor, Y. Bar-Cohen, J. O. Simpson and J. Smith, *Smart Mater. Struct.*, 1998, 7(6), R15.
18. S. Nemat-Nasser and J. Y. Li, *J. Appl. Phys.*, 2000, **87**(7), 3321.
19. L. M. Weiland and D. J. Leo, *J. Appl. Phys.*, 2005, **97**, 123530.
20. B. J. Akle and D. J. Leo, *Smart Mater. Struct.*, 2007, **16**, 1348.
21. M. Porfiri, *J. Appl. Phys.*, 2008, **104**, 104915.
22. Z. Chen and X. Tan, *IEEE/ASME Transactions on Mechatronics*, 2008, **13**(5), 519.
23. G.-H. Feng, *Comput. Mater. Sci.*, 2010, **50**, 158.
24. G.-H. Feng and J.-W. Tsai, *Polym. Eng. Sci.*, 2013, **53**(9), 2004.
25. M. Z. Bazant, K. Thornton and A. Ajdari, *Phys. Rev. E: Stat., Nonlinear, Soft Matter Phys.*, 2004, **70**, 021506.
26. F. Beunis, F. Strubbe, M. Marescaux, J. Beeckman, K. Neyts and A. R. M Verschueren, *Phys. Rev. E: Stat., Nonlinear, Soft Matter Phys.*, 2008, **78**, 011502.
27. F. Beunis, F. Strubbe, K. Neyts and A. R. M. Verschueren, *Appl. Phys. Lett.*, 2007, **90**, 182103.
28. R. Lumia and M. Shahinpoor, *J. Phys.: Conf. Ser.*, 2008, **127**(1).
29. R. K. Jain, S. Majumder and A. Dutta, *Smart Mater. Struct.*, 2012, **21**(7), 075004.
30. G.-H. Feng and J.-W. Tsai, *Biomed. Microdevices*, 2011, **13**(1), 169.
31. S. L. Li, W. Y. Kim, T. H. Cheng and I. K. Oh, *Smart Mater. Struct.*, 2011, **20**(3), 035008.

CHAPTER 19

Ionic Polymer Metal Composites: Recent Advances in Self-sensing Methods

MASOUD AMIRKHANI* AND PARISA BAKHTIARPOUR

Institut für Experimentelle Physik Universität Ulm, Albert-Einstein-Allee 11 89081 Ulm, Germany
*Email: masoud.amirkhani@uni-ulm.de

19.1 Introduction

An ionic polymer metal composite (IPMC) device converts electrical stimuli into mechanical movement and *vice versa*, and thus they can be used as both actuators and sensors.[1–3] IPMCs are perfluorinated sulfonic ionic polymer membranes sandwiched between two metallic electrodes and containing mobile cations and fixed anions. In the equilibrium state, ions are distributed evenly in the IPMC strip, but an electrical potential causes ion redistribution, followed by bending or general deformation of the strip as a consequence of electrostatic, osmotic, and elastic interaction forces. IPMCs have many advantages such as low activation voltages, large bending strains, and biocompatibility, the ability to function in both wet and dry conditions, and ease of miniaturization.[4] The application of IPMCs as smart materials in soft robotics is an emerging and cutting-edge technology with great potential to benefit aerospace, medical, and automotive industries. However, a smart system requires a self-sensing mechanism, which allows interaction with the environment and responding correspondingly. Therefore, the self-sensing

RSC Smart Materials No. 18
Ionic Polymer Metal Composites (IPMCs): Smart Multi-Functional Materials and Artificial Muscles, Volume 2
Edited by Mohsen Shahinpoor

Published by the Royal Society of Chemistry, www.rsc.org

ability for IPMCs is a crucial aspect in order to widen their application and for them to be classified as a smart material. The self-sensing abilities of IPMCs can be grouped into five classes, each of which exploits different properties of IPMCs. In the first class, when the IPMC is bent, an output voltage proportional to the amplitude and direction of displacement[5–10] is produced. This output voltage can serve both for energy harvesting[11] and sensing abilities. As in this technique the mechanical energy is converted into electrical energy, we call it mechanical-to-electrical transduction (MET). The second sensing technique is based on measuring the surface resistance (SR) of an IPMC during bending and deformation.[12,13] In the SR technique, the change of SR on each side of the IPMC is used to record the position of the cantilever. In the third method, the resonant frequency of the IPMC strip is used for the sensing ability.[14] In the fourth technique, an external element such as a strain gage is added to the IPMC actuator and used as the bending sensor.[15] However, strictly speaking, this method should not be considered to be a self-sensing method. For the first time, we introduce the fifth technique, which is a new technique and, instead of measuring the resistance of the surface of the IPMC, the resistance across the sample is measured. IPMCs can be considered to be a combination of resistances, capacitors, and diodes, which dictates a frequency dependence of the voltage–current curve. In this technique, which we call high-frequency resistance (HFR) sensing, the resistance behavior of the IPMC in the range of kilo-Hertz is exploited for sensing.

In this chapter, we will discuss the basics and the recent advances of MET and SR sensing methods, which are the most studied techniques, in addition to HFR.

19.2 MET Sensor

This is a widely used technique that is sensitive to tip displacement, velocity, force, and pressure.[1,3,5–10] A typical IPMC sample is a thin ionic polymer membrane consisting of free cations and fixed anions. Basically, mechanical stimulation of an IPMC causes polarization of electric charges, which in turn produce a measurable voltage on the electrodes (Figure 19.1). This is due to the redistribution of ions across the thickness of the membrane upon mechanical deformation. The sensing voltage is governed by electrostatic interactions, ionic migration, ionic diffusion, ionic convection and SR. Therefore, the sensing is related to the properties of the IPMC along the thickness and the length directions. In the following, we briefly discuss each of mentioned phenomena and their effects on sensing voltage.

In the absence of any mechanical or electrical stimuli and in the wet membrane of the IPMC, three electrostatic forces are present.[7] First is the electrostatic interaction between the oxygen molecules of water and cations such that each cation carries several water molecules. Second, the cations decorate fixed anion groups of polymer to neutralize the solution. Third is the dipole–dipole interaction, which leads to increasing the stretching of the polymer membrane.

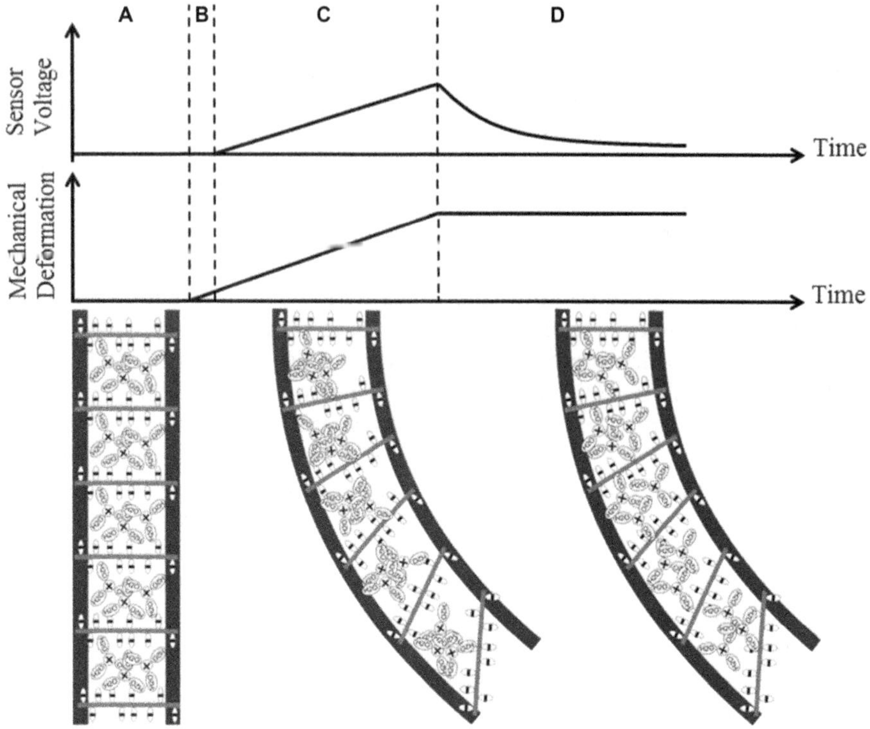

Figure 19.1 Side view illustration of an IPMC. The dark vertical and gray parallel lines represent electrodes and polymers with fixed cations, respectively. A: Before any mechanical deformation, anions are evenly distributed in the membrane. B: Upon applying mechanical deformation, there is a time delay between mechanical deformation and sensing of the voltage, which is related to the thickness of the IPMC and the velocity of deformation. This can produce a phase delay between deformation and the sensing signal. C: During mechanical deformation, anions are redistributed, causing increases in the sensing voltage. D: When the mechanical deformation is stopped, the sensing voltage will decay exponentially.

Now suppose a mechanical stimulus is applied to the IPMC. This forces the polymer to contract on one side and the opposite side to swell, which leads to migration of water from the contracted part to the expanded part of the membrane (Figure 19.1). As the water molecules are decorated with cations in spite of the attractive force between the mobile cations and the fixed anions, any water displacement results in displacement of cations. Furthermore, the initial contraction/swelling process produces inertia, thus causing extra cation migration to occur. The pressure gradient in the membrane is responsible for the convection of ions, thus the fluid in the membrane moves from high pressure to low pressure with a velocity that can be calculated using Darcy's law. Additionally, any imbalance in cation concentration leads to diffusion and redistribution of cations. It is shown that in order to predict the behavior of IPMCs and the produced voltage, one should also include the change in SR due to bending.[7]

We know that while a mechanical stimulus is applied to an IPMC, a detectable voltage on the electrodes is produced, but after stopping the stimulus, the voltage drop exponentially (Figure 19.1). Now the question is which phenomenon is responsible for increasing and decreasing the voltage during and after bending stops. It is almost impossible to provide a strict chronological order for each mentioned effect, but one can highlight the dominate effects in each stage of mechanical stimulus. Upon applying mechanical stress, in the first step, hydrate cations migrate toward the stretched part of polymer, and the imbalanced cation concentration leads to polarization, which eventually produces a voltage on the electrodes. In this first step, the swelling/contraction and the initial inertia of ions mostly define the strength of the measured voltage. However, due to changes in the concentration of cations and the electrostatic force between mobile cations and fixed anions, a back diffusion of cations occurs, which reduces the voltage. The effect of SR and convection has been discussed elsewhere.

The generated voltage has a linear relationship with bending. Bahramzadeh and Shahinpoor[5] found that lower velocities generate a smaller voltage with a lower increasing rate.[5] This can be due to two facts: one is due to low initial inertia, so the migration due to inertia is also lower; and another is due to additional time for the back diffusion of cations. The beauty of this technique is its versatility and sensitivity to pressure and deformation. However, as this technique is sensitive to almost any mechanical stimulus, it is thus difficult to distinguish between the voltage that is produced due to pressure or displacement. Furthermore, after stopping the external stimulus, the voltage decays to its original level, so one can only measure the relative bending, not overall bending.[7] This necessitates using a recording device to track each step of the stimulus. Generally, the voltage that is produced by this technique is one order of magnitude smaller than the voltage that is needed for actuating. So far, it is impossible to apply this technique for simultaneous sensing/actuating ability. However, the way around this problem is using two cantilevers of the IPMC strip in different formations, with one as a sensor and another one as an actuator (Figure 19.2).[16,17]

19.3 SR Sensor

As we mentioned before, each side of an IPMC is coated with a thin metallic layer. This metal layer is penetrated into the polymer membrane and contains micro-cracks on the surface. The mechanical deformation causes one side of the IPMC to swell and the surface cracks to widen, and on the other side where the membrane contracts, the cracks are squeezed together (Figure 19.3). It has been shown that if we model the surface as having simple resistance, the resistance on the swollen side becomes bigger, while on the contracted side of polymer it becomes smaller.[12,13] Therefore, the

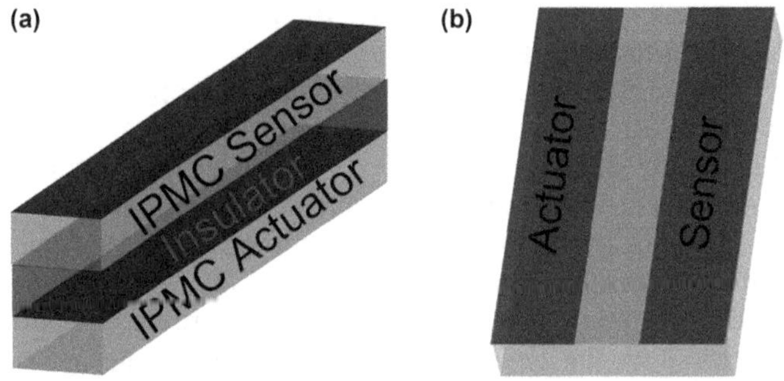

Figure 19.2 MET type of sensor, which can be used for simultaneous sensing and actuating abilities. (a) The sensor and actuator are stacked on top of each other and isolated by an isolator. (b) Using patterning, an IPMC is divided into two sections: sensor and actuator.

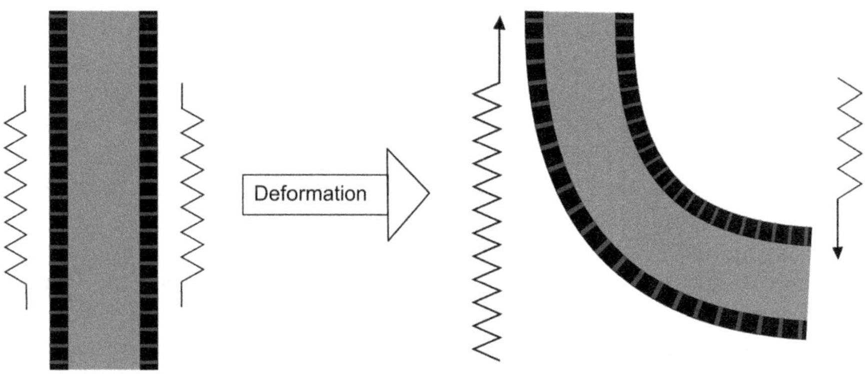

Figure 19.3 Side view illustration of an IPMC. The dark vertical and gray parallel lines in the electrode represent electrodes and cracks in the electrodes, respectively. The zigzag line represents the SR. (Left) A relaxed IPMC in which the crack distribution on both sides is the same, thus the resistance is equal. (Right) After deformation, one side of the IPMC is stretched and resistance becomes higher, while the other side is squeezed out and resistance decrease.

change in SR can be the basis of self-sensing IPMCs. This technique can be used while the actuator is on or off, so in theory it can measure the bending simultaneously while the actuator is on.

So far, two types of resistance-measuring techniques have been suggested. In the first technique,[12] the IPMC has a long clamp with four contacting points (two for each side) so a large portion of the IPMC is fixed between the clamp (Figure 19.4). Additionally, there are two extra contact points (one for each side) in the tip of the cantilever—these contacts can move freely. The fixed part of IPMC is used as the reference signal for the sensor. The actuating voltage is applied at point B, and then the voltages of four other points are measured (V_{A1}, V_{A2}, V_{C1}, and V_{C2}). By measuring the change in $V_{A1} - V_{C1}$

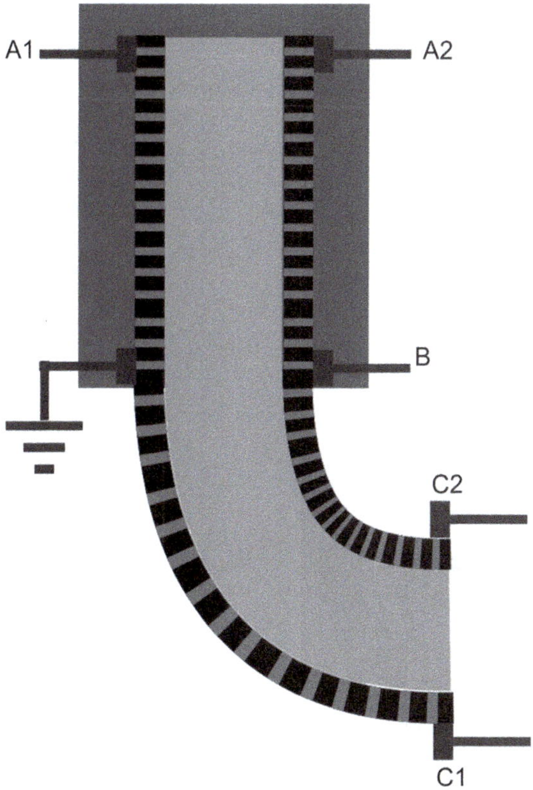

Figure 19.4 The difference between the voltages of the upper contact and lower contact on each side ($V_{A1} - V_{C1}$ and $V_{A2} - V_{C2}$) can be used as the sensor signal.

and $V_{A2} - V_{C2}$, the bending of the IMPC cantilever can be measured. Apart from simultaneous actuating and sensing capabilities, this technique has one other advantage, which is measuring the curvature regardless of the velocity of deformation. However, it has two disadvantages: one is the complication of adding electrodes to the tip of IPMC cantilever without compromising the actuating ability; the second is the existence of a fixed portion of IPMC between the clamp, which reduces the actuating ability and increases power consumption.

In the other method,[13] which exploits SR as a sensing method, the surface of the IPMC must be patterned in three different sections, which are the actuator, shield, and sensor (Figure 19.5).

The sensor and actuator section must be separate and shielded from each other. The shielding part is made to avoid cross-talk between the actuator and sensor sections. However, even when using the shielding section, cross-talk cannot be eliminated completely, so it is necessary to use additional mechanisms to cancel the cross-talk. If we assume that the cross-talk in each side is identical, one can remove it using a Wheatstone bridge.

Figure 19.5 Patterned surface for self-sensing using SR. The ground part of the sample is necessary to prevent cross-talk between the actuator and sensors.

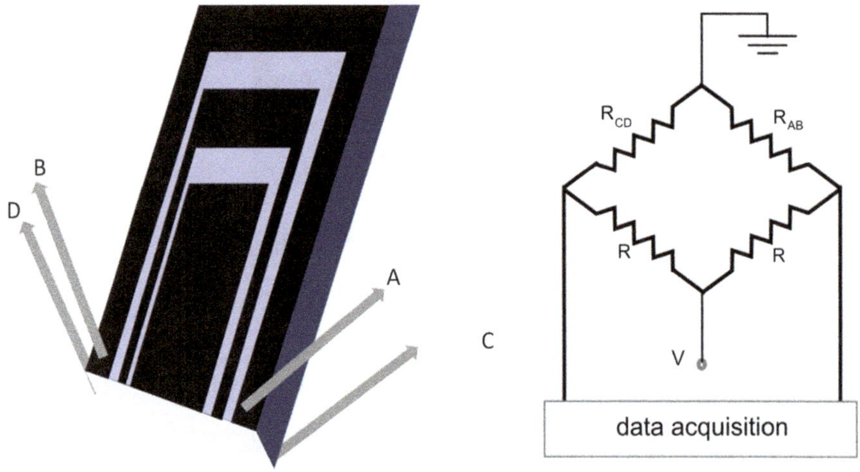

Figure 19.6 Wheatstone bridge for removing cross-talk between the actuator and sensor of a patterned IPMC. The voltage of the Wheatstone bridge (R_{AB} and R_{CD}) can be used as a motion detector. A, B, C and D are contact points for measuring resistance.

The disadvantages of this method are the complication of the sample fabrication due to patterning the surface and the small actuator area, which results in less bending. Additionally, the cross-talk between the actuator and sensor (Figure 19.6) cannot be prevented entirely, thus for lower bending, the sensor practically does not work.

19.4 HFR Sensor

IPMC cantilevers under a low-frequency driving voltage have a nonlinear response that is due to the electronic and ion current. However, for high frequencies (more than 1000 Hz), IPMCs behave with simple resistance and Ohm's law becomes valid. We exploit this key property of IPMCs to design a self-sensing system that works simultaneously with an actuator.

To understand the behavior of IPMCs we used an equivalent circuit. The two most common IPMC equivalent circuits are "RC" (resistor–capacitor)[18] and "RC + diode"[19] models. However, the RC model cannot predict the nonlinearity of IPMCs and the frequency dependency of the voltage and current. As the frequency response for the HFR sensing technique is an important aspect of IPMCs, thus the basis of our equivalent circuit is the "RC + diode" model. We used a modified "RC + diode" circuit to simulate the response of a HFR sensor while the actuating and sensing parameters were changed (the results are not shown here). Our equivalent circuit is illustrated in Figure 19.7.

Under the conditions of our experiments, this circuit responded more realistically to the voltage and frequency. Figure 19.8 illustrates the voltage–current curves for various frequencies. As one can see, IPMCs have a nonlinear behavior at low frequencies, but by increasing the frequency, the nonlinearity decreases. At around 1 kHz, the IPMC shows a linear behavior and follows Ohm's law, so for high-frequency voltages, IPMC elements can be substituted with simple resistance.

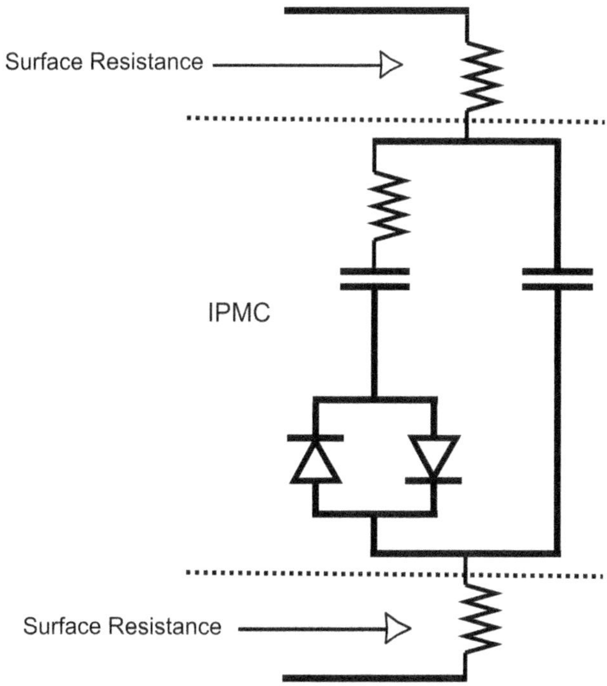

Figure 19.7 An equivalent circuit for an IPMC.

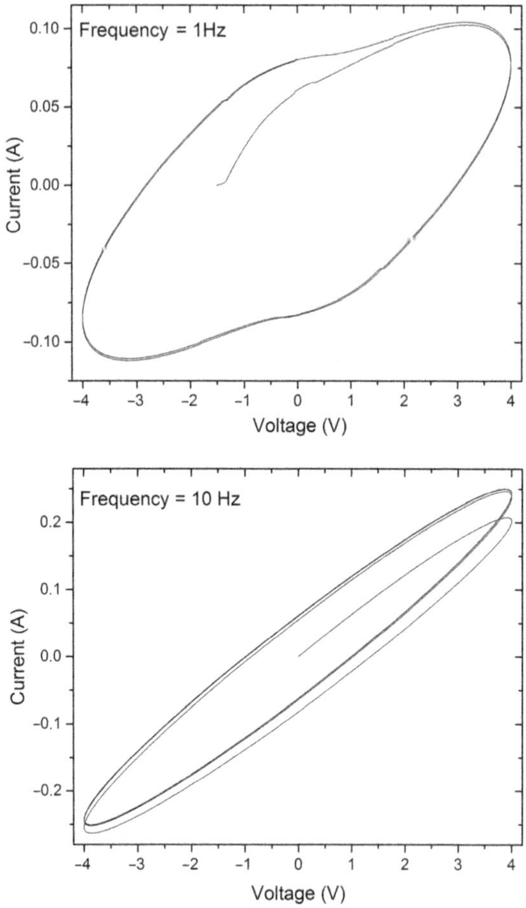

Figure 19.8 The simulation of IPMC responses for various frequencies using the equivalent circuit of Figure 19.7.

By knowing the frequency dependency of IPMCs, one can design a circuit to measure the change in the resistance across the IPMC. For high frequencies, we can model the sensing principle as a simple circuit that includes IPMC resistance and a small external resistance (around 1 Ohm). In Figure 19.9, the total resistance can be written as the following:

$$R_{tot} = R_{IPMC} + R_{ex} \tag{19.1}$$

The current that passes through the sample is the same for both resistances, so we can write eqn (19.1) as the following:

$$V_{tot} = V_{IPMC} + V_{ex} \tag{19.2}$$

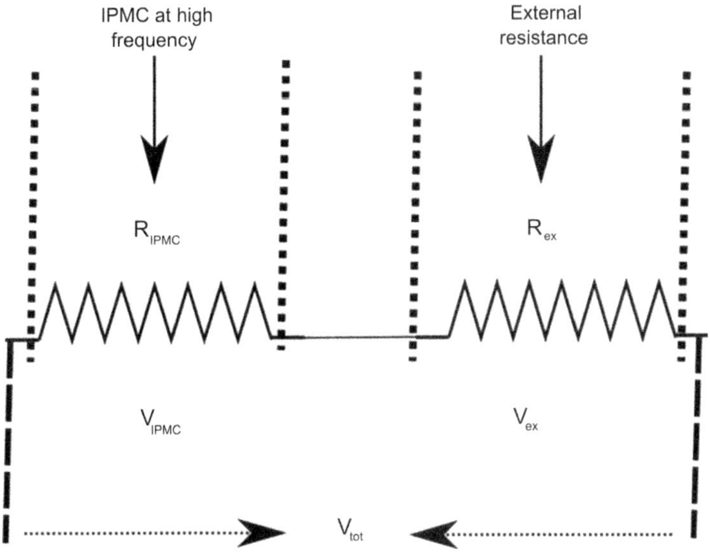

Figure 19.9 The basic circuit of the HFR sensing technique.

Eqn (19.2) shows that increases in V_{IPMC} lead to decreases in V_{ex} and *vice versa*; thus, by measuring V_{ex}, the change in the resistance of the IPMC can be evaluated.

19.4.1 Experiment

For simultaneous sensing and actuating abilities, two function generators must be used: one for high frequencies (sensing voltage) and another for low frequencies (actuating voltage). These function generators are in a parallel configuration, which means that they will interfere with each other. In order to avoid any interference between the function generators, we add two inductors to the low-frequency function generator and two capacitors to the high-frequency function generator. The inductors block the high-frequency voltage, while the low-frequency voltage can pass easily, and capacitors block the low-frequency voltage, while they act as simple resistance for the high-frequency voltage. We can write the impedance of the inductors and capacitors as the following:

$$Z_L = \mathrm{i}\omega L \quad \text{and} \quad Z_C = 1/\mathrm{i}\omega C \tag{19.3}$$

where Z_L, L, Z_C, and C are inductor impedance, inductance, capacitor impedance, and capacity, respectively. The total impedance is:

$$Z_{\text{total}} = Z_L + Z_C = \mathrm{i}\omega L + \frac{1}{\mathrm{i}\omega C} + \frac{1 - LC\omega^2}{\mathrm{i}\omega C} \tag{19.4}$$

To obtain perfect shielding, the total impedance for high (ω_H) and low (ω_L) frequencies must be the same:

$$\left|Z_{\text{total,High}}\right| = \left|\frac{1 - LC\omega_H^2}{i\omega_H C}\right| = \left|\frac{1 - LC\omega_L^2}{i\omega_L C}\right| = \left|Z_{\text{total,low}}\right| \tag{19.5}$$

Finally, it can be written:

$$LC = \frac{1}{\omega_L \omega_C} \tag{19.6}$$

Therefore, the final circuit contains 1 Ohm resistance, two voltage generators in distinct frequency ranges, two capacitors and two inductors to shield the function generators from each other, and finally a lock-in amplifier to measure the sensor signal. A shematic circuit is shown in Figure 19.10.

The resistance of the IPMC can be written using the following equation:

$$R = \rho\left(\frac{l}{A}\right) \tag{19.7}$$

where l, A, and ρ are the width, surafce area, and electrical resistivity across thet IPMC, respectively. During bending, the resistivity of the sample

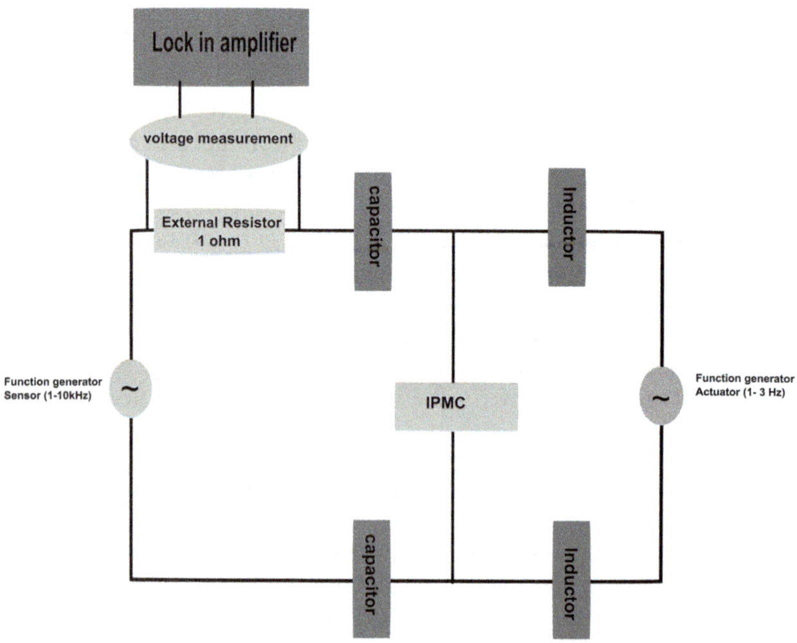

Figure 19.10 Schematic of a simultanous self-sensing and actuating circuit. Using two very distinctive frequencies in combination with a lock-in amplifier pervents any cross-talk between the sensor and actuator.

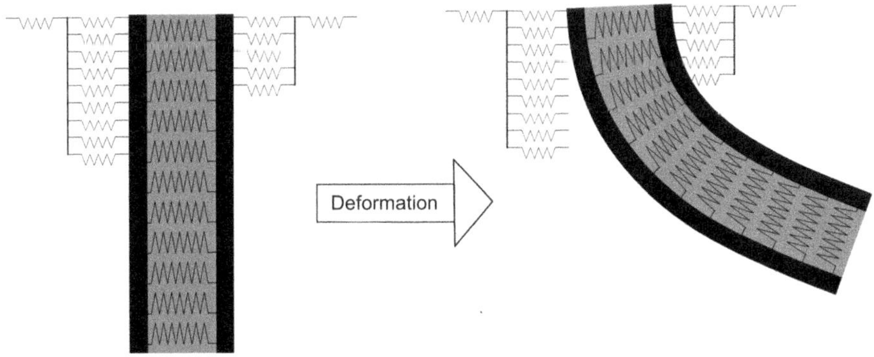

Figure 19.11 Schematic illustration of the asymmetrical electrodes. Here, for the sake of presentation, the sizes of the electrodes are exaggerated. In principle, the size of the big electrode is less than 5% of the cantilever.

changes, but at this stage, the sensitivity of HFR is not high enough to be measured. In order to solve this problem, we designed an asymmetric electrode instead of a symmetric one (Figure 19.11).

In this setup, the electrode on one side is always in contact with the IPMC, but the other electrode changes its contact area while the cantilever is bending. For this setup, we should rewrite eqn (19.1) as the following:

$$R_{\text{tot}} = R_{\text{IPMC}} + R_{\text{electrode}} + R_{\text{ex}} \tag{19.8}$$

One can model the electrode as n parallel electrodes with resistance of R; thus, the electrode can be written as the following:

$$\frac{1}{R_{\text{electrode}}} = \sum_{i=1}^{n} \frac{1}{R} = \frac{n}{R} \tag{19.9}$$

From eqn (19.9), we can conclude that a bigger contact leads to a smaller electrode resistance, which means that the voltage on the external resistance (R_{ex}) increases, and *vice versa*.

We used a Nafion-based IPMC with a platinum coating, which was made for us in the laboratory of Prof. Shahinpoor at Maine University.

19.4.2 Results and Discussion

Figure 19.12 shows the HFR sensing voltage in comparison with the laser sensor. The length of the bigger electrodes is around 5% and the width is around 10% of the size of the cantilever, and the electrode was made of thin copper. The thickness of the copper did not affect the signal except for in thicker one, which was a bit more rigid, so the bending was slightly smaller in one direction. Here, the sensor frequency and voltage for all tests were 10 kHz and 0.4 V, respectively. HFR sensor signals due to the actuating voltage and additional mechanical deformation correspond to the laser sensor (Figure 19.12), thus the HFR sensor is able to detect the position

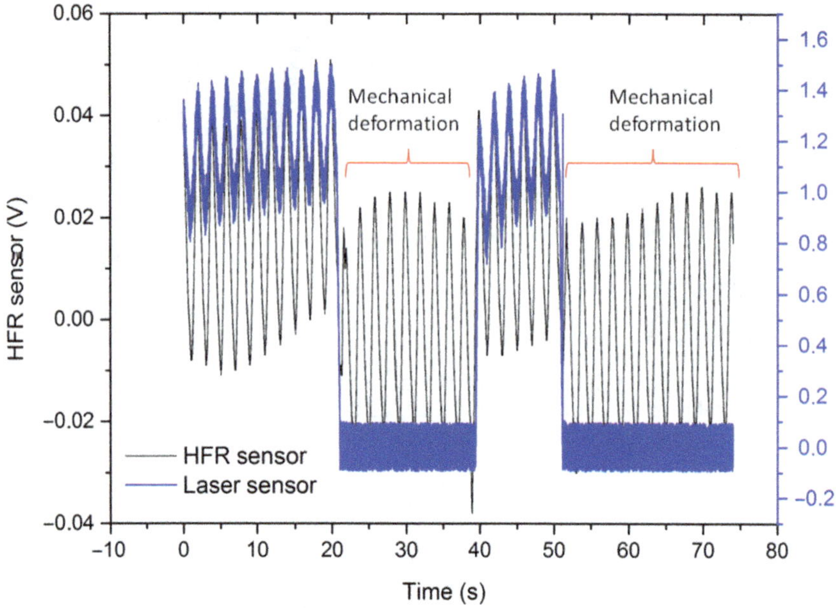

Figure 19.12 HFR signal and laser sensor while the actuator is running. There are two mechanical deformations of 3 mm, one starting at around 20 s and the other starting at around 50 s.

while actuation is running. However, one can say that the change of the sensing signal could be due to ion movement in the IPMC membrane, not the bending. This means that if we block the IPMC cantilever as the voltage is still running, the HFR sensor must show approximately the same signal.

To test whether the HFR sensor responds to the bending or actuating voltage, we blocked the cantilever and compared it with the free cantilever. The results of such a test are shown in Figure 19.13. We can see that the HFR sensor signal for the blocked sample is several-fold smaller than for the free cantilever. However, a careful check of the HFR sensor signal from the blocked sample reveals a slight dependence on the voltage of the actuator. This is because the cantilever is blocked at the tip, so there is still a tiny movement that can be detected by the HFR technique. As is clear from Figure 19.12, this technique measures the absolute position, not the relative position, because when the time is ~20 s or ~50 s, the cantilever is moved 3 mm, and after stopping the deformation, the level of signal did not change until the cantilever had returned to its original position. Therefore, Figures 19.12 and 19.13 confirm that the HFR sensing technique can measure the curvature of the IPMC.

For both MET (Section 19.2) and SR (Section 19.3), simultaneous sensing and actuating was accompanied by diminished actuating ability due to the decreasing size of the actuator caused by patterning. Nevertheless, in the HFR technique, the actuating ability remains totally untouched by the

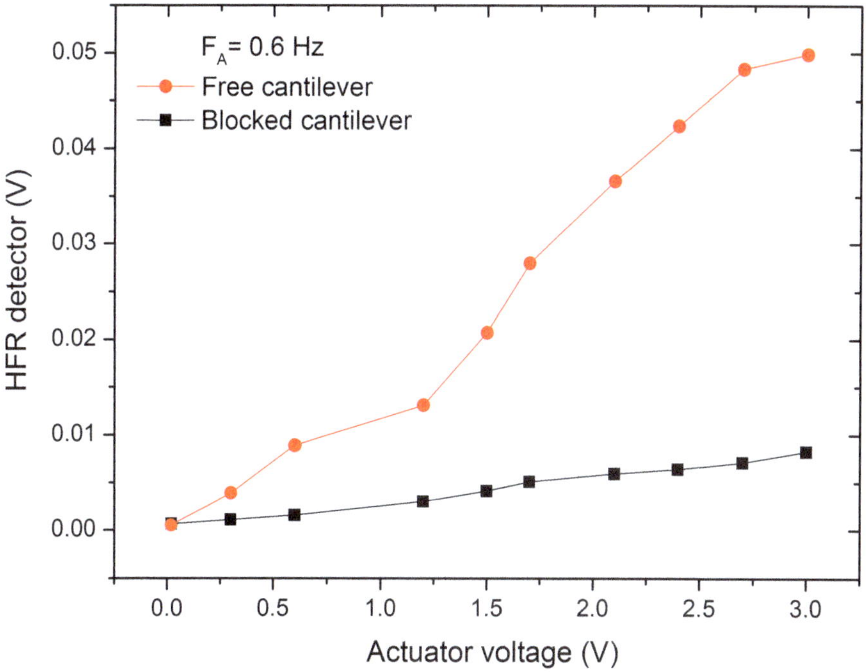

Figure 19.13 Comparison of the HFR sensor signal between blocked and non-blocked cantilevers. As we can see, the free sample has a much bigger signal than the blocked cantilever.

sensor (Figure 19.14). Figure 19.14 shows the laser signal while the HFR sensor was removed in comparison with a running HFR sensor. In principle, here two things might affect the bending: the asymmetric electrode and the high-frequency voltage. However, as we mentioned before, the longer electrode is soft and its length is less than 5% of the total length of the cantilever, so the bending is not blocked physically. Furthermore, the high-frequency voltage does not affect on the ion transportation in the membrane because cations cannot respond to such high frequencies, and their motion is derived only by the low-frequency voltage, which is the actuator voltage.

The bending of the cantilever is proportional to the actuator voltage; thus, we examined the sensitivity of the HFR sensor to various amounts of bending by changing the voltage of the actuator (Figure 19.15). As can be seen from Figure 19.15, the HFR sensor shows very good agreement with the laser sensor. We also examined the performance of the HFR sensors for various frequencies and voltages (Figure 19.16).

We should note that the laser voltage is linearly proportional to the tip displacement, and the HFR sensor voltage is also proportional to the laser sensor voltage. Therefore, for the range of frequencies and voltages used for the actuator, the HFR sensor shows an almost linear response to the bending.

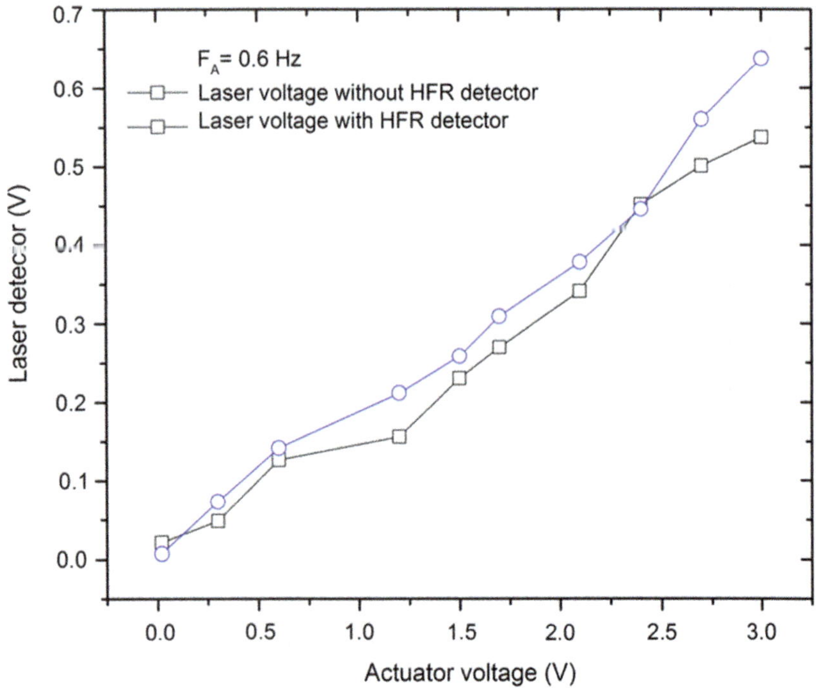

Figure 19.14 Comparison between bending for when the HFR detector is operating and when it is removed completely.

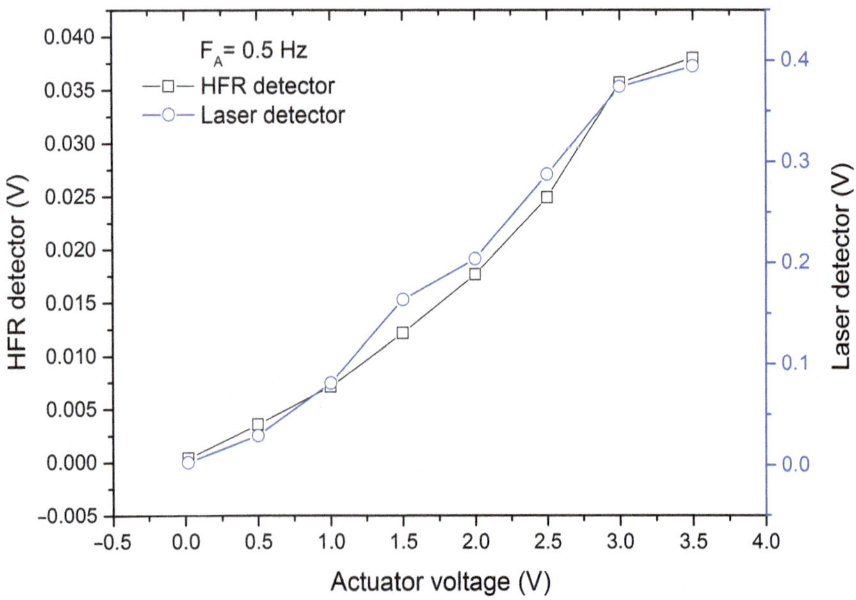

Figure 19.15 Sensitivity of the HFR sensor to different bending amplitudes.

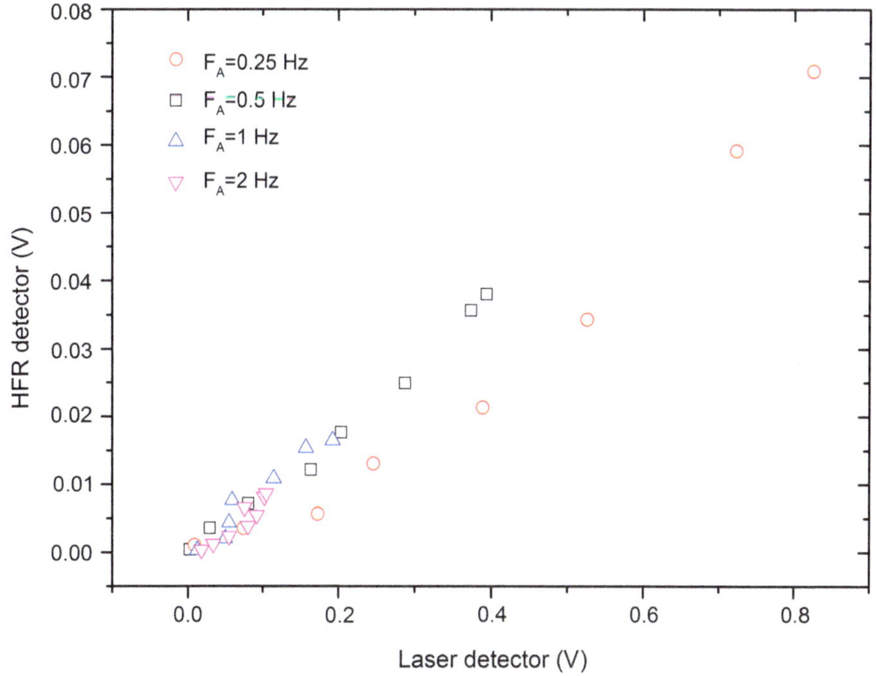

Figure 19.16 HFR sensing signal versus laser sensing signal.

19.5 Conclusion

In this chapter, we reviewed the recent advances in MET and SR sensors. Additionally, we present a proof of concept for a new technique of so-called HFR sensors. In the MET technique, the mechanical stimulus is translated into a measurable voltage. This technique is highly sensitive to acceleration, thus it is a very promising method for measuring impulse and impact. In spite of its ability to measure bending, it has some limitations for bending with a slow and constant velocity. The SR method exploits the change in the SR of the IPMC during bending; this technique is able to measure any bending regardless of velocity and acceleration. This method does not suffer from signal decay after stopping deformation as the MET sensor does, thus it measures the complete deformation cycle and no recording device is necessary. For the HFR sensing technique, the change across the IPMC cantilever was used to calculate the deformation. We illustrated that the deformation and sensing signals are proportional, and it is possible to measure the bending while the actuator is running. Both MET and SR sensors are able to perform simultaneous sensing and actuation, but realizing simultaneous measurements experimentally requires some considerations, such as of complicated fabrication processes and cross-talk between the actuator and sensor. Additionally, MET and SR methods impair actuating by decreasing the size of the actuating section and increasing the load on the actuator. The HFR sensing technique does not change the actuating ability and no additional procedures are needed for sample fabrication. However, this technique must undergo more extensive testing for various mechanical deformations in order to prove its ability as a reliable sensing method. In general, each of mentioned techniques has limitations and benefits; thus, depending on the type application, one should choose an appropriate technique.

References

1. Y. Bahramzadeh and M. Shahinpoor, *Soft Robotics*, 2014, **1**, 38.
2. K. J. Kim and S. Tadokoro, *Electroactive Polymers for Robotic Applications*. Springer, London, 2007.
3. J. Choonghee, D. Pugal, I. K. Oh, K. J. Kimc and K. Asaka, *Prog. Polym. Sci.*, 2013, **38**, 1037.
4. M. Shahinpoor and K. J Kim, *Smart Mater. Struct.*, 2005, **14**, 197.
5. Y. Bahramzadeh and M. Shahinpoor, *Smart Mater. Struct.*, 2011, **20**, 094011.
6. T. Ganley, D. L. S. Hung, G. Zhu and X. Tan, *IEEE/ASME Transactions on Mechatronics*, 2011, **16**, 80.
7. D. Ngo, C. Nam and K. K. Ahn, *Smart Mater. Struct.*, 2014, **23**, 025025.
8. E. Biddiss and T. Chau, *Med. Eng. Phys.*, 2006, **28**, 568.
9. L. Ferrara, M. Shahinpoor, K. J. Kim, B. Schreyer, A. Keshavarzi, E. Benzel and J. Lantz, *SPIE*, 1999, **3669**, 394.
10. K. Park, B. Lee, H.-M. Kim, K.-S. Choi, G. Hwang, G.-S. Byun and H.-K. Lee, *Int. J. Electrochem. Sci.*, 2013, **8**, 4098.
11. M. Aureli and M. Porfiri, *Continuum Mech. Thermodyn.*, 2013, **25**, 273.
12. A. Punning, M. Kruusmaa and A. Aabloo, *Sens. Actuators, A*, 2007, **136**, 656.
13. K. Kruusamae, P. Brunetto, S. Graziani, A. Punning, G. Di Pasqualeb and A. Aablooa, *Polym. Int.*, 2010, **59**, 300.
14. C. Bonomo, L. Fortuna, P. Giannone, S. Graziani and S. Strazzeri, *Smart Mater. Struct*, 2008, **17**, 015014.
15. K. K. Leang, Y. Shan, S. Song and K. J. Kim, *IEEE/ASME Transactions on Mechatronics*, 2012, **17**, 345.
16. H. Nakadoit, A. Sera, M. Yamakitatt, K. Asaka, Z.-W. Luot and K. Ito, Proceedings of International Conference on Mechatronics ICM 2007 4th IEEE International Conference, 2007, TuA1-B-5.
17. C. Bonomo, P. Brunetto and L. Fortuna, *IEEE Sens. J.*, 2008, **8**, 1486.
18. T. G. Noh, Y. Tak, J. D. Nam and H. Choi, *Electrochim. Acta*, 2007, **47**, 2341.
19. C. Bonomo and L. Fortuna, *Fellow, IEEE*, Pietro Giannone, and Salvatore Graziani, IEEE Transactions on Circuits and Systems—I: Regular Papers, 2006, 53, 338.

CHAPTER 20

A Continuum Multiphysics Theory for Electroactive Polymers and Ionic Polymer Metal Composites

JOHN G. MICHOPOULOS,*[a] MOHSEN SHAHINPOOR[b] AND ATHANASIOS ILIOPOULOS[c]

[a] Naval Research Laboratory, Computational Multiphysics Systems Lab Code 6394, Washington, DC 20374, USA; [b] University of Maine, Department of Mechanical Engineering, Orono, ME 04469, USA; [c] George Mason University, Computational Materials Science Center, Resident at NRL Code 6394, USA

20.1 Introduction

Progress in the manufacturing processes of various materials that are activated by excitation from multiple physical fields, such as electroactive polymers (EAPs) and ionic polymer metal composites (IPMCs), mainly for actuation applications such as for artificial muscles, has underlined the general need for the rigorous modeling of their behavior from a continuum-coupled multiphysics perspective.

On the one hand, the technological push for the multifunctional (*i.e.* actuation, sensing and energy harvesting), efficient and inexpensive design,

RSC Smart Materials No. 18
Ionic Polymer Metal Composites (IPMCs): Smart Multi-Functional Materials and Artificial Muscles, Volume 2
Edited by Mohsen Shahinpoor

Published by the Royal Society of Chemistry, www.rsc.org

prototyping and qualification of materials, devices and structures, and, on the other hand, the scientific pull of computational and mechatronic opportunities are the drivers motivating our continuous multiphysics behavioral modeling efforts.

The present chapter presents an overview of the development of a modeling effort that attempts not to ignore any coupled field and/or transport effects nor make any geometry dimensional simplifications other than those prescribed by the manufacturing process and the application specific context. The operational regime of our formulation is that of the macro length scale but is not limited to it. Consequently, the modeling effort can be classified as belonging to continuum multiphysics and mechanics. This effort is meant to represent an initial step towards modeling generalization for various length scales required for materials of this type.

Continuum systems under the influence of coupled multifield excitation or loading can exhibit static and dynamic behavior that can be highly variable in space and time across multiple scales, and that sometimes exhibits intense nonlinear behavior. Ionic polymers and composite materials and structural systems fall within this category of multiphysics systems when exposed to multiple physical excitations. Specific electromechanical modeling of such systems has already been attempted for simple low-dimensional geometries like membranes[1,2] and some one-dimensional systems in the form of simple strips.[3–10] In fact, a detailed exposition of the most comprehensive description of electromechanical models for simple strip configurations of IPMCs is given in Bar-Cohen[8] and especially in the chapter by Nemat-Nasser.[11] However, none of these contributions have considered the global continuous multiphysics perspective of more than two simultaneously acting fields, nor do they have derived multifield equations for arbitrarily shaped EAPs and IPMCs.

To address both of these issues, a multifield approach is required. Generalized approaches for deriving multiphysics theories have been developed both from thermodynamics and electrodynamics perspectives in 4D space–time.[12–17] All of the approaches consider the simultaneous presence of multiple fields but without the consideration of mass transport. To our knowledge, the first attempt to combine mass and heat transport with theories of deformable continua for the case of complex geometries was done for the case of isotropic hygrothermoelasticity[18] and the five-field theory of electromagnetic hygrothermoelasticity.[19]

In the context of solving boundary value problems in continuum mechanics, it is customary to define as field equations the set of partial differential equations (PDEs) that can be produced to describe field spatial and time evolution when all constitutive equations have been incorporated *via* term substitution. A brief overview of the generalized methodologies for achieving this goal has been presented more recently elsewhere.[20] The essence of the methodology for deriving such field evolution equations is based on utilizing the conservation laws of thermodynamics and electrodynamics that, when enabled with properly developed constitutive laws between the conjugate fields describing the behavior of multifield continua, yield an algebraically closed system of PDEs that are the field equations of the modeled system.

Furthermore, the problem to be addressed here lies in utilizing a field theory for such materials with the minimum amount of simplifications permitted by the feasibility of the computational implementation of solution schemes for the corresponding PDEs that govern the evolution of the corresponding state variables. Here we will consider the simultaneous presence of charge transport responsible for the ionic conductivity processes along with heat and mass transport (due to water or electrolyte diffusion) coupled with chemical reactions of the diffusing components in the material, all coupled with mechanical deformation. We have presented our preliminary formulation efforts for this problem from various perspectives elsewhere,[21–29] but in the present chapter, we are presenting a unifying overview of these efforts.

After the motivational impetus and the problem statement presented in this introduction, a brief description of the constitutive field theory (CFT) development process will be given in Section 20.2. Section 20.3 presents an application of this process for general anisotropic multicomponent hygrothermoelectroelasticity and derives the corresponding constitutive model. Section 20.4 further specializes this model with a simplification that applies to isotropic continua and produces field equations that maintain mechanical linearity and electrodynamic nonlinearity. Finally, conclusions will be given in the last section of this paper.

20.2 Overview of the Multifield and Constitutive Theory Framework

20.2.1 The Abstract Derivation Process

Every deformable solid under conditions of generalized loading that may include mass and heat fluxes, electric currents and mechanical loads or constrains can be considered as a system that is describable in space and time by the evolution of its state variables expressing the intensity of the involved fields. Some of these variables can be thought of as dependent or output parameters and some as independent or input parameters. Such a system can be idealized from two separate perspectives. One is the perspective of the systemic abstraction view of such a medium, as presented in Figure 20.1(a), and the other is its corresponding continuous multiphysics system, as shown in Figure 20.1(b).

The systemic abstraction allows for the decomposition of the bulk material state behavior and the structural state behavior, while the perspective of continuum mechanics (which is more traditional) is not equipped to make this differentiation, thus leading to confusion regarding what is externally observed during a structural material test. Usually, when researchers of material science refer to the term "material", they mostly mean a material

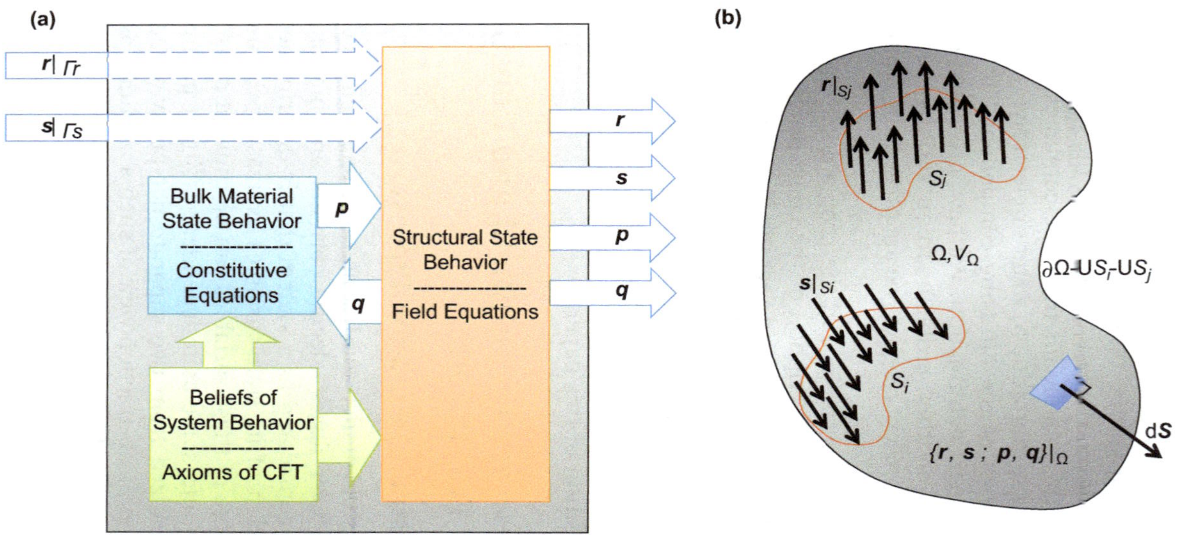

Figure 20.1 State variables under systemic (a) and continuum multiphysics (b) perspectives of a system under the influence of multiple generalized loads or fluxes and corresponding responses.

within a structure assuming, most of the time, a homogeneous state of strain. However, this is hardly ever possible because of the influence of the structural geometry on the observable and measured quantities. Similarly, when structural mechanics experts refer to the term "material", they think of a set of numbers (referring to the instantiation of the relevant material parameters) that influence how a structure behaves. However, constitutive behavior refers to the bulk material state behavior (in the sense of the behavior of the representative volume element behavior), while systemic behavior is the composition of both the bulk material state behavior and the structural behavior.

The discussion that follows applies to the bulk material state models, while it indirectly incorporates the geometry of the structure through the fact that we choose all thermodynamic quantities to be defined as densities per unit volume that, in order to satisfy the conservation laws of physics, eventually have to be integrated over the volume (or surface) where they are specified in the form of some boundary condition.

The bulk behavior of such a system is usually described as a set of relational restrictions among the state variables selected by the modeler as the ones pertinent to the situation at hand, given by:

$$\boldsymbol{F}(\boldsymbol{q},\xi,\boldsymbol{p}) = \boldsymbol{0} \tag{20.1}$$

where $\boldsymbol{q}$,ξ and $\boldsymbol{p}$ represent the input state variables, the internal state variables (if they exist) and the output state variables, respectively, all represented as components of the associated vectors (which is why italic bold symbols are used in eqn (20.1) and throughout this chapter). Those cases in which these relations can be solved with respect to the output variables are usually called constitutive relations of the form:

$$\boldsymbol{q} = \boldsymbol{C}(\xi,\boldsymbol{p}) \tag{20.2}$$

The vector $\boldsymbol{r}$ shown in Figure 20.1(b) does not participate in these expressions as they are only relevant to the description of the structural state behavior (*i.e.* for a deformable continuum, they would correspond to observable displacements associated through some kinematic relations with $\boldsymbol{P}$ that, in turn, would correspond to the strains associated with the constitutive relations).

The functionals $\boldsymbol{C}$ in eqn (20.2) represent an *a priori* definable multifunctional mapping of the form $\mathbb{R}^{\dim(\boldsymbol{p})} \times \mathbb{R}^{\dim(\xi)} \xrightarrow{C} \mathbb{R}^{\dim(\boldsymbol{q})}$. In most path history-independent state spaces, these functionals can be recovered by differentiation of an also *a priori* definable potential function $\Xi(\xi,\boldsymbol{p})$, usually definable as a thermodynamic potential, with respect to the internal (if present) and input (independent) state variables. This practice has been shown, followed by the early formulations of hyperelasticity,[13,30] where this potential was identified to be the strain energy density function. This potential function has to be constructed as a function of the input and the internal state variables and, if necessary, any time derivatives of them. This can be expressed by:

$$\boldsymbol{q} = \nabla_{\boldsymbol{p}}\, \Xi\left(\xi, \boldsymbol{p}, \frac{\partial^i \boldsymbol{p}}{\partial t^i}, \frac{\partial^i \xi}{\partial t^i}\right) \tag{20.3}$$

This formalism imposes a conjugation between input and output state variables in a way that allows us to form "cause–effect" pairs $\{\boldsymbol{q},\boldsymbol{p}\}$ and component form $\{q_i,p_i\}$ or conjugate variable pairs that have the property that their product has unit dimensionality of energy density per unit of volume or mass. Various researchers have introduced several choices for the potential function required for the constitutive relations in the context of the theory of continuous thermodynamics.[31–34] Some of these choices are internal energy, enthalpy, Helmholtz free energy and Gibbs free energy. However, these potentials are not independent from each other, as they are associated with appropriate Legendre transformations.[32]

In the context of continuous multiphysics, system behavior in terms of state evolution in most continuum systems is expressed in terms of solutions of PDEs that govern a special topological form on the fields described by the spatial and time distributions of the state variables and further restricts the potential values these variables can take; *i.e.*:

$$\aleph_i(\nabla^n \boldsymbol{q}, \boldsymbol{q}_{,t^m}, \ldots) = 0 \tag{20.4}$$

The traditional sources of such equations are the so-called "conservation" or "balance" laws of physics. These are the thermomechanical laws of conservation of mass, momentum, moment of momentum, energy, entropy flaw and the electrodynamic laws of conservation of electric displacement (Gauss–Faraday law), magnetic flux, electric charge, rotation of electric intensity (Faraday's law) and magnetic intensity (Ampere's law). These are also known as the axioms of continuous physics (ACP) that are not (formally) provable (in their most general form), but rather they are *a priori* beliefs that we accept to be true. Unfortunately, they are not always sufficient (algebraically closed) to completely determine the state variable field evolution and they are mostly used for ensuring that the corresponding CFTs maintain particular properties. It is exactly for this reason that the constitutive functionals are needed to obtain a complete set of useful algebraic equations. Introducing the constitutive equations in eqn (20.2) or (20.3) into eqn (20.4) and eliminating the independent variables or, more generally, half of the conjugate variables leads to a set of algebraically closed PDEs that constitute the so-called field equations of the system, and they have the form:

$$\wp_i(\boldsymbol{p}, \boldsymbol{p}_{,t^m}, \ldots) = 0 \tag{20.5}$$

where $\wp_i$ are differential operators acting on state variables $\boldsymbol{p}, \boldsymbol{p}_{,t^m}, \ldots$

Further simplifications and relational restrictions can be obtained by applying some or all of the additional axiomatic and meta-axiomatic restrictions that are traditionally called the axioms of constitutive theory.[15] The ACP have been historically expressed in either their global (integral) or the local (differential) forms, which are derived from the global form *via* the applications of the divergence (Gauss–Ostrogratsky) theorem.[15,34]

The global form of the conservation law can be written as:

$$\frac{\mathrm{d}}{\mathrm{d}t}\int_{V_\Omega} \Phi \mathrm{d}V = -\int_{\partial V_\Omega} \boldsymbol{F}\cdot \mathrm{d}\boldsymbol{S} - \int_{\partial V_\Omega} \Phi \mathbf{v}\cdot \mathrm{d}\boldsymbol{S} + \int_{V_\Omega} H \mathrm{d}V \tag{20.6}$$

where Φ is the conserved field within a volume V of domain Ω, $\boldsymbol{F}$ is the influx of Φ through the surface $\mathrm{d}\boldsymbol{S}$ and H is the amount of Φ produced in the body (source term). The quantity $\Phi\boldsymbol{v}$ in the second term of the right-hand side denotes the transport flux (quantity Φ per unit of time) that is moved by the velocity $\boldsymbol{v}$ through a part of the boundary. The respective local form of this equation is given as:

$$\frac{\partial \Phi}{\partial t} + \nabla \cdot (\boldsymbol{F} + \Phi \boldsymbol{v}) - H = 0 \tag{20.7}$$

The general process of simulating the behavior of a continuum system usually involves the solution of the PDEs describing its space and time evolution *via* application of a discretization method over the domain of their applicability and the subsequent solution of a set of ordinary differential and eventually algebraic equations. However, recent advances in material processing are suggesting that our ability to produce new material systems is outpacing our ability to model them.

20.2.2 Multiplicity of Thermodynamics

As indicated in the introduction, the formulation of a specific multiphysics theory capable of describing the interaction and the effects of the coexistence of multiple fields acting on a continuum system has been the focus of many approaches from various perspectives and driven by various applications needs. A brief historical overview of the various approaches is presented elsewhere.[20]

However, the two dominant approaches are rooted in the different representations of thermodynamics in the theory formation process. As stated eloquently by Maugin,[34] some of these approaches are more "traditional" and others are more adventurous. Some are based on microscale considerations while others discard any "molecular" or "atomic" length scale basis. Some exploit the experience from thermostatics while some simply ignore it. Some specify that their validity stands only for the case of being a little far-from-equilibrium, while others claim general validity well outside equilibrium. In the present chapter, we will focus on the two most prevalent approaches: the most adventurous presentation, which is that of *rational thermodynamics* (R.T.) and ignores altogether the acquired experience of thermostatics and claims larger validity anywhere outside equilibrium; and the most classical one, which is the *theory of irreversible processes* (T.I.P.), which does not ignore the experience acquired from the thermostatics and only operates in a region near equilibrium.

The T.I.P. is the most standard approach and is most widely accepted by physicists and physico-chemists. It is thoroughly described by various authors.[31–33] Relating to small deviations from equilibrium, it is supported by microscopic analysis proposed by Onsager[35,36] and Casimir.[32] The most important tenet of the T.I.P. is the axiom of local state that postulates that each part Ω of a material system S can be approximately considered, at each time t, as being in thermal equilibrium. Alternatively expressed, a thermodynamic process close to equilibrium can be viewed as a sequence of thermostatic equilibria. This allows us to endow entropy and temperature with their usual thermostatic definitions.

R.T. is the theory of phenomenological thermodynamics developed by Coleman, Noll, Truesdell, Eringen and co-workers in the 1960s.[12,13,15,37] It is based on the rational mechanics model of 18th and 19th century mathematicians (*i.e.* Lagrange and Cauchy) and the simple thermomechanics of Duhem (1911).[38] It openly ignores or bypasses the experience of thermostatics while it attempts to formally axiomatize thermodynamics. According to G. Maugin's interpretation,[34] its main postulates seem to be that those notions that could be defined precisely only in equilibrium for thermostatics exist *a priori* for any thermodynamic state, even largely outside equilibrium. This entropy and temperature are still granted to any state, so that the formal bases of R.T. are the *a priori* statements of the second law (entropy inequality) and the first law (energy conservation).

Both T.I.P. and R.T. accept the validity of the conservation laws (mass, momentum, moment of momentum and energy), as well as the entropy inequality. However, T.I.P. furthermore accepts the existence of entropy conservation, while R.T. does not require it nor uses it. To algebraically close the equations produced from the conservation principles, T.I.P. builds the constitutive equations between the conjugate variables expressing the fields present by differentiation of a thermodynamic potential, while R.T. begins from postulated forms of the constitutive functionals that are restricted to be consistent with Duhem's inequality.

In the present chapter, we will provide a derivation of a multiphysics theory for multifield continua under the scope of the T.I.P. only. Essentially, we will invoke the thermomechanics of electromagnetic continua, which is a branch of energetics that deals with the unification of continuum mechanics and electrodynamics of material media on top of the foundation provided by the general thermodynamics of irreversible processes near thermodynamic equilibrium.

20.3 Conservation Laws of Electrodynamics

20.3.1 Classic and Potential Formulations

The conservation laws of electrodynamics do not owe their existence to thermodynamic arguments, but they have been given in numerous formalisms. To ensure the proper definitions and their association with the rest of our development, we present here a brief exposition as it relates to the development of our theory.

From a macroscopic perspective, the conservation laws of electrodynamics are represented by Maxwell's system of four PDEs[39] given by a specialization of eqn (20.7), whereas the conserved quantities are the electric vector field $\boldsymbol{E}$, the magnetic flux density vector $\boldsymbol{B}$ (which sometimes is also called the magnetic field) and their conjugate fields represented by the electric and magnetic displacement vectors $\boldsymbol{D}$ and $\boldsymbol{H}$, respectively. For a large subset of the authors, $\boldsymbol{H}$ is also called the magnetic field (when $\boldsymbol{B}$ is called magnetic flux density). The associated form of the Maxwell system as it will be used in the present analysis is given by:

$$\nabla \cdot \boldsymbol{D} = \rho q \tag{20.8}$$

$$\nabla \cdot \boldsymbol{B} = 0 \tag{20.9}$$

$$\frac{\partial \boldsymbol{D}}{\partial t} - \nabla \times \boldsymbol{H} = -\boldsymbol{I} \tag{20.10}$$

$$\frac{\partial \boldsymbol{B}}{\partial t} + \nabla \times \boldsymbol{E} = 0 \tag{20.11}$$

These equations represent Gauss's law, the law of conservation of magnetic flux, Ampere's law and Faraday's law, respectively.[15] The quantities ρ, q and $\boldsymbol{I}$ represent the mass density, the electrical charge density per unit mass and the electric current density vector, respectively.

These four equations (*i.e.* eqn (20.8) through (20.11)) are expressed in terms of five vector field variables $\boldsymbol{E}$, $\boldsymbol{D}$, $\boldsymbol{H}$, $\boldsymbol{B}$ and $\boldsymbol{I}$ (which form the conjugate pairs $\{\boldsymbol{E}, \boldsymbol{D}\}$ and $\{\boldsymbol{H}, \boldsymbol{B}\}$), and therefore they are not algebraically closed. Some authors believe that the current conservation equation provides the required fifth equation, but because this equation is derivable form algebraic rewriting of eqn (20.8) through (20.11), it is not an independent equation.

In fact, if we take the divergence of both sides of Ampere's law (*i.e.* eqn (20.10)), and since the divergence of a curl of any vector is zero, we obtain:

$$\frac{\partial (\nabla \cdot \boldsymbol{D})}{\partial t} = -\nabla \cdot \boldsymbol{I} \tag{20.12}$$

Which, upon usage of Gauss's law (eqn (20.8)), reduces to the current continuity (or conservation) equation:

$$\rho \frac{\partial q}{\partial t} + \nabla \cdot \boldsymbol{I} = 0 \tag{20.13}$$

Therefore, additional relations need to be developed. This can be achieved by introducing the use of the electromagnetic constitutive relations. The most common (but not always applicable) form of these equations involves the postulate that a linear relationship between the electric and magnetic conjugate fields exists as follows:

$$\boldsymbol{D} = \boldsymbol{\varepsilon} \cdot \boldsymbol{E} \tag{20.14}$$

$$\boldsymbol{B} = \boldsymbol{\mu} \cdot \boldsymbol{H} \tag{20.15}$$

In eqn (20.14) and (20.15), the quantities $\boldsymbol{\varepsilon}$ and $\boldsymbol{\mu}$ represent the second-order dielectric (or permittivity) and magnetic permeability tensors, respectively, and they allow for the general case of describing an electromagnetically anisotropic medium. The bold upright font will be used for all tensors from now on. The dot product in these relations represents the multiplication of a second-order tensor with a first-order one (vector), and therefore it returns a vector. For isotropic materials, these two tensors can be expressed in terms of a scalar permittivity and permeability and the unit diagonal tensor $\mathbf{U}$ (not to be confused with the symbol for the set union operator), as follows:

$$\boldsymbol{\varepsilon} = \varepsilon \mathbf{U} \tag{20.16}$$

$$\boldsymbol{\mu} = \mu \mathbf{U} \tag{20.17}$$

For polarizable media, we need to introduce the electric and magnetic polarizations that are also defined by relations:

$$\boldsymbol{P} = \boldsymbol{D} - \boldsymbol{E} \tag{20.18}$$

$$\boldsymbol{M} = \boldsymbol{B} - \boldsymbol{H} \tag{20.19}$$

It is important to note that combining eqn (20.18) and (20.19) with eqn (20.14) and (20.15) reveals that the polarization vectors can be expressed as linear functions of the electric and magnetic fields, respectively, according to:

$$\boldsymbol{P} = (\boldsymbol{\varepsilon} - \mathbf{U}) \cdot \boldsymbol{E} = \boldsymbol{\kappa} \cdot \boldsymbol{E} \tag{20.20}$$

$$\boldsymbol{M} = (\boldsymbol{\mu} - \mathbf{U}) \cdot \boldsymbol{H} = \boldsymbol{\chi} \cdot \boldsymbol{H} \tag{20.21}$$

where $\boldsymbol{\kappa}$ and $\boldsymbol{\chi}$ are the electric and magnetic susceptibility second-order tensors, respectively, which can be further simplified for isotropic materials by utilizing eqn (20.16) and (20.17) for the case of materials with isotropic electric and magnetic response.

In some cases and for many reasons of practicality, the electric field and the magnetic flux density is also defined in terms of a scalar electric field ϕ and a vector magnetic potential $\boldsymbol{A}$ according to:

$$\boldsymbol{E} = -\nabla\varphi - \frac{\partial \boldsymbol{A}}{\partial t} \tag{20.22}$$

$$\boldsymbol{B} = \nabla \times \boldsymbol{A} \tag{20.23}$$

Vector calculus suggests that these two equations are realizable due to the validity of eqn (20.11) and (20.9), respectively. In fact, eqn (20.22) and (20.23) can be considered the dual forms of eqn (20.11) and (20.9).

By implication, inserting these two relations into eqn (20.8) and (20.10) along with the help of eqn (20.12), and after some algebraic manipulations, we can derive the following two relations:

$$\nabla \cdot \left(-\boldsymbol{\sigma} \cdot \frac{\partial \boldsymbol{A}}{\partial t} + (\boldsymbol{\sigma} \cdot \boldsymbol{v}) \times \nabla \times \boldsymbol{A} - \boldsymbol{\sigma} \cdot \nabla \varphi + \boldsymbol{I}_e \right) = 0 \tag{20.24}$$

$$\boldsymbol{\sigma} \cdot \frac{\partial \boldsymbol{A}}{\partial t} + \nabla \times \left(\boldsymbol{\mu}^{-1} \nabla \times \boldsymbol{A} \right) - (\boldsymbol{\sigma} \cdot \boldsymbol{v}) \times \nabla \times \boldsymbol{A} + \boldsymbol{\sigma} \cdot \nabla \varphi = \boldsymbol{I}_e \tag{20.25}$$

which represent the scalar and vector potential forms of the laws of Gauss and Ampere, respectively, and along with eqn (20.22) and (20.23) representing the potential formulations of electrodynamics. In these equations, we have introduced the second-order conductivity tensor $\boldsymbol{\sigma}$ and the external current density $\boldsymbol{I}_e$. These quantities are introduced from yet another linear constitutive relationship that appears as follows:

$$\boldsymbol{I} = \boldsymbol{\sigma} \cdot \boldsymbol{E} + \boldsymbol{I}_e \tag{20.26}$$

and expresses the generalized Ohm's law.

20.3.2 Electric Conductivity through Charge Relaxation

In the potential formulation of electrodynamics and through eqn (20.24) and (20.25), above we have introduced the second-order conductivity tensor $\boldsymbol{\sigma}$, but this quantity was not a part of the original equations of Maxwell. To reconcile this discrepancy between the original and the potential formalisms of electrodynamics, we need to refer to the relaxation theory of conductivity as it emerges from Gauss's law and the current continuity equation.

By introducing eqn (20.14) into Gauss's law given by eqn (20.8), we obtain:

$$\nabla \cdot (\boldsymbol{\varepsilon} \cdot \boldsymbol{E}) = \rho q \tag{20.27}$$

and combining this with the current continuity in eqn (20.13) along with the Ohm's law expressed by eqn (20.26), we obtain the following differential equation governing the electric charge density:

$$\frac{\partial q}{\partial t} + \boldsymbol{\sigma} : \boldsymbol{\varepsilon}^{-1} q = 0 \tag{20.28}$$

This equation has a solution of the form:

$$q(t) = q_0 \mathrm{e}^{-t/\tau} \tag{20.29}$$

where:

$$\tau = \boldsymbol{\varepsilon} : \boldsymbol{\sigma}^{-1} \tag{20.30}$$

is called the charge relaxation time. For a good conductor like copper, τ is of the order of 10^{-19} s, whereas for a good insulator like silica glass, it is of the order of 10^3 s, and for a pure insulator it becomes infinite.

When $\tau \gg t$, the external time scale is short compared to the charge relaxation time and the charges do not have time to redistribute, and therefore the medium operates under electrostatic conditions that essentially permit the eliminating of all time derivatives form our electrodynamics. Alternatively, when $\tau \ll t$, the external time scale is long compared to the charge relaxation time and the stationary solution of the continuity equation has been reached.

All electrodynamic fields required for the follow-up analysis have now been defined.

20.4 Transport of Multicomponent Mass, Heat and Electric Current in Deformable Continua

20.4.1 Mass, Charge and Current Density Conservation

We consider a material infused with a mixture of n (charged and uncharged) species that is placed in an electromagnetic field. The mass conservation law for each species, provided the system is close to mechanical equilibrium ($d\boldsymbol{v}/dt \simeq 0$) and also that the barycentric velocity itself is negligible ($\boldsymbol{v} \simeq 0$) can be written again with the help of eqn (20.7) as follows:

$$\rho \frac{\partial c_k}{\partial t} = -\nabla \cdot \boldsymbol{J}_k + \sum_{j=1}^{r} r_{kj} J_j, \ (k = 1, 2, \ldots, n) \tag{20.31}$$

where the second term of the right-hand side reflects the source term representing the mass production $r_{kj}J_j$ of species k per unit volume in the jth chemical reaction, with the coefficient r_{kj} being (when divided by the molecular mass M_k) proportional to the stoichiometric coefficient with which k appears in the chemical reaction j. In addition, coefficients c_k represent the mass fractions that are defined in terms of the species densities ρ_k as:

$$c_k = \frac{\rho_k}{\rho}, \left(\sum_{j=1}^{n} \rho_k = \rho \right) \tag{20.32}$$

and where $\boldsymbol{J}_k$ are the "diffusion flow fluxes"[32] defined by:

$$\boldsymbol{J}_k = \rho_k(\boldsymbol{v}_k - \boldsymbol{v}), \quad (k = 1, 2, \ldots, n) \tag{20.33}$$

with $\boldsymbol{v}_k$ being the velocity vector of species k and $\boldsymbol{v}$ being the velocity vector of the center of mass of the species defined by:

$$\boldsymbol{v} = \sum_{k=1}^{n} \rho_k \boldsymbol{v}_k / \rho \tag{20.34}$$

Assuming that the chemical reaction term can be neglected because there are no reactions going on and summing eqn (20.31) over all substances k yields the law of conservation of total mass:

$$\frac{\partial \rho}{\partial t} = -\nabla \cdot (\rho \boldsymbol{v}) \tag{20.35}$$

The total current density $\boldsymbol{I}$ can now be written in terms of the velocities of the species components:

$$\boldsymbol{I}=\sum_{k=1}^{n}\rho_k q_k \boldsymbol{v}_k=\rho q\boldsymbol{v}+\sum_{k=1}^{n}q_k\boldsymbol{J}_k \tag{20.36}$$

where eqn (20.33) has been introduced to obtain the right-hand side of eqn (20.36) and q_k is the electric charge per unit mass of component k and the total charge q per unit mass is defined as:

$$q=\rho^{-1}\sum_{k=1}^{n}\rho_k q_k=\sum_{k=1}^{n}c_k q_k \tag{20.37}$$

The second term of the right-hand side of eqn (20.36) is usually interpreted as the electric current density due to the relative motion of various species components. Thus, by introducing the replacement:

$$\boldsymbol{I}_c=\sum_{k=1}^{n}q_k\boldsymbol{J}_k \tag{20.38}$$

into eqn (20.36), we obtain:

$$\boldsymbol{I}=\boldsymbol{I}_\nu+\boldsymbol{I}_c \tag{20.39}$$

where the term $\boldsymbol{I}_\nu=\rho q\boldsymbol{v}$ represents the electric current due to convection and $\boldsymbol{I}_c$ represents the conduction current.

If we assume that there are no chemical reactions, combining eqn (20.31) and (20.37) yields the law of conservation of charge as follows:

$$\rho\frac{\partial q}{\partial t}=-\nabla\cdot\boldsymbol{I}_c \tag{20.40}$$

It is interesting to note that although the law of charge conservation has been derived here from the law of mass conservation of the species components, it can also be derived from a combination of the Maxwell equations as seen in Section 20.3.1, as is evident from eqn (20.13). Clearly, for the case of conductors, where the electric charge density must obey electric neutrality $\rho_e=\rho q=\rho\sum_{j=1}^{n}c_k q_k=0$, the conservation of charge equation in eqn (20.26) reduces to the divergence-free condition for the current density as follows:

$$\nabla\cdot\boldsymbol{I}_c=0 \tag{20.41}$$

20.4.2 Momentum Conservation

In classical Newtonian mechanics, momentum of a material system of mass m that is moving with a velocity $\boldsymbol{v}$ is defined as $\boldsymbol{p}=m\boldsymbol{v}$. The per unit volume conservation of momentum when the electromagnetic momentum $\boldsymbol{p}_{em} = \boldsymbol{E} \times \boldsymbol{H}$ is included can be written in terms of the mass density as:

$$\frac{\partial}{\partial t}(\rho\boldsymbol{v} + \boldsymbol{E} \times \boldsymbol{H}) = -\nabla \cdot (\rho\boldsymbol{v}\boldsymbol{v} + \boldsymbol{\tau}) + (\boldsymbol{F}_L + \boldsymbol{F}_b) \tag{20.42}$$

where $\rho\boldsymbol{v}\boldsymbol{v}$, $\boldsymbol{\tau}$, $\boldsymbol{F}_L$ and $\boldsymbol{F}_b$ are the convective part of the momentum flow tensor, the Cauchy stress tensor, the Lorentz force density per unit volume and the body force density per unit volume, respectively. The Lorentz force density is defined in terms of the electric field, the velocity, the current density, the magnetic flux density and the electric and the magnetic polarization as:

$$\boldsymbol{F}_L = \rho q(\boldsymbol{E} + \boldsymbol{v} \times \boldsymbol{B}) + \boldsymbol{I} \times \boldsymbol{B} + (\nabla\boldsymbol{E}) \cdot \boldsymbol{P} + (\nabla\boldsymbol{B}) \cdot \boldsymbol{M} + \frac{\mathrm{d}}{\mathrm{d}t}(\boldsymbol{P} \times \boldsymbol{B}) - \frac{\mathrm{d}}{\mathrm{d}t}(\boldsymbol{M} \times \boldsymbol{E}) \tag{20.43}$$

When this expression for the Lorentz force density is introduced into eqn (20.42), it yields:

$$\begin{aligned}\frac{\partial}{\partial t}(\rho\boldsymbol{v}) = & -\nabla \cdot \boldsymbol{\tau} + \rho q\boldsymbol{E} + \boldsymbol{I} \times \boldsymbol{B} + (\nabla\boldsymbol{E}) \cdot \boldsymbol{P} + (\nabla\boldsymbol{B}) \cdot \boldsymbol{M} + \frac{\mathrm{d}}{\mathrm{d}t}(\boldsymbol{P} \times \boldsymbol{B}) \\ & - \frac{\mathrm{d}}{\mathrm{d}t}(\boldsymbol{M} \times \boldsymbol{E}) + \boldsymbol{F}_b\end{aligned} \tag{20.44}$$

This equation, in addition to representing the momentum conservation, also represents the usual equations of motion known as force balance equations (when the vectors are expanded to their component form).

20.4.3 Energy Conservation

The energy conservation equation can be obtained again by introducing the relevant energy quantities in the generic conservation equation in eqn (20.7) that takes the form:

$$\frac{\partial E}{\partial t} = -\nabla \cdot \boldsymbol{J}_E + Q \tag{20.45}$$

where E, $\boldsymbol{J}_E$ and Q are the total internal energy density, the associated total energy current and the energy generated in the body (source term), respectively. The total energy can be written in terms of the internal energy density U, the kinetic energy of matter and the stored electromagnetic energy as follows:

$$E = U + \frac{1}{2}(\rho v^2 + \boldsymbol{D} \cdot \boldsymbol{E} + \boldsymbol{B} \cdot \boldsymbol{H} - \boldsymbol{P} \cdot \boldsymbol{E} - \boldsymbol{M} \cdot \boldsymbol{B}) + 2(\boldsymbol{E} \times \boldsymbol{H}) \tag{20.46}$$

The energy current can be defined as follows:

$$\boldsymbol{J}_E = \boldsymbol{J}_q + \left(\frac{1}{2}\rho v^2\boldsymbol{v} + U\boldsymbol{v} + \boldsymbol{\tau} \cdot \boldsymbol{v} - (\boldsymbol{P} \cdot \boldsymbol{E} + \boldsymbol{M} \cdot \boldsymbol{B})\boldsymbol{v} + \boldsymbol{E} \times \boldsymbol{H}\right) \tag{20.47}$$

where $\boldsymbol{J}_q$ represents the heat flux crossing the boundary.

After multiplying eqn (20.44) by $\boldsymbol{v}$, introducing eqn (20.46) and (20.47) into eqn (20.36) and incorporating the electromagnetic energy balance expressed by the Poynting theorem:

$$\boldsymbol{E}\cdot\frac{\partial\boldsymbol{D}}{\partial t}+\boldsymbol{H}\cdot\frac{\partial\boldsymbol{B}}{\partial t}=-\nabla\cdot(\boldsymbol{E}\times\boldsymbol{H})-\boldsymbol{I}\cdot\boldsymbol{E} \tag{20.48}$$

and after the well-known replacement $\nabla\boldsymbol{v}=\frac{d\gamma}{dt}$ and some algebraic manipulations, we obtain the conservation of internal energy in the form:

$$\frac{\partial U}{\partial t}=-\nabla\cdot\boldsymbol{J}_q-\boldsymbol{\tau}:\frac{\mathrm{d}\gamma}{\mathrm{d}t}+\boldsymbol{I}\cdot\boldsymbol{E}+\boldsymbol{E}\cdot\frac{\mathrm{d}\boldsymbol{P}}{\mathrm{d}t}+\boldsymbol{B}\cdot\frac{\mathrm{d}\boldsymbol{M}}{\mathrm{d}t}+Q \tag{20.49}$$

This equation represents the third conservation law that is referred to as the energy conservation law and sometimes it is also referred to as the first law of thermodynamics.

20.4.4 Entropy Conservation and the Second Law

A significant difference between T.I.P. and R.T. is the fact that in T.I.P. the conservation principles extend beyond the mass, momentum and energy conservation, and they include the conservation of entropy, while R.T. do not require it. A requirement of T.I.P. for the derivation of the entropy conservation equations is the choice for the definition of the Gibbs equation[32] that essentially collects all of the entropy producing and consuming sources in a per unit time sense. By following the program described in de Groot and Mazur[32] for polarizable media without chemical reactions between the species, we can adopt the form:

$$\frac{\mathrm{d}S}{\mathrm{d}t}=\frac{1}{T}\frac{\mathrm{d}U}{\mathrm{d}t}+\frac{Q}{T}+\frac{\boldsymbol{\tau}}{T}:\frac{\mathrm{d}\gamma}{\mathrm{d}t}-\frac{\boldsymbol{E}}{T}\cdot\frac{\mathrm{d}\boldsymbol{P}}{\mathrm{d}t}-\frac{\boldsymbol{B}}{T}\cdot\frac{\mathrm{d}\boldsymbol{M}}{\mathrm{d}t}-\sum_{i=1}^{n}\frac{\mu_i}{T}\frac{\mathrm{d}C_i}{\mathrm{d}t} \tag{20.50}$$

where S, γ, μ_i and C_i are the entropy per unit volume, the strain tensor, the chemical potential of component i in the system and the associated mass concentration of the species component i, respectively. Introducing into eqn (20.50) the expressions from the previous equations, we can finally obtain:

$$\begin{aligned}\frac{\mathrm{d}S}{\mathrm{d}t}=&-\nabla\cdot\frac{1}{T}\left(\boldsymbol{J}_q-\sum_{k=1}^{n}\mu_k\boldsymbol{J}_k\right)-\frac{1}{T^2}\boldsymbol{J}_q\cdot\nabla T-\frac{1}{T}\sum_{k=1}^{n}\boldsymbol{J}_k\cdot\left(T\nabla\frac{\mu_k}{T}-q_k\boldsymbol{\kappa}^{-1}\cdot\boldsymbol{P}\right)\\&-\frac{Q}{T}-\frac{\boldsymbol{\tau}}{T}:\frac{\mathrm{d}\gamma}{\mathrm{d}t}-\frac{1}{T}\frac{\mathrm{d}\boldsymbol{P}}{\mathrm{d}t}\cdot\left(\boldsymbol{\kappa}^{-1}\cdot\boldsymbol{P}-\boldsymbol{E}\right)-\frac{1}{T}\frac{\mathrm{d}\boldsymbol{M}}{\mathrm{d}t}\cdot\left((\boldsymbol{\chi}-\mathsf{U})\cdot\boldsymbol{\chi}^{-1}\cdot\boldsymbol{M}-\boldsymbol{B}\right)\end{aligned} \tag{20.51}$$

The last two terms in this equation are expressing the electric and magnetic relaxation due to the respective polarizations, and when the medium is not polarizable, they vanish due to the nullification of the differences inside the parentheses. For the special case when the material is a conductor (not electric polarization) and it is not magnetic, eqn (20.51) reduces to:

$$\frac{dS}{dt} = -\nabla \cdot \frac{1}{T}\left(\boldsymbol{J}_q - \sum_{k=1}^{n} \mu_k \boldsymbol{J}_k\right) - \frac{1}{T^2}\boldsymbol{J}_q \cdot \nabla T - \frac{1}{T}\sum_{k=1}^{n} \boldsymbol{J}_k \cdot \left(T\nabla \frac{\mu_k}{T} - q_k \mathrm{E}\right) - \frac{Q}{T} - \frac{\tau}{T} : \frac{d\gamma}{dt} \tag{20.52}$$

which is the same as eqn (XIII.37) in de Groot and Mazur.[32]

It is easy to confirm that by introducing the entropy current flow in the form:

$$\boldsymbol{J}_S = \frac{1}{T}\left(\boldsymbol{J}_q - \sum_{k=1}^{n} \mu_k \boldsymbol{J}_k\right) \tag{20.53}$$

and the entropy source term (or dissipation function):

$$\psi = \frac{1}{T^2}\boldsymbol{J}_q \cdot \nabla T - \frac{1}{T}\sum_{k=1}^{n} \boldsymbol{J}_k \cdot \left(T\nabla \frac{\mu_k}{T} - q_k \boldsymbol{\kappa}^{-1} \cdot \boldsymbol{P}\right) - \frac{Q}{T} - \frac{\tau}{T} : \frac{d\gamma}{dt} - \frac{1}{T}\frac{d\boldsymbol{P}}{dt} \cdot \left(\boldsymbol{\kappa}^{-1} \cdot \boldsymbol{P} - \boldsymbol{E}\right) - \frac{1}{T}\frac{d\boldsymbol{M}}{dt} \cdot \left((\chi - \mathsf{U}) \cdot \chi^{-1} \cdot \boldsymbol{M} - \boldsymbol{B}\right) \tag{20.54}$$

eqn (20.51) takes the original form of the conservation law expressed by eqn (20.7) as follows:

$$\frac{\partial S}{\partial t} = -\nabla \cdot \boldsymbol{J}_S + \psi \tag{20.55}$$

This equation represents the fourth equation of T.I.P., but in addition to it, the second law of thermodynamics requires that the dissipation function is always positive definite, *i.e.*:

$$\psi \geq 0 \tag{20.56}$$

which, after eqn (20.53), can be expanded to:

$$\frac{1}{T^2}\boldsymbol{J}_q \cdot \nabla T - \frac{1}{T}\sum_{k=1}^{n} \boldsymbol{J}_k \cdot \left(T\nabla \frac{\mu_k}{T} - q_k \boldsymbol{\kappa}^{-1} \cdot \boldsymbol{P}\right) - \frac{Q}{T} - \frac{\tau}{T} : \frac{d\gamma}{dt} - \frac{1}{T}\frac{d\boldsymbol{P}}{dt} \cdot \left(\boldsymbol{\kappa}^{-1} \cdot \boldsymbol{P} - \boldsymbol{E}\right) - \frac{1}{T}\frac{d\boldsymbol{M}}{dt} \cdot \left((\chi - \mathsf{U}) \cdot \chi^{-1} \cdot \boldsymbol{M} - \boldsymbol{B}\right) \geq 0 \tag{20.57}$$

In the T.I.P., both eqn (20.55) and (20.57) must hold simulateneously.[32]

Because of eqn (20.20) and (20.36), eqn (20.57) can be written to include the electric conduction current as follows:

$$\frac{1}{T^2}\boldsymbol{J}_q\cdot\nabla T-\frac{1}{T}\sum_{k=1}^{n}\boldsymbol{J}_k\cdot\nabla\mu_k+\frac{1}{T}\boldsymbol{I}\cdot\boldsymbol{E}-\frac{Q}{T}-\frac{\boldsymbol{\tau}}{T}:\frac{\mathrm{d}\boldsymbol{\gamma}}{\mathrm{d}t}$$
$$-\frac{1}{T}\frac{\mathrm{d}\boldsymbol{P}}{\mathrm{d}t}\cdot\left(\boldsymbol{\kappa}^{-1}\cdot\boldsymbol{P}-\boldsymbol{E}\right)-\frac{1}{T}\frac{\mathrm{d}\boldsymbol{M}}{\mathrm{d}t}\cdot\left((\boldsymbol{\chi}-\mathsf{U})\cdot\boldsymbol{\chi}^{-1}\cdot\boldsymbol{M}-\boldsymbol{B}\right)\geq 0 \tag{20.58}$$

This expression is a more well-known expression of the second law of thermodynamics.

20.5 Development of Constitutive Theory

While the conservation laws given by the $n+3$ equations in eqn (20.31), (20.44), (20.49) and (20.54) are providing the necessary equational infrastructure describing the evolution behavior of their associated fields, they are not algebraically closed since we have more unknowns than equations. Therefore, additional equations have to be generated to achieve the goal of algebraic closure. For this purpose, specific constitutive equation coupling the conjugate pair field variables must be generated.

To this end, and according to the T.I.P., the existence of a thermodynamic potential is postulated. The choices for such a potential function are extended among all those that participate in the Legendre transformations between the various combinations of thermodynamic potentials.[32,33]

Thus, in what follows, we introduce the free energy F and the enthalpy H relative to the internal energy and entropy *via* the expressions:

$$F=U-ST \tag{20.59}$$

and:

$$H=F-\boldsymbol{E}\cdot\boldsymbol{P}-\boldsymbol{H}\cdot\boldsymbol{M}-\sum_{k=1}^{n}\mu_k C_k \tag{20.60}$$

where C_k represents the concentration of species k in units of mass per unit volume and is defined as $C_k=\rho c_k=\rho_k$.

Combining eqn (20.59) and (20.60) yields an expression of the internal energy in terms of enthalpy as follows:

$$U=H+ST+\boldsymbol{E}\cdot\boldsymbol{P}+\boldsymbol{H}\cdot\boldsymbol{M}+\sum_{k=1}^{n}\mu_k C_k \tag{20.61}$$

which in turn enables the rewriting of the energy equation in eqn (20.49) in the form:

$$\frac{\partial H}{\partial t}=-\nabla\cdot\boldsymbol{J}_q+\boldsymbol{\tau}:\frac{\mathrm{d}\boldsymbol{\gamma}}{\mathrm{d}t}-S\cdot\frac{\mathrm{d}T}{\mathrm{d}t}-T\cdot\frac{\mathrm{d}S}{\mathrm{d}t}+\boldsymbol{I}\cdot\boldsymbol{E}-\boldsymbol{E}\cdot\frac{\mathrm{d}\boldsymbol{P}}{\mathrm{d}t}-\boldsymbol{H}\cdot\frac{\mathrm{d}\boldsymbol{M}}{\mathrm{d}t}$$
$$-\sum_{k=1}^{n}\mu_k\frac{dC_k}{dt}+Q \tag{20.62}$$

By eliminating the heat source term $\frac{Q}{T}$ from the inequality in eqn (20.58) *via* eqn (20.62), we obtain the transformed inequality:

$$
\begin{aligned}
&-\left(\frac{\mathrm{d}H}{\mathrm{d}t}+\frac{\mathrm{d}T}{\mathrm{d}t}\right)-\frac{1}{T}\boldsymbol{J}_q\cdot\nabla T+\boldsymbol{\tau}:\frac{\mathrm{d}\gamma}{\mathrm{d}t}-\sum_{k=1}^{n}\boldsymbol{J}_k\cdot\nabla\mu_k+\boldsymbol{I}\cdot\boldsymbol{E}\\
&-\boldsymbol{P}\cdot\left(\boldsymbol{\kappa}^{-1}\cdot\frac{\mathrm{d}\boldsymbol{P}}{\mathrm{d}t}-\frac{\mathrm{d}\boldsymbol{E}}{\mathrm{d}t}\right)-\boldsymbol{M}\cdot\left((\boldsymbol{\chi}-\mathsf{U})\cdot\boldsymbol{\chi}^{-1}\cdot\frac{\mathrm{d}\boldsymbol{M}}{\mathrm{d}t}-\frac{\mathrm{d}\boldsymbol{H}}{\mathrm{d}t}\right)\geq 0
\end{aligned}
\tag{20.63}
$$

We now assume that the enthalpy function depends on the possible state field variables as follows:

$$H=H(\gamma,\boldsymbol{E},\boldsymbol{H},T,\nabla T,C_k)\tag{20.64}$$

This allows us to write the time derivative of the enthalpy as follows:

$$\frac{\partial H}{\partial t}=\frac{\partial H}{\partial\gamma}\frac{\partial\gamma}{\partial t}+\frac{\partial H}{\partial\boldsymbol{E}}\frac{\partial\boldsymbol{E}}{\partial t}+\frac{\partial H}{\partial\boldsymbol{H}}\frac{\partial\boldsymbol{H}}{\partial t}+\frac{\partial H}{\partial T}\frac{\partial T}{\partial t}+\frac{\partial H}{\partial\nabla T}\frac{\partial\nabla T}{\partial t}+\sum_{k=1}^{n}\frac{\partial H}{\partial C_k}\frac{\partial C_k}{\partial t}\tag{20.65}$$

Combining eqn (20.65) and (20.63) produces:

$$
\begin{aligned}
&-\left(\frac{\partial H}{\partial t}+S\right)\frac{\mathrm{d}T}{\mathrm{d}t}-\frac{\partial H}{\partial\nabla T}\nabla T-\frac{1}{T}\boldsymbol{J}_q\cdot\nabla T-\sum_{k=1}^{n}\boldsymbol{J}_k\cdot\nabla\mu_k\\
&+\left(\boldsymbol{\tau}-\frac{\partial H}{\partial\gamma}\right):\frac{\mathrm{d}\gamma}{\mathrm{d}t}-\sum_{k=1}^{n}\left(\mu_k+\frac{\partial H}{\partial C_k}\right)-\left(\boldsymbol{P}+\frac{\partial H}{\partial\boldsymbol{E}}\right)\cdot\left(\boldsymbol{\kappa}^{-1}\cdot\frac{\mathrm{d}\boldsymbol{P}}{\mathrm{d}t}-\frac{\mathrm{d}\boldsymbol{E}}{\mathrm{d}t}\right)\\
&-\left(\boldsymbol{M}+\frac{\partial H}{\partial\boldsymbol{M}}\right)\cdot\left((\boldsymbol{\chi}-\mathsf{U})\cdot\boldsymbol{\chi}^{-1}\cdot\frac{\mathrm{d}\boldsymbol{M}}{\mathrm{d}t}-\frac{\mathrm{d}\boldsymbol{H}}{\mathrm{d}t}\right)\geq 0
\end{aligned}
\tag{20.66}
$$

This inequality should be satisfied for all variations of the variables in eqn (20.63) and therefore the coefficients of each term must vanish simultaneously. This leads to the actualization of eqn (20.3), which can be written in terms of the components of the tensor, associated vectors and scalars as follows:

$$\tau_{ij}=\frac{\partial H}{\partial\gamma_{ij}},\quad P_i=-\frac{\partial H}{\partial E_i},\quad M_i=-\frac{\partial H}{\partial H_i},\quad S=-\frac{\partial H}{\partial T},\quad\frac{\partial H}{\partial T_{,i}}=0,\quad\mu_k=-\frac{\partial H}{\partial C_k}\tag{20.67}$$

The last of the sub-equations in eqn (20.67) indicates that the H is independent of the temperature gradient. The equations in eqn (20.67) reduce the inequality in eqn (20.77) to:

$$-\frac{1}{T}\boldsymbol{J}_q\cdot\nabla T-\sum_{k=1}^{n}\boldsymbol{J}_k\cdot\nabla\mu_k\geq 0\tag{20.68}$$

which can be satisfied when:

$$\boldsymbol{J}_q=-\mathbf{k}\cdot\nabla T\tag{20.69}$$

$$\boldsymbol{J}_k=-\mathbf{d}_k\cdot\nabla C_k-q_kC_k\mathbf{d}_k\cdot\nabla\phi\tag{20.70}$$

Eqn (20.69) expresses the familiar Fourier's law, where **k** is the second-order heat conductivity tensor. The first term of eqn (20.70) expresses Fick's first law, where $\mathbf{d}_k$ are the mass diffusivity second-order tensors for each species k. The second term expresses the mass current due to electromigration of the species that are electrically charged and tend to move towards the opposite polarity electrodes (if any exist). It should be noted here that eqn (20.68) provides the thermodynamic justification for adopting both Fourier's and Fick's laws as described by eqn (20.69) and (20.70), respectively.

Expanding the enthalpy H into a Taylor series up to the second order around the origin of the state space (*i.e.* $\mathsf{H}=\mathsf{H}_0$, $\tau_{ij}=\gamma_{ij}=0$, $E_i=H_i=0$, $T=T_0$, $C_k=C_{k0}$) yields:

$$\begin{aligned}\mathsf{H}=\mathsf{H}_0&+\frac{1}{2}c_{ijkl}\gamma_{ij}\gamma_{kl}-e_{kij}\gamma_{ij}E_k-h_{kij}\gamma_{ij}H_k+\frac{1}{2}\varepsilon_{ij}E_iE_j+\frac{1}{2}\mu_{ij}H_iH_j\\&-\alpha_{ij}\gamma_{ij}(T-T_0)-p_iE_i(T-T_0)-q_iH_i(T-T_0)-\frac{\rho c_p}{2T_0}(T-T_0)^2\\&-\sum_{k=1}^{n}\beta_{(k)}ij\gamma_{ij}(C_k-C_{k0})-\sum_{k=1}^{n}r_{(k)}iE_i(C_k-C_{k0})-\sum_{k=1}^{n}s_{(k)}iH_i(C_k-C_{k0})\\&-\sum_{k=1}^{n}\frac{d_k}{2}(C_k-C_{k0})^2-\sum_{k=1}^{n}w_k(T-T_0)(C_k-C_{k0})\end{aligned}\tag{20.71}$$

Applying the equations in eqn (20.68) to eqn (20.71) yields the following set of constitutive equations:

$$\tau_{ij}=c_{ijkl}\gamma_{kl}-e_{kij}E_k-h_{kij}H_k-\alpha_{ij}(T-T_0)-\sum_{k=1}^{n}\beta_{(k)}ij(C_k-C_{k0})\tag{20.72a}$$

$$P_i=-e_{kij}\gamma_{ij}+\varepsilon_{ij}E_j-p_i(T-T_0)-\sum_{k=1}^{n}r_{(k)}i(C_k-C_{k0})\tag{20.72b}$$

$$M_i=-h_{kij}\gamma_{ij}+\mu_{ij}H_j-q_i(T-T_0)-\sum_{k=1}^{n}s_{(k)}i(C_k-C_{k0})\tag{20.72c}$$

$$S=-\alpha_{ij}\gamma_{ij}-p_iE_i-q_iH_i-\frac{\rho c_\gamma}{T_0}(T-T_0)-\sum_{k=1}^{n}w_k(C_k-C_{k0})\tag{20.72d}$$

$$\mu_k=-\beta_{(k)ij}\gamma_{ij}-r_{(k)i}E_i-s_{(k)i}H_i+d_k(C_k-C_{k0})+w_k(T-T_0),\quad k=1,\ldots,n\tag{20.72e}$$

In all of the constitutive equations in eqn (20.71), the terms are written in the Einstein notation for the components of the tensors and vectors, and therefore they expand to the respective sums, and the indices range from 1 to 3, except for the species concentration terms, which are governed by the summation symbol. The indices in parentheses vary with the summation index that counts the species, and the parentheses are used to distinguish them from the indices indicating vector or tensor components. The parentheses are not used when the species index is the only one.

20.6 General Field Evolution Equations

By introducing these constitutive equations into the PDEs expressing the conservation laws given by eqn (20.31), (20.44), (20.55) and (20.61) with the help of eqn (20.51), and after using Onsager's reciprocity relations and extensive algebraic manipulations, we can derive the following field evolution equations:

$$\frac{\partial C_k}{\partial t} = \sum_{j=1}^{n} \nabla \cdot \left(\boldsymbol{\eta}_{kj} \cdot \nabla C_k + q_k C_k \boldsymbol{\eta}_{kj} \cdot \nabla \varphi \right) + \nabla \cdot (\mathbf{k} \cdot \nabla T) + \zeta_k \nabla \cdot (\boldsymbol{\beta} \nabla \boldsymbol{u}), \quad k = 1, 2, \ldots, n \tag{20.73}$$

$$\rho \frac{\partial^2 \boldsymbol{u}}{\partial t^2} = \nabla \cdot (\mathbf{C} : \nabla \boldsymbol{u}) - \nabla \cdot (\boldsymbol{\alpha} \nabla T) - \sum_{k=1}^{n} \nabla \cdot (\boldsymbol{\beta} \nabla C_k + \bar{\boldsymbol{\beta}} C_k \nabla \varphi) + \rho q \boldsymbol{E} + \boldsymbol{I} \times \boldsymbol{B} + (\nabla \boldsymbol{E}) \cdot \boldsymbol{P} + (\nabla \boldsymbol{B}) \cdot \boldsymbol{M} + \frac{\mathrm{d}}{\mathrm{d}t}(\boldsymbol{P} \times \boldsymbol{B}) - \frac{\mathrm{d}}{\mathrm{d}t}(\boldsymbol{M} \times \boldsymbol{E}) + \boldsymbol{F}_b \tag{20.74}$$

$$\rho C_p \frac{\partial T}{\partial t} = \sum_{j=1}^{n} \nabla \cdot \left(\xi \boldsymbol{\eta}_j \cdot \nabla C_k + C_k \bar{\boldsymbol{\eta}}_j \cdot \nabla \varphi \right) + \nabla \cdot (\mathbf{k} \cdot \nabla T) + \zeta \nabla \cdot \left(\boldsymbol{\beta} \nabla \frac{\partial \boldsymbol{u}}{\partial t} \right) + \boldsymbol{I} \cdot \boldsymbol{E}, \tag{20.75}$$

$$\frac{\partial S}{\partial t} = -\vec{\nabla} \cdot \frac{1}{T} \left(\mathbf{k} \cdot \nabla T - \sum_{k=1}^{n} \mu_k (\boldsymbol{\eta}_k \cdot \nabla C_k + q_k C_k \boldsymbol{\eta}_k \cdot \nabla \varphi) \right) - \frac{1}{T} \sum_{k=1}^{n} (\boldsymbol{\eta}_k \cdot \nabla C_k + q_k C_k \boldsymbol{\eta}_k \cdot \nabla \varphi) \cdot \left(T \nabla \frac{\mu_k}{T} \right) + \frac{1}{T} \boldsymbol{I} \cdot \boldsymbol{E} - \frac{Q}{T} - \frac{\boldsymbol{\tau}}{T} : \frac{\mathrm{d}\gamma}{\mathrm{d}t} - \frac{1}{T} \frac{\mathrm{d}\boldsymbol{P}}{\mathrm{d}t} \cdot \left(\boldsymbol{\kappa}^{-1} \cdot \boldsymbol{P} - \boldsymbol{E} \right) - \frac{1}{T} \frac{\mathrm{d}\boldsymbol{M}}{\mathrm{d}t} \cdot \left((\boldsymbol{\chi} - \mathbf{U}) \cdot \boldsymbol{\chi}^{-1} \cdot \boldsymbol{M} - \boldsymbol{B} \right) \tag{20.76}$$

Eqn (20.73) through (20.75) with eqn (20.8) through (20.11), along with eqn (20.14), (20.15), (20.18) and (20.19) from electrodynamics, define a complete set of algebraically closed system of equations. Eqn (20.76) is not necessary for the algebraic completion of the system but is given for the sake of generality as entropy is one of the scalar fields characterizing the multi-physics behavior of homogeneous and anisotropic continua under the influence of multicomponent species mass transport and heat transport along with electrodynamic fields and mechanical loads.

20.7 Specific Field Evolution Equations

To address the case of a relevant reduction of the developed theory, for the case of EAP/IPMC media, we consider absence of polarizations and magnetic field, as well as absence of piezoelectric, piezomagnetic, thermoelectric, thermomagnetic and hygromagnetic effects (*i.e.* $\mu_{ij}=e_{kij}=h_{kij}=p_i=q_i=s_{(k)i}=0$) and also assume homogeneous isotropy of all relevant constitutive behaviors. In addition, we consider mass transport of an electroneutral multicomponent system of charged and uncharged components. For this case, the entropy production source term can be simplified as follows:

$$\psi=-\frac{1}{T^2}\boldsymbol{J}_q\cdot\nabla T-\frac{1}{T}\sum_{k=1}^{1}\boldsymbol{J}_k\cdot\left(T\nabla\frac{\mu_k}{T}\right)+\frac{1}{T}\boldsymbol{I}\cdot\boldsymbol{E}-\frac{Q}{T}-\frac{\tau}{T}:\frac{\mathrm{d}\gamma}{\mathrm{d}t} \tag{20.77}$$

while the entropy flux term is still expressed by eqn (20.53) and consequently the conservation of entropy equation reduces to an alternate from of eqn (20.52) that contains the electric current density, as follows:

$$\begin{aligned}\frac{\partial S}{\partial t}=&-\vec{\nabla}\cdot\frac{1}{T}\left(\boldsymbol{J}_q-\sum_{k=1}^{1}\mu_k\boldsymbol{J}_k\right)-\frac{1}{T^2}\boldsymbol{J}_q\cdot\nabla T-\frac{1}{T}\sum_{k=1}^{1}\boldsymbol{J}_k\cdot\left(T\nabla\frac{\mu_k}{T}\right)\\&+\frac{1}{T}\boldsymbol{I}\cdot\boldsymbol{E}-\frac{Q}{T}-\frac{\tau}{T}:\frac{\mathrm{d}\gamma}{\mathrm{d}t}\end{aligned} \tag{20.78}$$

On the account of the isotropic elastic medium, the components of the fourth-order Hooke's tensor reduce to those expressed on the two independent Lamé parameters λ, μ as follows:

$$c_{ijkl}=\left(\lambda+\frac{2}{3}\mu\right)\delta_{ij}\delta_{kl}+\mu\left(\delta_{ik}\delta_{jl}+\delta_{il}\delta_{jk}-\frac{2}{3}\delta_{ij}\delta_{kl}\right) \tag{20.79}$$

Introducing all of the above mentioned assumptions and this relation in eqn (20.71) reduces the enthalpy to the simpler form:

$$\begin{aligned}\mathsf{H}=&\mathsf{H}_0+\mu\gamma_{ij}\gamma_{ij}+\frac{\lambda}{2}\gamma_{kk}\gamma_{nn}+\alpha\gamma_{kk}(T-T_0)-\frac{\rho c_p}{2T_0}(T-T_0)^2-\sum_{j=1}^{n}\beta_j\gamma_{kk}(C_j-C_{j0})\\&-\sum_{k=1}^{n}\frac{d_k}{2}(C_k-C_{k0})^2-\sum_{k=1}^{n}w_k(T-T_0)(C_k-C_{k0})\end{aligned} \tag{20.80}$$

Through application of the equations in eqn (20.67)—which are still valid—this last expression of the enthalpy function leads to the following constitutive equations:

$$\tau_{ij} = 2\mu\gamma_{ij} + \lambda\gamma_{kk}\delta_{ij} - \alpha(T - T_0)\delta_{ij} - \sum_{k=1}^{n}\beta_k(C_k - C_{k0})\delta_{ij} \tag{20.81a}$$

$$P_i = \varepsilon_{ij}E_j - p_i(T - T_0) - \sum_{k=1}^{n} r_{ik}(C_k - C_{k0}) \tag{20.81b}$$

$$S = -\alpha\gamma_{kk} - \frac{\rho c_\gamma}{T_0}(T - T_0) - \sum_{k=1}^{n} w_k(C_k - C_{k0}) \tag{20.81c}$$

$$\mu_k = -\beta_k\gamma_{kk} - d_k(C_k - C_{k0}) - w_k(T - T_0), \quad k = 1, \ldots, n \tag{20.81d}$$

The corresponding field equations are now being reduced to:

$$\frac{\partial C_k}{\partial t} = \sum_{j=1}^{n}\eta_{kj}(\nabla^2 C_k + q_k C_k \nabla^2\varphi) + k'\nabla^2 T + \zeta_k\beta\nabla\cdot(\nabla\boldsymbol{u}), \quad k = 1, 2, \ldots, n \tag{20.82}$$

$$\rho\frac{\partial^2\boldsymbol{u}}{\partial t^2} = \mu\nabla^2\boldsymbol{u} - (\lambda + \mu)\nabla\left(\vec{\nabla}\boldsymbol{u}\right) - \alpha\nabla T - \sum_{k=1}^{n}(\beta_k\nabla C_k) + \rho q\boldsymbol{E} + \boldsymbol{F}_b \tag{20.83}$$

$$\rho C_p\frac{\partial T}{\partial t} = k'\nabla^2 T + \sum_{k=1}^{n}\eta_k(\nabla^2 C_k + q_k C_k\nabla^2\varphi) - \frac{\alpha T_0}{k'}\nabla\cdot\left(\frac{\partial\boldsymbol{u}}{\partial t}\right) + \boldsymbol{I}\cdot\boldsymbol{E} + \mathrm{Q} \tag{20.84}$$

where η_{kj}, k', ζ_k are the effective mass diffusivities, the effective heat conductivity and the effective coefficient expressing the coupling between mass transport and deformation displacement, respectively. The $n+2$ field equations in eqn (20.82) through (20.84) have $n+2+2$ unknowns and therefore we need two more equations to obtain an algebraically closed system. For that purpose, we assume the quasi-static reduction of the equations of electromagnetics that, by neglecting the magnetic fields and for the case of a conducting medium, reduce to the current conservation equation in eqn (20.41), which, after utilizing the generalized Ohm's law:

$$\boldsymbol{I} = \sigma\boldsymbol{E} \tag{20.85}$$

with σ being the electric conductivity, and introducing the electric constitutive equation in eqn (20.22), itself reduces to the standard Laplace equation for the scalar potential, as follows:

$$-\vec{\nabla}\cdot(\sigma\boldsymbol{E}) = -\vec{\nabla}\cdot(-\sigma\nabla\varphi) = \sigma\nabla^2\varphi = 0 \tag{20.86}$$

Therefore, eqn (20.82) through (20.86) and eqn (20.22) govern the n-species multicomponent behavior of an EAP/IPMC system, and although they represent a specialization of the general equations described in Section 20.6, they can be further specialized by deciding what and how many species one may want to involve.

20.8 Application to a Bi-component Electrohygrothermoelastic Medium

For the case of modeling EAP/IPMC materials, we now focus on the context of their application. We take the more general CFT developed up to now and specialize it for this particular application. We have assumed a general application context that fits the actuation utilization of EAP/IPMC materials. In this context, the presence of the following fields has been considered:

- Mass transport driven from the diffusive processes of two liquid substances, neutral water and an electrolyte (either polar water or free components from the utilized perfluorinated hybrid polymer). The state variables in the continuum that describe the distribution of these substances in the material are the corresponding mass concentrations C_1 and C_2 for neutral water and electrolyte, respectively.
- There is a temperature field state variable $\theta = T - T_0$ expressing the difference between the initial temperature of the system at any point and the current temperature caused from both the heat influx in the system and all of the contributing irreversible internal processes (sources) including heat conduction, mass transport, charge transport and strain gradients.
- There is an electric vector field state variable $\boldsymbol{E}$ with components E_i corresponding to the applied macroscopic electric field. Alternatively, the associated scalar potential ϕ can be used because of the reduced form of eqn (20.22), which is $\boldsymbol{E} = -\nabla\phi$.
- There is a second-order strain tensor field distribution expressed by its individual components γ_{ij}.

The corresponding conjugate state variables are the chemical potentials μ_1, μ_2, the entropy S, the electric field displacement D_i and the stress tensor components τ_{ij}.

Replacing the Lamé coefficients with the traditional engineering constants though the well-known transformations:[40]

$$\mu = G = \frac{E}{2(1+\nu_p)}, \quad \lambda = \frac{E\nu_p}{(1+\nu_p)(1-2\nu_p)} \tag{20.87}$$

where E, ν_p are the Young's modulus and the Poisson's ratio, respectively, allows us to rewrite eqn (20.81a) for the case the bi-component medium in the Duhamel–Neumann constitutive form:

$$\tau_{ij} = 2G\left\{\gamma_{ij} + \frac{\nu_p}{1-2\nu_p}\gamma_{kk}\delta_{ij}\right\} - 2G\left\{\frac{1+\nu_p}{1-2\nu_p}\left[\alpha(T-T_0)+\sum_{k=1}^{2}\beta_k[(C_k-C_{k0})+q_k(C_k-C_{k0})\varphi]\right]\delta_{ij}\right\} \tag{20.88}$$

The inverse of this equation is:

$$\gamma_{ij} = \frac{1}{2G}\left(\tau_{ij} - \frac{\nu_p}{1+\nu_p}\sigma_{kk}\delta_{ij}\right) + \left[\alpha(T-T_0) + \sum_{k=1}^{2}\beta_k[(C_k-C_{k0})+q_k(C_k-C_{k0})\varphi]\right]\delta_{ij} \tag{20.89}$$

In these last two equations, we have made the substitution $\beta_k \to \beta_k(1+q_k\phi)$ to account for the electromigrative effects of mass transport in the constitutive law by making the hygro-expansion coefficients β_k depend on the electric potential ϕ when it is present.

Introducing the form of the stress and strain constitutive form into the system of field eqn (20.82) through (20.84) and dropping the extra source terms as no extra sources can be justified for EAP/IPMC applications enables the expression of them in terms of the Cauchy stress tensor components as follows:

$$\frac{\partial C_j}{\partial t} = \sum_{i=1}^{2}\eta_{ji}(\nabla^2 C_j + q_j C_j \nabla^2\varphi) + k'\nabla^2\theta + \zeta_j\beta_j\nabla^2\tau_{kk}, \quad j=1,2 \tag{20.90}$$

$$\rho\frac{\partial^2 u_i}{\partial t^2} = \nabla\cdot\tau_{ij} - \alpha\nabla\theta - \sum_{k=1}^{2}\beta_k(\nabla C_k + q_k C_k \nabla\varphi) \tag{20.91}$$

$$\rho C_p\frac{\partial\theta}{\partial t} - \sum_{j=1}^{2}\eta_j T_0\frac{\partial C_j}{\partial t} - \alpha T_0\frac{\partial\tau_{kk}}{\partial t} - \rho q T_0\frac{\partial\varphi}{\partial t} = k'\nabla^2\theta + \sum_{k=1}^{n}\eta'_k(\nabla^2 C_k + q_k C_k\nabla^2\varphi) \tag{20.92}$$

Twice differentiating the stress tensor components as expressed by eqn (20.88) yields:

$$\tau_{ij,ij} = 2G\frac{1-\nu_p}{1-2\nu_p}\nabla^2\gamma_{kk} - \frac{E}{1-2\nu_p}\left[\alpha\nabla^2\theta + \sum_{k=1}^{2}\beta_k(\nabla^2 C_k + q_k C_k\nabla^2\varphi)\right] \tag{20.93}$$

The equation of motion as expressed by eqn (20.91) can now be expressed in terms of stresses *via* eqn (20.93) as follows:

$$\frac{1}{u_{c2}}\frac{1-\nu_p}{1+\nu_p}\frac{\partial^2\tau_{kk}}{\partial t^2}+\frac{6G}{u_{c2}}\frac{1-\nu_p}{1-2\nu_p}\left[\alpha\frac{\partial^2\theta}{\partial t^2}+\sum_{k=1}^{2}\beta_k\frac{\partial^2 C_k}{\partial t^2}\right]$$
$$=\frac{1-\nu_p}{1+\nu_p}\nabla^2\tau_{kk}+4G\left[\alpha\nabla^2\theta+\sum_{k=1}^{2}\beta_k\left(\nabla^2 C_k+q_kC_k\nabla^2\varphi\right)\right] \tag{20.94}$$

in which we have inserted $u_{c2}=\{2G(1-\nu_p)/[\rho(1-2\nu_p)]\}^{1/2}$ to represent the speed of the irrotational stress waves (also known as P-waves, dilatational waves or longitudinal waves). However, for quasi-static processes where u_{c2} is much larger than the speed of mechanical and electrical loading, as well the speed of diffusion for heat and mass transport, the left-hand side of eqn (20.94) can be neglected and thus it reduces to:

$$\frac{1-\nu_p}{2E}\nabla^2\tau_{kk}+\alpha\nabla^2\theta+\sum_{k=1}^{2}\beta_k\left(\nabla^2 C_k+q_kC_k\nabla^2\varphi\right)=0 \tag{20.95}$$

A possible solution of this equation with respect to τ_{kk} can be constructed to have the form:

$$\tau_{kk}=-\frac{2E}{1-\nu_p}\left[\alpha\theta+\sum_{k=1}^{2}\beta_k[(C_k-C_{ko})+q_kC_k\varphi]\right]+\Psi \tag{20.96}$$

with the new function Ψ satisfying Laplace's equation:

$$\nabla^2\Psi=0 \tag{20.97}$$

By utilizing eqn (20.86) and (20.96), we can now rearrange eqn (20.90), (20.92) and (20.95) into the following forms:

$$\frac{\partial C_1}{\partial t}=D_{m11}\left(\nabla^2C_1+q_1C_1\nabla^2\varphi\right)+D_{m12}\left(\nabla^2C_2+q_2C_2\nabla^2\varphi\right)+\lambda_1D_h\nabla^2\theta, \tag{20.98}$$

$$\frac{\partial C_2}{\partial t}=D_{m21}\left(\nabla^2C_1+q_1C_1\nabla^2\varphi\right)+D_{m22}\left(\nabla^2C_2+q_2C_2\nabla^2\varphi\right)+\lambda_2D_h\nabla^2\theta, \tag{20.99}$$

$$\frac{\partial\theta}{\partial t}=D_h\nabla^2\theta+\nu_1D_{m1}\left(\nabla^2C_1+q_1C_1\nabla^2\varphi\right)+\nu_2D_{m2}\left(\nabla^2C_2+q_2C_2\nabla^2\varphi\right)-N_c\frac{\partial\Psi}{\partial t} \tag{20.100}$$

It is worthwhile observing here that if we ignore the thermal effects, eqn (20.98) and (20.99) can be reduced to the diffusion equation developed by the simpler diffusion electromigration system developed elsewhere.[9,10]

We can further simplify the system by introducing the substitution:

$$R = \theta + N_c \Psi \tag{20.101}$$

in which we set:

$$\Psi = \tau_{kk} + \frac{2E}{1-\nu_p}\left[\alpha\theta + \sum_{k=1}^{2} \beta_k[(C_k - C_{ko}) + q_k C_k \varphi]\right] \tag{20.102}$$

in a manner that satisfies Laplace's equation in eqn (20.97).

We can further rearrange eqn (20.98) through (20.100) by proper algebraic manipulations as follows:

$$D_1\left(\nabla^2 C_1 + q_1 C_1 \nabla^2 \varphi\right) = \frac{\partial C_1}{\partial t} + \kappa_1 \frac{\partial C_2}{\partial t} - \lambda_1 \frac{\partial R}{\partial t} \tag{20.103}$$

$$D_2\left(\nabla^2 C_2 + q_2 C_2 \nabla^2 \varphi\right) = \frac{\partial C_2}{\partial t} + \kappa_2 \frac{\partial C_1}{\partial t} - \lambda_1 \frac{\partial R}{\partial t} \tag{20.104}$$

$$\mathscr{D}\nabla^2 R = \frac{\partial R}{\partial t} - \nu_1 \frac{\partial C_1}{\partial t} - \nu_2 \frac{\partial C_2}{\partial t} \tag{20.105}$$

In all of these equations, the diffusion coefficients $D_1, D_2, \mathscr{D}$ of the left-hand sides of these equations, as well as the weight coefficients $\kappa_1, \kappa_2, \lambda_1, \lambda_2, \nu_1, \nu_2$ of the time derivative terms of the right-hand sides of the same equations, are phenomenological quantities associated with Onsager's reciprocity relationship and the relevant Onsager coefficients. All these coefficients can be considered properties of the material and therefore need to be determined experimentally for each specific system.

Finally, it should be noted that eqn (20.103) through (20.105), along with eqn (20.97) and eqn (20.86), represent the reduced system of field equations for the bi-component electrohygrothermoelastic medium.

20.9 Conclusions

We initiated this chapter by introducing the general process of forming multiphysics theories and their basis of the two major schools of thought relative to thermodynamics (T.I.P. and R.T.).

We have subsequently presented a description of a general theoretical framework for a continuum multiphysics theory describing the behavior of deformable anisotropic continua under the influence of multicomponent species mass transport, heat transport and electromagnetic fields. We first presented the conservation principles of electrodynamics of continua and then the conservation laws according to the T.I.P., as well as the process for deriving the associated CFT required for the algebraic closure of the system of PDEs that eventually lead to the field equations describing the multiphysics behavior of the continuum system.

We finally applied a specialization of the developed theory for the case of the isotropic non-magnetic bi-component system that appears to be the most suitable description for the case of EAPs or ICMP materials.

Acknowledgements

The authors acknowledge the support by the Office of Naval Research through the 6.1 core funding of the Naval Research Laboratory.

References

1. W. Kuhn, B. Hargitay, A. Katchalsky and H. Eisenberg, *Nature*, 1950, **165**, 514–516.
2. A. J. Grodzinsky and J. R. Melcher, *IEEE Trans. Biomed. Eng.*, 1976, **BME-23**, 421–433.
3. I. V. Yannas and A. J. Grodzinsky, *J. Mechanochem. Cell Motil.*, 1973, **2**, 113–125.
4. P. G. de Gennes, K. Okumura, M. Shahinpoor and K. J. Kim, *Europhys. Lett.*, 2000, **50**, 513–518.
5. M. Shahinpoor, in *1999 Symposium on Smart Structures and Materials*, ed. Y. Bar-Cohen, International Society for Optics and Photonics, 1999, pp. 109–120.
6. H. B. Schreyer, M. Shahinpoor and K. J. Kim, in *1999 Symposium on Smart Structures and Materials*, ed. Y. Bar-Cohen, International Society for Optics and Photonics, 1999, pp. 192–198.
7. M. Shahinpoor, in *SPIE's 7th Annual International Symposium on Smart Structures and Materials*, ed. V. V. Varadan, International Society for Optics and Photonics, 2000, pp. 310–320.
8. Y. Bar-Cohen, *Electroactive Polymer (EAP) Actuators as Artificial Muscles: Reality, Potential, and Challenges*, SPIE Press, 2004.
9. D. J. Leo, K. Farinholt and T. Wallmersperger, *Smart Struct. Mater.*, 2005, **5759**, 170–181.
10. W. J. Yoon, P. G. Reinhall and E. J. Seibel, *Sens. Actuators, A*, 2007, **133**, 506–517.
11. S. Nemat-Nasser and C. W. Thomas, in *Electroactive Polymer (EAP) Actuators as Artificial Muscles: Reality, Potential, and Challenges*, ed. Y. Bar-Cohen, SPIE Press, 2004, p. 765.
12. C. Truesdell and R. Toupin, in *Principles of Classical Mechanics and Field Theory*, ed. S. Flügge, Springer-Verlag, Berlin, Germany, 1960, vol. 1, pp. 226–858.
13. C. Truesdell and W. Noll, in *Encyclopedia of Physics/Handbuch der Physik*, Springer-Verlag, Berlin, Germany, 1965, vol. 3.
14. A. E. Green and P. M. Naghdi, *Proc. R. Soc. London, Ser. A*, 1995, **448**, 335–356.
15. A. C. Eringen and G. A. Maugin, *Electrodynamics of Continua I. Foundations and Solid Media*, Springer-Verlag, New York, NY, USA, 1990.

16. H. F. Tiersten, *Int. J. Eng. Sci.*, 1971, **9**, 587.
17. H. Parkus, *Magneto-Thermoelasticity*, Springer Verlag, Wien, Germany, 1972, vol. 118.
18. G. C. Sih, J. G. Michopoulos and S. C. Chou, *Hygrothermoelasticity*, Martinus Nijhoff Publishers (now Kluwer Academic), 1986.
19. J. G. Michopoulos and G. C. Sih, *Coupled Theory of Temperature Moisture Deformation and Electromagnetic Fields*, 1984.
20. J. G. Michopoulos, C. Farhat and J. Fish, *J. Comput. Inf. Sci. Eng.*, 2005, **5**, 198.
21. J. Michopoulos and M. Shahinpoor, *Proceedings of the First World Congress on Biomimetics and Artificial Muscles*, Albuquerque New Mexico, 2002.
22. J. G. Michopoulos, *ASME 2003 Int.*, 2003, 1 A, 343–351.
23. J. G. Michopoulos, *Vol. 1 23rd Comput. Inf. Eng. Conf. Parts A B*, 2003, 343–351.
24. J. Michopoulos, *Comput. Sci.*, 2004, **3039**, 621–628.
25. J. Michopoulos and M. Shahinpoor, in *Second World Congress on Biomemetics, Artificial Muscles and Nano-Bio, Albuqerque, New Mexico, USA*, 2004.
26. J. G. Michopoulos, *Smart Struct. Mater*, 2004, **5387**, 12–23.
27. J. Michopoulos and M. Shahinpoor, in *Proceedings of ASME-IMECE 2005, ASME International Mechanical Engineering Congress and RD&D Exposition, Orlando, Florida*, 2005, pp. IMECE2005-82426.
28. J. G. Michopoulos, *Int. J. Multiscale Comput. Eng.*, 2006, **4**, 265–279.
29. J. G. Michopoulos and M. Shahinpoor, *Comput. Sci. 2006*, 2006, **3992 LNCS**, 131–138.
30. C. Truesdell, *J. Ration. Mech. Anal.*, 1952, **1**(125–171), 173–300.
31. I. Prigogine, *Etude Thermodynamique des Phenomenes Irreversible*, Desoer, Liege, 1947.
32. S. R. de Groot and P. Mazur, *Non-Equilibrium Thermodynamics*, North Holland Publ. Co., Amsterdam, 1962.
33. R. Haase, *Thermodynamics of Irreversible Processes*, Addison Wesley Publishing Co. Inc., Reading, Mass., 1969.
34. G. A. Maugin, *The Thermomechanics of Nonlinear Irreversible Behaviors: An Introduction*, World Scientific, 1999.
35. L. Onsager, *Phys. Rev.*, 1931, **37**, 405.
36. L. Onsager, *Phys. Rev.*, 1931, **38**, 2265–2279.
37. B. D. Coleman and W. Noll, *Arch. Ration. Mech. Anal.*, 1963, **13**, 167–178.
38. P. Duhem, *Traite d'Energetique*, Gauthier-Villars, Paris, 1911.
39. J. Maxwell, *A treatise on Electricity and Magnetism*, 1881, vol. II.
40. L. D. Landau and E. M. Lifshitz, *Theory of Elasticity*, Pergamon Press Ltd., Oxford, UK, 1970.

CHAPTER 21

Multiphysics Modeling of Nonlinear Ionic Polymer Metal Composite Plates

JOHN G. MICHOPOULOS,*[a] MOSHEN SHAHINPOOR[b] AND ATHANASIOS ILIOPOULOS[c]

[a] Naval Research Laboratory, Computational Multiphysics Systems Lab Code 6394, Washington, DC 20374, USA; [b] University of Maine, Department of Mechanical Engineering, Orono, ME 04469, USA; [c] George Mason University, Computational Materials Science Center, Resident at NRL Code 6394, USA

21.1 Introduction

As discussed in the previous chapter[1] of this volume, the equation of mechanical equilibrium in eqn (20.95) is a direct consequence of having applied the law of conservation of momentum and having assumed that the speed of the mechanical waves is much higher than the quasi-static dynamic multiphysics loading that is usually relevant for the electromechanical actuation needed for various artificial muscle applications. This equation has also been further formulated under the premise that the infinitesimally small strain assumption is justifiable, along with the small displacement assumption. In fact, most electromechanical formulations of such structures in the past have always been one-dimensional (but not necessarily justifiably) and implicitly or explicitly assuming Kirchhoff's small strain bending theory.[2–9] However, based on the fact that ionic polymer metal composite

RSC Smart Materials No. 18
Ionic Polymer Metal Composites (IPMCs): Smart Multi-Functional Materials and Artificial Muscles, Volume 2
Edited by Mohsen Shahinpoor

Published by the Royal Society of Chemistry, www.rsc.org

(IPMC) structures and devices are almost always characterized by very large deformations (especially out of their initial plane), it is naturally desirable to at least remove the assumption of the small displacements. To address this issue, we have developed an analytical framework that contains the derivation of the set of appropriate partial differential equations (PDEs) that govern the large deflection of plates activated by multiphysics fields.[10–14] Efforts to present numerical and analytical solutions of the derived system of PDEs have also been presented by the authors.[11,12,15–18] In the present chapter, we provide an overview that summarizes the most important results of our efforts and also present some new results as well.

In the next section, we present the derivation of the generalized von Karman equations for the multiphysics case. The chapter that follows shows some general results of applying the finite element method (FEM) for numerically solving this system. In the subsequent section, we present a data-driven inverse method that used experimental measurements of deflections to identify an analytical solution to the system. The chapter closes with a final section containing the conclusions.

21.2 Derivation of the Generalized von Karman Equations

The operating specifications for a large-deflection multiphysics theory of plates with large deformations can be defined as follows:

- The plate of thickness h is thin relative to its characteristic length dimension L (which is usually the largest dimension of the plate; *i.e.* $h/L \ll 1$).
- The magnitude of the deflection is large and of the order of the plate thickness or larger (*i.e.* $|w| \geq O(h)$).
- The magnitude of the deflection is much smaller than the largest plate dimension L (*i.e.* $|w| \geq L$).
- The slope of the deflected surface is small (*i.e.* $|\partial w/\partial x| \ll 1$, $|\partial w/\partial y| \ll 1$).
- The tangential displacements u and v are infinitesimal; only the non-linear terms that depend on $\partial w/\partial x$, $\partial w/\partial y$ are retained. All the rest are neglected.
- All strains are small and therefore linear constitutive laws are still valid between strains and stresses.
- Kirchhoff's hypotheses are still holding (*i.e.* the traction on surfaces parallel to the mid-surface are small, strains vary linearly with z, the distance from the mid-surface within the plate remains invariant and the normal to the mid-surface remains normal during loading).
- The material remains mechanically isotropic during deformation, though it can be graded from any other point of view.

At the outset, we need to involve the stress constitutive response for a system of interest. For a multicomponent species mass transport system under simultaneous electrothermoelastic conditions, we can rewrite eqn (20.88) of the previous chapter in this volume[1] in the form:

$$\tau_{ij} = \frac{E}{1-\nu_p^2}\left\{(1-\nu_p)\gamma_{ij} + \left[\nu_p\gamma_{kk} - (1+\nu_p)\left[\alpha(T-T_0) - \sum_{k=1}^{n}\beta_k(C_k - C_{ko})(1+q_k\varphi)\right]\right]\delta_{ij}\right\} \tag{21.1}$$

which can further be simplified as:

$$\tau_{ij} = \frac{E}{1-\nu_p^2}\left\{(1-\nu_p)\gamma_{ij} + \left[\nu_p\gamma_{kk} - (1+\nu_p)\sum_{k=1}^{2n+1}X_{(k)ij}\right]\right\} \tag{21.2}$$

where the following substitutions have been employed:

$$\begin{aligned} &X_{(1)ij} = \alpha(T-T_0)\delta_{ij}, \\ &X_{(k)ij} = \beta_{k-1}(C_{k-1} - C_0)\delta_{ij}, \quad k = 2, \ldots, n+1 \\ &X_{(k)ij} = \beta_{k-n-1}(C_{k-n-1} - C_0)q_{k-n-1}\varphi\delta_{ij}, \quad k = n+2, \ldots, 2n+1 \end{aligned} \tag{21.3}$$

The inverse form of eqn (21.2) represents the Duhamel–Neumann relations for the strain components as follows:

$$\gamma_{ij} = \frac{1}{E}\left[(1+\nu_p)\tau_{ij} - \nu_p\tau_{kk}\delta_{ij}\right] + \sum_{k=1}^{2n+1}X_{(k)ij} \tag{21.4}$$

We now follow the derivation process presented by Nowacki[19] and introduce the substitution:

$$X_{(k)ij} = Y_k\delta_{ij} \tag{21.5}$$

and then introduce an additive decomposition of all fields in eqn (21.1) as described by:

$$Y_k(x, y, z) = Y_{ko}(x, y) + zY_{k1}(x, y) \tag{21.6}$$

such that the dependence through the thickness direction is given a linear distribution of the thickness direction of the second component $Y_{k1}(x, y)$ that, as with the first component $Y_{ko}(x, y)$, depend solely on the x and y coordinates of the plane where the plate/membrane structure is defined. As a consequence of this decomposition, eqn (21.4) can now be written as:

$$\gamma_{ij} = \frac{1}{E}\left[(1+\nu_p)\tau_{ij} - \nu_p\tau_{kk}\delta_{ij}\right] + \sum_{k=1}^{2n+1}(Y_{ko} + zY_{k1})\delta_{ij}, \quad i,j = 1,2 \tag{21.7}$$

Similarly, the components of strain γ_{ij} can be decomposed to two parts $\gamma_{ij}^{(1)}(x, y)$ and $\gamma_{ij}^{(2)}(x, y, z)$ according to:

$$\gamma_{ij}(x,y,z) = \gamma_{ij}^{(1)}(x,y) + \gamma_{ij}^{(2)}(x,y,z) = \gamma_{ij}^{(1)}(x,y) - z\frac{\partial^2 w}{\partial x_i \partial x_j} \tag{21.8}$$

where w is the deflection of the plate. Introducing this into eqn (21.7) and solving the result with respect to the stress tensor components yields:

$$\tau_{ij} = \tau_{ij}^{(1)} + \tau_{ij}^{(2)} \tag{21.9}$$

With:

$$\tau_{ij}^{(1)} = \frac{E}{1-\nu_p^2}\left\{(1-\nu_p)\gamma_{ij}^{(1)} + \left[\nu_p \gamma_{kk}^{(1)} - (1+\nu_p)\sum_{k=1}^{2n+1} Y_{k0}\delta_{ij}\right]\right\} \tag{21.10}$$

$$\tau_{ij}^{(2)} = \frac{Ez}{1-\nu_p^2}\left\{(1-\nu_p)w_{,ij} + \left[\nu_p w_{,kk} - (1+\nu_p)\sum_{k=1}^{2n+1} Y_{k1}\delta_{ij}\right]\right\} \tag{21.11}$$

To derive a set of mechanical equilibrium equations valid for these conditions, we consider the specific equilibrium of a plate element of dimensions dx, dy, h along the axes x, y, z, respectively, as shown in Figure 21.1. From Figure 21.1(a), which shows the force equilibrium

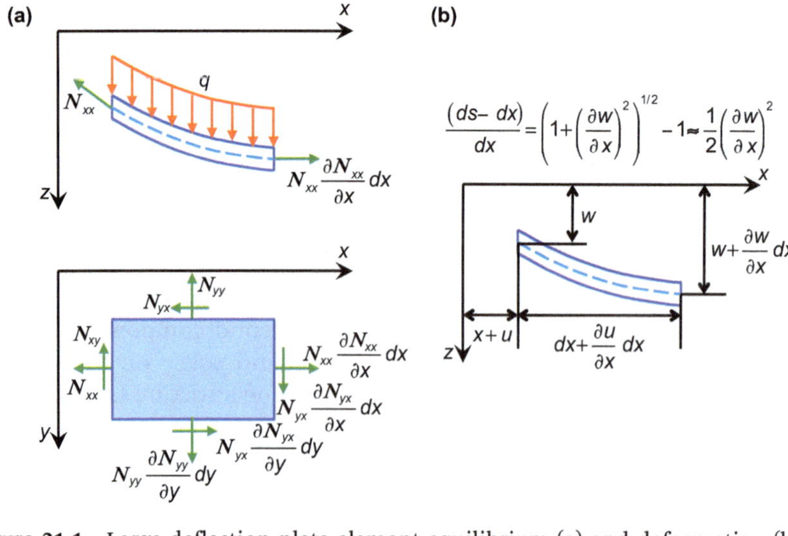

Figure 21.1 Large deflection plate element equilibrium (a) and deformation (b).

conditions of the infinitesimal plate element, we can derive the general balance of moments equation:

$$\frac{\partial^2 M_x}{\partial x^2} + 2\frac{\partial^2 M_{xy}}{\partial x \partial y} + \frac{\partial^2 M_y}{\partial y^2} = -q \\ -\frac{\partial}{\partial x}\left(N_x \frac{\partial w}{\partial x} + N_{xy}\frac{\partial w}{\partial y}\right) - \frac{\partial}{\partial y}\left(N_{xy}\frac{\partial w}{\partial x} + N_y \frac{\partial w}{\partial y}\right) \tag{21.12}$$

Under the substitution $(x, y, z) \rightarrow (x_1, x_2, x_3)$ and the derivative notation equivalence defined by $a_{,i} \leftrightarrow \frac{\partial a}{\partial x_i}$, we can write eqn (21.12) in the contracted form:

$$M_{ij,ij} + N_{ij}w_{,ij} = -q \tag{21.13}$$

where M_{ij}, N_{ij}, w, q are the bending and torsional moments, the longitudinal forces, the plate deflection and the distributed transverse load on the plate, respectively.

The in-plane forces can be evaluated by utilizing eqn (21.10) of the previous chapter in this volume[1] for the in-plane form of the stresses and integrating them across the thickness of plate element to derive:

$$N_{ij} = \int_{-h/2}^{h/2} \tau_{ij}^{(1)} dx_3 = D\left\{(1-\nu_p)\gamma_{ij}^{(1)} + \left[\nu_p \gamma_{kk}^{(1)}\delta_{ij} - (1+\nu_p)\sum_{k=1}^{2n+1} Y_{k0}\delta_{ij}\right]\right\} \tag{21.14}$$

Similarly, the moments are given by:

$$M_{ij} = \int_{-h/2}^{h/2} \tau_{ij}x_3 dx_3 = -N\left\{(1-\nu_p)w_{,ij} + \left[\nu_p w_{,kk} - (1+\nu_p)\sum_{k=1}^{2n+1} Y_{k1}\right]\delta_{ij}\right\} \tag{21.15}$$

with the flexural rigidities given by:

$$N = \frac{Eh^3}{12(1-\nu^2)}, \quad D = \frac{Eh}{1-\nu^2} \tag{21.16}$$

By expressing the forces N_{ij} in terms of the Airy stress function, we can write:

$$N_{ij} = \tau_{ij}h = h(\nabla^2\delta_{ij} - \partial_i\partial_j)F \tag{21.17}$$

where we have introduced the well-known equivalence notation $\partial_i \leftrightarrow \frac{\partial}{\partial x_i}$.

By introducing eqn (21.14), (21.15) and (21.17) into eqn (21.13), we obtain:

$$\nabla^2\nabla^2 w + (1+\nu)\sum_{k=1}^{2n+1}\nabla^2 Y_{k1} = \frac{h}{N}\left(\frac{q}{h} + F_{,22}w_{,11} - 2F_{,12}w_{,12} + F_{,11}w_{,22}\right) \tag{21.18}$$

This equation is the first of two PDEs that appear to be generalizations of the Foppl–von Karman system of PDEs for large-deflection plates.[20,21]

In deriving this multiphysics plate deflection equation in eqn (21.18), we did not need to make an assumption about the geometric nonlinearity of the problem due to the large deflection. However, the total extension (strain) $\varepsilon_{ii}^{(1)}$ of a linear element in terms of a direction i when, at the same time, there is a deflection w involved, as shown in Figure 21.1(b), is given by the nonlinear expression:

$$\varepsilon_{ii}^{(1)} = u_{i,i} + \frac{1}{2}(w_{,i})^2 \tag{21.19}$$

Also, the shear component is given by:

$$\varepsilon_{ij}^{(1)} = \frac{1}{2}(u_{i,j} + u_{j,i}) + \frac{1}{2}(w_{,i}w_{,j}) \tag{21.20}$$

We now utilize the strain compatibility for a plate element in the x–y (or x_1–x_2) plane, which can be expressed as:

$$\varepsilon_{11,22}^{(1)} + \varepsilon_{22,11}^{(1)} = 2\varepsilon_{12,12}^{(1)} \tag{21.21}$$

and along with eqn (21.19) and (21.20) yields:

$$\varepsilon_{11,22}^{(1)} + \varepsilon_{22,11}^{(1)} - 2\varepsilon_{12,12}^{(1)} = (w_{,12})^2 - \frac{1}{2}w_{,11}w_{,22} \tag{21.22}$$

Introducing eqn (21.4) into eqn (21.22) and expressing the in-plane forces in terms of the Airy stress function as defined in eqn (21.17) yields the form:

$$\nabla^2\nabla^2 F + E\sum_{k=1}^{2n+1}\nabla^2 Y_{k1} = E\left[(w_{,12})^2 - w_{,11}w_{,22}\right] \tag{21.23}$$

which represents the second PDE, which is a multiphysics generalization of the Foppl–von Karman PDEs. This equation, too, is only different from the classical second von Karman equation for large-deflection plates in that it contains the second term in the left-hand side, representing the in-plane contribution of the additional fields.

To achieve algebraic closure, the system of eqn (21.18) and (21.23) needs to be supplemented from the PDEs governing the generalized scalar fields Y_k, which, through eqn (21.3) and (21.5), can be expressed as follows:

$$\begin{aligned} &Y_1 = \alpha(T - T_0), \\ &Y_k = \beta_{k-1}(C_{k-1} - C_0), \quad k = 2, \ldots, n+1 \\ &Y_k = \beta_{k-n-1}(C_{k-n-1} - C_0)q_{k-n-1}\varphi, \quad k = n+2, \ldots, 2n+1 \end{aligned} \tag{21.24}$$

while eqn (21.6) in conjunction with eqn (21.24) allows the definition of the quantities:

$$
\begin{aligned}
&Y_{11} = \alpha(T^{(1)} - T_0), \\
&Y_{k1} = \beta_{k-1}(C_{k-1}^{(1)} - C_0^{(1)}), \quad k = 2, \dots, n+1 \\
&Y_{k1} = \beta_{k-n-1}(C_{k-n-1}^{(1)} - C_0^{(1)}) q_{k-n-1}\varphi^{(1)}, \quad k = n+2, \dots, 2n+1
\end{aligned}
\tag{21.25}
$$

for which the definition introduced by eqn (21.6) has been extended to the set of the respective scalar field variables as follows:

$$
\begin{aligned}
&T(x, y, z) = T^{(0)}(x, y) + zT^{(1)}(x, y) \\
&C_k(x, y, z) = C_k^{(0)}(x, y) + zC_k^{(1)}(x, y) \quad k = 1, \dots, n
\end{aligned}
\tag{21.26}
$$

Under this formalism, the bi-component system of heat and mass transport equations in eqn (20.98), (20.99) and (20.100) that were introduced in the previous chapter[1] transforms to:

$$
\begin{aligned}
\frac{1}{\alpha}\frac{\partial Y_{11}}{\partial t} + N_c \frac{\partial \Psi}{\partial t} = {} & \frac{D_h}{\alpha}\nabla^2 Y_{11} + \nu_1 D_{m1}\left(\frac{1}{\beta_1}\nabla^2 Y_{21} + \frac{1}{\beta_1}\nabla^2 Y_{41}\right) \\
& + \nu_2 D_{m2}\left(\frac{1}{\beta_2}\nabla^2 Y_{31} + \frac{1}{\beta_2}\nabla^2 Y_{51}\right)
\end{aligned}
\tag{21.27}
$$

$$
\begin{aligned}
\frac{1}{\beta_1}\frac{\partial Y_{21}}{\partial t} = {} & D_{m11}\left(\frac{1}{\beta_1}\nabla^2 Y_{21} + \frac{1}{\beta_1}\nabla^2 Y_{41}\right) \\
& + D_{m12}\left(\frac{1}{\beta_2}\nabla^2 Y_{31} + \frac{1}{\beta_2}\nabla^2 Y_{51}\right) + \frac{\lambda_1 D_h}{\alpha}\nabla^2 Y_{11},
\end{aligned}
\tag{21.28}
$$

$$
\begin{aligned}
\frac{1}{\beta_2}\frac{\partial Y_{31}}{\partial t} = {} & D_{m21}\left(\frac{1}{\beta_1}\nabla^2 Y_{21} + \frac{1}{\beta_1}\nabla^2 Y_{41}\right) \\
& + D_{m22}\left(\frac{1}{\beta_2}\nabla^2 Y_{31} + \frac{1}{\beta_2}\nabla^2 Y_{51}\right) + \frac{\lambda_2 D_h}{\alpha}\nabla^2 Y_{11},
\end{aligned}
\tag{21.29}
$$

Since we have derived the generalized von Karman equations for time-independent problems, we can eliminate the $\frac{\partial \Psi}{\partial t}$ term from eqn (21.27). In addition, if we let the reciprocals of the hygro-expansion constants and the thermal expansion constant be absorbed on redefined constants, then the last three equations can be reformulated as:

$$
\frac{\partial Y_{11}}{\partial t} = D_h \nabla^2 Y_{11} + \nu_1 D'_{m1}\left(\nabla^2 Y_{21} + \nabla^2 Y_{41}\right) + \nu_2 D'_{m2}\left(\nabla^2 Y_{31} + \nabla^2 Y_{51}\right) \tag{21.30}
$$

$$\frac{\partial Y_{21}}{\partial t} = \lambda_1 D_h' \nabla^2 Y_{11} + D_{m11}\left(\nabla^2 Y_{21} + \nabla^2 Y_{41}\right) + D_{m12}'\left(\nabla^2 Y_{31} + \nabla^2 Y_{51}\right) \quad (21.31)$$

$$\frac{\partial Y_{31}}{\partial t} = \lambda_2 D_h'' \nabla^2 Y_{11} + D_{m21}'\left(\nabla^2 Y_{21} + \nabla^2 Y_{41}\right) + D_{m22}\left(\nabla^2 Y_{31} + \nabla^2 Y_{51}\right) \quad (21.32)$$

with:

$$D_{m1}' = \alpha D_{m1} / \beta_1, D_{m2}' = \alpha D_{m2} / \beta_2, D_h' = \beta_1 D_h, D_{m12}' = \beta_1 D_{m12} / \beta_2, D_h'' = \beta_2 D_h$$

These three equations can be expressed compactly as follows:

$$\frac{\partial Y_{j1}}{\partial t} = \sum_{i=1}^{5} D_{ji} \nabla^2 Y_{i1}, \quad j = 1, 2, 3 \quad (21.33)$$

The sixth and final equation comes by the electric potential equation that, for the case of electroneutrality, can be written as:

$$\nabla^2 Y_{41} = \nabla^2 Y_{51} = 0 \quad (21.34)$$

Thus, the system of equations in eqn (21.18), (21.23), (21.30)–(21.32) and (21.34) forms the complete system for the large-deflection IPMC plates under bi-component mass transport, heat transport and electric field.

It is worth noting that the equations in eqn (21.33) are generalizable for the case of n-component systems to:

$$\frac{\partial Y_{j1}}{\partial t} = \sum_{i=1}^{2n+1} D_{ji} \nabla^2 Y_{i1}, \quad j = 1, ..., n + 1 \quad (21.35)$$

This system generates one PDE for the heat conduction and n equations, one for each species of mass conservation.

21.3 Special Cases

For the case of a bi-component electrohygrothermoelastic system, the mass transport equations in eqn (20.90) of the previous chapter in this volume reduce to the form:

$$\frac{\partial C_j^{(1)}}{\partial t} = \sum_{i=1}^{2} \eta_{ji}\left(\nabla^2 C_j^{(1)} + q_j C_j^{(1)} \nabla^2 \varphi\right) + k' \nabla^2 T^{(1)}, \quad j = 1, 2 \quad (21.36)$$

while the corresponding energy conservation equation in eqn (20.92) reduces to:

$$\rho C_p \frac{\partial T^{(1)}}{\partial t} = k' \nabla^2 T^{(1)} + \sum_{k=1}^{2} \eta_k'\left(\nabla^2 C_k^{(1)} + q_k C_k^{(1)} \nabla^2 \varphi\right) \quad (21.37)$$

and the electric potential equation is recovered by introducing $\boldsymbol{E} = -\nabla\varphi$ into eqn (20.27) of the previous chapter in this volume,[1] along with the equation:

$$\rho_e = \rho q = \rho \sum_{j=1}^{n} c_k q_k$$

$$\nabla^2\varphi = -\frac{\rho}{\varepsilon}\sum_{j=1}^{2} c_k q_k \tag{21.38}$$

or in the case of electroneutrality:

$$\nabla^2\varphi = 0 \tag{21.39}$$

The corresponding von Karman equations are reduced to:

$$\nabla^2\nabla^2 w + (1+\nu)\left[\beta_1\left(\nabla^2 C_1^{(1)} + C_1^{(1)} q_1 \nabla^2\varphi\right) + \beta_2\left(\nabla^2 C_2^{(1)} + C_2^{(1)} q_2 \nabla^2\varphi\right) + \alpha\nabla^2 T\right]$$
$$= \frac{h}{N}\left(\frac{q}{h} + F_{,22}w_{,11} - 2F_{,12}w_{,12} + F_{,11}w_{,22}\right) \tag{21.40}$$

$$\nabla^2\nabla^2 F + E\left[\beta_1\left(\nabla^2 C_1^{(1)} + C_1^{(1)} q_1 \nabla^2\varphi\right) + \beta_2\left(\nabla^2 C_2^{(1)} + C_2^{(1)} q_2 \nabla^2\varphi\right) + \alpha\nabla^2 T\right]$$
$$= E\left[(w_{,12})^2 - w_{,11}w_{,22}\right] \tag{21.41}$$

Eqn (21.36)–(21.37) and (21.39)–(21.41) represent a system of six PDEs with six unknowns and therefore it is algebraically closed.

On the other hand, the simplest case is that of a one-species or one-component system. In this case, the heat conduction equation in eqn (21.37) reduces to:

$$\rho C_p \frac{\partial T^{(1)}}{\partial t} = k'\nabla^2 T^{(1)} + \eta_1'\left(\nabla^2 C_1^{(1)} + q_1 C_1^{(1)}\nabla^2\varphi\right) \tag{21.42}$$

The mass transport equation in eqn (21.36) reduces to:

$$\frac{\partial C_1^{(1)}}{\partial t} = k'\nabla^2 T^{(1)} + \eta_{ji}\left(\nabla^2 C_1^{(1)} + q_1 C_1^{(1)}\nabla^2\varphi\right) \tag{21.43}$$

The complete set of equations for one component in terms of the alternate representation given by eqn (21.18), (21.23), (21.30)–(21.32) and (21.34) reduces to the system:

$$\nabla^2\nabla^2 w + (1+\nu)\left(\nabla^2 Y_{11} + \nabla^2 Y_{21} + \nabla^2 Y_{41}\right)$$
$$= \frac{h}{N}\left(\frac{q}{h} + F_{,22}w_{,11} - 2F_{,12}w_{,12} + F_{,11}w_{,22}\right) \tag{21.44}$$

$$\nabla^2\nabla^2F + E(\nabla^2Y_{11} + \nabla^2Y_{21} + \nabla^2Y_{41}) = E[(w_{,12})^2 - w_{,11}w_{,22}] \quad (21.45)$$

$$\frac{\partial Y_{11}}{\partial t} = D_h\nabla^2Y_{11} + \nu_1 D'_{m1}(\nabla^2Y_{21} + \nabla^2Y_{41}) \quad (21.46)$$

$$\frac{\partial Y_{21}}{\partial t} = \lambda_1 D'_h\nabla^2Y_{11} + D_{m11}(\nabla^2Y_{21} + \nabla^2Y_{41}) \quad (21.47)$$

$$\nabla^2Y_{41} = 0 \quad (21.48)$$

Further simplifications can only be achieved by simplifying the physics. For example, in the case when there is no heat transfer, this system further reduces to:

$$\nabla^2\nabla^2w + (1+\nu)(\nabla^2Y_{21} + \nabla^2Y_{41}) = \frac{h}{N}\left(\frac{q}{h} + F_{,22}w_{,11} - 2F_{,12}w_{,12} + F_{,11}w_{,22}\right) \quad (21.49)$$

$$\nabla^2\nabla^2F + E(\nabla^2Y_{21} + \nabla^2Y_{41}) = E[(w_{,12})^2 - w_{,11}w_{,22}] \quad (21.50)$$

$$\frac{\partial Y_{21}}{\partial t} = D_{m11}(\nabla^2Y_{21} + \nabla^2Y_{41}) \quad (21.51)$$

$$\nabla^2Y_{41} = 0 \quad (21.52)$$

Alternatively, if mass transport is neglected, then the same system reduces to:

$$\nabla^2\nabla^2w + (1+\nu)(\nabla^2Y_{11}) = \frac{h}{N}\left(\frac{q}{h} + F_{,22}w_{,11} - 2F_{,12}w_{,12} + F_{,11}w_{,22}\right) \quad (21.53)$$

$$\nabla^2\nabla^2F + E(\nabla^2Y_{11}) = E[(w_{,12})^2 - w_{,11}w_{,22}] \quad (21.54)$$

$$\frac{\partial Y_{11}}{\partial t} = D_h\nabla^2Y_{11} \quad (21.55)$$

In the case of just electromechanical coupling, the system reduces to:

$$\nabla^2\nabla^2w + (1+\nu)\nabla^2Y_{41} = \frac{h}{N}\left(\frac{q}{h} + F_{,22}w_{,11} - 2F_{,12}w_{,12} + F_{,11}w_{,22}\right) \quad (21.56)$$

$$\nabla^2\nabla^2F + E(\nabla^2Y_{41}) = E[(w_{,12})^2 - w_{,11}w_{,22}] \quad (21.57)$$

$$\nabla^2Y_{41} = 0 \quad (21.58)$$

Based on this analysis, it is apparent that the system of the generalized von Karman equations in its simplest form still requires one extra PDE to describe the evolution of the additional field, whether that is the heat transport, the mass transport or the electric field presence.

21.4 Numerical Solution of a Special Case

There is no known closed-form solution of the generalized von Karman equations, not even for the simpler original ones.[20] The only option available is an approximate solution based either on utilizing the sum of products of orthogonal functions[22] or numerical discretization via finite difference or finite element analysis (FEA) or other numerical methods. Flexibility of existing codes along with the astonishing progress of computational technologies during recent years has made FEA our choice for the present task. Any code that implements a "discontinuous Galerkin" FEA approach such as "freeFEM++",[23] "flexPDE"[24] or "COMSOL Multiphysics"[25] can be used for this task. Because of its adaptive mesh capability, efficient problem definition script language and excellent graphics capabilities, we selected flexPDE to solve the appropriate system PDEs for the problem of an IPMC plate as indicated in our earlier work.[11–13,15,26] However, for the purpose of demonstrating the feasibility of solving the IPMC plate problem in a variety of computational environments, here we will present the results of solving the system of relevant PDEs by using COMSOL Multiphysics.[25]

Without loss of generality, the specific problem selected to be solved numerically involves the generalized von Karman equations with one extra field. In particular, we present the solution of the heat transport case described by eqn (21.53)–(21.55). However, since COSMOL Multiphysics (as well as most codes) is not capable of accepting the definition of higher-than-second-order PDEs, we introduce the quantities $w' = \nabla^2 w$ and $F' = \nabla^2 F$ that permit the reduction of the system of the two fourth-order and one second-order equations in eqn (21.53)–(21.55) to the system of five second-order PDEs as follows:

$$\nabla^2 w = w' \tag{21.59}$$

$$\nabla^2 w' + (1+\nu)\left(\nabla^2 Y_{11}\right) = \frac{h}{N}\left(\frac{q}{h} + F_{,22}w_{,11} - 2F_{,12}w_{,12} + F_{,11}w_{,22}\right) \tag{21.60}$$

$$\nabla^2 F = F' \tag{21.61}$$

$$\nabla^2 F' + E(\nabla^2 Y_{11}) = E[(w_{,12})^2 - w_{,11}w_{,22}] \tag{21.62}$$

$$\frac{\partial Y_{11}}{\partial t} = D_h \nabla^2 Y_{11} \tag{21.63}$$

The computational domain will be considered to correspond to that of a rectangular plate with the short dimension being 0.10 m and with a dimensional aspect ratio of 1.618 (as a first-order approximation of the golden ratio), while the thickness was assigned to be 0.0015 m. The Young's modulus and Poisson's ratio assigned to the problem were $E = 270$ MPa and $\nu = 0.3$. In order to invoke the buckling behavior, we also set $Y_{11} = T$, which, through eqn (21.25), essentially enforces a value of $\alpha = 1\ \text{K}^{-1}$.

Here, we will only present the solution of a simply supported plate along all the edges with lateral mechanical load q given by:

$$q = q_0 \sin(t) \tag{21.64}$$

with $q_0 = 1N/m^2$. We also consider Neumann boundary conditions $\boldsymbol{n} \cdot (\nabla Y_{11}) = g - gY_{11}$ along all edges with:

$$g = g_0[1 + \sin(t)] \tag{21.65}$$

and $g_0 = 2$. This choice was motivated by the goal of investigating the time delays due to the diffusive processes and the verification of the reversible action on the deflection field as a function of the applied stimulus. In the first case, this field is applied along the entire boundary. Figure 21.2 represents the time evolution of both boundary conditions along with their specific values for the four distinct times at which we will present representative results.

The solution results for the three main fields of deflection, the Airy stress function and the temperature for the case of $D_h = 0.01$ at the four time instances of 1, 2, 3 and 4 s are shown in Figure 21.3.

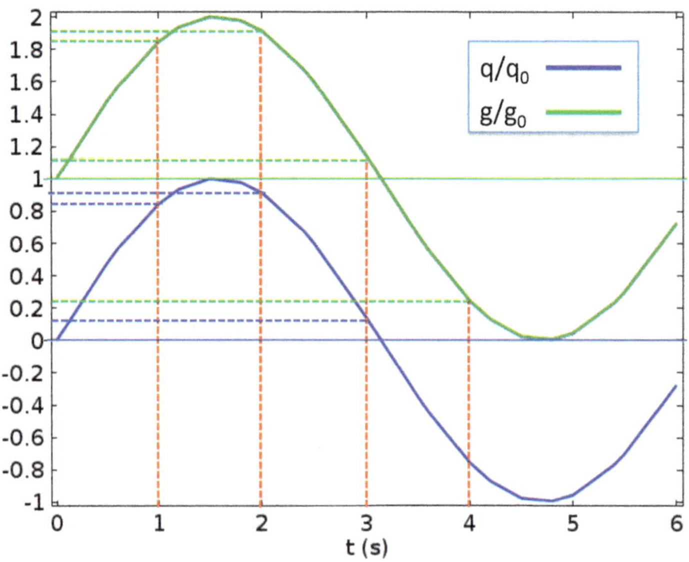

Figure 21.2 Normalized transverse load q/q_0 and heat flux g/g_0 applied at the edges of the plate.

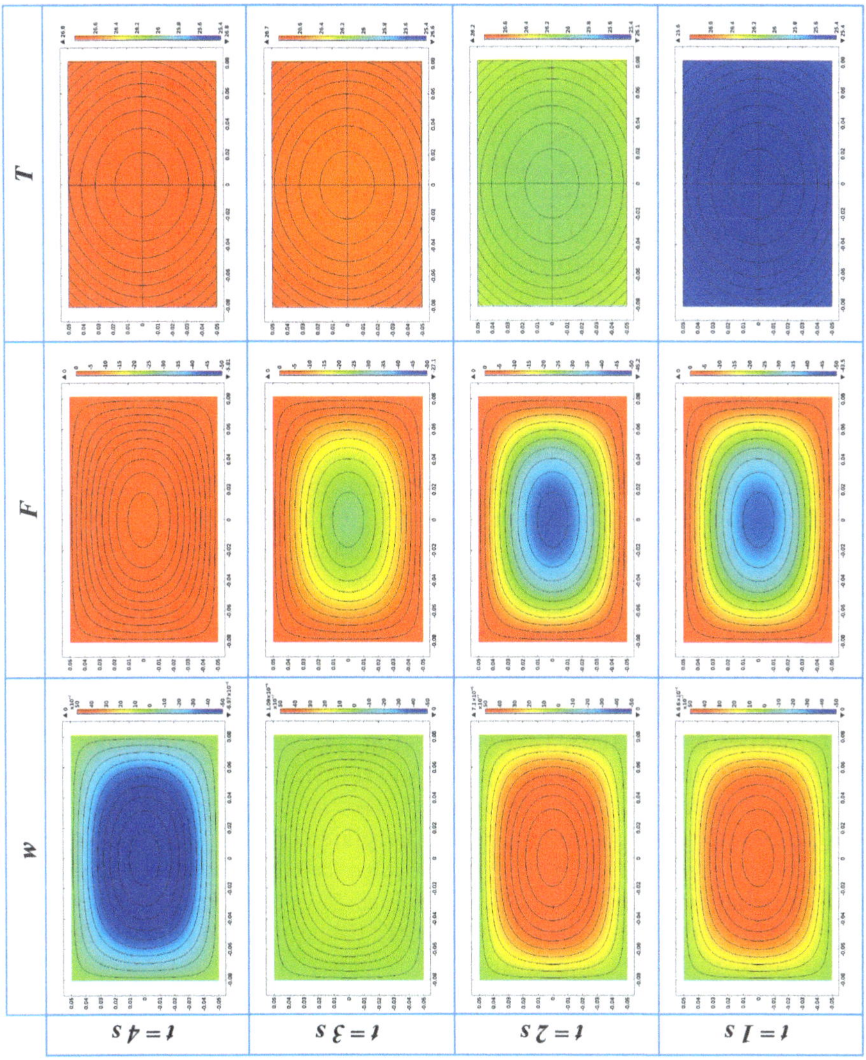

Figure 21.3 Snapshots of the evolution of the solution fields w, F and T for diffusivity $D_h = 0.01$ m^2 s^{-1} for times $t = 1$, 2, 3 and 4 s. The domain axes represent the spatial coordinates and the numbers reflect length in m.

As anticipated, the deflection field follows the variation of the lateral load distribution q and is in phase with it. Another noticeable result is that the evolution of the temperature field seems not to present a noticeable variation in space as the diffusivity is high enough to allow the heat flux to generate an almost uniform field throughout the domain.

A completely different situation emerges for the case of $D_h = 0.00001$, as is presented in Figure 21.4. The most striking difference here is that the buckling behavior of the plate has been captured, and wrinkles appear at the four corners. The extremum of the deflection is no longer at the mid-point of the plate as in the previous case, but instead, multiple extrema appear near the four corners that give rise to the wrinkles.

Another observable characteristic from Figure 21.4 is that the plate is not deforming in a manner that is in phase with the lateral load pressure as in the previous case (Figure 21.3), but rather its behavior is modulated by the temperature conduction from the boundaries. In fact, observing the evolution of the temperature field as a function of time (right column of images in Figure 21.4) shows that the heat conduction is much slower, and a higher temperature appears near the boundaries that slowly diffuses into the rest of the domain while the maximum temperature decreases as a function of increasing time, while most of the central area of the plate domain remains at the initial temperature of 25 °C.

To underline the drastic difference between the unstable (buckled) and stable behaviors of the plate, we present in Figure 21.5 an isometric 3D view of the plate at $t = 2$ s for the case of $D_h = 0.00001$ in Figure 21.5(a) and $D_h = 0.01$ in Figure 21.5(b).

To obtain a better idea of the transition from the stable to unstable (buckling) behavior of the plate, we present in Figure 21.6 the evolution of the deflection field for the same four time instances for values of $D_h = 1\text{e}{-6}$, 1e–5 and 1e–4, and in Figure 21.7, we present the evolution of the deflection field for values of $D_h = 1\text{e}{-3}$, 1e–2 and 1e–1 at the same time increments.

These two figures clearly indicate that the transition from one extremum to multiple extrema occurs in the region $1\text{e}{-4} < D_h < 1\text{e}{-2}$. In fact, if we plot the time evolution of the mid-point of the plate for various values of the diffusivity as shown in Figure 21.8, we also see both the out-of-phase buckling behavior of the mid-point for values $D_h = 1\text{e}{-6}$, 1e–5 and 1e–4 in Figure 21.8(a) and for $D_h = 1\text{e}{-3}$ in Figure 21.8(b), and the in-phase behavior for $D_h = 1\text{e}{-3}$, 1e–2, 1e–1 and 1e–0.

According to the behavior of the mid-point of the plate, it appears that the buckling instability appears for a value in the region $1\text{e}{-3} < D_h < 1\text{e}{-2}$.

The temperature evolution as a function of time for four distinct points of the plate (the mid-point of the plate, the mid-points of the edges and the corner point) and four distinct values of the diffusivity $D_h = 1\text{e}{-6}$, 1e–5, 1e–4 and 1e–3 is presented in Figure 21.9.

This figure clearly shows that, for all diffusivities, the temperature of the corner point is the highest. For the case of $D_h = 1\text{e}{-6}$ and 1e–5, we see the that the evolutions of the mid-point of the edges are identical as they fall on

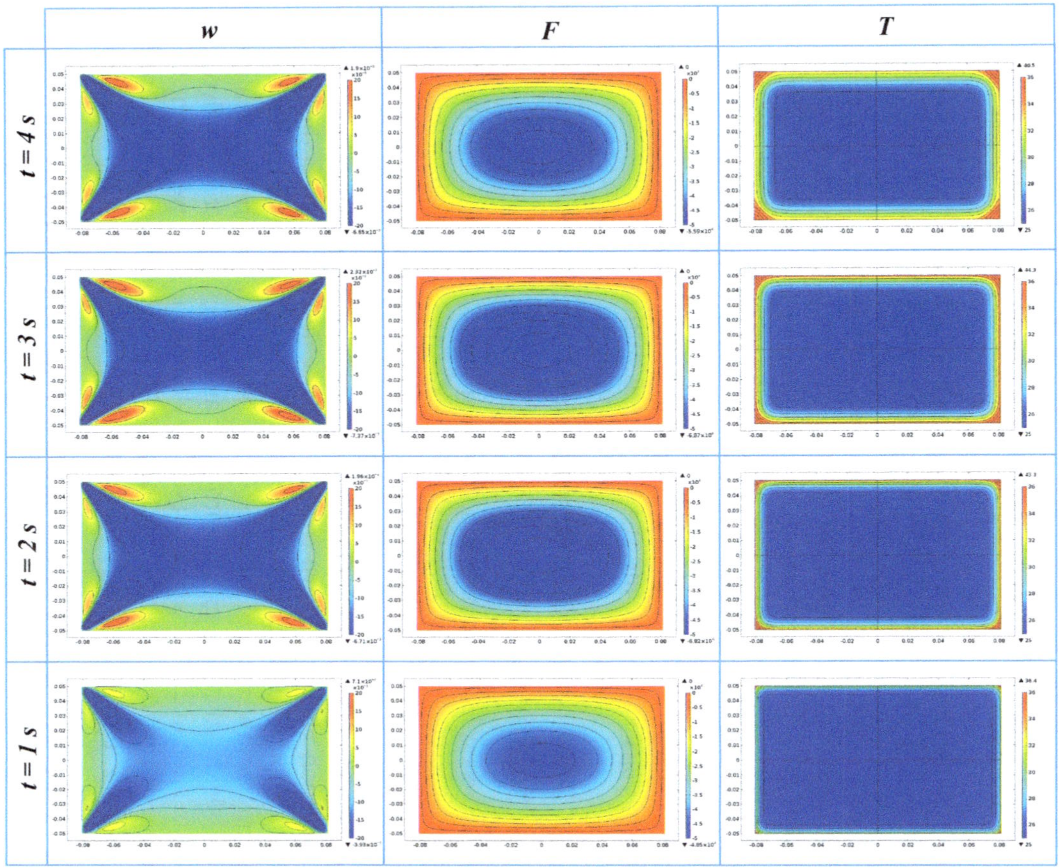

Figure 21.4 Snapshots of the evolution of the solution fields w, F and T for diffusivity $D_h = 0.00001$ m^2 s^{-1} for times $t = 1$, 2, 3 and 4 s. The domain axes represent the spatial coordinates and the numbers reflect length in m.

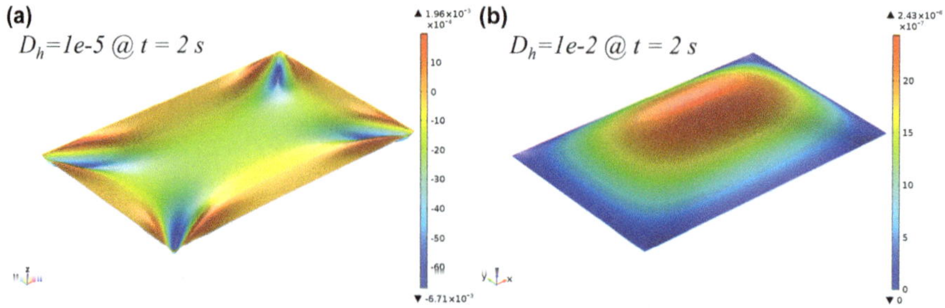

Figure 21.5 Isometric 3D view of the deflection field w in m for $D_h = 0.01\ \mathrm{m^2\ s^{-1}}$ (a) and for $D_h = 0.00001\ \mathrm{m^2\ s^{-1}}$ (b) at time $t = 2$ s.

top of each other, while at the same time the mid-point remains at the initial constant temperature of 25 °C. As the diffusivity increases to $D_h = 1\text{e}{-4}\ \mathrm{m^2\ s^{-1}}$, the mid-points of the edges appear to exhibit different histories after 1.9 s and the mid-points begin to increase their temperatures at about the same time. Furthermore, as the diffusivity increases to $D_h = 1\text{e}{-3}\ \mathrm{m^2\ s^{-1}}$, these trends become more dominant and the differences in the evolution of all temperatures become smaller and they all seem to coalesce, especially as the diffusivity increases further.

21.5 Data-driven Construction of Analytical Solutions

In order to explore obtaining a solution without solving numerically the multiphysics PDEs describing the dynamics of the system altogether, we demonstrate here a data-driven approach for constructing analytical solutions. The proposed solution fields are approximated by sums of basis functions borrowed from well-known single-physics approximations. The identification of the unknown coefficients is achieved by casting the problem as an optimization problem where an objective function is constructed that expresses the norm of the error vector between the experimental values of the fields involved in the problem and their respective analytical approximations constructed as previously described. First, we present the experimental procedure and then the global optimization procedure for a square IPMC plate.

21.5.1 Experimental Procedure for Data Collection

To acquire preliminary experimental data reflecting the electrically activated bending of IPMC plates, an IPMC square specimen was cut with dimensions 40 mm×40 mm×0.3 mm.

Figure 21.10(a) presents a view of the IPMC plate mounted on a conductive frame that was used to apply voltage boundary conditions along the entire boundary. In Figure 21.10(b), the experimental setup is shown, including a load cell used to determine the deflection at the mid-point of the plate

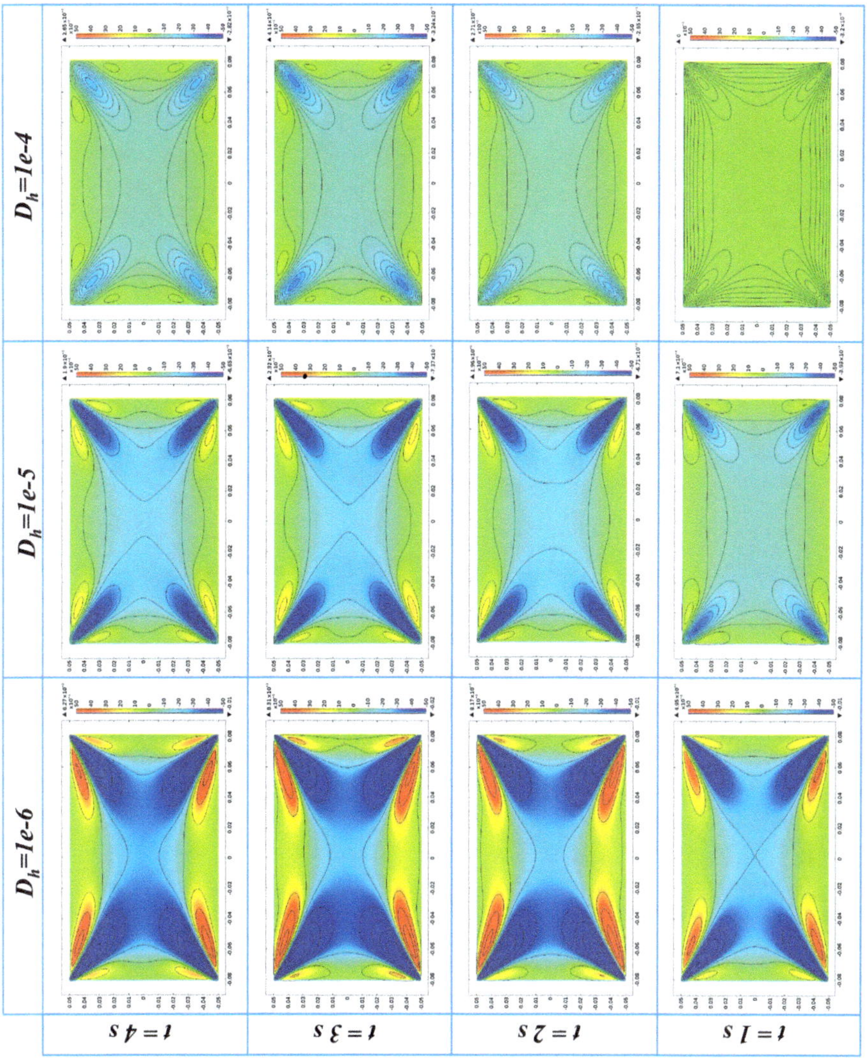

Figure 21.6 Snapshots of the evolution of the solution field w for diffusivity values $D_h = 0.000001$, 0.00001 and 0.0001 m^2 s^{-1} for times $t = 1$, 2, 3 and 4 s. The domain axes represent the spatial coordinates and the numbers reflect length in m.

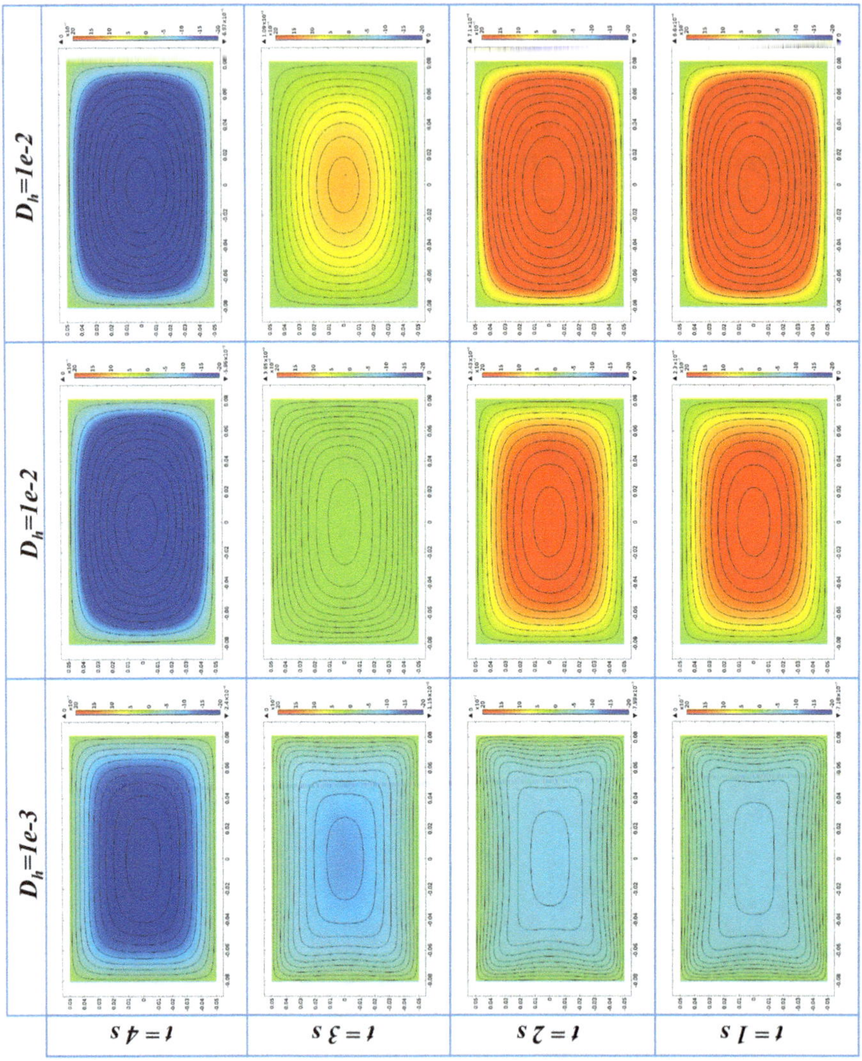

Figure 21.7 Snapshots of the evolution of the solution field w for diffusivity values $D_h = 0.001$, 0.01 and 0. 1 m^2 s^{-1} for times $t = 1$, 2, 3 and 4 s. The domain axes represent the spatial coordinates and the numbers reflect length in m.

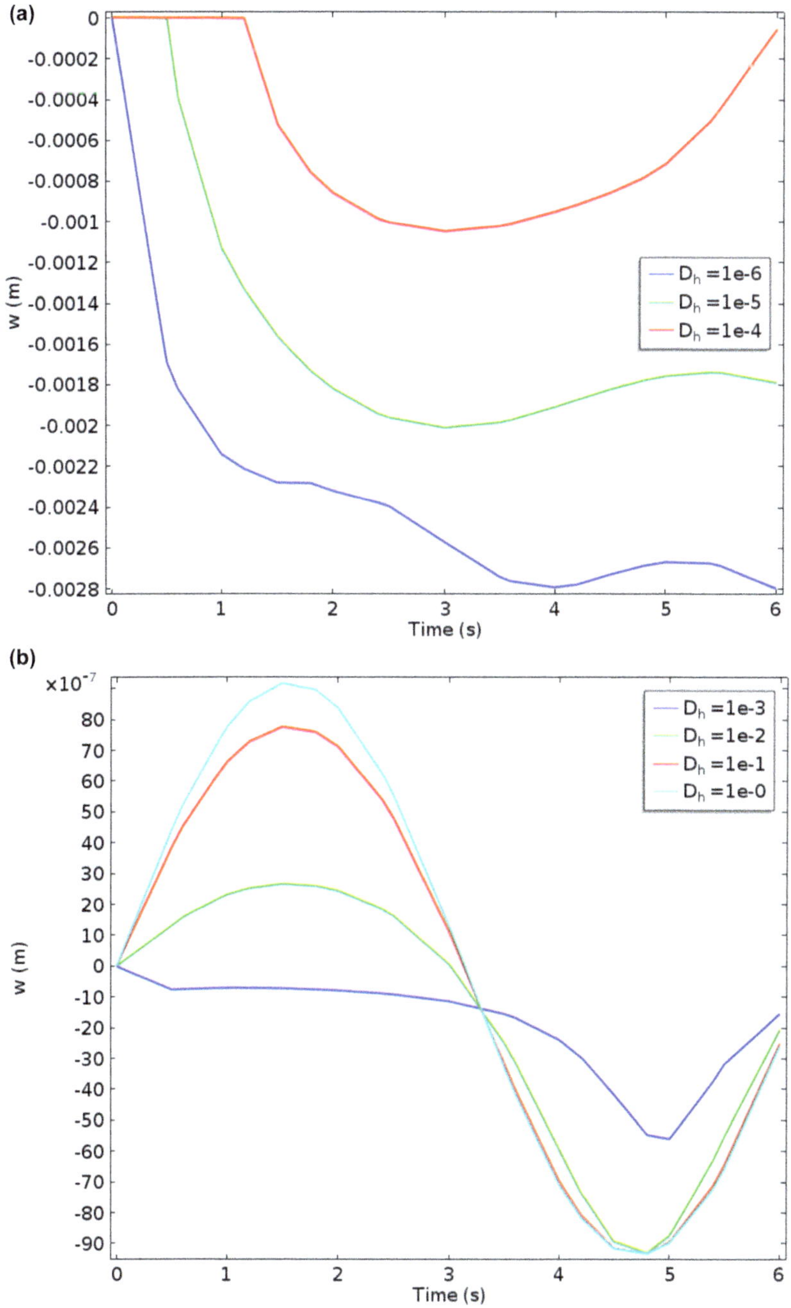

Figure 21.8 Evolution of the solution field w at the mid-point of the plate for diffusivity values $D_h = 0.000001$, 0.00001 and 0.0001 $m^2\ s^{-1}$ (a) and for $D_h = 0.001$, 0.01, 0.1 and 1.0 $m^2\ s^{-1}$ (b).

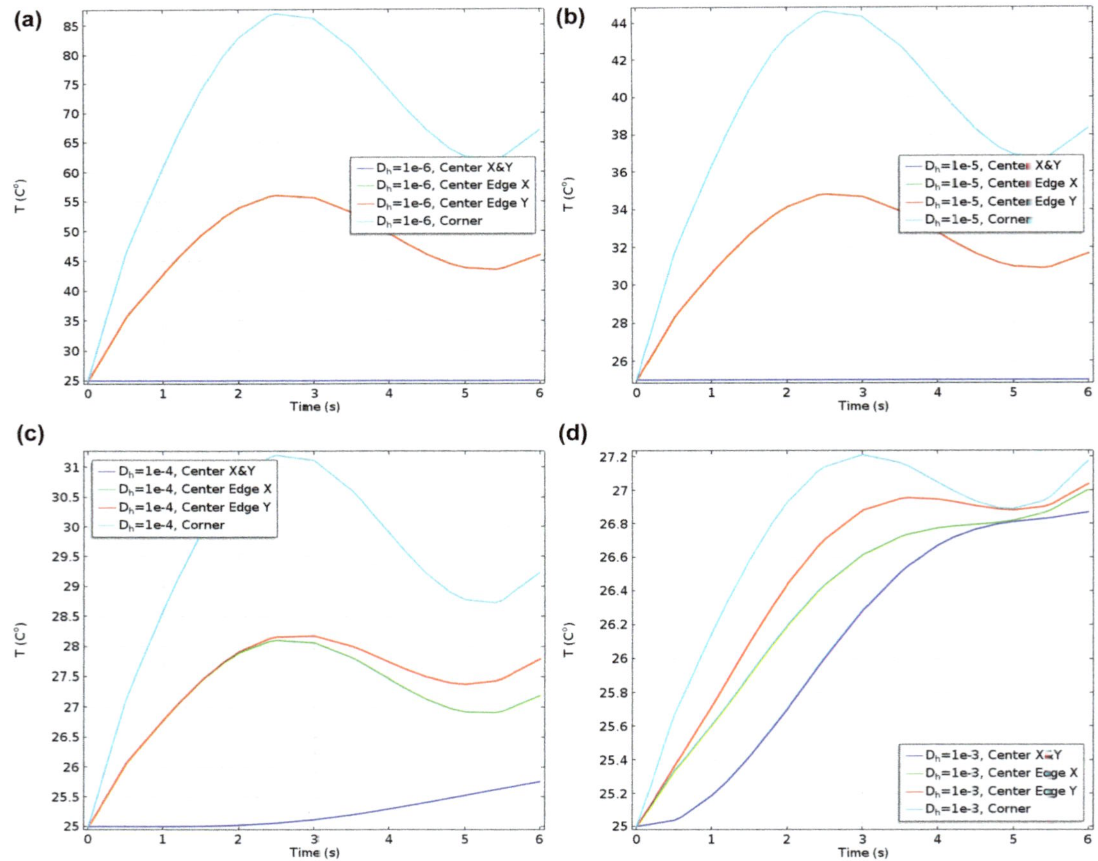

Figure 21.9 Evolution of the solution field T at the center of the plate, the two midpoints of the edges and the corner of the plate for diffusivity values $D_h = 0.000001\ \text{m}^2\ \text{s}^{-1}$ (a), for $D_h = 0.00001\ \text{m}^2\ \text{s}^{-1}$ (b), for $D_h = 0.0001\ \text{m}^2\ \text{s}^{-1}$ (c) and for $D_h = 0.001\ \text{m}^2\ \text{s}^{-1}$ (d).

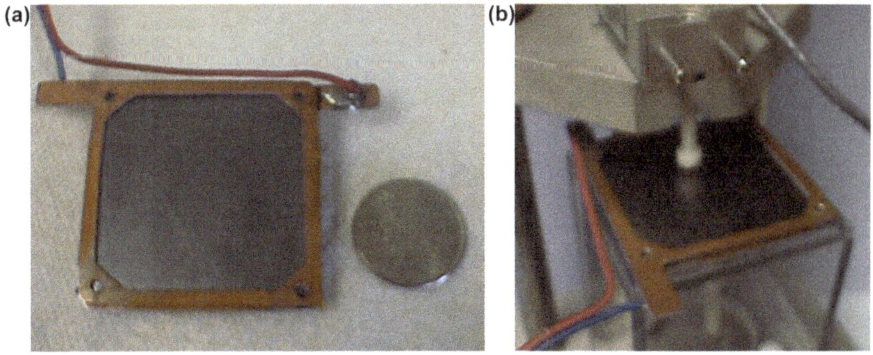

Figure 21.10 Close-up view of a square IPMC plate mounted on a conductive frame (a), and experimental setup of a loading frame with a loading cell in place (b).

surface. The load cell was mounted on a precision screw-type platform capable of applying measurable displacement towards and away from the center of the plate. Consecutive turns of the screw were applied, all corresponding to a displacement of about 1 micron. For each one of these turns, when zero contact force was first observed (the load cell was not touching the plate) the distance travelled by the load cell is considered to be equal to the deflection of the plate.

Figure 21.11(a) shows the experimental results of applying a sinusoidally varying voltage and the corresponding deflection at the middle of the plate as a function of time. Figure 21.11(b) shows the voltage *vs.* the mid-point deflection for the first 2 s. These are the data to be used in the process that follows.

21.5.2 Design Optimization for the Analytical Approximation of Simulated Behavior

The inverse approach character of the proposed methodology is based on minimizing—in the least-squares sense—the objective function:

$$f^{o}(c_{ij}) = \left\{ \sum_{i=1}^{n} [w_i^s(c_{1j}) - w_i^e]^2 + \sum_{i=1}^{n} [F_i^s(c_{2j}) - F_i^e]^2 + \sum_{i=1}^{n} [V_i^s(c_{3j}) - V_i^e]^2 \right\} \tag{21.66}$$

where $w_i^s(c_{1j}), F_i^s(c_{2j}), V_i^s(c_{3j})$ are the unknown variables of the simulated fields corresponding to the deflection, the Airy stress function and the voltage distributions, respectively, and w_i^e, F_i^e, V_i^e are the corresponding experimental values of the same variables, respectively. Here, the unknowns to be determined by the minimization of the objective function defined by eqn (21.66) are the free coefficients c_{ij}.

In our preliminary experimental implementation, the Airy stress function and the voltage distribution over the domain of the space variables (x,y) were

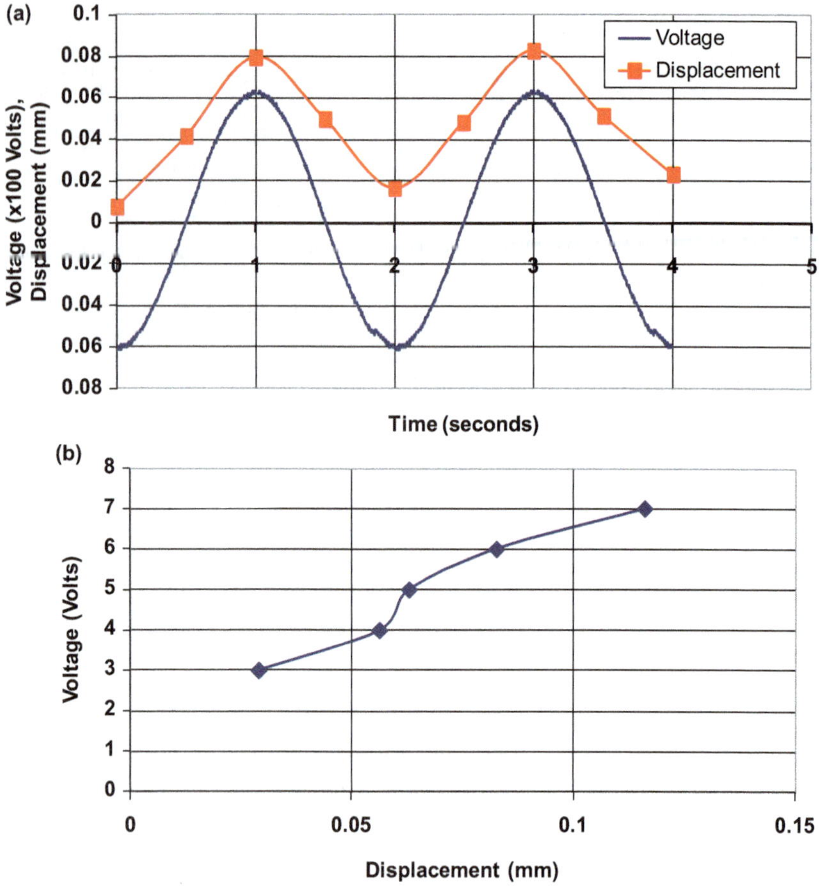

Figure 21.11 Voltage and mid-point deflection *vs.* time (a) and voltage *vs.* mid-point deflection (b).

not measured. Therefore, only the first term of eqn (21.66) (corresponding to the deflections) was used in the actual numerical analysis, since it was the only one for which we could obtain experimental measurements. This implies that $c_{ij} = c_{1j} = c_j$.

We constructed the simulated deflection solution $w_i^s(c_{1j}\,;\, x_i, y_i)$ indexed for specific locations $(x_i,\ y_i)$ as an additive composition of just three basis functions τ_j with $(j = 1,\ 2,\ 3)$, which satisfy the boundary conditions of zero deflection along the edges according to:

$$w_i^s(x_i, y_i) = \frac{4q^{\mathrm{equiv}}a^4}{\pi^5 D}\left[c_1\tau_1(x_i, y_i) + c_2\tau_2(x_i, y_i) + c_3\tau_3(x_i, y_i)\right] + w_i^{\mathrm{offset}} \quad (21.67)$$

where D is the flexural rigidity of the plate, q^{equiv} is an equivalent lateral load distribution per unit of area of a small-deflection plate that generates deflections identical to our multifield plate and w_i^{offset} was added to capture the possible existence of any initial deflection. The three free coefficients c_i

weighting the basis functions, are the unknowns (design variables) to be determined.

Based on the analytical solutions in the form of infinite series constructed for approximating a solution satisfying the bi-harmonic equation governing the single physics of small-deflection bending of rectangular plates given in the past, we chose one of them to construct our basis functions out of the first three terms[27] according to:

$$\tau_m(x_i,y_i) = \cos\frac{m\pi x_i}{a}\left(1-\frac{\alpha_m\tanh\alpha_m+2}{2\cosh\alpha_m}\cosh\frac{m\pi y_i}{a} + \frac{1}{2\cosh\alpha_m}\frac{m\pi y_i}{a}\sinh\frac{m\pi y_i}{a}\right) \quad (21.68)$$

with $\alpha_m = m\pi b/2a$ and where a, b are the dimensions of the plate along the x and y axes, respectively.

Since the experimental data for the deflection are collected only at the mid-point of the plate, we used another solution of an equivalent single physics-approximating solution of the bi-harmonic equation to generate pseudo-experimental data for the rest of the points on the plate according to:

$$w_i^e(x_i,y_i) = \frac{16q^{\text{equiv}}}{\pi^6 D}\sum_{m=1}^{100}\sum_{n=1}^{100}\frac{a^4b^4}{mn(b^2m^2+a^2n^2)^2}\sin\frac{m\pi(2x_i+a)}{2a}\sin\frac{m\pi(2y_i+b)}{2b} + w_i^{\text{offset}} + w_i^{\text{gaussian\_noise}}, \quad (21.69)$$

where $w_i^{\text{gaussian\_noise}}$ represents a term (injecting noise) for the purpose of emulating the error due to the experimental data-acquisition methodology. It is worth mentioning here that the reason we use only the first 100 terms of the double series in the above relation is due to the fact that we are concerned with capturing the deflection of the plate in an approximate manner. The injected noise is expected and has been chosen to be of equal or higher magnitude than the approximation error introduced from not considering more terms.

To eliminate the presence of the equivalent mechanical load distribution from eqn (21.67) and (21.69), the voltage *vs.* mid-point (maximum) deflection observed in Figure 21.10(b) has been approximated by the following second-order polynomial:

$$w^{\max} = 000000137903V_o^2 + 000000635696V_o + 0000000234504 \quad (21.70)$$

By equating this deflection with that provided by eqn (21.67) at the mid-point and solving the resulting equation with respect to q^{equiv}, we obtain the following voltage-dependent solution:

$$q^{\text{equiv}} = \frac{\pi^5 D}{4a^4}\frac{(000000137903V_o^2 + 000000635696V_o + 0000000234504 - w_i^{\text{offset}})}{[c_1\tau_1(0,0)+c_2\tau_2(0,0)+c_3\tau_3(0,0)]}. \quad (21.71)$$

Table 21.1 Computed solution coefficients C_i and their comparison with known terms of the single-physics solution.

	c_1	c_2	c_3
Known values	1.00000	0.00411523	0.00032
Solution values	1.00000	0.00432335	0.000310941
Deviation	0%	5.1%	2.8%

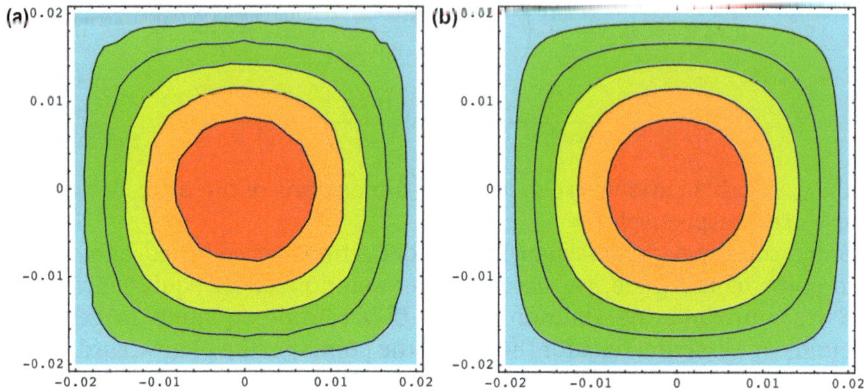

Figure 21.12 Deflection field distributions for the pseudo-experimental case (a) and the analytically computed case (b). The domain axes represent the spatial coordinates and the numbers reflect length in m.

Eqn (21.67) through (21.71) fully define the quantities participating in eqn (21.66) and therefore determination of the unknown parameters c_j can now be achieved by using any optimization methodology for minimizing the objective function defined by eqn (21.66). In the context of this work, the implementation of the optimization procedure was formulated in Mathematica[28] *via* the algorithms available for global optimization by using the Nelder–Mead method. The resulting solution is presented in Table 21.1 and the three coefficients are compared with the known values of the first three terms of the infinite series that approximates the bi-harmonic solution.

Now that the design variables c_i have been estimated, the analytical form of the deflection field as expressed by eqn (21.67) is fully determined. For comparison purposes, Figure 21.12 shows the experimental (Figure 21.12(a)) and simulated (Figure 21.12(b)) distributions of the displacement field over the square plate at the same arbitrary time increments.

Clearly, the computationally determined simulated solution tracks the pseudo-experimental one very well. The jaggedness of the contours in the pseudo- experimental deflection distributions shown in Figure 21.11(a) is due to the noise introduced in eqn (21.69).

21.6 Conclusions

In the present chapter, we derived the generalized form of the von Karman equations for multifield conditions and described various special cases of them. A demonstration of how to solve a representative system of such equations via the use of the FEA method was also presented for a rectangular simply supported plate exposed to transverse mechanical load and to heat fluxes at the supporting edges. Both mechanical and heat loads were defined as functions of time. We presented representative results of the solution of the system of the proper PDEs and demonstrated and identified both the stable and unstable (buckling) regimes of the plates. As an alternative to the FEA method, we also presented a data-driven method that uses a design optimization approach for solving an inverse problem, by exploiting partial experimental data, to determine the unknown coefficients of the basis functions that form an analytical approximation of the solution fields.

Acknowledgements

The authors acknowledge the support by the Office of Naval Research through the 6.1 core funding of the Naval Research Laboratory.

References

1. J. G. Michopoulos, M. Shahinpoor and A. Iliopoulos, ed. M. Shahinpoor, Royal Society of London, 2015, pp. XX–XX.
2. W. Kuhn, B. Hargitay, A. Katchalsky and H. Eisenberg, *Nature*, 1950, **165**, 514–516.
3. I. V. Yannas and A. J. Grodzinsky, *J. Mechanochem. Cell Motil.*, 1973, **2**, 113–125.
4. A. J. Grodzinsky and J. R. Melcher, *IEEE Trans. Biomed. Eng.*, 1976, **BME-23**, 421–433.
5. M. Shahinpoor, in *1999 Symposium on Smart Structures and Materials*, ed. Y. Bar-Cohen, International Society for Optics and Photonics, 1999, pp. 109–121.
6. H. B. Schreyer, M. Shahinpoor and K. J. Kim, in *1999 Symposium on Smart Structures and Materials*, ed. Y. Bar-Cohen, International Society for Optics and Photonics, 1999, pp. 192–198.
7. M. Shahinpoor, in *SPIE's 7th Annual International Symposium on Smart Structures and Materials*, ed. V. V. Varadan, International Society for Optics and Photonics, 2000, pp. 310–320.
8. Y. Bar-Cohen, *Electroactive Polymer (EAP) Actuators as Artificial Muscles: Reality, Potential, and Challenges*, SPIE Press, 2004.
9. S. Nemat-Nasser and C. W. Thomas, in *Electroactive Polymer (EAP) Actuators as Artificial Muscles: Reality, Potential, and Challenges*, ed. Y. Bar-Cohen, SPIE Press, 2004, p. 765.
10. J. Michopoulos and M. Shahinpoor, in *Proceedings of the First World Congress on Biomimetics and Artificial Muscles*, Albuquerque, New Mexico, 2002.

11. J. Michopoulos and M. Shahinpoor, in *Second World Congress on Biomemetics, Artificial Muscles and Nano-Bio*, Albuqerque, New Mexico, USA, 2004.
12. J. Michopoulos and M. Shahinpoor, in *Proceedings of ASME-IMECE 2005, ASME International Mechanical Engineering Congress and RD&D Exposition, Orlando, Florida*, 2005, pp. IMECE2005–82426.
13. J. G. Michopoulos, *Int. J. Multiscale Comput. Eng.*, 2006, **4**, 265–279.
14. J. G. Michopoulos, *Smart Struct. Mater.*, 2004, **5387**, 12–23.
15. J. G. Michopoulos, *ASME 2003 Int.*, 2003, 1A, 343–351.
16. J. G. Michopoulos and M. Shahinpoor, *Comput. Sci.*, 2006, **3992 LNCS**, 131–138.
17. J. G. Michopoulos, *23rd Comput. Inf. Eng. Conf. Parts A B*, 2003, **1**, 343–351.
18. J. Michopoulos, *Comput. Sci.*, 2004, **3039**, 621–628.
19. W. Nowacki, *Thermoelasticity*, Pergamon Press, Oxford, UK, 2nd edn, 1986.
20. T. Von Kármán, *Festigkeitsprobleme im maschinenbau*, [publisher not identified], Leipzig, 1910.
21. M. Lewicka, L. Mahadevan and M. R. Pakzad, *Proc. R. Soc. A*, 2011, **467**, 402–426.
22. S. Levy, *Bending of Rectangular Plates with Large Deflections*, 1942.
23. F. Hecht, *J. Numer. Math.*, 2012, **20**, 251–266.
24. FlexPDE 6, PDE Solutions Inc., 2014.
25. COMSOL.
26. J. Michopoulos and S. M., in *NASA-FEMCI 2003 Workshop*, 2003.
27. S. Timoshenko and S. Woinowsky-Krieger, *Theory of Plates and Shells*, Mcgraw-Hill, 1959.
28. S. Wolfram, *The Mathematica Book*, Cambridge University Press, New York, NY, USA, 4th edn 1999.

CHAPTER 22

Ionic Polymer Metal Composites as Dexterous Manipulators and Haptic Feedback/Tactile Sensors for Minimally Invasive Robotic Surgery

MOHSEN SHAHINPOOR

Department of Mechanical Engineering, University of Maine, Orono, Maine 04469, USA
Email: shah@maine.edu

22.1 Introduction

Robotic surgery has already created a paradigm shift in medical surgical procedures and will continue to expand to all surgical and microsurgical procedures. There is no doubt that in doing so, robotic surgical systems such as the da Vinci will become much more intelligent and sophisticated with the integration, implementation and synergy of new intelligent material systems that will make surgical tools and equipment more functional and more intelligent in biomimetic sensing and actuation and kinesthetic interaction with organs during robotic surgery.

RSC Smart Materials No. 18
Ionic Polymer Metal Composites (IPMCs): Smart Multi-Functional Materials and Artificial Muscles, Volume 2
Edited by Mohsen Shahinpoor

Published by the Royal Society of Chemistry, www.rsc.org

The current robotic surgical systems evolved from laparoscopic surgical procedures and made it possible for surgeons to perform surgery away from the patient with much more concentration and ease. However, what was lost in this transition by the surgeons was the feeling sensation of tissues and organs and kinesthetic force feedback during surgery. It is interesting to note that even during laparoscopic surgery, the surgeons can still feel and sense the tissues and organs they are handling and operating on with laparoscopic/endoscopic tools during surgery and feel the kinesthetic forces at work. However, the kinesthetic force feedback was replaced with visual feedback during robotic surgery. It is to be noted that some of this kinesthetic force feedback was lost in the transition from open to laparoscopic surgery due to trocar friction and varying lever arms. However, with smart materials such as ionic polymer metal composites (IPMCs) and appropriate calibration and tuning, one may be able to recover the kinesthetic force feedback during surgery using IPMCs. While one cannot possibly minimize the importance of visual feedback during robotic surgery,[1–4] this chapter focuses on smart and soft haptic/tactile feedback. Specifically, the scope of the chapter is limited to realistic haptic/tactile feedback (human-like manual surgical dexterity) to surgeons during robotic surgery by means of specially configured arrays of IPMC actuators and sensors integrated with surgical end-effectors for haptic feedback/tactile sensing. IPMCs are great for such robotic force feedback applications because they work perfectly well in the wet human body environment and generate millivolt-level sensing signals for kinesthetic force feedback. We believe that considering IPMCs for haptic and kinesthetic force feedback is novel. This chapter discusses the applications of intelligent materials and artificial muscles to robotic surgery in connection with haptic, tactile and kinesthetic force feedback to surgeons during robotic surgery.

22.2 Introduction to Smart Materials and Artificial Muscles

In this chapter, we will introduce a number of smart materials and artificial muscles that can be employed during robotic surgery to enhance the quality of surgical operations with robotic structures and provide force, haptic and kinesthetic feedback to surgeons during robotic surgery. Smart materials are generally defined as multi-functional materials that can perform sensing, energy harvesting and actuation, in addition to providing additional signals pertaining to their environmental disturbances and changes. Due to the requirement of biocompatibility, not all smart materials can be easily used in robotic surgery. For example, piezoceramic materials, such as lead zirconate titanate, are not recommended for use in a biomedical environment due to the presence of lead-related materials. However, piezopolymeric materials, such as polyvinylidene fluoride, may be used. Here, biomimetic IPMCs are described as electroactive polymers (EAPs) and artificial muscles for haptic feedback and tactile sensing during robot-assisted surgery. However, other artificial muscles may also be suitable for such applications pending biocompatibility, voltage and activation requirements. These will include conductive and conjugated polymers as EAPs and artificial muscles, metal hydride artificial muscles, chemoelectromechanical contractile artificial muscles such as polyacrylonitrile fibrous gels, biopolymeric artificial muscles such as chitosan gels, magnetic gels, shape memory alloys and shape memory polymers.

At this stage, it is appropriate to acknowledge the great work accomplished in developing tactile sensing and haptic feedback in recent years, enabling a smooth transition to applications of IPMCs in tactile sensing and haptic feedback.

22.3 Haptic/Tactile Feedback Technology Overview

Many technologies have been used to develop haptic and tactile feedback systems for surgeons during robotic surgery. Each will be described, along with their research and developments. Salisbury, Cutkosky and Okamura are the pioneers of haptic/tactile feedback in robotic surgery and have contributed tremendously to this technology. Salisbury and his students at MIT began developing innovative human–machine interfaces and haptics in the early 1980s. Salisbury should be credited for the creation of the first dexterous three-fingered Salisbury hand[5] and his pioneering efforts to make it highly dexterous, as well as his revolutionary robotic surgery research and development[6] through haptic robotic dexterous hand designs. These designs eventually led to the formation of Integrated Surgical Systems (now called Intuitive Surgical Inc.), which created the da Vinci Surgical System®, classified as a master–slave surgical system. Jacobson *et al.*[7] should also be given credit for the creation of another dexterous four-fingered hand, the "Utah/MIT Dexterous Hand", in the early 1980s. Following these developments, haptic research progressed significantly, as can be seen from early research by Okamura *et al.*,[8] who presented an overview of dexterous manipulation. Furthermore, Okamura and Cutkosky[9] reported their work on feature detection for haptic exploration with robotic fingers. Later, Okamura *et al.*[10] presented a very compelling paper on using a force-feedback joystick to control dynamic systems. Barbagli and Salisbury studied the effects of sensor/actuator asymmetries in haptic interfaces.[11] Griffin *et al.*[12] further discussed feedback strategies for shared control in dexterous telemanipulation. Prasad *et al.*[13] further developed a modular two DOF (degree of freedom) force-sensing instrument for laparoscopic surgery. Following that effort, Salisbury *et al.*[14] made advances towards haptic rendering in 2004. Okamura then presented some methods for haptic feedback in teleoperated robot-assisted surgery,[15] and along with Bethea, Kitagawa, Fitton, Cattaneo and Gott, presented a number of applications of haptic feedback in robotic surgery.[16] Kitagawa *et al.*[17] made further advances towards understanding the effects of sensory substitution on suture-manipulation forces for robotic surgical systems. Abbott *et al.*[18] were able to develop a number of haptic virtual fixtures for robot-assisted manipulation with applications to minimally-invasive robotic surgery (MIRS). This was followed later by an Okamura paper on haptic feedback in robot-assisted minimally invasive surgery.[19] Misra *et al.*[20] further discussed the importance of organ geometry and boundary constraints for the planning of medical interventions. Haptic and tactile feedback in robotic surgery received additional boosts through the original work of Okamura *et al.*,[21] who further developed haptics for robot-assisted minimally invasive surgery. Force feedback was recognized to be of extreme importance in MIRS, and thus Okamura *et al.*[22] presented the essence of force feedback and sensory substitution for robot-assisted surgery. Some of the recent work in connection with the advancement of haptic/tactile feedback in MIRS can be traced in the publications of Okamura *et al.*,[23] Yamamoto *et al.*[24] and Stanley *et al.*[25] on haptic jamming and a recent patent disclosure on advanced haptics by Guerin *et al.*[26] The pioneering work of Dutson and his research team at the UCLA Center for Advanced Surgical and Interventional Technology (CASIT) should be briefly mentioned. They are developing a pneumatic balloon-based tactile feedback system that measures forces at the distal end of the robotic grasper *via* a force sensor array made with *FlexiForce*® sensors, and the forces from each sensor element are translated into proportional pressures that are applied to the surgeon's fingers *via* balloon actuator arrays. This system is currently under development for use with the da Vinci robotic surgical system.[27,28] *FlexiForce*® sensors mounted at the distal end of robotic grasper forceps proportionately translate pressures applied to the surgeon's fingers *via* balloon actuator arrays. Researchers have already demonstrated a reduced grasping force using this system. Dargahi *et al.*[29] have presented a new idea for the modeling and testing of an endoscopic piezoelectric-based tactile sensor. Some other interesting and innovative research on haptic/tactile feedback in robotic surgery is the advancement made by Kuchenberger,[30–32] whereby Kuchenbecker's VerroTouch System, for example, restores the sense of touch and vibration lost in robotic surgery, enabling the surgeon to feel the texture of rough surfaces or the start and end of contact with tissue. Similarly, Tadano *et al.*,[33] of the Precision and Intelligence Laboratory, Tokyo Institute of Technology and Tokyo Medical and Dental University, have developed a pneumatic robot with haptic feedback called IBIS IV. Steinberg of Cambridge Research and Development in Nashua has introduced a haptic interface that allows users to feel the touch and pressure applied by a robot while it works. The system uses a pneumatic air pressure-measuring device attached to the back of the head and relays steady pressure that the user can feel as a robot touches or grabs an object.[34] A new surgical robot with haptic feedback recently developed in Europe at the Technical University of Eindhoven is called "Sofie", a haptic feedback robot designed and developed by van Den Beden.[35] One should also mention new haptic devices such as the PHANTOM Omni® haptic joystick device from SensAble technologies[36] and the development of a new Tactile Capsule which restores surgeons' sense of touch during laparoscopy by a team of engineers from the

Science and Technology of Robotics in Medicine (STORM) Laboratory at Vanderbilt University.[37] For a state-of-the-art paper on force and tactile sensing for minimally invasive surgery, see Puangmali *et al.*[38]

Having fully explored and understood the state of the art in haptic/tactile feedback sensing, we can now introduce the new smart biomimetic multifunctional IPMCs into this critical field for MIRS. It must be mentioned that IPMCs as sensors and haptic actuators are far more appropriate than *FlexiForce*® because they can also sense bending, twisting, rolling, tensioning, twirling and turning and they are biocompatible, while *FlexiForce*® can only measure the normal pressure in a flat configuration. It uses a piezoresistive material that can only sense the normal force. The harder you press, the lower the sensor's resistance. The sensor itself is thin and flexible, but the resistance does not change while being flexed. In that sense, the introduction of the versatile IPMCs to MIRS as a new family of haptic/tactile feedback sensors is highly desirable and timely. The literature on employing soft smart materials and IPMCs for haptic/tactile feedback sensing in general robotics will be discussed later when IPMC actuation and sensing phenomena are discussed.

In connection with some background on robotic surgery relevant to this chapter, there exists a very rich body of literature on robotic surgery and, in particular, there are recent books and patents that cover all aspects of robotic surgery from basic design and kinematics to surgical work spaces and clinical aspects. In particular, Gharagozloo and Najam[39] wrote the first book on robotic surgery from a clinical perspective in 2008. Dasgupta *et al.*[40] presented clinical views of and practices in urologic robotic surgery. Rosen, Hannaford and Satava[41] wrote the first book on robotic surgery devoted to systems applications and vision rather than a purely clinical approach. Navaranjan *et al.*[42] wrote a book on robotic surgery in 2012 with an emphasis on general considerations for this emerging field. Sood and Leichtle[43] wrote a book on the essentials of robotic surgery in 2013. Shahinpoor and Gheshmi[44] recently published a book on robotic surgery focusing on using smart materials, robotic structures and artificial muscles with additional emphasis on the design, kinematics and work spaces of surgical robots. Changqing[45] also recently published a book on robotic cardiac surgery with an emphasis on its clinical aspects and complications. Shahinpoor and Gheshmi[46,47] have also presented novel designs for ophthalmic robotic microsurgery as well as robotic laparoscopic surgery.

22.4 IPMC Manufacturing and Biocompatibility

The prelude to manufacturing IPMCs was manufacturing electrically controllable ionic gels or synthetic (artificial) muscles from polyelectrolytes as reported by Adolf *et al.*[48] and Shahinpoor.[49] In order to integrate electrodes with ionic gels, a number of manufacturing approaches based on an oxidation/reduction (redox) chemical procedure that leads to an electroless chemical plating of ionic membranes were developed as reported in Oguru *et al.*[50] and Shahinpoor and Mojarrad,[51] who present detailed original manufacturing procedures for IPMCs. Kim and Shahinpoor[52–54] also presented detailed manufacturing methodologies for IPMCs based on Taguchi's manufacturing optimization procedure. Shahinpoor and Kim,[55] Segalman *et al.*[56] and Shahinpoor[57] presented a number of plausible models for electrically controllable soft biomimetic polymeric artificial muscles as early as 1992. One of the basic polyelectrolytes used in manufacturing IPMCs is perfluorinated sulfonic membranes, a form of which is commercially known as Nafion™ and is manufactured by DuPont. These materials are basically biocompatible, as reported by Kim *et al.*[58] and Dickert *et al.*[59] Thus, plating these membranes with biocompatible metals such as platinum (Pt) or gold will lead to biocompatible IPMCs as biomimetic muscle-like soft actuators and sensors for biomedical applications and, in particular, "robotic surgery". This will open an exciting area in robotic surgery for realistic haptic/tactile feedback sensing by surgeons. These polymeric muscles are manufactured by chemical processes[51–55] in which a noble metal, such as Pt or gold, is deposited near a boundary as embedded electrodes that penetrate in a fractal and dendritic manner within the molecular network of the base ionic polymer. IPMCs chemically plated with Pt form a sandwich of Pt–Nafion–Pt, where the Pt is often only a few tens of microns thick (typically 10–20 microns for a $1\,\text{cm}\times4\,\text{cm}\times0.2\,\text{mm}$ IPMC strip). IPMC artificial muscles are capable of absorbing polar fluids such as water because of the presence of Nafion. The absorbed water somewhat gels up on internal cations as hydrated molecules. Thus, IPMCs are capable of operating in very humid or wet environments, underwater, in blood and also in air. In this context, they are quite suitable for surgical operations in a human body.

22.4.1 IPMC Biomimetic Robotic Actuation

IPMCs are active multifunctional smart materials (*i.e.*, they deform significantly when excited by a relatively low voltage and generate voltage when deformed). Grodzinsky[60] and Yannas and Grodzinsky[61] were the first to present a continuum model for the electrochemistry of the deformation of a charged polyelectrolyte membrane. However, it was Shahinpoor *et al.*[62] who introduced IPMCs as smart multifunctional EAPs and artificial muscles in 1998, followed by four review articles on the state-of-the-art IPMCs by Kim and Shahinpoor in recent years.[63–66] The application of these active polymers to artificial muscles can be traced to Shahinpoor *et al.*[67] and Nemat-Nasser.[68] Due to space limitations, the discussion on IPMC actuation will be very limited. Interested readers are invited to read the review references.[63–67]

As will be explained in this and the following two sections, one can bond (solder) a pair of electrodes with attached lead wires to the top and bottom of the IPMC sheet and quickly observe that, upon application of a low voltage, the sheet deforms accordingly. If the voltage is dynamic, say sinusoidal, the bending becomes oscillatory. IPMCs have a very large bandwidth to kilo-Hertz both in actuation and sensing. For a full theoretical description of the charge dynamics and electromechanics of IPMCs, see de Gennes *et al.*[69] and Porfiri.[70–72] Once an electric field is imposed on an IPMC cantilever, the conjugated and hydrated cations rearrange to accommodate the local electric field and thus the network deform or bend in a spectacular manner under a small electric field such as tens of volts per millimeter. de Gennes *et al.*[69] have presented the standard Onsager formulation of the underlying principle of IPMC actuation/sensing phenomena using linear irreversible thermodynamics.

Charge dynamics modeling and Poisson–Nernst–Planck formulation can also be applied, as discussed by Porfiri[70–72] and Bahramzadeh.[73]

In order to assess the electrical properties of the IPMCs, the standard AC impedance method that can reveal the equivalent electric circuit of IPMCs has been adopted. Typical results are reported in Shahinpoor *et al.*,[67] Bahramzadeh[73] and Bonomo *et al.*[74] Overall, it is interesting to note that the IPMCs are nearly resistive (>50 Ω) in the high-frequency range and fairly capacitive (>100 μF) in the low-frequency range. IPMCs generally have a surface resistance (R_{SS}) of a few Ohms per centimeter, a near-boundary resistance (R_S) of a few tens of Ohms per centimeter and a cross-resistance (R_P) of a few hundreds of Ohms per millimeter, with a typical cross-capacitance (C_g) of a few hundreds of microfarads per millimeter. This approach is based upon the experimental observation of the considerable surface electrode resistance based on an equivalent circuit model (see Shahinpoor *et al.*[67]). Another important feature of the IPMCs is the molecular diffusion and transfer of cations and their hydrated water molecules as they move between the porous electrodes.

22.4.2 IPMC Versatile Sensing Feedback

IPMCs are excellent versatile sensors for complex deformations and kinesthetics of internal organs and tissues. As reported above, they can cooperatively actuate and sense during their operation inside the body, and this makes them unique as haptic actuators and sensors for robotic surgery. Upon application of an electric field (imposed voltage across the membrane), the cations migrate towards the cathode and thus cause a stress gradient that deforms IPMC in an actuator mode. On the other hand, mechanically deforming the IPMC forces the cations to redistribute and move across the thickness, and this produces an electric field between the electrodes based on the Poisson–Nernst–Planck phenomena (see Porfiri[70–72]). Shahinpoor[75,76] and Mojarrad *et al.*[77] had introduced the physical phenomenon of the "flexogelectric" effect in connection with the dynamic sensing of ionic polymeric gels, where manually bending or twisting the IPMCs resulted in a measurable electric field. Sadeghipour *et al.*[78] had reported earlier that Nafion in a hydrogen environment and sandwiched between two flat electrodes could be used as a vibration energy damper. Shahinpoor and Mojarrad[79] introduced the ionic polymers as cooperative sensors and actuators in 2002. Additional contributions on IPMC sensing followed with the work of Ferrara *et al.*[80] on measuring the force and pressure between vertebrates, Bonomo *et al.*[81] on IPMC general sensing, Henderson *et al.*[82] on the near-DC sensing capabilities of IPMCs, Farinholt *et al.*[83] on the relationship between charge and deformation in IPMCs, Chen *et al.*[84] on IPMC sensing underwater and Bahramzadeh *et al.*[85] on curvature sensing by IPMCs. Bonomo *et al.*[86] modeled IPMC sensors, finding good correspondence between their model and experimental results. Recently, Yamakita *et al.*[87] used the position sensing capabilities of IPMCs to develop a closed position controller using H_∞ theory (see Zames[88]). Doping the IPMC material with specific counter-ions affects the performance of the material, emphasizing either the sensing or actuation properties.

22.4.3 IPMC-Based Haptic/Tactile Feedback Sensing Technology

In order to initiate a transformative and translational haptic/tactile feedback sensing technology in robotic surgery in comparison to generic haptic/tactile feedback sensing technology, the IPMCs are introduced. The following contributions are of importance to the proposed effort. As early as 2000, Konyo *et al.*[89] discussed artificial tactile feel displays using soft gel actuators. Based on the concept of smart pads introduced by Mazzone *et al.*,[90] Nakano *et al.*[91] introduced a novel soft patch array of IPMCs that can be used simultaneously as an input and output device for designing, presenting or recognizing objects in three-dimensional space. Shahinpoor further introduced the soft organ and tissue compression by IPMCs[92] by introducing multi-fingered soft robotic hands as shown in Figure 22.1(a) and (b).

The idea of creating a dexterous five-fingered hand made entirely with IPMCs and emulating the real five-fingered hand of a surgeon and generating haptic feedback has been explored by the author, as shown in Figures 22.2–22.4. Here, each finger basically consists of stacked up IPMC strips wired directly for finger actuation. It is also possible to create multi-fingered IPMC hands to be used as soft biomimetic biocompatible catchers/grabbers of a thrombus (a blood clot formed *in situ* within the

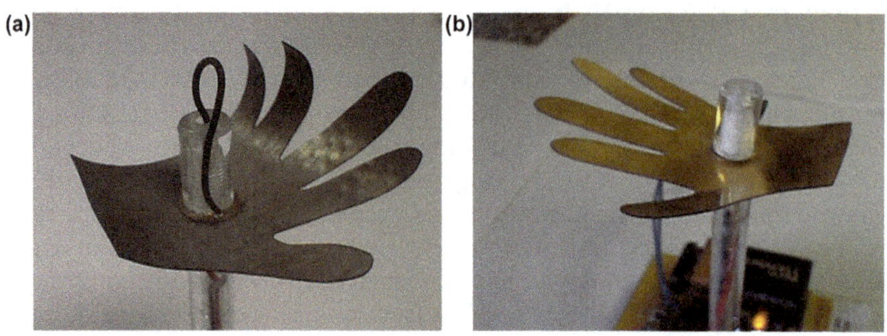

Figure 22.1 Two examples (Pt [a] and gold [b]) of dexterous soft biomimetic five-fingered IPMC hands with actuation and haptic feedback sensing capabilities controlled by just a couple of electrodes.

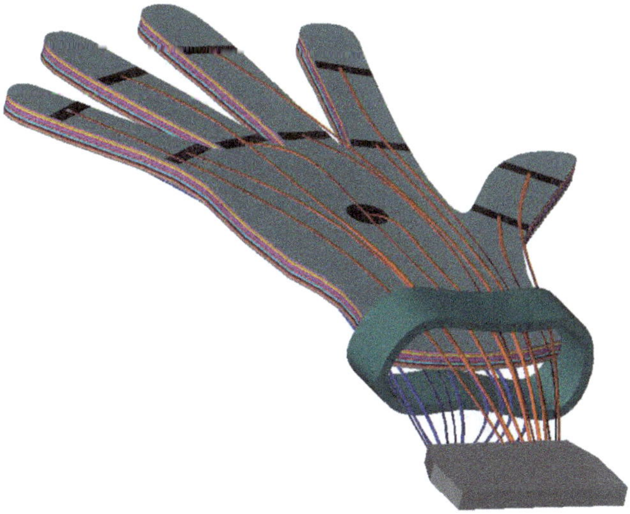

Figure 22.2 Concept drawing of an electrically powered biomimetic soft dexterous hand emulating a surgeons' hand by stacked IPMC fingers.

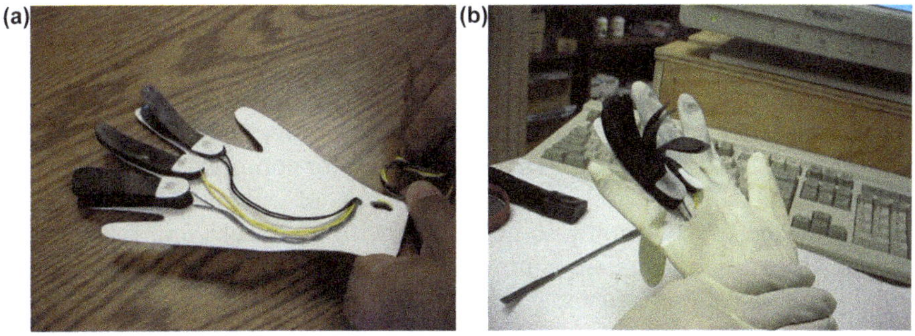

Figure 22.3 Fabricated biomimetic soft dexterous hand (a) and encapsulating it in a latex glove (b) to emulate a surgeons' hand by stacking IPMC fingers and powering them by electrical wire connections.

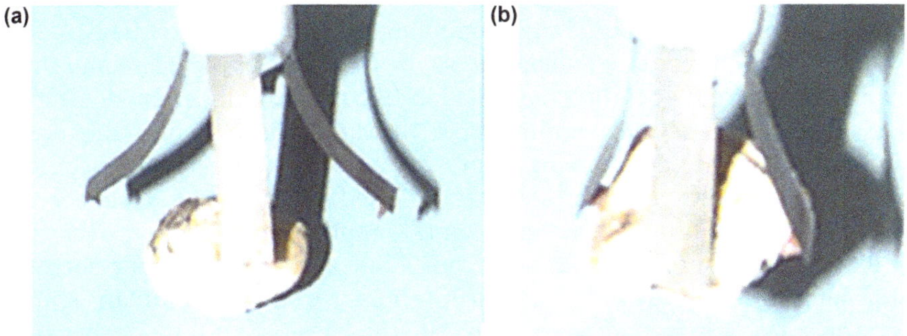

Figure 22.4 Four-fingered IPMC grabber to be used as a soft biomimetic biocompatible catcher/grabber of blood clots or other loose/lost items and parts during endovascular robotic surgery: (a) approaching; (b) grabbing.

vascular system of the body and impeding blood flow), blood clots or other loose/lost items and parts during endovascular robotic surgery, as shown in Figure 22.4.

Later, Citerin and Kheddar presented their research on EAP actuators for tactile display.[93] Konyo and Tadokoro later reported an IPMC-based tactile display for pressure and texture presentation on a human finger.[94] Recently, Jain *et al.*[95] further contributed to the design and control of an electromyography (EMG)-driven IPMC-based dexterous robotic finger and Aw *et al.*[96] presented their results on an IPMC actuated robotic surgery end-effector. Chatterjee *et al.*[97] introduced a miniaturized IPMC-based five-fingered dexterous hand. Figures 22.1–22.5 depict examples of such dexterous soft biomimetic multi-fingered IPMC hands with haptic actuation and sensing capabilities. However, we will focus on using various configurations of IPMCs for haptic feedback and tactile sensing during robotic surgery.

22.5 Applications of IPMCs for Robotic Surgery

22.5.1 Brief Introduction to IPMCs as Multifunctional Materials

IPMCs or ionic polymer conductor nano-composites are chemically plated ionic polymers manufactured by a redox operation with a noble metal such as Pt or gold to keep them biocompatible. Refer to the seminal publication by Shahinpoor, Kim and Mojarrad[67] for comprehensive coverage of the various properties and applications of such materials. The basic material is commonly ionic Teflon with relatively few fixed ionic groups. Once an electric field is imposed on such a network, the conjugated and hydrated cations rearrange to accommodate the local electric field and thus the network deforms, and in the simplest of cases, such as in thin membrane sheets, spectacular bending is observed (Figure 22.5) under small electric fields such as tens of volts per millimeter.

Typical experimental deflection curves are depicted in Figure 22.6. Once an electric field is imposed on an IPMC cantilever, in the polymeric network the hydrated cations migrate to accommodate the local electric field.

This creates a pressure gradient across the thickness of the beam and thus the beam undergoes bending deformation (Figure 22.6) under small electric fields such as tens of volts per millimeter. The variations in tip blocking forces and deflections are depicted in Figure 22.7. IPMCs can generate electrical power like an electromechanical battery if flexed, bent, twisted or squeezed.

Keshavarzi, Shahinpoor, Kim and Lantz[98] applied the sensing capabilities of IPMCs to measure blood pressure, pulse rate and rhythm during surgery. Motivated by the idea of measuring pressure in the human spine, Ferrara *et al.*[80] applied pressure across the thickness of an IPMC strip while measuring the output voltage. Typically, flexing of such material in a

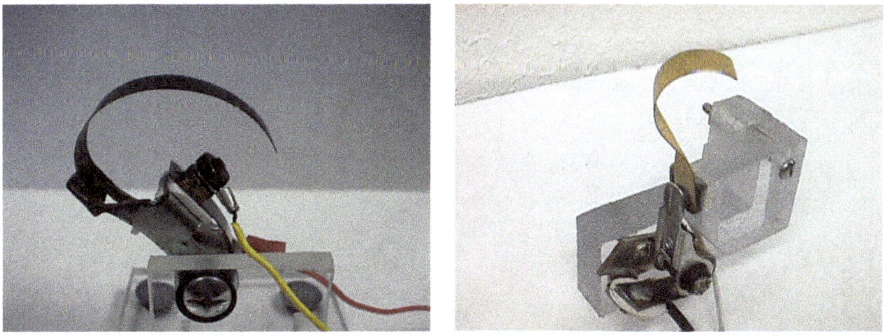

Figure 22.5 Typical deformation of strips (10 mm×80 mm×0.34 mm) of IPMCs under a step voltage of 4 V: pure bending (left); bending and twisting (right).

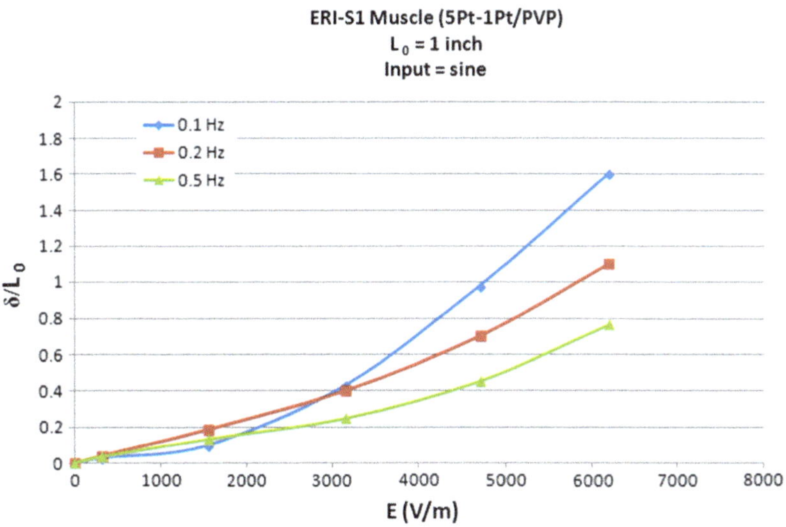

Figure 22.6 Typical frequency-dependent actuation of IPMCs in a cantilever configuration.

cantilever form sets them into a damped vibration mode that can generate a similar damped signal in the form of electrical power (voltage or current), as shown in Figure 22.8.

It will be shown in the next section that the output voltage of an IPMC strip is a function of the curvature and its rate of change. While an output voltage of 1–2 mV can be derived by dynamic sensing from a sample with dimensions of 10 mm×30 mm×0.2 mm, the achievable voltage is smaller in a quasi-static sensing case using the same sample. Small samples (1 cm×4 cm×0.2 mm) of IPMCs generate signals in the range of a few millivolts in the presence of a typical 90° bending deformation. IPMC

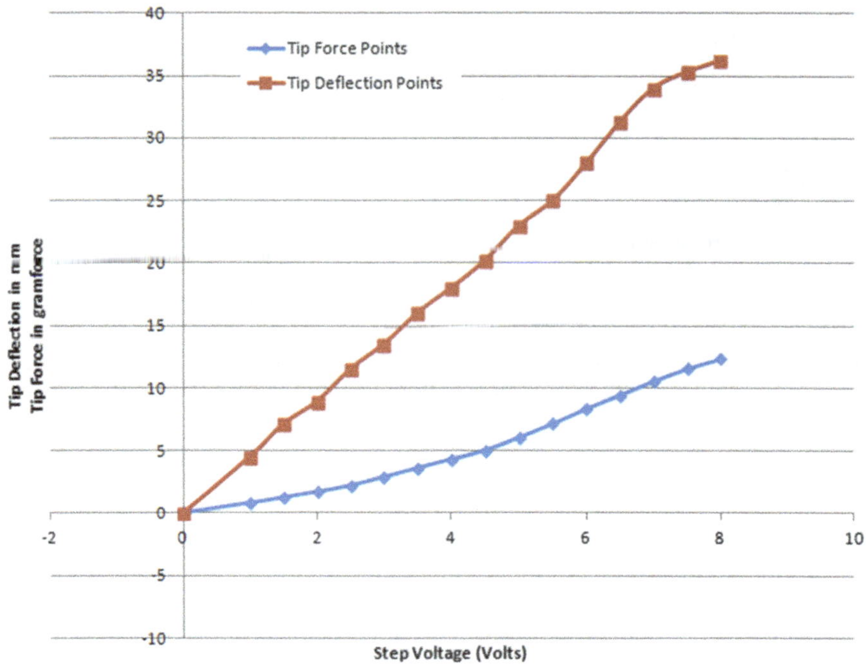

Figure 22.7 Variations in tip blocking forces and the associated deflections for a 1 cm × 5 cm × 0.3 mm IPMC Pt–Pd sample in a cantilever configuration.

sensors have proven to be highly sensitive to applied deformation over a large frequency range.

The experimental results showed that an almost linear relationship exists between the voltage output and the imposed displacement of the tip of the IPMNC (ionic polymer metal nano-composites) sensor. As far as force generation is concerned, IPMNCs generally have a very high force density. Figure 22.9 below displays the cantilever and load cell configuration for measuring the tip blocking force of typical samples of IPMNCs.

22.6 Feasibility of Providing Kinesthetic Force Feedback to Surgeons during Robotic Surgery by EAP Sensors (IPMCs)

One can integrate IPMCs as EAP nano-sensors and nano-actuators with robotic surgical end-effectors such as intuitive grasping forceps to provide kinesthetic force/torque feedback to surgeons during robotic surgery. The approach is to employ IPMCs as surgeons' feel/haptic sensors for the kinesthetics of internal organs in interaction with various surgical robotic end-effectors. This aims to integrate the kinesthetic force feedback signals with joysticks and foot pedals in order for surgeons to receive kinesthetic force feedback in real time during robotic surgery. To achieve these

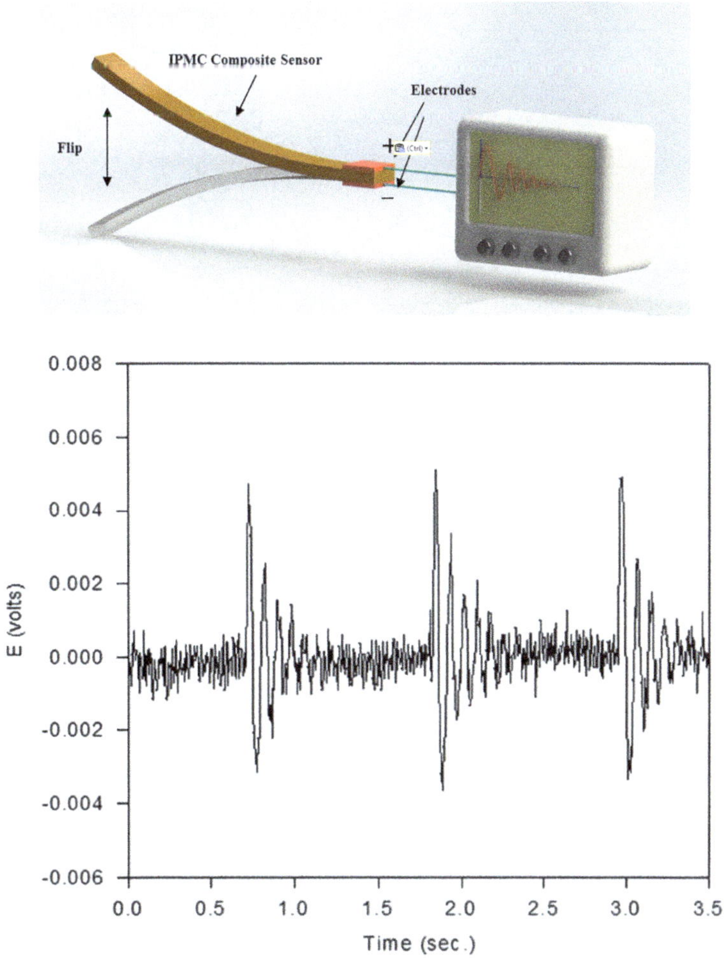

Figure 22.8 A typical voltage response of an IPMC strip (1 cm × 4 cm × 0.2 mm) under oscillatory mechanical excitations (top) and damped oscillatory output voltages (bottom).

objectives, suitable IPMC strips and loops are attached to the end-effectors and wired through the end-effectors to the electronics providing kinesthetic force or torque feedback to surgeons. To provide the force or torque sensation to the surgeons, the specific IPMC sensor, say a tip bender, will be subjected to a bending force to create an output signal in millivolts. This signal will then be amplified electronically and fed into a linear actuator or servo motor to generate the same force or torque and affect the operational forces/torques in the joysticks or foot pedals used by surgeons. If the tip of a 3 cm × 1 cm × 0.2 mm IPMC strip experiences a blocking force by contacting an organ/tissue during surgery, it will develop a blocking kinesthetic force of about 20 g. The same strip in a cantilever configuration, if moved by the

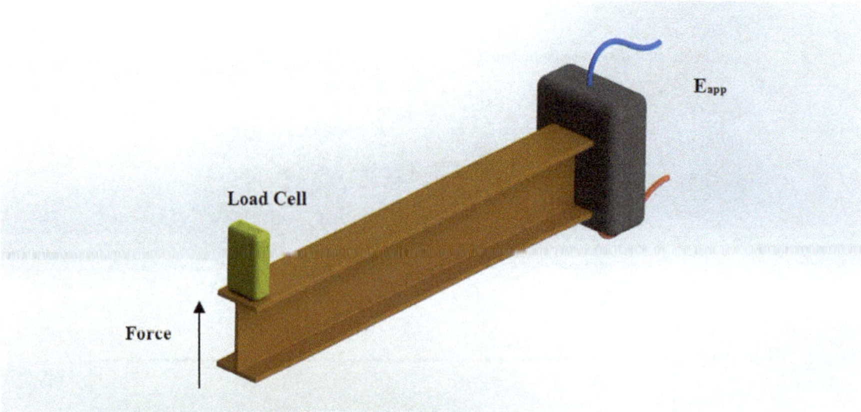

Figure 22.9 Cantilever and load cell configuration for measuring the tip blocking force of IPMNC samples.

kinesthetics of internal organs and end-effectors during surgery, will generate about 4 mV, which can be correlated to the kinesthetic forces of the internal organ movement during robotic surgery. Our experimentally obtained data indicate that IPMCs may provide dynamic feedback of kinesthetic forces to surgeons during surgery and gradually train surgeons to feel the kinesthetics and kinesthetic forces and torques applied to internal organs and tissues during surgery.

The fact that IPMCs can work well in the wet human body environment during robotic surgery will be an advantage for developing this technology. IPMCs are basically biocompatible because they are Teflon-based plastics in a nano-composite form with a noble metal such as Pt or gold. They are electrically safe and self-powered and do not need any source of voltage or current to provide kinesthetic force feedback in millivolts, and they are autoclaveable and hardly wear out. IPMCs also work well in an electrocautery environment because IPMCs are basically ionic Teflon with Pt electrodes that can withstand the high temperature of cauterization without melting or burning during surgery.

The current robotic surgical systems evolved from laparoscopic surgical procedures and made it possible for surgeons to perform surgery away from the patient with much greater concentration and ease. However, what was lost in this transition by the surgeons was the feeling sensation of tissues and organs and kinesthetic force feedback during surgery. It is interesting to note that even during laparoscopic surgery, surgeons can still feel and sense the tissues and organs they are handling and operating on with laparoscopic/endoscopic tools during surgery and feel the kinesthetic forces at work. However, kinesthetic force feedback was replaced with visual feedback during robotic surgery. It is to be noted that some of this kinesthetic force feedback was lost in the transition from open to laparoscopic surgery due to trocar friction and varying lever arms. However, with smart materials such as

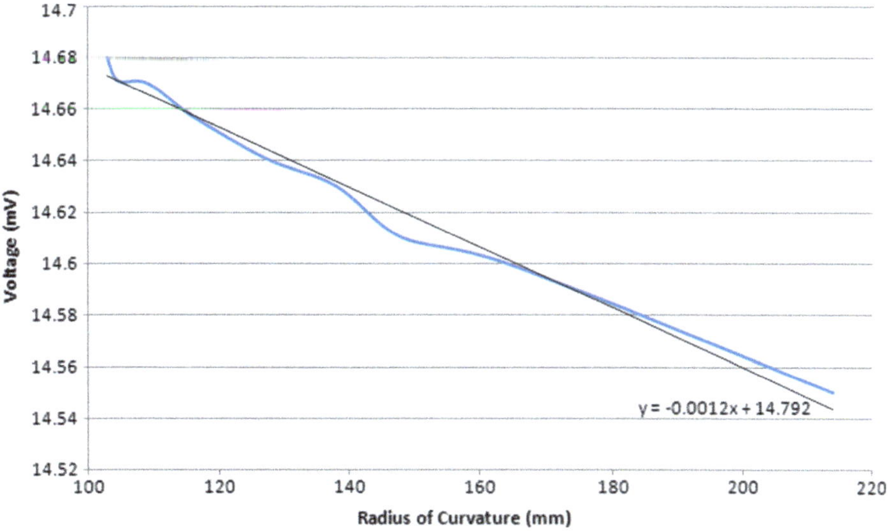

Figure 22.10 Typical sensing response of IPMC strips in bending. The sensing response in millivolts can be correlated to tip kinesthetic force.

IPMCs and appropriate calibration and tuning, one may be able to recover the kinesthetic force feedback during surgery. IPMCs are great for such robotic force feedback applications because they work perfectly well in the wet human body environment and generate millivolt-level sensing signal for kinesthetic force feedback. Figure 22.10 depicts the typical sensing output of IPMC strips in bending and how the bending force is measured by a load cell.

Note that the output voltage of an IPMC strip is a function of the curvature and its rate of change. While an output voltage of 1–2 mV can be derived by dynamic sensing from a sample with dimensions of 10 mm × 30 mm × 0.2 mm, the achievable voltage is smaller in a quasi-static sensing case using the same sample. Figure 22.11 depicts the experimental setup for measuring the bending curvature of IPMCs *versus* electrical parameters such as voltage and transient current.

IPMCs generate signals in the range of a few millivolts in the presence of deformation. IPMC sensors have been proven to be highly sensitive to applied deformation over a large frequency range. Figure 22.12 shows the general responses of IPMC sensors to fast excitations followed by slow bending accompanied by high-frequency noise.

In order to obtain kinesthetic force feedback information from IPMCs, two calibration procedures are required. To this end, the calibration of deformation of IPMC sensors with respect to a generated signal and then the bending force of IPMC actuators can be easily related to the observed deformation. It should be emphasized that the results of using IPMCs for kinesthetic force feedback to surgeon during robotic surgery are very promising, in that kinesthetic force feedback to surgeons may be possible using IPMCs in bending, twisting, loop or compression loading. In the next

(a)

(b)

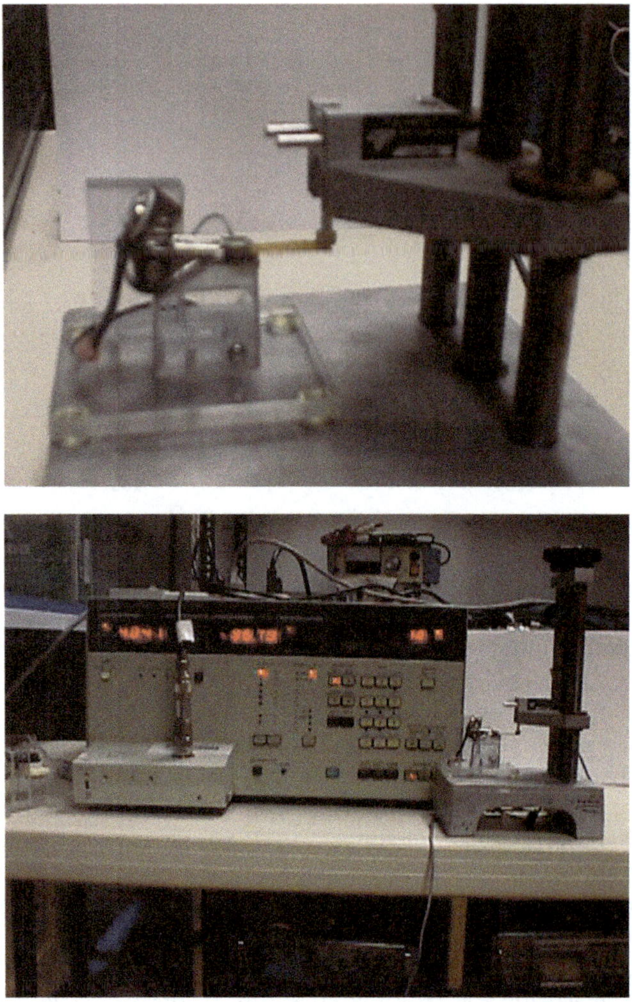

Figure 22.11 Experimental setup (a) for the measurement of the bending force of an IPMC cantilever using a high-resolution load cell and impedance analyzer (b).

section, we will elaborate on some additional experimental data that may help us to achieve our objectives.

22.7 Integration of IPMCs with Robotic End-effectors for Kinesthetic Force Feedback to Surgeons during Robotic Surgery by EAP Sensors (IPMCs)

Ionic polymeric composites (IPMCs) with distributed nano-sensing and nano-actuation can be employed in robotic surgery in order to provide kinesthetic force feedback to surgeons. Several apparatuses for the modeling

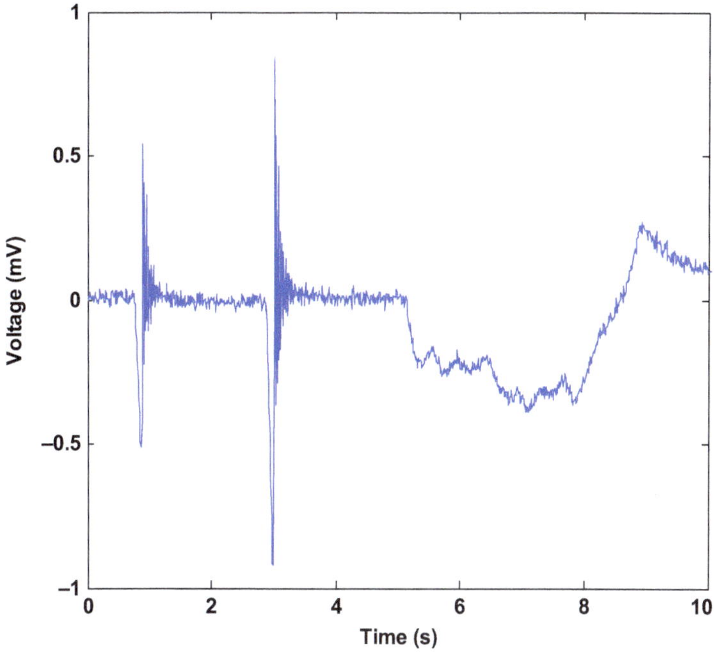

Figure 22.12 General response of an IPMC sensor to fast excitations followed by slow bending.

and testing of the various IPMC artificial muscles are described to show the viability of the application of electroactive IPMCs for providing surgeons with kinesthetic force feedback during robotic surgery. Here, we present some of the data generated by placing small strips of IPMCs on the contact face of grasping forceps as shown in Figure 22.13. One can also explore the bending and loop configurations for IPMCs (Figure 22.14) to interact with bodily organs and tissues during robotic surgery and provide us with kinesthetic force signals due to the bending of the IPMC strips or deformation of the loop, which provides a voltage signal that can be correlated to the force exerted.

Kinesthetic force feedback signals in the range of a few millivolt can be generated by touching soft human body anatomical plastic organs and can be correlated with the grasping forces of surgeons. Thus, one will be able to correlate the signal generated by the IPMC sensor mounted on an end-effector to the kinesthetic organ force experienced by the IPMC strip. In order to translate the kinesthetic force feedback voltage signal from the IPMC strips to kinesthetic force feedback to surgeons' hands during surgery, the voltage signal should be amplified electronically and applied to a servo integrated with joysticks to simultaneously provide the surgeons with haptic/tactile feedback during surgery. Figures 22.15–22.17 show some typical sensing signals from IPMC strips attached to surgical robotic end-effectors in compression, bending and loop configurations. By compression, we mean direct compression of IPMCs with a normal load such as in grasping forceps

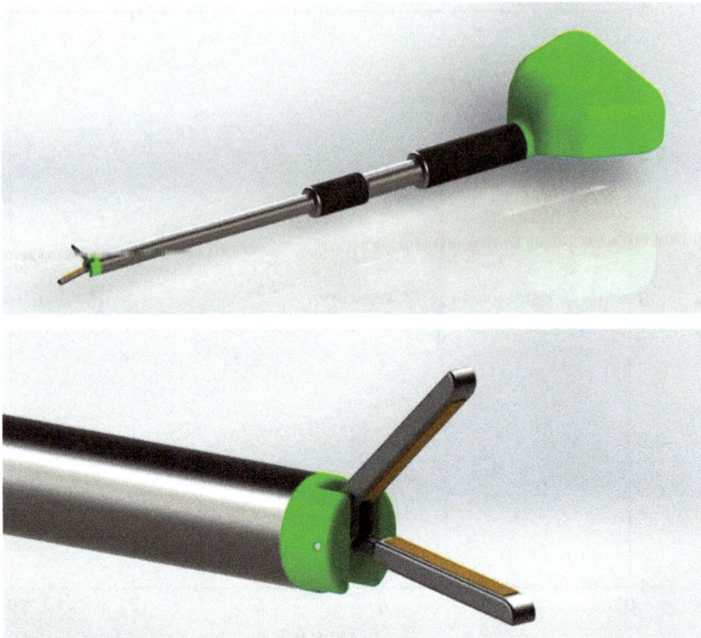

Figure 22.13 IPMC strips mounted on the faces of grasping forceps and wired into the electronics: normal view (top); expanded tip view (bottom).

(Figure 22.18). Note that the voltage output can be correlated with the kinesthetic normal force applied to the strip electronically, as described below.

The relation between the bending moment M and the radius of the curvature ρ of the neutral axis of the beam is as follows: $M=\frac{EI}{\rho}$, where ρ is the radius of the curvature and $I=\frac{1}{12}b.h^3$. Note that based on the dimensions given in Figure 22.16 (10 mm×30 mm×0.2 mm), $I=\frac{1}{12}\times 0.01\times 0.0002^3=6.67\times 10^{-15}$ m^4 and with $E=1200$ MPa for IPMCs and considering pure bending, one can calculate the required force for bending: $F=\frac{M}{L}=\frac{M}{0.03}=33.3\ M$. Thus, based on the curvatures measured in Figure 22.10, one can calculate the corresponding M and calculate the equivalent force experienced by bending the IPMC during surgery. For example, for $\rho=100$ mm, $M=81\times 10^{-6}$ Nm, and $F=2.74\times 10^{-3}$ $N=2.74$ mN.

Thus, in order to make robotic surgery more intelligent and haptics-based, one should equip surgical robotic end-effectors with smart nano-composites such as IPMCs capable of force, haptic and impedance sensing that can be fed back to the surgeon. Thus, one will be able to correlate the signal out of the IPMC mounted on an end-effector to the kinesthetic force experienced by the IPMC strip. In order to translate the kinesthetic force feedback voltage

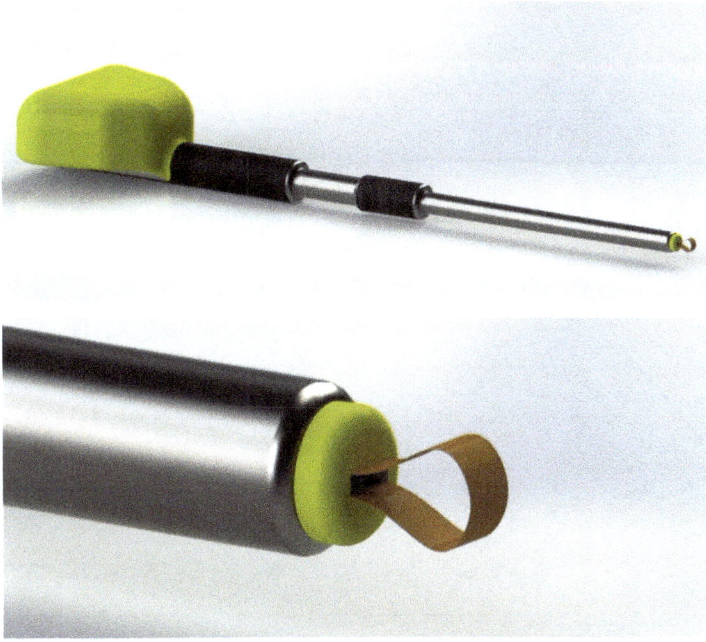

Figure 22.14 Kinesthetic force feedback IPMC loop (top) for haptic interaction with bodily organs, with expanded tip view (bottom).

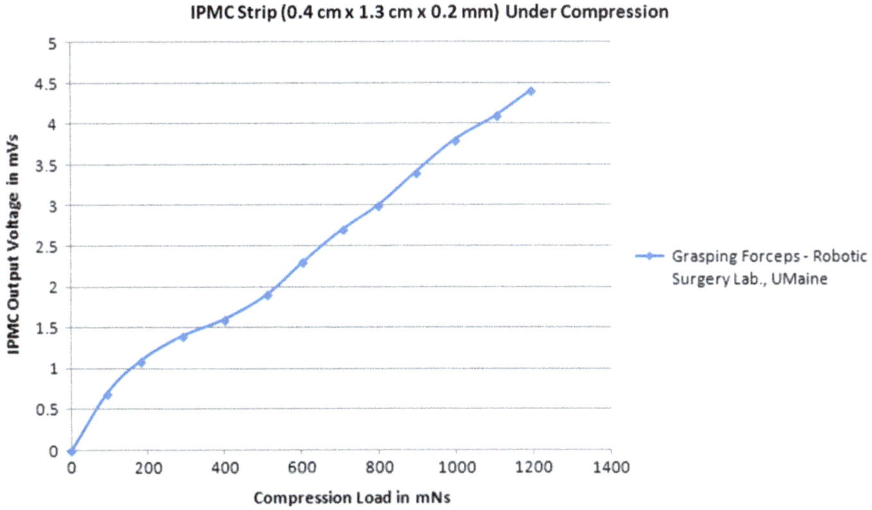

Figure 22.15 Typical strong sensing signal in millivolts from an IPMC strip in compression mode (direct compression of IPMCs with a normal load). The voltage output can be correlated with the kinesthetic normal force applied to the strip by body organs during robotic surgery.

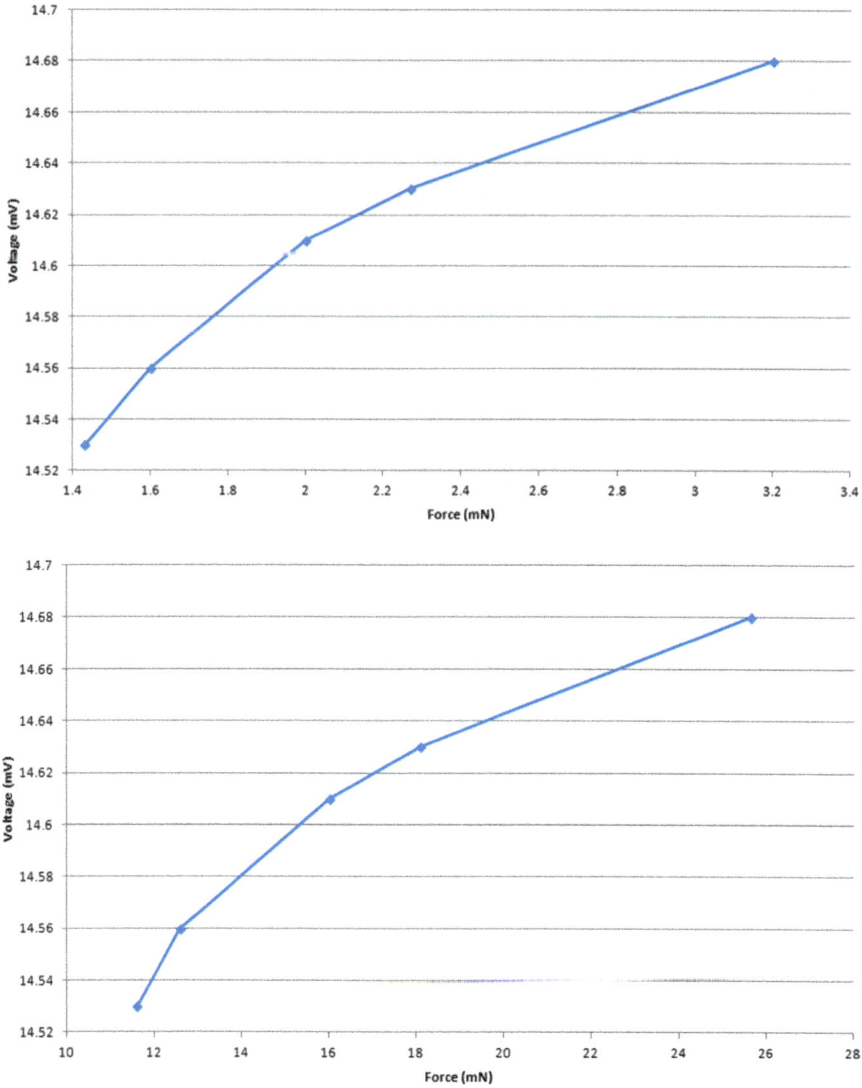

Figure 22.16 Typical force sensing signal in milli-Newtons of an IPMC strip (top: 10 mm×30 mm×0.2 mm; bottom: 10 mm×30 mm×0.4 mm) in bending/twisting mode (bending or twisting of IPMCs due to kinesthetic interaction of da Vinci end-effectors and plastic body organs). The voltage outputs were correlated with the kinesthetic force applied to the strip.

signal from the IPMC strips into kinesthetic force feedback to surgeons' hands during surgery, the voltage signal will be amplified electronically and applied to a servo integrated with joysticks to simultaneously provide the surgeons with kinesthetic force feedback during surgery. For other kinesthetic configurations of IPMCs, such as bending kinesthetics or loop kinesthetic, similar operations are applied. Figure 22.19(a) and (b) depict the general experimental setup for correlating the sensing voltage signal coming

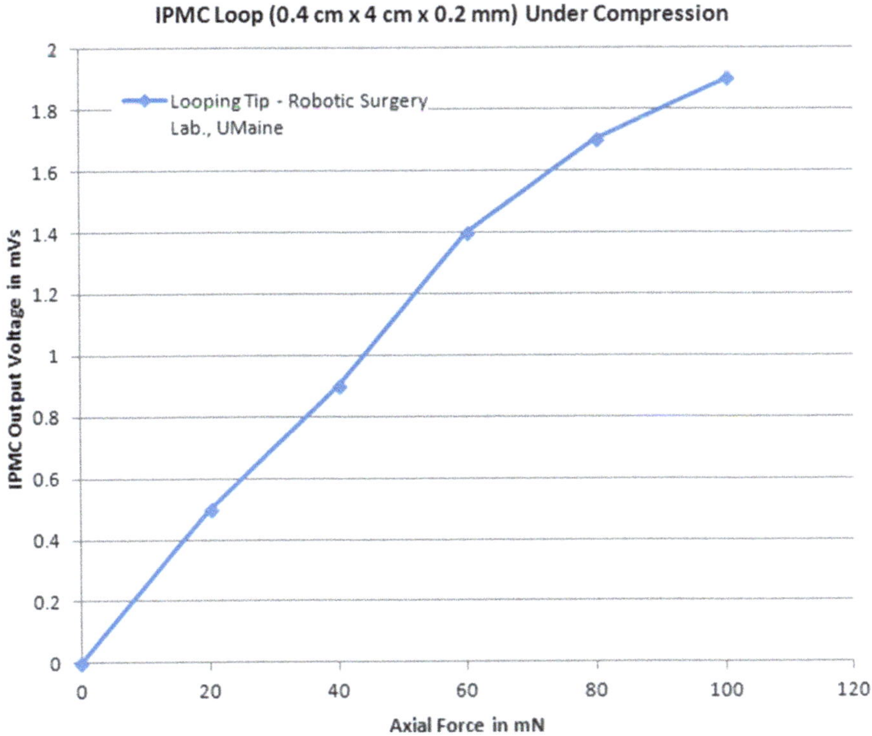

Figure 22.17 Typical sensing signal in millivolts from the IPMC loop configuration (loop deformation of IPMCs due to kinesthetic interaction of robotic end-effectors and plastic body organs).

from the bending or loop configurations of IPMCs in kinesthetic interaction with end-effectors and organs with the actual kinesthetic forces at work in Newtons.

The IPMC bending sensor is attached to a cantilever beam so that the IPMC strip follows the imposed curvature of the beam shape. One end of the sample is clamped to the fixed end of the beam. Both the curvature and rate of change of the curvature are controlled using a servo motor that bends the tip of the beam. An AX-12 Dynamixel servo motor with a step size of 0.29° was used to control the tip bending of the cantilever beam. The servo was also controlled through LabVIEW software. For voltage measurement, an NI-9219 A/D data acquisition module was used and the data were processed in LabVIEW.

To reduce the signal noise, a band pass filter was used to filter low (less than 0.01 Hz) and high (greater than 5 Hz) frequencies. The output voltage was amplified by a factor of 10 for easier processing of the data. In order to calibrate the sensor, different curvature inputs were applied to the cantilever at different rates.

Incorporating haptic force feedback may also enable expansion of robotic surgery to other surgical procedures, such as intraocular surgery or microsurgery, which are difficult to perform without a sense of touch or force feedback integrated with surgical robotic end-effectors. These robotic

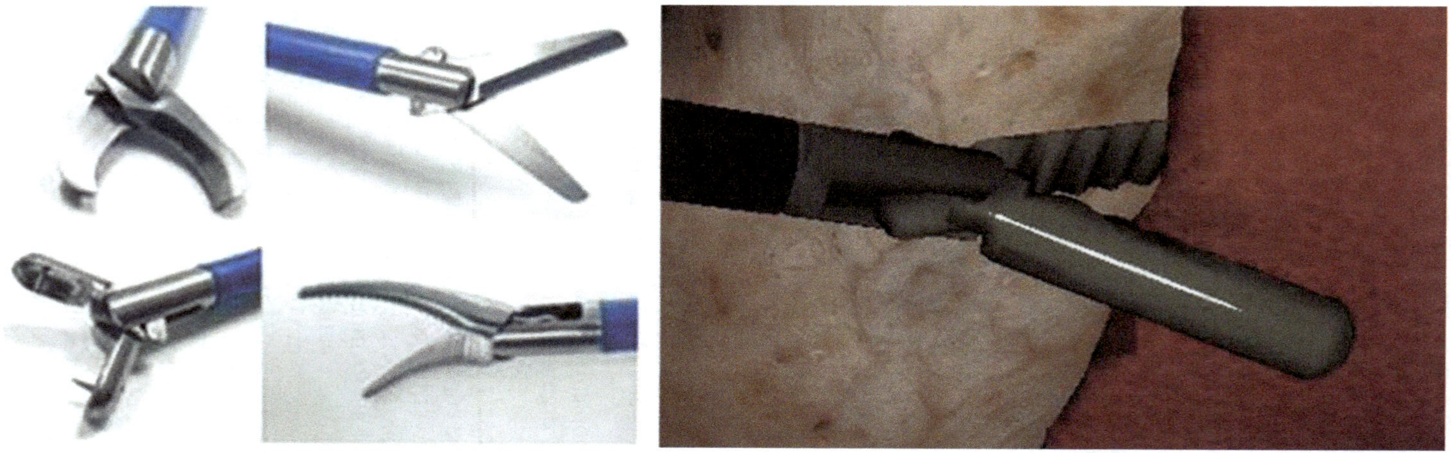

Figure 22.18 Some typical end-effectors (left) that can be integrated with various configurations of IPMCs for sensing the contact surfaces and kinesthetic forces experienced by surgeons (right).

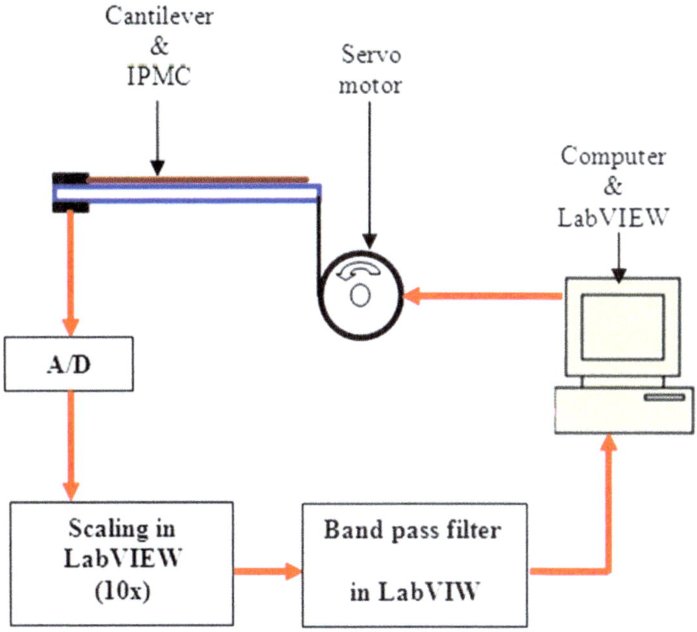

Figure 22.19a Schematic of the experimental setup to measure the dynamic kinesthetic output voltages of IPMC strips in bending and to electronically correlate these voltages to the forces generated *via* torque measurements of the servo motors.

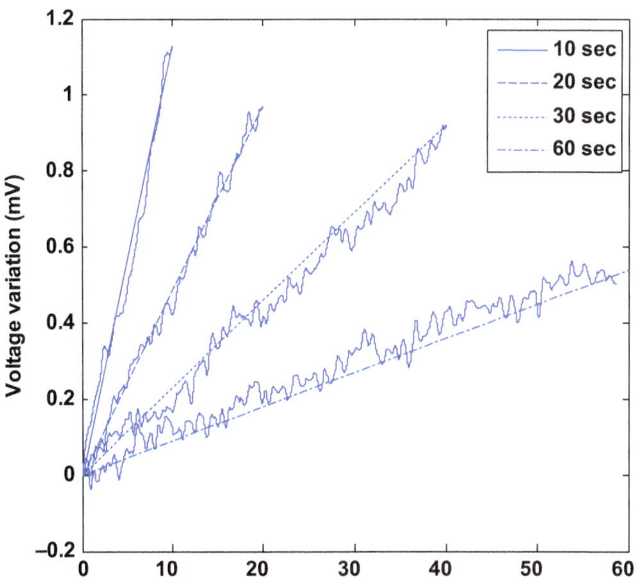

Figure 22.19b Ramp responses of an IPMC sensor at four different curvature rates of 10 s. 20 s, 40 s and 60 s. The IPMC strip was dynamically bent to amount to 10 mm, which is equal to a 500 mm radius of curvature.

end-effectors, such as needle holders, grasping forceps, dissecting forceps, scissors, biopsy spoons, retractors, electrosurgical tips and retractors and uterine manipulators, play a fundamental role in robotic surgery. The intent of this chapter was to address more advanced surgical operations in which the surgical robotic end-effectors are integrated with IPMCs. It is to be noted that some of this kinesthetic force feedback was lost in the transition from open to laparoscopic surgery due to trocar friction and varying lever arms. However, with smart materials such as IPMCs and appropriate calibration and tuning, one may be able to recover kinesthetic force feedback during surgery using IPMCs. IPMCs are great for such robotic force feedback applications because they work perfectly well in the wet human body environment and generate millivolt-level sensing signals for kinesthetic force feedback.

22.8 IPMC-Based Haptic/Tactile Feedback Technology

In order to initiate a transformative and translational haptic/tactile feedback sensing technology in robotic surgery in comparison to the generic haptic/tactile feedback sensing technology that was fully described in Section 22.1, IPMCs are introduced. The following contributions are of importance to the proposed effort. Figures 22.1–22.3 depict some examples of such dexterous, soft, biomimetic, five-fingered IPMC hands with haptic actuation and sensing capabilities. Here are developed various configurations of IPMCs for haptic/tactile feedback sensing. These are shown in Figure 22.20 with various

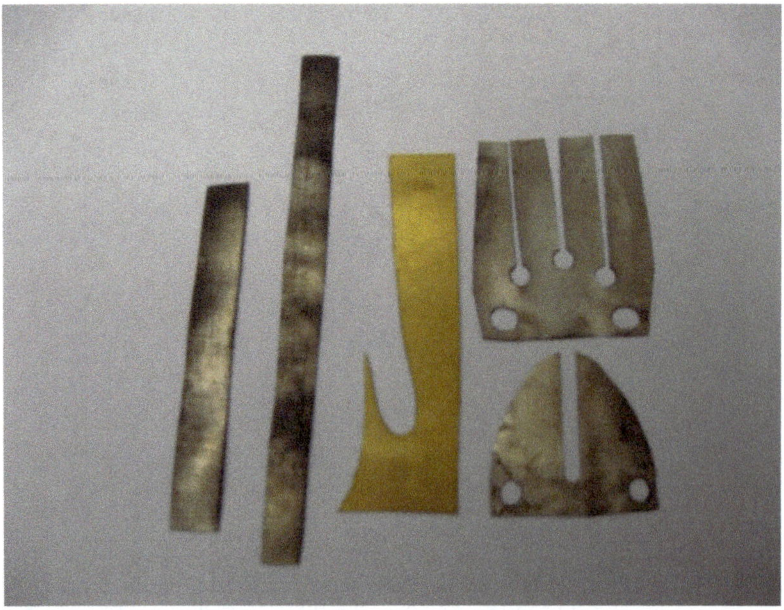

Figure 22.20 Various two-dimensional geometries of IPMCs (Pt and gold plating).

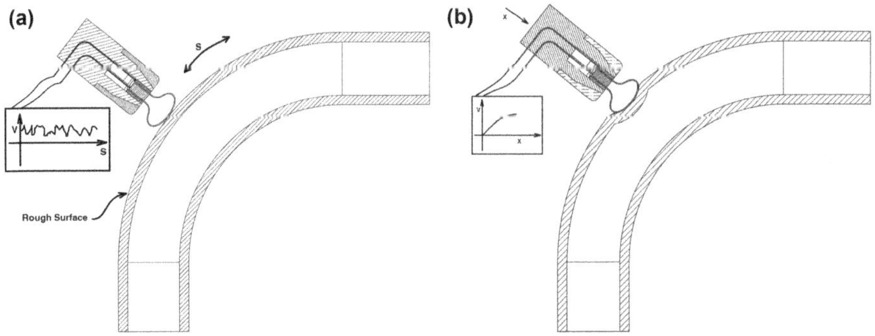

Figure 22.21 IPMC haptic/tactile feedback loop configurations (a) moved softly on the skin of an organ tissue boundary, generating tactile data, or (b) pressed softly against an organ tissue boundary, generating haptic feedback signals.

geometries of bendable, twistable, twirlable, squeezable, multi-fingered dexterous hands plated with biocompatible electrodes of Pt and gold.

22.9 Configuration of IPMC Haptic Feedback/Tactile Loop Sensing Elements with Robotic Surgical End-effectors

IPMC-based haptic/feedback and tactile sensors can be integrated with multi-fingered grippers and dexterous hand configurations and end-effectors such as needle holders, suture holders, grasping forceps, dissecting forceps, scissors, biopsy spoons, retractors, electrosurgical tips and retractors, thermo-surgical tips and refractors and uterine manipulators (Figure 22.21).

Acknowledgements

This work was partially supported by Environmental Robots Inc.

References

1. C. E. Reiley, T. Akinbiyi, D. Burschka, D. C. Chang, A. M. Okamura and D. D. Yuh, *J. Thorac. cardiovasc. Surg.*, 2008, **135**(1), 196–202.
2. L. Callan, N. Chen and S. Pautler, *UWO Med. J.*, 2013, **82**(1), 22–24.
3. *Surgical Robotics: Systems Applications and Visions* ed. J. Rosen, B. Hannaford and R. M. Satava, Springer Publishers, New York, Dordrecht, Heidelberg, London, 2011.
4. M. E. Hagen, J. J. Meehan, I. Inan and P. Morel, *Surg. Endosc.*, 2008, **22**(6), 1505–1508.
5. J. K. Salisbury and J. J. Craig, *Int. J. Robot. Res.*, 1982, **1**(1), 4–17.

6. http://www.intuitivesurgical.com/company/history/.
7. S. C. Jacobsen, J. E. Wood, D. F. Knutti and K. B. Biggers, *Int. J. Robot. Res.*, 1984, **3**(4), 21–50.
8. A. M. Okamura, N. Smaby and M. R. Cutkosky, Proceedings of International Conference of Robotics and Automation, ICRA'2000, IEEE International Conference, ICRA.
9. A. M. Okamura and M. R. Cutkosky, *Int. J. Robot. Res.*, 2001, **20**(12), 925–938.
10. A. M. Okamura, C. Richard and M. R. Cutkosky, *J. Eng. Educ.*, 2002, **91**(3), 345–349.
11. F. Barbagli and K. Salisbury, Proceedings of the 11th Symposium on Haptic Interfaces for Virtual Environment and Teleoperator Systems, Los Angeles, CA, March 2003, pp. 140–147.
12. W. B. Griffin, W. R. Provancher and M. R. Cutkosky, Proceedings of the IEEE/RSJ International Conference on Intelligent Robots and Systems, Las Vegas, NV, October 2003, 2003, Vol. 3, pp. 2791–2796.
13. S. K. Prasad, M. Kitagawa, G. S. Fischer, J. Zand, M. A. Talamini, R. H. Taylor and A. M. Okamura, *Medical Image Computing and Computer-Assisted Intervention-MICCAI*, 2003, 279–286.
14. K. Salisbury, F. Conti and F. Barbagli, *IEEE Comput. Graphics Appl.*, 2004, **24**(2), 24–32.
15. A. M. Okamura, *Indust. Robot.*, 2004, **31**(6), 499–508.
16. B. T. Bethea, A. M. Okamura, M. Kitagawa, T. P. Fitton, S. M. Cattaneo, V. L. Gott, W. A. Baumgartiner and D. D. Yu, *J. Laparoendosc. Adv. Surg. Tech.*, 2004, **14**(3), 191–195.
17. M. Kitagawa, D. Dokko, A. M. Okamura and D. D. Yuh, *J. Thorac. Cardiovasc. Surg.*, 2005, **129**(1), 151–158.
18. J. J. Abbott, P. Marayong and A. M. Okamura, *J. Robot. Res.*, 2007, STAR 28, 49–64.
19. A. M. Okamura, *Curr. Opin. Urolog.*, 2009, **19**(1), 102110.
20. S. Misra, K. J. Macura, K. T. Ramesh and A. M. Okamura, *Med. Eng. Phys.*, 2009, **31**(2), 195–206.
21. A. M. Okamura, L. N. Verner, C. E. Reiley and M. Mahvash, *J. Robot. Res.*, 2011, **STAR 66**, 361–372.
22. A. M. Okamura, L. N. Verner, T. Yamamoto, J. C. Gwilliam and P. G. Griffiths, *Surg. Robot.*, 2011, 419–448.
23. A. M. Okamura, C. Basdogan, S. Baillie and W. S. Harwin, *IEEE Trans. Haptics*, 2011, **4**(3), 153–154.
24. T. Yamamoto, N. Abolhassani, S. Jung, A. M. Okamura and T. N. Judkins, *Int. J. Med. Robot. Comp. Assist. Surg.*, 2012, **8**(1), 45–56.
25. A. A. Stanley, J. C. Gwilliam, A. M. Okamura, Proceedings of World Haptics Conference (WHC), 2013, pp. 25–30.
26. K. R. Guerin, A. Okamura, M. Rotella, X. He and J. Bohren, "System and Method for Sensor Fusion of Single Range Camera Data and Inertial Measurement for Motion Capture", US Patent App. 13/780,252, (2013).

27. M. L. Franco, C. H. King, M. O. Culjat, C. E. Lewis, J. W. Bisley, E. C. Holmes, W. S. Grundfest and E. P. Dutson, *Int. J. Med. Robot. Comp. Assist. Surg.*, 2009, **5**(1), 13–19.
28. C. H. King, M. O. Culjat, M. L. Franco, J. W. Bisley, G. P. Carman, E. P. Dutson and W. S. Grundfest, *IEEE Trans. Haptics*, 2009, **2**(1), 52–56.
29. J. Dargahi, R. Sedaghati, H. Singh and S. Najarian, *Mechatronics*, 2007, **17**(8), 462–467.
30. W. McMahan, E. D. Gomez, L. Chen, K. Bark, J. C. Nappo, E. I. Koch, D. I. Lee, K. R. Dumon, N. N. Williams and K. J. Kuchenbecker, *J. Robot. Surg.*, 2013, 23–32.
31. K. Bark, W. McMahan, A. Remington, J. Gewirtz, A. Wedmid, D. I. Lee and K. J. Kuchenbecker, *Surg. Endosc.*, 2012, **27**(2), 656–664.
32. W. McMahan, J. Gewirtz, D. Standish, P. Martin, J. A. Kunkel, M. Lilavois, A. Wedmid, D. I. Lee and K. J. Kuchenbecker, *IEEE Transactions on Haptics, Special Issue on Haptics in Medicine and Clinical Skill Acquisition*, 2011, **4**(3), 210–220.
33. K. Tadano, K. Kawashima, K. Kojima and N. Tanaka, *J. Robot. Mechatron.*, 2010, **22**(2), 179–194.
34. http://allaboutroboticsurgery.com/images/NH-Union-Leader-Haptic-Article.pdf.
35. L. J. M. van den Bedem, R. Hendrix, P. C. J. N. Rosielle, M. Steinbuch, H. Nijmeijer, Proceedings of the 2009 IEEE International Conference on Mechatronics and Automation, Changchun, China, pp. 60–65.
36. http://support.sensable.com/documents/documents/PDD_InstallGuide.pdf.
37. http://www.medgadget.com/2013/10/tactile-capsule-restores-surgeons-sense-of-touch-during-laparoscopy.html.
38. P. Puangmali, K. Althoefer and L. D. Seneviratne, *IEEE Sens. J.*, 2008, **8**(4), 371–381.
39. F. Gharagozloo and F. Najam, *Robotic Surgery*, McGraw Hill Publishers, London and New York, 1st edn, 2008.
40. P. Dasgupta, J. O. Peabody and M. Menon, *Urologic Robotic Surgery in Clinical Practice*, Springer publishers, London and New York, 1st edn, 2009.
41. J. Rosen, B. Hannaford and R. Satava, *Surgical Robotics: Systems Applications and Visions*, Springer Publishers, London and New York, 2010.
42. P. Navaranjan, R. Williams and M. Neufeld, *Robotic Surgery – A General Guide to an Emerging Field*, Kindle eBooks Publishers, 2012.
43. M. Sood and S. Leichtle, *Essentials of Robotic Surgery*, Spry Publishing Company, 2013.
44. M. Shahinpoor and S. Gheshmi, *Robotic Surgery: Smart Materials, Robotic Structures, and Artificial Muscles*, PAN Stanford Publishing Company, London and New York, 2014.
45. *Robotic Cardiac Surgery*, ed. G. Changqing, Springer Publishing Company, London and New York, 2014.

46. M. Shahinpoor and S. Gheshmi, "Intraocular Robotic Surgery System", Provisional Patent, UMaine, US Patent Office No.- 61/716,718, Filed October 22, 2012, now patent pending.
47. M. Shahinpoor and S. Gheshmi, "Space-Saving Laparoscopic Surgical Slave Robot", Provisional Patent, UMaine, US Patent Office No.- 61/718,822, Filed October 24, 2012, now patent pending.
48. D. Adolf, M. Shahinpoor, D. Segalman and W. Witkowski, "Electrically Controlled Polymeric Gel Actuators", (world's first patent on synthetic artificial muscles), *US Patent Office,* US Patent No. 5,250,167, Issued October, 5, 1993.
49. M. Shahinpoor, "Spring-Loaded Ionic Polymeric Gel Linear Actuator", *US Patent Office*, US Patent No. 5,389,222, Issued February 14, 1995.
50. K. Oguro, H. Takenaka and Y. Kawami, "Actuator Element", US patent Office No. 5,268,082, Issued December 7, 1993.
51. M. Shahinpoor and M. Mojarrad, "Soft Actuators and Artificial Muscles", *US Patent Office,* United States Patent 6,109,852, Issued August 29, 2000.
52. K. J. Kim and M. Shahinpoor, Proceeding of SPIE 8th Annual International Symposium on Smart Structures and Materials, Newport Beach, California, 2001, Vol. 4329-(58).
53. K. J. Kim and M. Shahinpoor, *Polymer*, 2002, **43**(3), 797–802.
54. K. J. Kim and M. Shahinpoor, *Smart Materials and Structures (SMS)*, Institute of Physics Publication, 2003, vol. 12, issue 1, pp. 65–79.
55. M. Shahinpoor and K. J. Kim, "Method of Fabricating a Dry Electro-Active Polymeric Synthetic Muscle", US Patent Office, Patent No. 7, 276, 090, Issued October 2, 2007.
56. D. Segalman, W. Witkowski, D. Adolf and M. Shahinpoor, *Int. J. Smart Mater. Struct.*, 1992, **1**(1), 44–54.
57. M. Shahinpoor, *Int. J. Smart Mater. Struct.*, 1992, **1**(1), 91–94.
58. G. Kim, H. Kim, I. J. Kim, J. R. Kim, J. I. Lee and M. Ree, *J. Biomater. Sci., Polym. Ed.*, 2009, **20**(12), 1687–1707.
59. U. Latif, F. L. Dickert, R. G. Blach and H. D. Feucht, *J. Chem. Soc. Pak.*, 2013, **35**(1), 17–22.
60. A. J. Grodzinsky, "Electromechanics of Deformable Polyelectrolyte Membranes", Department of Electrical Engineering, MIT, Sc.D. Dissertation, June 1974.
61. I. V. Yannis and A. J. Grodzinsky, *J. Mechanochem. Cell Motil.*, 1973, **2**, 113–125.
62. M. Shahinpoor, Y. Bar-Cohen, J. Simpson and J. Smith, *Smart Mater. Struct. J.*, 1998, 7, R15–R30.
63. M. Shahinpoor and K. J. Kim, *Smart Mater. Struct. Int. J.*, 2001, **10**, 819–833.
64. K. J. Kim and M. Shahinpoor, *Smart Materials and Structures (SMS)*, Institute of Physics Publication, 2003, vol. 12, issue 1, pp. 65–79.
65. M. Shahinpoor and K. J. Kim, *Smart Mater. Struct. Int. J.*, 2004, **13**(4), 1362–1388.

66. M. Shahinpoor and K. J. Kim, *Smart Mater. Struct. Int. J.*, 2005, **14**(1), 197–214.
67. M. Shahinpoor, K. J. Kim and M. Mojarrad, *Artificial Muscles: Applications of Advanced Polymeric Nano-Composites*, Taylor and Francis Publishers, London and New York, 1st edn, 2007.
68. S. Nemat-Nasser, *J. Appl. Phys.*, 2002, **92**, 2899.
69. P. G. de Gennes, K. Okumura, M. Shahinpoor and K. J. Kim, *Europhys. Lett.*, 2000, **50**(4), 513–518.
70. M. Porfiri, *J. Appl. Phys.*, 2008, **104**(10), 104915.
71. M. Porfiri, *Smart Mater. Struct.*, 2009, **18**(1), 015016.
72. M. Porfiri, *Phys. Rev. E: Stat., Nonlinear, Soft Matter Phys.*, 2009, **79**(4), 041503.
73. Y. Bahramzadeh, "Multiphysics Modeling and Simulation of Dynamic Curvature Sensing in Ionic Polymer Metal Composites (IPMCs) with Application in Soft Robotics", Doctoral Dissertation, Dept. of Mechanical Engineering, University of Maine, December 2013.
74. C. Bonomo, L. Fortuna, P. Giannone and S. Graziani, *IEEE Trans. Circuits Syst. I, Fundam. Theory Appl.*, 2006, **53**, 338–350.
75. M. Shahinpoor, Proc. SPIE (1995) North American Conference on Smart Structures Materials, February 28-March 2 (1995), San Diego, California, 1995, vol. 2441, paper no. 05.
76. M. Shahinpoor, ICIM'96, and Third European Conference on Smart Structures and Materials, Lyon, France, 1996, pp. 1006–1011.
77. M. Mojarrad and M. Shahinpoor, Proceedings of the 1997 SPIE Conference, 1997, No. 3042: pp. 52–60.
78. K. Sadeghipour, R. Salomon and S. Neogi, *Smart Mater. Struct.*, 1992, **1**, 172–179.
79. M. Shahinpoor and M. Mojarrad, "Ionic Polymer Sensors and Actuators", *US Patent Office,* No. **6,475,639**, Issued November 5, 2002.
80. L. Ferrara, M. Shahinpoor, K. J. Kim, B. Schreyer, A. Keshavarzi, E. Benzel, J. Lantz, Proceedings of the SPIE Conference, 1999, No., 3669: pp. 394–401.
81. C. Bonomo, L. Fortuna, P. Giannone, S. Graziani and S. Strazzeri, *Smart Mater. Struct.*, 2006, **15**, 749–758.
82. B. K. Henderson, S. Lane, M. Shahinpoor, K. J. Kim, and D. Leo, Proceeding of AIAA Space 2001 Conference and Exposition, Albuquerque, New Mexico, 2001, AIAA 2001-4600.
83. K. Farinholt, K. Newbury, M. Bennet and D. Leo, First World Congress on Biominetics and Artificial Muscles, Albuquerque, NM, 9–11 December 2002.
84. Z. Chen, X. Tan, A. Will and C. Ziel, *Smart Mater. Struct.*, 2007, **16**, 1477–1488.
85. Y. Bahramzadeh, and M. Shahinpoor, Proceedings of SPIE, 2011, No. 7976, pp. 761–768.
86. C. Bonomo, L. Fortuna, P. Giannone and S. Graziani, *Sens. Actuators, A*, 2005, **123-4**, 146–54.

87. M. Yamakita, A. Sera, N. Kamamichi, K. Asaka and Z. W. Luo, Proceedings. 2006 Conference on International Robotics and Automation Orlando, FL, USA: IEEE, 2006, pp. 1834–1839.
88. G. Zames, *IEEE Trans. Autom. Control*, 1981, **26**(2), 301–320.
89. M. Konyo, S. Tadokoro, T. Takamori and K. Oguru, Proceedings of IEEE International Conference on Robotics and Automation, April, 2000, pp. 3416–3421.
90. A. Mazzone, C. P. Spagno and A. M. Kunz, Proceedings of the ACM Symposium on Virtual Reality Software and Technology, VRST 2003, Osaka, Japan, October 2003, pp. 1–3.
91. M. Nakano, A. Mazzone, F. Piffaretti, R. Gassert, M. Nakao, M. Nakao and H. Bleuler, Proc. SPIE 5759, Smart Structures and Materials 2005: Electroactive Polymer Actuators and Devices (EAPAD), 331.
92. M. Shahinpoor, "Electrically-Controllable Multi-Fingered Resilient Heart Compression Devices", CIP to US Patent No. Number 6, 464, 655, *US Patent Office*, Patent No.7, 198, 594, Issued April 3, 2007.
93. J. Citerin and A. Kheddar, Electroactive Polymer Actuator for tactile Display, in *The Sense of Touch and Its Rendering: Progress in Haptics Research*, ed. A. Bicchi, M. Buss, M. O. Ernst, A. Peer, STAR 45, Springer-Verlag, Berlin-Heidelberg, 2008, ch. 6, pp. 131–154.
94. M. Konyo and S. Tadokoro, IPMC Based Tactile Dislay for Pressure and Texture Presentation on a Human Finger, in *Biomedical Applications of Electroactive Polymer Actuators*, ed. F. Carpi and E. Smela, John Wiley and Sons publishers, London and New York, 2009, ch. 8, pp. 161–174.
95. R. K. Jain, S. Datta and S. Majumder, "Design and Control of an EMG Driven IPMC Based Artificial Muscle Finger", in Chapter 15, © 2012 Jain *et al.*, licensee. InTech. This is an open access chapter distributed under the terms of the Creative Commons Attribution License (http://creativecommons.org/licenses/by/3.0), which permits unrestricted use, distribution, and reproduction in any medium, provided the original work is properly cited, (2012).
96. K. Aw, L. Fu and A. McDaid, *Int. J. Smart Nano Mater.*, Taylor and Francis, 2013, 4 (3), 1–11.
97. D. Chatterjee, N. Hanumaiah, Y. Bahramzadeh and M. Shahinpoor, *Advanced Materials Research*, Trans Tech Publications, Switzerland, 2013, vol. 740, pp. 492–495.
98. A. Keshavarzi, M. Shahinpoor, K. J. Kim, and J. Lantz, Proc. SPIE Smart Materials and Structures Conference, March 1-5, 1999, New Port Beach, California, Publication No. SPIE 3669-36.

CHAPTER 23

Ionic Polymer Metal Composites as Soft Biomimetic Robotic Artificial Muscles

MOHSEN SHAHINPOOR

Department of Mechanical Engineering, University of Maine, Orono, Maine 04469, USA
Email: shah@maine.edu

23.1 Introduction

Biomimetic smart material systems and structures have become important in recent years due to their potential engineering applications. Accordingly, based on such materials, their structures and their integration with appropriate sensors and actuators, novel applications useful for a large number of engineering applications have emerged. In that context, electroactive polymers (EAPs) and, in particular, ionic polymer metal composites (IPMCs) introduced in 1998 by Shahinpoor, Bar-Cohen, Harrison and Smith[1] have now become a family of powerful electronic smart materials with a variety of actuation, sensing and energy harvesting possibilities for engineering, industrial and medical applications.

The reader is referred to four review articles on these materials by Shahinpoor and co-workers.[2–5] Biomimetic robotic IPMC systems present unique opportunities for a number of advanced applications. There have been numerous attempts to dynamically model and analyze biomimetic robotic IPMC systems as can be seen as early as in 1991 in the research works

RSC Smart Materials No. 18
Ionic Polymer Metal Composites (IPMCs): Smart Multi-Functional Materials and Artificial Muscles, Volume 2
Edited by Mohsen Shahinpoor

Published by the Royal Society of Chemistry, www.rsc.org

of Shahinpoor[6,7] on the conceptual design, kinematics and dynamics of swimming robotic structures and ionic polymeric muscles, of Mojarrad and Shahinpoor[8,9] on noiseless propulsion for swimming robotic structures using polyelectrolyte ion-exchange membranes and artificial muscles, of Chamorro[10] on swimming robotic structures equipped with IPMC artificial muscles, of Mojarrad[11,12] on ionic polymeric gels as smart materials and artificial muscles for biomimetic swimming robotic applications, and of Nakabo, Mukai and Asaka[13] on biomimetic soft robots using IPMCs. For botanical (plant-like) biomimetic robotic structures, the reader is referred to Shahinpoor and Thompson's[14,15] work on the Venus flytrap (*Dionaea muscipula*) and the waterwheel plant as smart carnivorous botanical structures with built-in sensors and actuators, and Shahinpoor's[16–18] work on robotic Venus flytraps made with IPMCs. Similar biomimetic robotic muscular systems mimicking jellyfish, eels, rays, anemone and even dolphins and sharks can be fabricated using IPMCs. The reader is referred to some general presentations on these concepts by Kim, Tan, Choi and Pugal[19] on biomimetic robotic artificial muscles and Shahinpoor[20–24] on IPMC-based muscular biomimetic nano-sensors and nano-actuators. Furthermore, Kottke, Partridge and Shahinpoor[25] discuss the potential neural activation of IPMCs as artificial muscles for biomimetic robotic actuation.

There have also been numerous attempts to analyses the dynamic behavior of IPMCs as biomimetic robotic artificial muscles as described by Toi and Kang,[26] Kothera,[27] Shahinpoor,[28,29] Porfiri[30,31] and Bahramzadeh.[32,33] Recent progress on the processing and development of various multiphysics-activated materials such as EAPs and IPMCs as distributed nano-sensors, nano-actuators and artificial muscles and their applications has underlined the general need for rigorous and complete modeling of their behaviors from a continuum-coupled multiphysics perspective. This chapter presents the fundamentals of biomimetic robotic artificial muscles, including coupled fields and/or transport effects. This chapter further summarizes the modeling of biomimetic robots based on bio-inspired actual botanical or biological entities' designs as observed in carnivorous plants (*Droseraceae* family), such as the Venus flytrap (*Dionaea muscipula Ellis*) or the waterwheel plant (*Aldrovanda vesiculosa*), as well as swimming marine biostructures such as the jellyfish.

23.2 IPMC Manufacturing and Biocompatibility for Biomimetic Robotic Applications

As described in Chapter 23 IPMCs are manufactured by a REDOX operation which leads to electroless chemical plating. For detailed IPMC manufacturing procedures refer to Shahinpoor and Mojarrad,[34] Adolf *et al.*[35], Oguru *et al.*[36] and Shahinpoor[37] and Kim and Shahinpoor.[3,38–40] One of the basic polyelectrolytes used in manufacturing IPMCs is perfluorinated sulfonic membranes, a form of which is commercially known as Nafion™ and is manufactured by DuPont. These materials are basically biocompatible, as reported by Kim *et al.*[41] and Lath, *et al.*[42] Thus, plating these

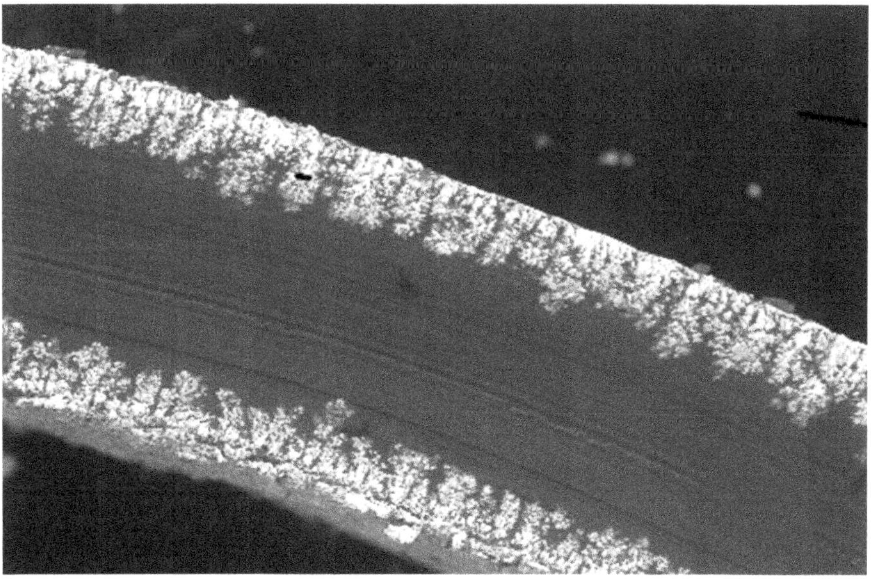

Figure 23.1 Dendritic/fractal penetration of the conductive phase into the ionic molecular network of IPMCs.

membranes with biocompatible metals such as platinum (Pt) or gold (Au) will lead to biocompatible IPMCs as biomimetic muscle-like soft actuators and sensors for biomedical applications. These polymeric muscles are manufactured by a REDOX technique in which the ionic polymer is first oxidized by a metallic salt such as tetra ammine platinum chloride hydrate ($Pt(NH_3)_4Cl_2 \cdot xH_2O$) and is then reduced by sodium borohydride ($NaBH_4$) in a chemical process in which a noble metal, such as platinum or gold, is deposited near a boundary as embedded electrodes that penetrate in a fractal and dendritic manner (Figure 23.1) within the molecular network of the base ionic polymer.

Segalman *et al.*[43] and Shahinpoor[44] had presented models for electrically controllable soft biomimetic polymeric artificial muscles as early as 1992. The basic polyelectrolyte used in manufacturing IPMCs have been perfluorinated sulfonic or carboxylic ion containing membranes, a sulfonic form of which is commercially known as Nafion™ and is manufactured by DuPont. IPMCs chemically plated with Pt form a sandwich of Pt–Nafion–Pt, where the Pt is often only a few tens of microns thick as shown in Figure 23.1 (typically 10–20 microns for a 1 cm×4 cm×0.2 mm IPMC strip).

23.3 IPMC Actuation as Biomimetic Robotic Artificial Muscles

Shahinpoor *et al.*[1] introduced IPMCs as smart multifunctional EAPs and artificial muscles in 1998, followed by four review articles on the state-of-the-art IPMCs by Kim and Shahinpoor in recent years.[2–5] Shahinpoor

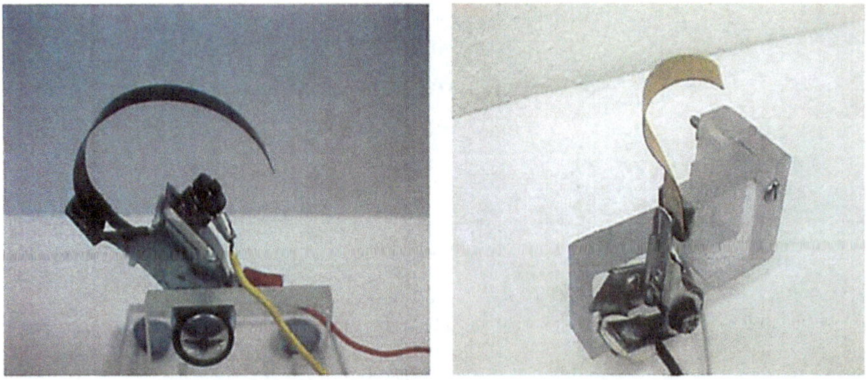

Figure 23.2 Bending deformation of a 1 cm × 8 cm × 0.2 mm IPMC strip under a low voltage (4 V) with a symmetrically placed pair of Pt electrodes (left) and a similar IPMC strip with an asymmetrically placed pair of gold electrodes (right).

et al.[45] and Nemat-Nasser[46] have discussed the applications of these active polymers as artificial muscles. Metal plated IPMCs strips deform spectacularly, upon application of a low voltage, (Figure 23.2). IPMCs are active multifunctional smart materials (*i.e.* they deform significantly when excited by a relatively low voltage and generate voltage when deformed).

Figure 23.2 displays the typical deformations of strips of IPMCs under a low voltage. Note that the bending is towards the anode electrode, and as voltage is increased, the bending increases. If the voltage is dynamic, say sinusoidal, the bending becomes oscillatory. IPMCs have a very large bandwidth to kilo-Hertz both in actuation and sensing. For a full theoretical description of the charge dynamics and electromechanics of IPMCs, see de Gennes *et al.*[47] and Porfiri.[30,31,48] Once an electric field is imposed on an IPMC cantilever, the conjugated and hydrated cations rearrange to accommodate the local electric field and thus the network deforms or bends in a spectacular manner (Figure 23.2) under a small electric field such as tens of volts per millimeter. de Gennes *et al.*[47] have presented the standard Onsager formulation on the underlining principles of IPMC actuation/sensing phenomena using linear irreversible thermodynamics (LITD).

Charge dynamics modeling and PNP formulations can also be applied, as discussed by Porfiri[30,31,48] and Bahramzadeh.[49]

23.4 Some Electrical Properties of IPMCs as Biomimetic Robotic Artificial Muscles

In order to assess the electrical properties of the IPMCs, the standard AC impedance method that can reveal the equivalent electric circuits of IPMCs has been adopted. Typical results are reported in Shahinpoor *et al.*,[47] Bahramzadeh[49] and Bonomo *et al.*[50] Overall, it is interesting to note that the IPMC is nearly resistive (>50 Ω) in the high-frequency range and fairly

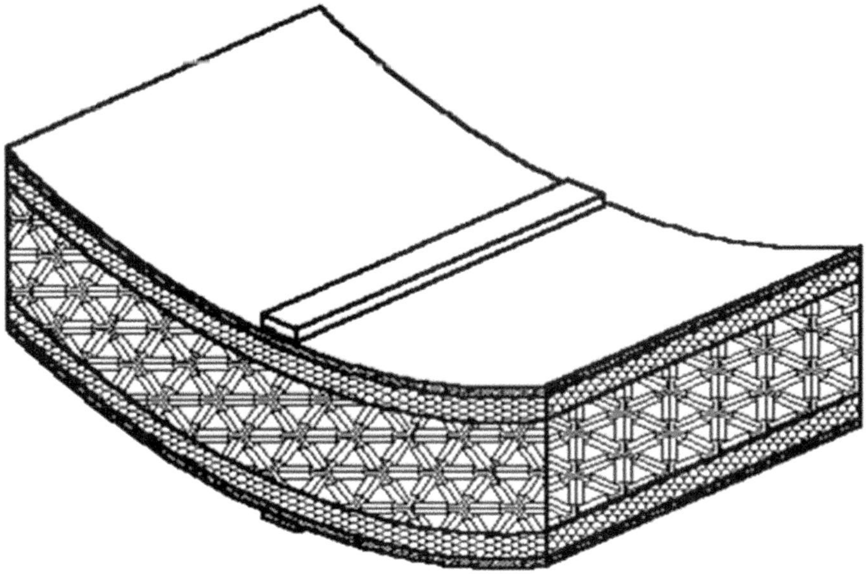

Figure 23.3 Cartoon of an IPMC structural configuration displaying the activation electrodes (middle of the top and bottom), surface electrodes (black; Pt), near-boundary electrodes (little circles) and internal capacitive structure with impedance (molecular network).

capacitive (>100 μF) in the low-frequency range. IPMCs generally have a surface resistance (R_{SS}) of about a few Ohms per centimeter, a near-boundary resistance (R_S) of a few tens of Ohms per centimeter and a cross-resistance (R_P) of a few hundreds of Ohms per millimeter, with a typical cross-capacitance (C_g) of a few hundreds of microfarads per millimeter. This approach is based upon the experimental observation of the considerable surface electrode resistance based on an equivalent circuit model (see Shahinpoor *et al.*[45]). Figure 23.3 depicts a general material configuration based on the performance requirements.

The next step is to place external electrodes (Figure 23.3) on the material and apply an electric field to force the cations to migrate away from the anodic internal electrodes from one region of the material toward the cathodic regions of the polymer and thus cause a corresponding deformation as depicted in Figure 23.2.

23.5 IPMCs as Versatile Sensors for Biomimetic Robotic Sensing

IPMCs are excellent versatile sensors for complex deformations and kinesthetics of internal organs and tissues. As reported above, they can co-operatively actuate and sense during their operation inside the body and this makes them unique as haptic actuators and sensors for robotic surgery. Upon application of an electric field (imposed voltage across the membrane), the cations migrate towards the cathode and thus cause a stress gradient that deforms an IPMC in an actuator mode. On the other hand, mechanically deforming the IPMC forces the cations to redistribute and move across the thickness, and this produces an electric field between the electrodes based on the PNP phenomena (see Porfiri[30,31,48]). Shahinpoor[2,51,52] and Mojarrad *et al.*[53] had introduced the physical phenomenon of the "flexogelectric" effect in connection with the dynamic sensing of ionic polymeric gels, where manually bending or twisting the IPMCs resulted in a measurable electric field. Sadeghipour *et al.*[54] had reported earlier that Nafion in a hydrogen environment and sandwiched between two flat electrodes could be used as a vibration energy damper. Shahinpoor and Mojarrad[12] introduced the ionic polymers as cooperative sensors and actuators in 2002. Additional contributions on IPMC sensing followed with the work of Ferrara *et al.*[55] on measuring the force and pressure between vertebrates, Bonomo *et al.*[56] on IPMC general sensing, Henderson *et al.*[57] on the near-DC sensing capabilities of IPMCs, Farinholt *et al.*[58] on the relationship between charge and deformation in IPMCs, Chen *et al.*[59] on IPMC sensing underwater and Bahramzadeh *et al.*[32,33] on curvature sensing by IPMCs. Bonomo *et al.*[60] modeled IPMC sensors, finding good correspondence between their model and experimental results. Recently, Yamakita *et al.*[61] used the position sensing capabilities of IPMCs to develop a closed-position controller using H_∞ theory (see Zames[62]). Doping the IPMC material with specific counter-ions affects the performance of the material, emphasizing either the sensing or actuation properties. Lei, Lim and Tan[63] have studied marine environmental sensing by biomimetic robotic fish equipped with IPMCs. Figure 23.4 depicts the typical dynamic sensing of IPMCs suitable for haptic/tactile feedback sensing.

23.6 Underlying Fundamentals of Biomimetic Robotic Actuation and Sensing in IPMCs

As discussed before, another important feature of the IPMCs is the molecular diffusion and transfer of cations and their hydrated water molecules between the cathode and the anode electrodes as they move between the porous electrodes. Figure 23.5 depicts both actuation and self-powered sensing mechanisms.

Note in Figure 23.5 that the hydrated cations migrate towards the cathode electrode and cause the muscle to bend towards the anode. Thus, the cationic migration under the influence of an imposed electric field causes the IPMCs to bend and develop a dynamic curvature in their structures and thus mimic the fin undulation of many marine creatures. Furthermore, the fact that IPMCs can operate very well underwater makes it advantageous to use IPMCs for fish fin dynamic undulation as a biomimetic robotic muscle.

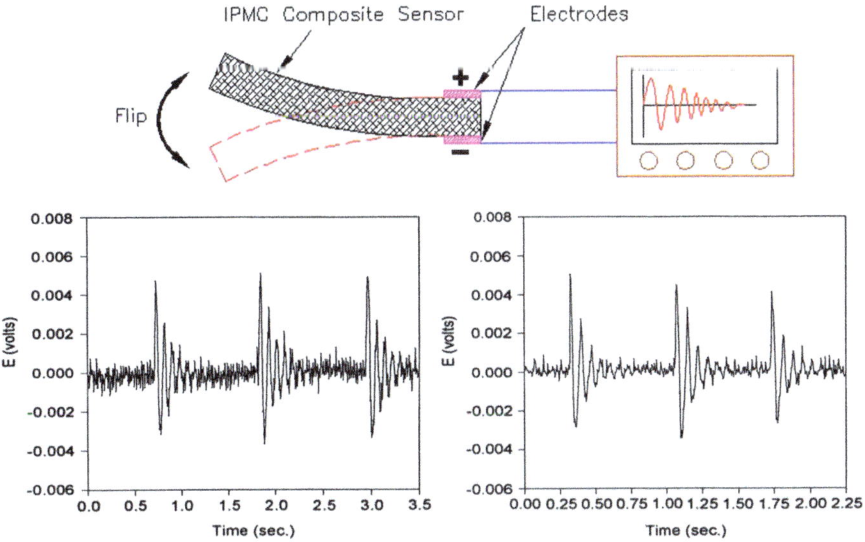

Figure 23.4 Dynamic sensing IPMCs: (a) sensing setup; (b) sensing signal (20 mm × 8 mm × 0.2 mm); (c) sensing signal (25 mm × 8 mm × 0.15 mm).

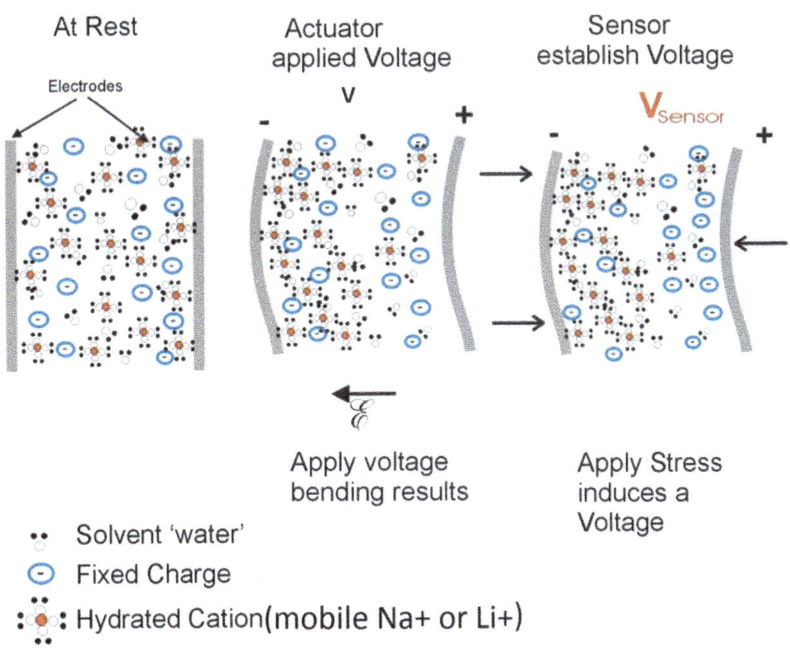

Figure 23.5 Sensing and actuation mechanisms in IPMCs due to internal ionic redistribution either due to mechanical deformation (sensing and energy harvesting, from left to right) or an imposed electric field (actuation, from right to left).

Based on these mechanisms of biomimetic robotic actuation and sensing in IPMCs, various kinds of biomimetic robotic fish fins (Figure 23.6), biomimetic robotic jellyfish (Figures 23.7 and 23.8), biomimetic robotic Venus flytraps (Figure 23.9), biomimetic robotic flying pairs of IPMC wings and artificial bats (Figure 23.10) and other types of nastic plant motion can be designed, fabricated and made operational for a variety of applications. In particular, biomimetic robotic marine structures can be designed and

Figure 23.6 An assortment of IPMC biomimetic robotic fish caudal fins and their associated fish.

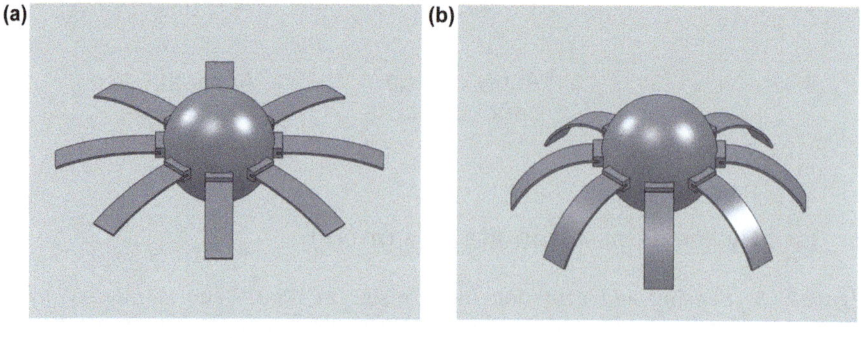

Figure 23.7 General design of an IPMC-based jellyfish with initial bending of the IPMC fins (a) and more bending of the IPMC fins (b).

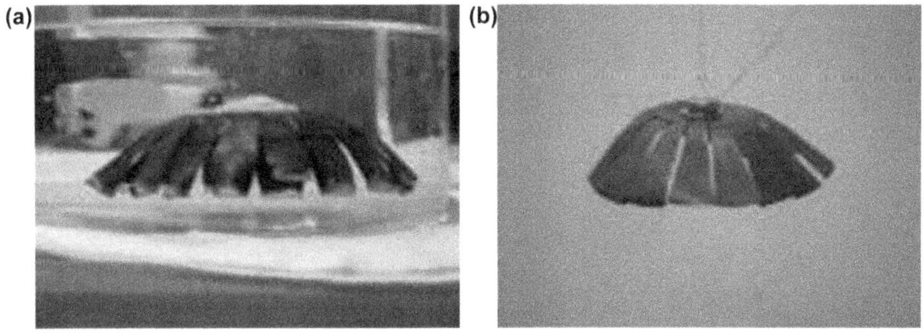

Figure 23.8 A biomimetic robotic jelly fish swimming underwater (a) and a jellyfish-like robot suspended in the air (b).

fabricated by IPMCs to operate underwater and monitor marine conditions and schools of fish and other marine structures.

As discussed by Shahinpoor,[16,18] the closing mechanism of the lobes of a Venus flytrap is very similar to the electromechanical bending in IPMCs due to cationic migration. In the case of Venus flytrap lobes, the migration of $\mathbf{Ca^{2+}}$ cations causes the lobes to quickly close and trap the pray, as shown in Figure 23.9(a) and (b). Figure 23.11 displays the similar $\mathbf{Ca^{2+}}$ migrations in the lobes of a Venus flytrap, forcing them to close.

Inspired by these mechanisms, three-dimensional (3D) versions of biomimetic robotic traps with multiple lobes may be designed and operated, as shown in Figure 23.12.

Rapid muscular movements in carnivorous plants, such as the Venus flytrap, which are triggered by antenna-like sensors (trigger hairs), present an opportunity to study distributed biomolecular motors. Carnivorous plants, such as the Venus flytrap, possess built-in intelligence (trigger hairs) as a strategy to capture prey, which can be turned on in a controlled manner. In the case of the Venus flytrap, the prey that is lured by the sweet nectar in the Venus flytrap's pair of jaw-like lobes has to flip and move the trigger hairs, which are colorless, bristle-like and pointed. The dynamically moved trigger hairs then electro-elastically send an electric signal to the internal ions in the lobe to migrate outwardly for the jaw-like lobes, causing them to close rapidly so as to capture the prey. The manner in which the Venus flytrap lobes bend inward to capture the prey shows a remarkable similarity to typical IPMC bending in an electric field. Furthermore, the mechanoelectrical sensing characteristics of IPMCs also show a remarkable resemblance to the mechanoelectrical trigger hairs on the lobes of the Venus flytrap. Thus, one can integrate IPMC lobes with a common electrode in the middle of one end of the lobes to act like a spine and use IPMC bristles as trigger fingers to sense the intrusion of a fly or insect to send a sensing signal to a solid state relay, which then triggers the actuation circuit of the IPMC lobes to rapidly bend toward each other and close. The two lobes that form the trap are attached to the midrib common electrode, which is conveniently termed the spine. The upper surface of each lobe is dished, and

Figure 23.9 Venus flytrap ((a) open lobes; (b) closed lobes) with trigger hairs and an actual biomimetic robotic Venus flytrap with sensing whiskers made from IPMCs (c).

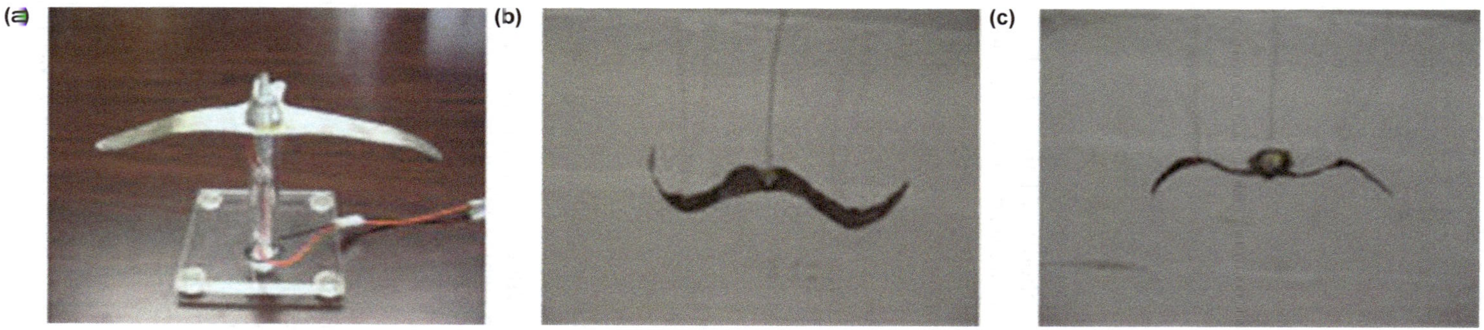

Figure 23.10 Biomimetic robotic flying pair of IPMC wings (a) and IPMC artificial bats (b, c) flying in air.

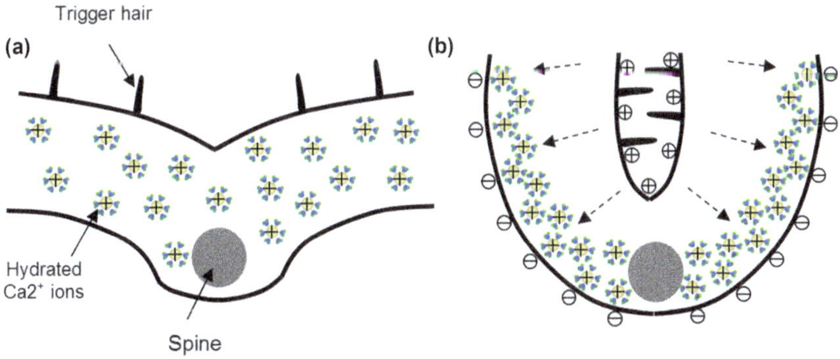

Figure 23.11 Basic Ca^{2+} cation migration mechanism induced by trigger hairs: (a) initial state, lobes open with trigger hairs; (b) final state after lobes have closed.

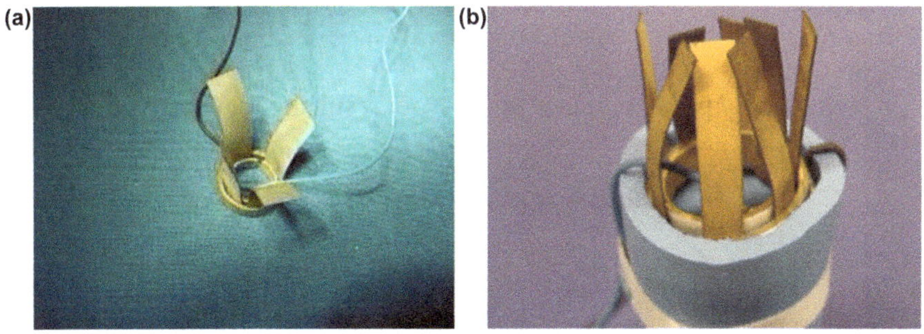

Figure 23.12 Three-dimensional versions of biomimetic robotic traps with four lobes (a) and eight lobes (b).

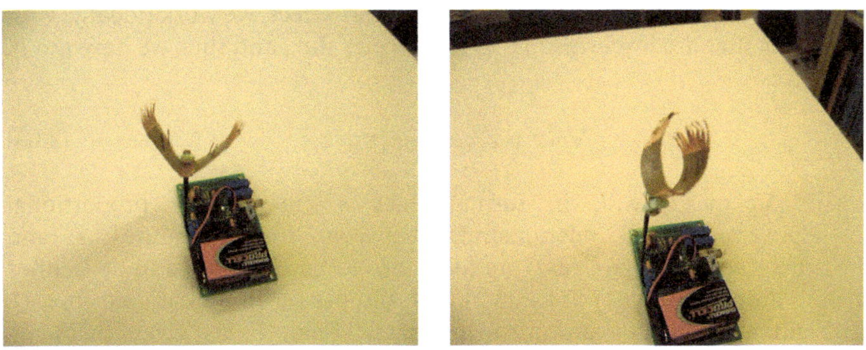

Figure 23.13 Final biomimetic robotic Venus flytrap: (left) trap open; (right) trap closing.

spaced along the free margins of the lobes are some 15–20 prong-like teeth. These are tough and pointed, and are inclined at an inward angle so that when the trap is sprung shut, they will interlock. Figure 23.13 depicts the final biomimetic robotic Venus flytrap (VFT) that was built and tested.

23.7 Modeling of Biomimetic Robotic Actuation and Sensing in IPMCs

de Gennes, Okumura, Shahinpoor and Kim[47] presented the first phenomenological modeling of sensing and actuation in IPMCs based on LITD and an equilibrium of forces and fluxes. Due to space limitations, a brief discussion on the LITD theory follows. There are ionic fluxes and forces at work within the IPMCs. Once an electric field is imposed on such a network, the conjugated and hydrated cations rearrange to accommodate the local electric field and thus the network deforms. In the simplest of cases, such as in thin membrane sheets, spectacular bending is observed under small electric fields, such as a few tens of kilovolts per meter. The underlying principle of the IPMC's actuation and sensing capabilities can be described by the standard Onsager formulation using LITD. When *static conditions* are imposed, a simple description of *mechanoelectric effect* is possible based upon two forms of transport: *electron transport* (with a current density $\underset{\sim}{J}$) and *ion transport* (with a flux $\underset{\sim}{Q}$). The conjugate forces include the electric field $\underset{\sim}{E}$ and the pressure gradient $-\underset{\sim}{\nabla}p$ across the pairs of electrodes. The resulting equations have the concise form of:

$$\underset{\sim}{J}(x,y,z,t) = \sigma \underset{\sim}{E}(x,y,z,t) - L_{12} \underset{\sim}{\nabla} p(x,y,z,t) \tag{23.1}$$

$$\underset{\sim}{Q}(x,y,z,t) = L_{21} \underset{\sim}{E}(x,y,z,t) - K \underset{\sim}{\nabla} p(x,y,z,t) \tag{23.2}$$

where σ and K are the material electric conductance and the Darcy permeability, respectively. A cross-coefficient is usually $L = L_{12} = L_{21}$. The simplicity of the above equations provides a compact view of the underlying principles of actuation, transduction and sensing of the ionic polymer nano-composites. When we measure the *direct* effect, we work (ideally) with electrodes that are impermeable to ion species flux, and thus we have $Q = 0$. This gives:

$$\underset{\sim}{\nabla} p(x,y,z,t) = \frac{L}{K} \underset{\sim}{E}(x,y,z,t) \tag{23.3}$$

This $\underset{\sim}{\nabla}p(x,y,z,t)$ will, in turn, induce a curvature $\underset{\sim}{\kappa}$ proportional to $\underset{\sim}{\nabla}p(x,y,z,t)$. The relationships between the curvature $\underset{\sim}{\kappa}$ and pressure gradient $\underset{\sim}{\nabla}p(x, y, z, t)$ are fully derived and described in de Gennes *et al.*[49] From eqn (23.1) and (23.2), one imposes a finite cationic flux, Q. This situation creates an intrinsic electric field, $\vec{E}$, as given below in eqn (23.4):

$$\vec{E} = \frac{L}{\sigma} \nabla p = \frac{12(1-\nu_p)}{(1-2\nu_p)} \left\{ \frac{L}{\sigma h^3} \right\} \Gamma \tag{23.4}$$

Note that the notations ν_p, h, and Γ are the Poisson ratio, the strip thickness and an imposed torque at the built-in end produced by a force F, applied to the free end and multiplied by the free length of the

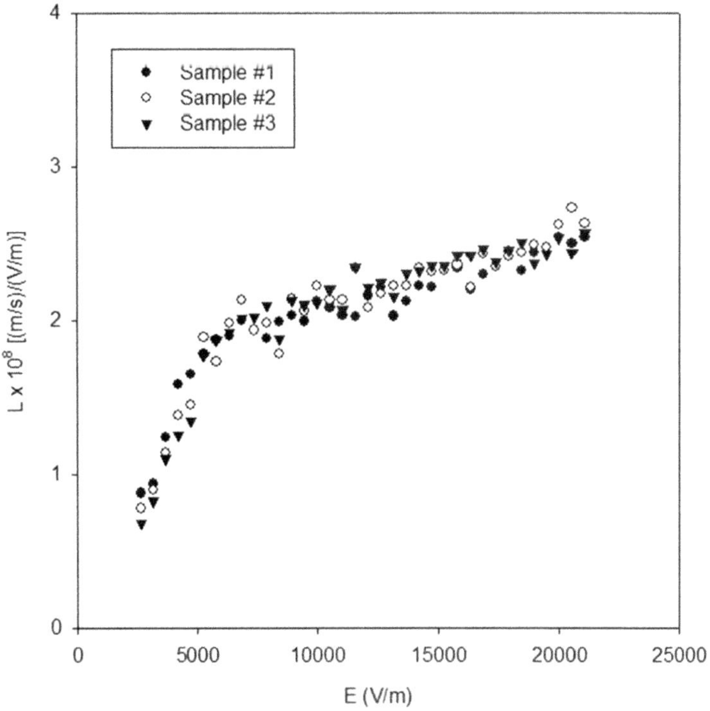

Figure 23.14 Experimental determination of the Onsager coefficient L using three different samples.

strip l_g, respectively. Note that, in this case, the Onsager coefficient L is indeed a function of the imposed electric field $\underset{\sim}{E}$, as shown in Figure 23.14.

The pressure gradient $\underset{\sim}{\nabla} p(x,y,z,t)$ will, in turn, induce a curvature $\underset{\sim}{\kappa}$ proportional to $\underset{\sim}{\nabla} p(x,y,z,t)$. The relationships between the curvature $\underset{\sim}{\kappa}$ and pressure gradient $\underset{\sim}{\nabla} p(x,y,z,t)$ are well known. Let us just mention that $\underset{\sim}{\kappa} = \mathbf{M}(\underset{\sim}{E})/\mathrm{YI}$, where $\mathbf{M}(\underset{\sim}{E})$ is the local induced bending moment and is a function of the imposed electric field $\underset{\sim}{E}$, Y is the Young's modulus (elastic stiffness) of the strip, which is a function of the hydration of the IPMC, and I is the moment of inertia of the strip. Note that, locally, $\mathbf{M}(\underset{\sim}{E})$ is related to the pressure gradient such that in a simplified scalar format:

$$\nabla p(x,y,z,t)\ (\mathbf{M}/\mathrm{I}) = \mathrm{Y}\ \underset{\sim}{\kappa} \tag{23.5}$$

where $\nabla p(x,y,z,t)$ is the pressure gradient or the difference between the tensile and the compressive stresses in the uppermost remote surfaces of the IPMC strip. From eqn (23.5), it is clear that the vector form of curvature $\underset{\sim}{kE}$ is related to the imposed electric field $\underset{\sim}{E}$ by:

$$\underset{\sim}{\kappa}_E = (\mathrm{L}(\underset{\sim}{E})/\mathrm{KY})\underset{\sim}{E} \tag{23.6}$$

Based on this simplified model, the tip bending deflection δ_{max} of an IPMC strip of length l_g should be almost linearly related to the imposed electric field due to the fact that:

$$\kappa_E \cong \left[2\delta_{max} \Big/ \left(l_g^2 + \delta^2_{max}\right)\right] \cong 2\delta_{max}/l_g^2 \cong (\mathrm{L}(E)/\mathrm{KY})E \tag{23.7}$$

The pertinent parameters have been experimentally measured to be $K = \sim 10^{-18}$ m^2 CP^{-1} and $\sigma \sim 1$A mV^{-1} or S m^{-1}. The Onsager coefficient L based on our experimental measurements is of the order of 10^{-8} m^2 V-s^{-1}. The simplicity of the above equations provides a compact view of the underlying principles of actuation, transduction and sensing of the IPMCs and will guide the design methodologies of performing experiments to further guide this theory. On the other front of theoretical modeling, Corry *et al.*,[64] Porfiri,[30,31] Bahramzadeh *et al.*[32,33,49] and Aureli *et al.*[65] have proposed the use of PNP theory to study charge dynamics and ion transport within IPMCs. As an external voltage is applied at both sides of an IPMC membrane, an electric field gradient across the membrane is induced. This is in accordance with the Nernst–Planck equations such that:

$$J = -D\left[\nabla \mathrm{C} + \frac{\mathrm{zF^*}}{\mathrm{RT}}\mathrm{C} \cdot \nabla \mathrm{V}\right] \tag{23.8}$$

where J is the flux of ionic species in mol (m^{2}s)$^{-1}$, C is the concentration of ionic species in mol m^{-3}, V is the electric potential field in volts, D is the diffusion coefficient in m^2 s^{-1}, z is the valence of ionic species, F* is the Faraday constant, R is the universal gas constant and T is the absolute temperature in degrees Kelvin. The exact mechanism that causes ion diffusion due to mechanical stimulation has been investigated by Porfiri[30,31] by considering the charge dynamics and micro-mechanics of ion diffusion in ion channels of porous Nafion membrane. The most general governing equations for charge kinetic of ionic polymers are the PNP equations. These equations are:

$$\frac{\partial C}{\partial t} = -D\nabla \cdot \left[\nabla \mathrm{C} + \frac{\mathrm{zF^*}}{\mathrm{RT}}\mathrm{C} \cdot \nabla \mathrm{V}\right], \tag{23.9}$$

$$\nabla^2 V + \frac{\rho}{\varepsilon} = 0, \tag{23.10}$$

where $\rho = F * (c^+ - c^-)$. Generally, the thickness of polymer membrane is significantly smaller than the other two dimensions, and it is a reasonable assumption that the ion diffusion is dominant over the thickness of the membrane with respect to two other dimensions, so the above equations can be expressed as a function of x (thickness across the membrane).

23.8 Some Experimental Results

The experimental results for the mechanoelectrical voltage generation of IPMCs in a flexing mode are shown in Figure 23.15. Figure 23.15 depicts the

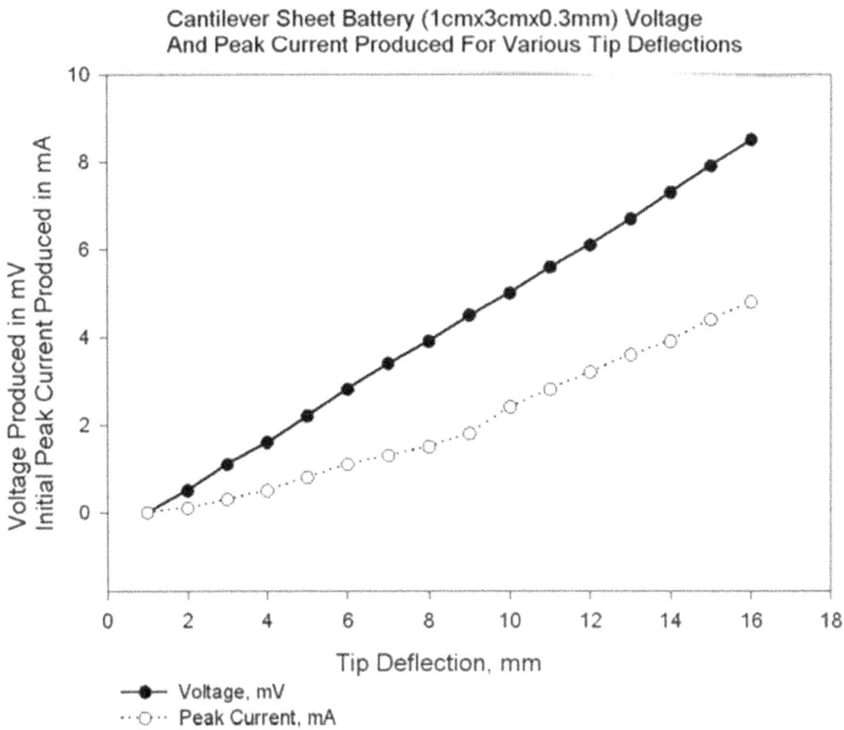

Figure 23.15 Typical voltage/current output of IPMC samples.

current and the power outputs for a sample of thin sheets of IPMCs. The experimental results depicted in Figure 23.15 show that an almost linear relationship exists between the voltage output and the imposed displacement of the tip of the IMPC sensor (Figure 23.15).

Note that once an electric field is imposed on such a network, the conjugated and hydrated cations rearrange to accommodate the local electric field and thus the network deforms, and in the simplest of cases, such as in thin membrane sheets, spectacular bending is observed under small electric fields such as tens of volts per millimeter. If the voltage signal or the electric field applied to IPMC strips is an alternating voltage, the IPMC strip will oscillate with the frequency imposed on it by the applied voltage. If the frequency of the applied field matches the natural frequency of vibrations of the strip in a cantilever form, the strip will become excited and will resonate. Figures 23.16 and 23.17 depict typical force and deflection characteristics of cantilever samples of IPMC composites.

As far as force generation is concerned, IPMCs generally have a very high force density of about 40, as depicted in Figure 23.16. Figure 23.17 depicts how robust the biomimetic sensing of IPMCs is in the sense of manually flipping the IPMC a couple of times, like a trigger finger in the lobes of a Venus flytrap, followed by one slow bending, in real time, to generate a different output signal.

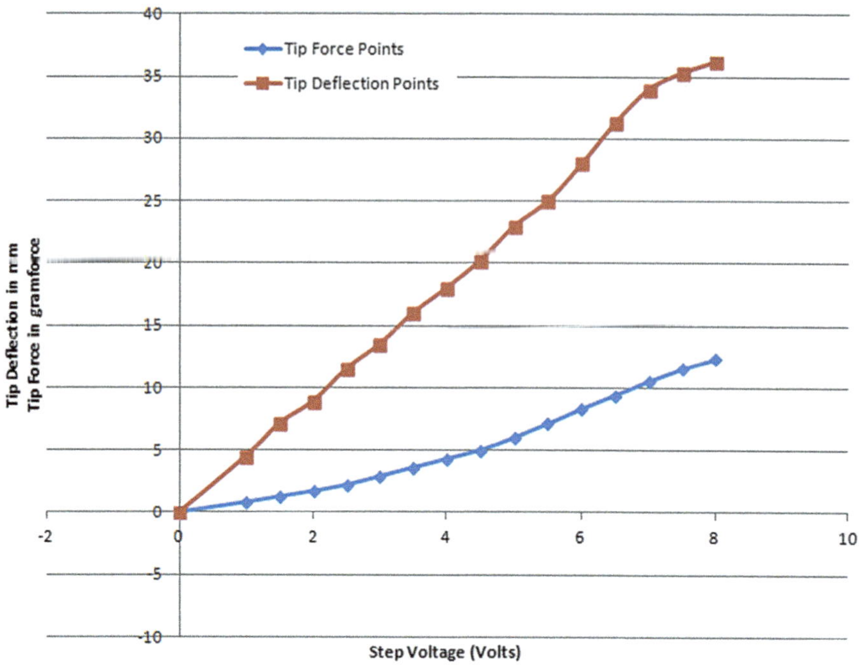

Figure 23.16 Variation of tip blocking force and the associated deflection if allowed to move versus the applied step voltage for a 1 cm × 5 cm × 0.3 mm IPMC Pt–Pd sample weighing about 0.25 g in a cantilever configuration generating up to 10 g of blocking force, giving rise to a force density of about 40.

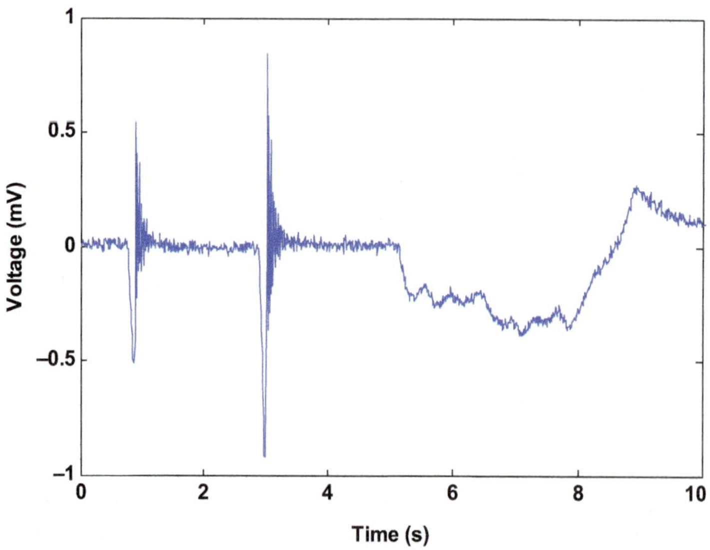

Figure 23.17 Response of an IPMC trigger finger in the artificial Venus flytrap to two fast excitations followed by one slow bending.

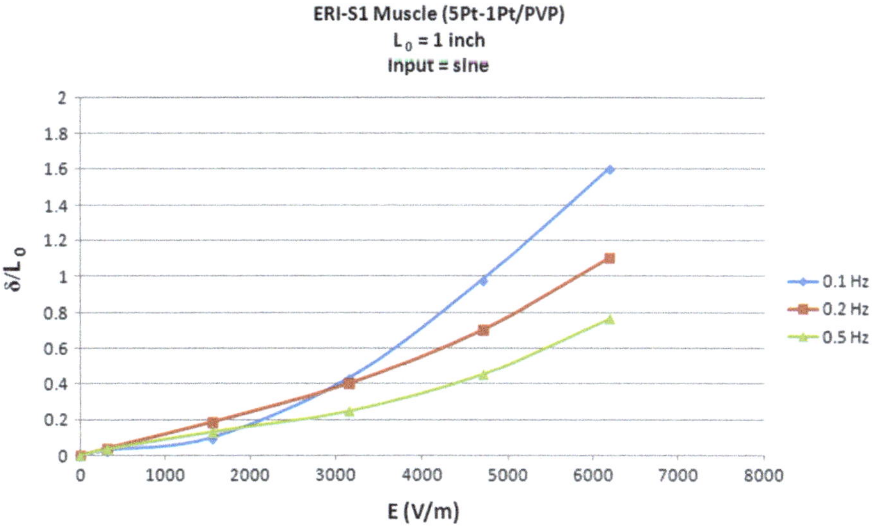

Figure 23.18 Non-dimensional deflection of IPMC cantilever strips (30 mm × 5 mm × 0.2 mm) to a square-wave electric field of various frequencies.

Figure 23.18 depicts the deflection time responses of IPMC strips to sinusoidal voltages of various frequencies.

23.9 Multicomponent Theories of Biomimetic Robotic Actuation and Sensing in IPMCs

For the case of three-dimensional general multicomponent hygro-thermo-electro-elasticity modeling of IPMCs, we have to focus on the context of their actuation and sensing applications. In this context, the presence of the following fields has been considered:

- There is mass transport driven from the diffusive processes of two liquid substances, neutral water and an electrolyte (either polar water or free components from the utilized perfluorinated hybrid polymer). The state variables in the continuum that describe the distribution of these substances in the material are the corresponding mass concentrations ${}^1c$ and ${}^2c$ for neutral water and electrolyte, respectively.
- There is charge transport from the electric current diffusive processes induced on the ionic species from the presence of the macroscopic electric field. The state of charge transport evolution is described in terms of the distinct distributions of anion n_+, cation n_- and undissociated n_u molecule populations that result from the reactive dissociation of the electrolyte.
- There is a temperature field state variable θ expressing the difference between the initial temperature of the system at any point and the current temperature caused from both heat influx in the system and all

the contributing irreversible internal processes (sources), including heat conduction, mass transport, charge transport and strain gradients.

- There is an electric vector field state variable $\vec{E}$ corresponding to the macroscopic long-range electric field applied.
- There is a strain tensor field distribution expressed by its individual components γ_{ij}.

The corresponding conjugate state variables are the chemical potentials ${}^1\mu$ and ${}^2\mu$, the currents I_+, I_- and I_u, the entropy S, the electric field displacement $\vec{D}$ and the stress tensor components τ_{ij}.

Applying the process described in the previous section for the case of homogeneous mechanically isotropic systems, the following set of field governing partial differential equations (PDEs) is obtained:

$$ {}^ic_{,t} = d_{ci}\nabla^2({}^ic) + d_{ci}\nabla^2\theta + \gamma_{ci}\nabla^2\tau'_{kk} + h_{ci}\nabla^2 V, \quad i=1,2 \text{ (mass conservation)} \tag{23.11} $$

$$ \frac{1-\nu}{2E}\nabla^2\tau'_{kk} + \alpha_0\nabla^2\theta + {}^i\beta_0\nabla^2({}^ic) + p_0\nabla^2 V = 0, \quad i=1,2 \text{ (equilibrium)} \tag{23.12} $$

$$ \begin{aligned} &c_T\theta,t - T_0({}^id)^ic_{,t} - T_0\alpha_0\tau'_{kk,t} - T_0 g_j E_{j,t} \\ &\quad = d'_{Ti}\nabla^2({}^ic) + d_T\nabla^2\theta + d'_c\nabla^2\tau'_{kk} + h_T\nabla^2 V - \frac{1}{T_0^2}\vec{q}\vec{\nabla}\theta \\ &\qquad - \frac{1}{T_0}\sum_{k=1}^{2}\vec{J}_k\vec{\nabla}\mu_k + \frac{1}{T_0}\vec{I}\vec{\nabla}V - \frac{1}{T}\tau'_{kk}\nabla\gamma'_{kk} \quad i=1,2 \text{ (heat conduction)} \end{aligned} \tag{23.13} $$

$$ \begin{aligned} &\varepsilon_0\nabla^2 V + p_i\gamma'_{kk,i} - g_i\theta_{,i} - {}^jh_i({}^jc_{,i}) \\ &\quad = q_c(n_- - n_+ - n_u) \quad j=1,2 \text{ (electric displacement)} \end{aligned} \tag{23.14} $$

$$ \begin{aligned} &\vec{\nabla}\cdot\vec{I} + q_c\frac{\partial(n_- - n_+ - n_u)}{\partial t} = 0, \\ &\vec{\nabla}\cdot\vec{I}_u = -b, \vec{\nabla}\cdot\vec{I}_+ = -n_+b, \quad \vec{\nabla}\cdot\vec{I}_- = -n_-b, \text{ (current continuity)} \end{aligned} \tag{23.15} $$

This system of nine PDEs is complemented by the following constitutive equations:

$$ {}^j\mu = -{}^j\beta_0\gamma_{kk} - {}^jh_iE_i - \frac{c_C}{C_0}({}^jc) - {}^2d\theta, \quad j=1,2 \text{ (chemical potentials)} \tag{23.16} $$

$$\begin{aligned} \tau'_{ij} &= \frac{E}{1+\nu}\left\{\gamma_{ij} + \frac{\nu}{1-2\nu'}\left[\gamma_{kk} + \frac{1+\nu}{\nu'}\left(p_k E_k - \alpha_0\theta - {}^{i}\beta_0({}^{i}c)\right)\right]\delta_{ij}\right\} \\ &\quad + t_{ij} \quad i=1,2 \text{ (stresses)} \\ \gamma_{ij} &= \frac{1+\nu}{E}\left(\tau'_{ij} - \frac{\nu\tau'_{kk}}{1+\nu}\delta_{ij}\right) \\ &\quad - [p_k E_k - \alpha_0\theta - {}^{i}\beta_0({}^{i}c)]\delta_{ij} \quad i=1,2 \text{ (strains)} \end{aligned} \tag{23.17}$$

$$S = -\alpha_0\gamma_{kk} - g_i E_i - \frac{c_T}{T_0}\theta - {}^{1}d({}^{1}c) - {}^{2}d({}^{2}c)(\text{entropy}) \tag{23.18}$$

$$D_i = \varepsilon_0 E_i + p_i\gamma_{kk} - g_i\theta - {}^{1}h_i({}^{1}c) - {}^{2}h_i({}^{2}c)(\text{electric displacement}) \tag{23.19}$$

Note that the nonlinear electrodynamic stress tensor components are given by:

$$\begin{aligned} t_{ij} &= \frac{1}{2}\varepsilon_0 E_i E_i \delta_{ij} + [p_i\gamma_{kk} - g_i\theta - {}^{1}h_i({}^{1}c) - {}^{2}h_i({}^{2}c)]E_i\delta_{ij} \\ t_{kk} &= \frac{1}{2}\varepsilon_0 E_i E_i + [p_i\gamma_{kk} - g_i\theta - {}^{1}h_i({}^{1}c) - {}^{2}h_i({}^{2}c)]E_i \end{aligned} \tag{23.20}$$

and the phenomenological fluxes are given by:

$$\begin{aligned} q_i &= -k_0\theta_{,i} - {}^{j}l'_0({}^{j}\mu_{,i}) - \kappa'_0 E_i - \lambda'_0\gamma_{kk,i}, \\ J_i &= -k'_0\theta_{,i} - {}^{j}l_0({}^{j}\mu_{,i}) - \kappa''_0 E_i - \lambda''_0\gamma_{kk,i}, \quad j=1,2 \\ I_i &= -k'''_0\theta_{,i} - {}^{j}l''({}^{j}\mu_{,i}) - \kappa_0 E_i - \lambda'''_0\gamma_{kk,i} \end{aligned} \tag{23.21}$$

It is important to realize here that all coefficients of the monomial forms participating in these equations are material constants that are sometimes related with each other,[22] but they are nevertheless material constants to be determined.

This concludes our modelling of biomimetic robotic IPMC artificial muscles.

23.10 Conclusions

This chapter described the initial development of IPMC-based biomimetic robotic smart materials and, in particular, caudal fins for robotic fish propulsion, followed by IPMC-based biomimetic robotic jellyfish and flying bats, and then on to biomimetic robotic carnivorous plants, particularly Venus flytraps. A phenomenological model of the underlying sensing and actuation mechanisms in biomimetic robotics was then presented, which was based on LITD with two driving forces—an electric field and a solvent pressure gradient—and two fluxes—electric current density and the ionic solvent flux. Furthermore, a continuum model based on PNP equations was described that dealt with the charge dynamics of cationic migration within the molecular network of IPMCs to generate deformation due to an imposed electric field or generated dynamic output voltage and transient current upon experiencing mechanical deformation, force, torque, strain and stress. The mechanoelectrical sensing characteristics of IPMCs also showed a remarkable resemblance to mechanoelectrical trigger hairs on the lobes of the Venus flytrap. Thus, IPMCs were fabricated in a configuration similar to the lobes or traps of the Venus flytrap with a common electrode in the middle of one end of the lobes to act like a spine, and IPMC bristles were used as trigger fingers to sense the intrusion of a fly or insect and subsequently send a sensing signal to a solid state relay, which then triggered the actuation circuit of the IPMC lobes to rapidly bend toward each other and close. The chapter ended with the irreversible thermodynamic, PNP and multi-component modeling of IPMCs as biomimetic robotic artificial muscles and materials with embedded sensing and actuation capabilities.

Acknowledgements

This work was partially supported by Environmental Robots Inc.

References

1. M. Shahinpoor, Y. Bar-Cohen, J. Simpson and J. Smith, *Smart Mater. Struct. Int. J.*, 1998, 7, R15–R30.
2. M. Shahinpoor and K. J. Kim, *Smart Mater. Struct. Int. J.*, 2001, **10**(4), 819–833.
3. K. J. Kim and M. Shahinpoor, 2003, *Smart Mater. Struct. Int. J.*, (SMS), Institute of Physics Publication, 2003, vol. 12, No. 1, pp. 65–79.
4. M. Shahinpoor and K. J. Kim, *Smart Mater. Struct. Int. J.*, , 2004, **13**(4), 1362–1388.
5. M. Shahinpoor and K. J. Kim, *Smart Mater. Struct. Int. J., 2005*, 2005, **14**(1), 197–214.
6. M. Shahinpoor, 1991, Proc. ADPA/AIAA/ASME/SPIE Conf. on Active Materials & Adaptive Structures, Alexandria, VA, November, 1991, pp. 222–229.
7. M. Shahinpoor, *Int. J. Smart Mater. Struct.*, 1992, **1**(1), 91–94.
8. M. Mojarrad and M. Shahinpoor, 1996, Proc. SPIE 1996 North American Conference on Smart Structures and Materials, February 27–29, 1996, San Diego, California, vol. 2716, paper no. 27.
9. M. Mojarrad and M. Shahinpoor, 1997, Proceedings of 1997 IEEE International Conference on Robotics and Automation 20–25 Apr 1997, Vol. 3, pp. s2152–2157.

10. G. Chamorro, 2000, Swimming Robotic Structures Equipped with IPMC Artificial Muscles, Department of Mechanical Engineering, University of New Mexico, Albuquerque, New Mexico, M.Sc., May 2000.
11. M. Mojarrad, 2001, Study of Ionic Polymeric Gels As Smart Materials and Artificial Muscles for Biomimetic Swimming Robotic Applications, Department of Mechanical Engineering, University of New Mexico, Albuquerque, New Mexico, Ph.D., December, 2001.
12. M. Shahinpoor and M. Mojarrad, 2002, Ionic Polymer Sensors and Actuators, *U. S.* 6475639, Issued November 5.
13. Y. Nakabo, T. Mukai, K. Asaka, *Electroactive Polymers for Robotic Applications*, Springer Publishers, London and New York, 2007, pp. 165–198.
14. M. Shahinpoor and M. S. Thompson, Proc. 1994 Int. Conf. on Intelligent Materials, ICIM'94, June 1994, Williamsburg, VA, pp. 1086–1094.
15. M. Shahinpoor and M. S. Thompson, *J. Mater. Sci. Eng.*, 1995, **C2**, 229–233.
16. M. Shahinpoor, Proceedings of 4th International Conference on Artificial Muscles, 5th International Congress on Biomimetics, Artificial Muscles and Nano-Bio (Nano-Bio 2009), Seri Life Science Center, Osaka, Japan, November 25–28.
17. M. Shahinpoor, *Bioinspiration and Biomimetics*, Institute of Physics (IOP) Publishing Ltd., London, UK, 2011, vol. 6, 046004, pp. 1–11.
18. M. Shahinpoor, 2013, Proceedings of the 7th International Congress on Biomimetics, Artificial Muscles and Nano-Bio (Nano-Bio 2013), Jeju Island, South Korea, August 26–30.
19. K. J. Kim, X. Tan, H. R. Choi and D. Pugal, *Biomimetic Robotic Artificial Muscles*, Worlfd Scientific Publishing, United Kingdom, 2013.
20. M. Shahinpoor, Biomimetic Nananosensors and Nanoactuators, in *Biomimetics and Bioinspired Nanomaterials*, ed. C. S. S. R. Kumar, John Wiley and Sons Publishers, Wiley-VCH Verlag, GmnH & Co. KGaA, Boschstrase 12, Weinheim, Germany, 2010, vol. 7, ch. 8, p. 2830301.
21. M. Shahinpoor, Muscular Biopolymers, in *Topics in Engineered Biomimecry: Biomimetics, Bioinspiration and Bioreplication*, ed. A. Lakhtakia and R.-J. Martin-Palma, Elsevier publishers, Waltham, MA, USA, 2013.
22. M. Shahinpoor, *Electrochim. Acta*, 2003, **48**(14–16), 2343–2353.
23. M. Shahinpoor, 2007, Proceedings of the 4th International Congress on Biomimetics, Artificial Muscles and Nano-Bio 2007, Cartagena, Spain, Europe, November 6–8.
24. M. Shahinpoor, J. Phys. CS, Institute of Physics, *J. Physics: Conf. Ser.*, 2009, vol. 127, pp. 20–28.
25. E. A. Kottke, L. D. Partridge and M. Shahinpoor, Proceedings of the Second World Congress On Biomimetics and Artificial Muscle (Biomimetics and Nano-Bio 2004), December 5–8, 2004, Albuquerque Convention Center, Albuquerque, New Mexico, USA.
26. Y. Toi and S. S. Kang, SEISAN KENKYU, Institute of Industrial Science, University of Tokyo, 2003, Vol. 55, No. 5, pp. 449–452.
27. C. S. Kothera, *J. Int. Mater. Syst. Struct.*, 2007, **18**(3), 219–234.

28. M. Shahinpoor, Proceedings of the 22nd, International Congress on Computer-Assisted Radiology and Surgery (CARS 2008), Barcelona, Spain, June 23–28.
29. M. Shahinpoor, Proceedings of the 6th International Congress on Biomimetics, Artificial Muscles and Nano-Bio (Nano-Bio 2011), Paris, France, October 25–27.
30. M. Porfiri, *J. Appl. Phys.*, 2008, **104**(10), 104915.
31. M. Porfiri, *Smart Mater. Struct. J.*, 2009, **18**(1), 16–26.
32. Y. Bahramzadeh and M. Shahinpoor, 2011, Proceedings of SPIE, No. 7976, pp. 761–768.
33. Y. Bahramzadeh and M. Shahinpoor, *Smart Mater. Struct. Int. J.*, 2011, **20**(9), 7, 094011.
34. M. Shahinpoor and M. Mojarrad, 2000, Soft Actuators and Artificial Muscles, *U. S. Pat.* 6109852, Issued August 29.
35. D. Adolf, M. Shahinpoor, D. Segalman and W. Witkowski, Electrically Controlled Polymeric Gel Actuators, (world's first patent on synthetic artificial muscles), *U. S. Pat.* 5250167, Issued October, 5, 1993.
36. K. Oguro, H. Takenaka and Y. Kawami, Actuator Element, *US Patent,* 5 268 082, 1993.
37. M. Shahinpoor, Spring-Loaded Ionic Polymeric Gel Linear Actuator, *US Patent*, 5 389 222, 1995.
38. K. J. Kim and M. Shahinpoor, Proceeding of SPIE 8th Annual International Symposium on Smart Structures and Materials, Newport Beach, California, 2001, Vol. 4329-(58).
39. K. J. Kim and M. Shahinpoor, *Polym. J.*, 2002, **43/3**, 797–802.
40. M. Shahinpoor and K. J. Kim, 2007, Method of Fabricating a Dry Electro-Active Polymeric Synthetic Muscle, *U. S. Pat.* 7276090, Issued October 2.
41. G. Kim, H. Kim, I. J. Kim, J. R. Kim, J. I. Lee and M. Ree, *J Biomater. Sci., Polym. ed.*, 2009, **20**(12), 1687–1707.
42. U. Latif, F. L. Dickert, R. G. Blach and H. D. Feucht, *J. Chem. Soc. Pak.*, 2013, **35**(1), 17–22.
43. D. Segalman, W. Witkowski, D. Adolf and M. Shahinpoor, *Int. J. Smart Mater. Struct.*, 1992, **1**(1), 44–54.
44. M. Shahinpoor, Proc. 1996, SPIE 2779, Third International Conference on Intelligent Materials, ICIM'96, and Third European Conference on Smart Structures and Materials, edited by: Pierre F. Gobin; Jacques Tatibouet, 1996, pp. 1006–1011.
45. M. Shahinpoor, K. J. Kim and M. Mojarrad, *Artificial Muscles: Applications of Advanced Polymeric Nano-Composites*, Taylor and Francis Publishers, London and New York, First edn, 2007.
46. S. Nemat-Nasser, *J. Appl. Phys.*, 2002, **92**, 2899.
47. P. G. de Gennes, K. Okumura, M. Shahinpoor and K. J. Kim, *Europhys. Lett.*, 2000, **50**(4), 513–518.
48. M. Porfiri, *Phys. Rev. E: Stat. Phys., Plasmas, Fluids, Relat. Interdiscip. Top.*, 2009, **79**(4), 041503.

49. Y. Bahramzadeh, 2013, Multiphysics Modeling and Simulation of Dynamic Curvature Sensing in Ionic Polymer Metal Composites (IPMCs) with Application in Soft Robotics, Doctoral Dissertation, Dept. of Mechanical Engineering, University of Maine, December 2013.
50. C. Bonomo, L. Fortuna, P. Giannone and S. Graziani, 2006, IEEE Transactions on Circuits and Systems 1: Fundamental Theory and Applications, 2006, vol. 53, pp. 338–350.
51. M. Shahinpoor, 1995, Proc. SPIE (1995) North American Conference on Smart Structures and Materials, February 28 March 2 (1995), San Diego, California, 1995, vol. 2441, paper no. 05.
52. M. Shahinpoor, Proc., Los Alamos National Laboratories, Workshop on Self-Assembling and Biomimetic Materials, December 15–17, (1997), Los Alamos, NM 1997.
53. M. Mojarrad and M. Shahinpoor, 1997, Proceedings of the 1997 SPIE Conference, No. 3042, pp. 52–60.
54. K. Sadeghipour, R Salomon and S. Neogi, *Smart Mater. Struct.*, 1992, **1**, 172–179.
55. L. Ferrara, M. Shahinpoor, K. J. Kim, B. Schreyer, A. Keshavarzi, E. Benzel, J. Lantz, Proceedings of the SPIE Conference, 1999, No. 3669, pp. 394–401.
56. C. Bonomo, L. Fortuna, P. Giannone, S. Graziani and S. Strazzeri, *J. Smart Mater. Struct.*, 2006, **15**, 749–758.
57. B. K. Henderson, S. Lane, M. Shahinpoor, K. J. Kim and D. Leo, Proceeding of AIAA Space 2001 Conference and Exposition, Albuquerque, New Mexico, AIAA 2001-4600 August, 2001.
58. K. Farinholt, K. Newbury, M. Bennet and D. Leo, First World Congress on Biominetics and Artificial Muscles, Albuquerque, NM, 9-11 December 2002.
59. Z. Chen, X. Tan, A. Will and C. Ziel, *Smart Mater. Struct.*, 2007, **16**, 1477–1488.
60. C. Bonomo, L Fortuna, P. Giannone and S. Graziani, *Sens. Actuators, A*, 2005, **123–4**, 146–54.
61. M. Yamakita, A. Sera, N. Kamamichi, K. Asaka and Z. W. Luo, Proceedings 2006 Conference on International Robotics and Automation Orlando, FL, USA: IEEE, 2006, pp. 1834–9.
62. G. Zames, *IEEE Trans. Automatic Control*, 1981, **26**(2), 301–320.
63. H. Lei, C. Lim and X. Tan, *J. Intell. Mater. Syst. Struct*, 2013, **24**, 1557.
64. B. Corry, S. Kuyucak and S. H. Chung, *Biophys. J.*, 2000, **78**, 2364–2381.
65. M. Aureli, V. Kopman and M. Porfiri, *IEEE/ASME Transactions on Mechatronics*, 2010, **15**(4), 603–614.

CHAPTER 24

Ionic Electroactive Actuators with Giant Electromechanical Responses

YUE ZHOU,*[a] MEHDI GHAFFARI,*[b] CHAD WELSH[a] AND Q. M. ZHANG[a,b]

[a] Department of Electrical Engineering, Pennsylvania State University, University Park, Pennsylvania 16802, USA; [b] Department of Materials Science and Engineering, Pennsylvania State University, University Park, Pennsylvania 16802, USA
*Email: yzz131@psu.edu; mxg1019@psu.edu

24.1 Aligned Nanoporous Microwave-exfoliated Graphite Oxide Actuators with Ultra-high Strain and Elastic Energy Density Induced under a Few Volts

24.1.1 Background

Materials that have the ability to produce large actuation forces are critical for many different purposes such as artificial muscles, precise motion and position control, and micro-electromechanical systems.[1–9] These materials are much more effective and useful if the large actuation force can be generated using little voltage potential, which then allows these actuator materials to be integrated in a wide range of complex micro-electronic devices.

RSC Smart Materials No. 18
Ionic Polymer Metal Composites (IPMCs): Smart Multi-Functional Materials and Artificial Muscles, Volume 2
Edited by Mohsen Shahinpoor

Published by the Royal Society of Chemistry, www.rsc.org

Large strain generation is not only critical in these actuators, but also materials that are able to generate large force are crucial. If large strain is generated but the material is unable to generate enough force, the material will be ineffective. For example, there are ionic polymer actuators such as gels and soft polymers that are able to produce >50% strain; however, the elastic modulus of these gel actuators is a very small (<1 MPa).[10–12] Therefore, it is critical to select materials that are able to generate a useful force magnitude for various applications.

Graphene is a single-sheet atomic layer conductor with good mechanical properties and it is highly electrically conductive. This material offers good qualities for use with ionic electrolytes in order to achieve large actuation with a small applied potential. Furthermore, graphene can be activated to produce nanomorphology that enhances the properties of graphene as an actuator material. Activated "nanoporous" microwave-exfoliated graphene oxide (A-aMEGO) actuators are developed in this section and large electro-actuation strain (>50%) under a few volts is observed as a result of using activated graphene. High force exertion ($\sim$1.5 J cc^{-1}) is also observed using the A-aMEGO. Due to the high degree of alignment of the nanopores in the graphene sheets, the strain is in the direction of the surface normal to the graphene sheets, as shown in Figure 24.1a and b.

The principle mechanism for the A-aMEGO operation as shown in Figure 24.1a and b is believed to be due the ingression and/or depletion of mobile ions in the nanopores between the graphene sheets to compensate for charge change when the electrodes are under an external electric potential. The change in position of the mobile ions results in a change in the inter-sheet spacing distance, and thus increased thickness or strain is produced (see Figure 24.1a and b). A large strain (>50%) can be achieved if the distance between the adjacent wall is similar to the change in the inter-sheet spacing. The activation of MEGO with KOH has allowed for a Brunauer–Emmett–Teller specific surface area of 3100 m^2 g^{-1} due to the fact that nanopores range between 1 and 4 nm in diameter. This activated material was coined aMEGO.[13] This graphene material can be aligned and assembled to form an ionic actuator device that produces large strain induced under a low voltage.

The graphene sheet expansion is mainly in the direction perpendicular to the sheet surface. To generate large strain, sheets should be stacked on top of one another to obtain a double-layer configuration (see Figure 24.1b). This stacking formation can be realized by using a vacuum-assisted method to stack the graphene sheets on top of one another.[14–17] By using this technique, A-aMEGO actuators with thicknesses of 50 μm were produced and the electrical and mechanical properties were measured. The vacuum-assisted assembly results show highly aligned nanomorphology and the desired layering of the graphene sheets (see Figure 24.1c). The dried A-aMEGO sheets have a density of 1.25 g cc^{-1}, whereas graphite has a density of 2.2 g cc^{-1}. This indicates that the pore volume fraction is 43% for the films consisting of A-aMEGO sheets.

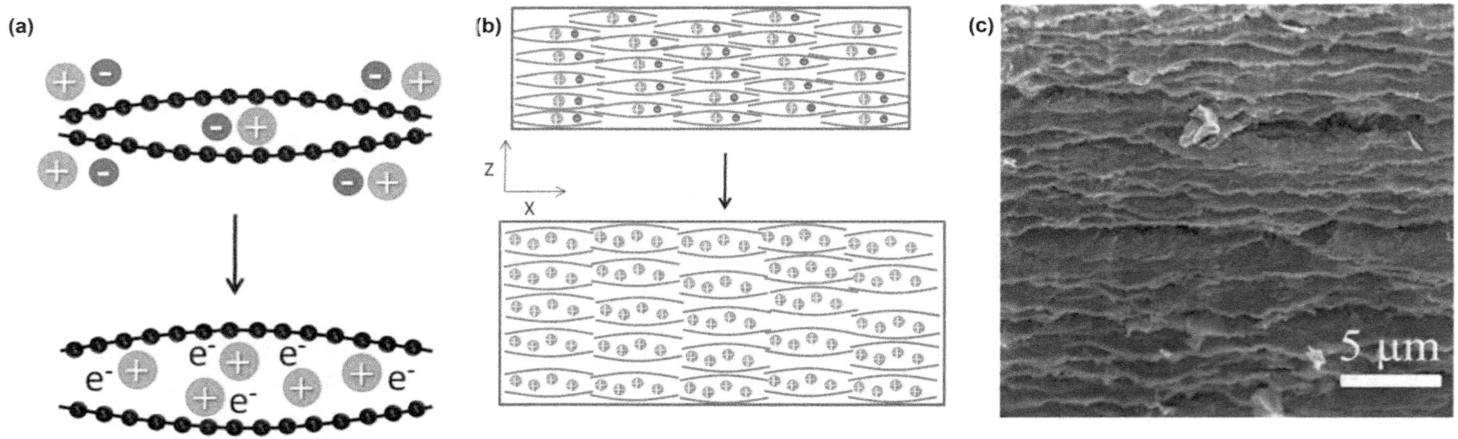

Figure 24.1 (a) Illustration of the ingression of mobile ions between graphene sheets under an applied potential. The mobile ions (cations) accumulate near the electron charges (e^-) on the cathode (graphene sheets). (b) Illustration of highly arranged aMEGO sheets without an applied voltage and with an applied voltage. Strain is observed in the *z*-direction or thickness direction in this configuration. (c) Scanning electron microscope cross-section image of A-aMEGO sheets.

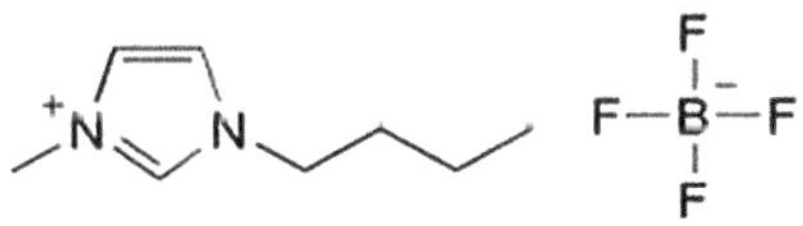

9.7 × 7.0 × 3.3 3.1 × 3.1 × 3.1

Figure 24.2 Chemical structure and dimensions (Å) of $BMIM^+$ and BF_4^- ions.

Two polymer binders are investigated in this section. One of the binders is poly(tetrafluoroethylene) (PTFE), which has a dielectric constant of 2.1, and the other binder is a polar fluoropolymer poly(vinylidene fluoride/chlorotrifluoroethylenen) (P[VDF-CTFE]) (91/9 mol% composition), which has a dielectric constant of 12.[18–20] It is suggested that higher-dielectric-constant polymers may enhance the mobile ion concentration in the graphene composite electrode due to the higher capacitance values in the system.[19]

Ionic liquid (IL) is critical for the A-aMEGO actuator, since it supplies the mobile ions that are depicted in Figure 24.1. It is also known that a proper mixture of an IL and a molecular liquid (ML) can improve the overall ion conduction of the pure IL.[22] The IL selected for the A-aMEGO actuator was 1-butyl-3-methylimidazolium tetrafluoroborate ($[BMIM][BF_4]$), which is an imidazolium-based IL.[21] Imidazolium-based ILs are typically chosen for electroactive devices since this IL has a wide electrochemical window and high ionic conductivity.[4,7,22] $BMIM^+$ is the cation and is slightly larger than the BF_4^- anion (see Figure 24.2). The ML mixed with the pure $[BMIM][BF_4]$ is acetonitrile (AN) (2 M). This mixture will improve the ion conduction of the pure electrolyte.

24.1.2 Experimental Preparation and Characterization

A vacuum-assisted assembly process was used to align the nanoporous aMEGO sheets.[14–17] The aMEGO (10 mg) sheets were dispersed in *N,N*-dimethylformamide (DMF; 5 ml). This mixture was filtered through a porous alumina membrane with an average pore size of 0.02 μm. The filtering of this dispersed mixture allowed for the proper alignment of the aMEGO sheets.

To prepare PTFE as a 10 wt% binder for the 90 wt% A-aMEGO, the A-aMEGO sheets were placed in a dish containing the 10 wt% PTFE dispersed in water and diluted with isopropyl alcohol, and enough time was allowed for the PTFE chains to diffuse into the sample. Different immersion times were evaluated to observe the uniformity of diffusion of the PTFE, and it was determined that after 1 week, a homogenous dispersion of the PTFE remained in the A-aMEGO layers. The amount of PTFE binder can be controlled by varying the amount of PTFE in the sample.

The P(VDF-CTFE) nanocomposite was prepared in a one-step process. A solution of P(VDF-CTFE) in DMF (0.2 wt%) along with a dispersion of aMEGO (10 mg) in DMF (2 ml) was prepared. These solutions were mixed together in a 1:1 ratio and sonicated to produce a homogeneous dispersion.

The final solution allowed for a ratio of 90 wt% of aMEGO to 10 wt% of P(VDF-CTFE). Again, the binder concentration can be adjusted by changing the ratio of aMEGO to P(VDF-CTFE).

After filtration, two samples with dimensions of 2 mm×2 mm and thicknesses of 50 μm were immersed in the selected electrolyte. The electro-actuation strain of the A-aMEGO actuators was tested using the pure IL along with 2 M [BMIM][BF_4]/AN, which resulted in a 60–62 wt% of the electrolyte in the nanocomposite. After the samples were completely infiltrated by the electrolyte, the electro-actuation strain testing and cyclic voltammetry (CV) properties were characterized. The nanocomposite samples had a 25% increase in thickness and less than 3% increases in the lateral dimensions after being immersed in the electrolyte.

The nanocomposite samples were bonded to a Teflon sample holder that had two square-shaped dishes with side lengths of 3 mm separated by a 1 mm wide channel. Each nanocomposite sample served as an electrode while the 1 mm channel allowed for ion conduction between the electrodes. Two stainless steel pins (1.5 mm diameter) were precisely controlled with micromanipulators and positioned into light contact with the nanocomposite samples. A lightweight aluminum reflector plate was glued to one of the stainless steel pins in order to measure displacement with a fiber optic sensor. The electro-actuation strains in the anode and cathode were characterized using the MTI 2100 Fotonic fiber optic displacement sensor, which has a resolution of 25 nm. The capacitance of the ionic nanocomposite actuators was measured with the Verstat 4 potentiostat.

24.1.3 Electro-actuation Strain, Specific Capacitance, and Elastic Energy Density

The electro-actuation strains of both the anode and cathode are presented for the graphene actuators. The cyclic electro-actuation strain of the cathode at a scan rate of 50 mV s^{-1} from 0 to 4 V for A-aMEGO/P(VDF-CTFE) with [BMIM][BF_4] electrolyte generated large strain (see Figure 24.3). The strain value was calculated using eqn (24.1):

$$\text{Strain} = \left(\frac{\text{Final Thickness} - \text{Initial Thickness}}{\text{Initial Thickness}}\right) \tag{24.1}$$

The maximum strain for the A-aMEGO/P(VDF-CTFE) nanocomposite was 56.6% at 4 V. The strain in the *x*-direction (see Figure 24.1b) was investigated as well, and it showed strain values of less than 3%, which is significantly smaller than the strain in the *z*-direction. This large strain anisotropy is caused by the highly aligned nanocomposite morphology (see Figure 24.1c). Strain induced in the perpendicular direction of the desired actuation can adversely affect the performance of actuators in practical applications that are based on strain along a single direction.[23,24]

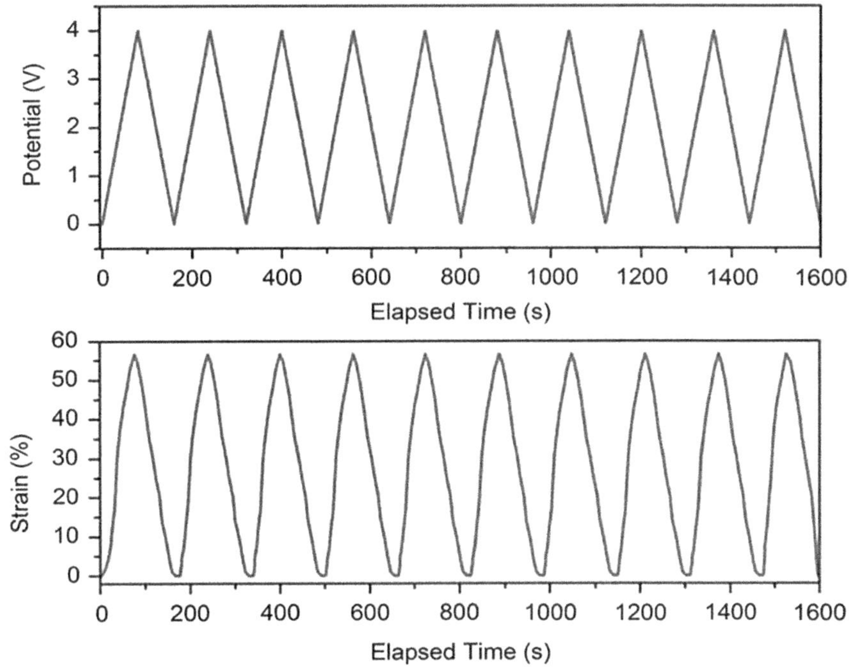

Figure 24.3 Cyclic actuation of the A-aMEGO/P(VDF-CTFE) nanocomposite cathode (bottom) with a [BMIM][BF_4]/AN electrolyte. A near-linear response is seen with the applied voltage.

The strain results for the cathode showed larger strain than the anode, which is due to the difference in cation and anion size. The peak strain results for the cathode at a scan rate of 50 mV s^{-1} from 2 V to 4 V are summarized (see Figure 24.4a). The best electro-actuation was observed with the IL/ML electrolyte mixture. As was mentioned previously, the A-aMEGO/P(VDF-CTFE) produced the best strain of 56.6% at 4 V, while the lowest strain at 4 V was the pure IL at 36.2%. These results show that the IL/ML mixture provides the best strain results, which indicates that the ionic mobile conductivity is improved and better than for pure IL. This increased mobility with the IL/ML leads to a greater amount of ions collected at each of the A-aMEGO electrodes, and consequently, this causes larger strain.

Another factor to consider with actuators is the response time of the actuator. Scan rates at 50, 100, and 500 mV s^{-1} were tested under a potential of 4 V (see Figure 24.4b). As is expected with ionic devices, the increased operation frequency leads to decreased actuation response. The A-aMEGO/P(VDF-CTFE) with the IL/ML electrolyte still produces a strain of 20.3% at 500 mV s^{-1}. Furthermore, the strain response of these actuators can be increased by reducing the length that the mobile ions need to travel. The mobile ion movement occurs through drift, where the time constant is proportional to distance, and also through diffusion, where the time

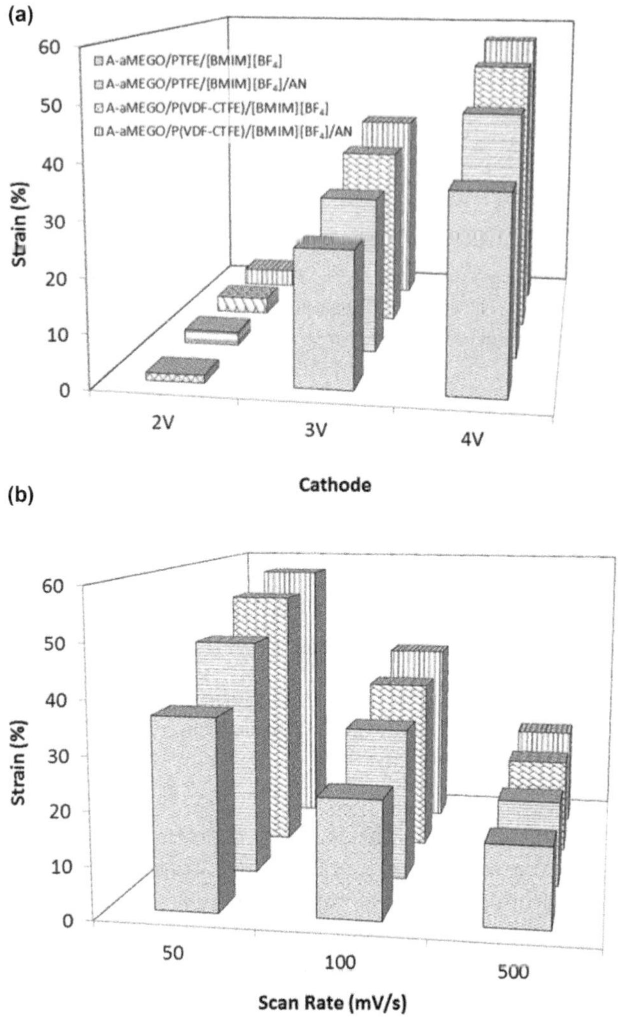

Figure 24.4 (a) Electromechanical strain response of the A-aMEGO/polymer nanocomposite cathode at a 50 mV s^{-1} scan rate under different voltages. (b) Electromechanical strain response of the nanocomposite cathode at 4 V for varied scan rates.

constant is proportional to the square root of distance. By using an alternative electrode design, the drift and diffusion time of the ions can be improved (see Section 24.3).

The potentiostat was used to measure the specific capacitance of the A-aMEGO actuators. The ratio of electro-actuation strain on the cathode over the specific capacitance was calculated for the scan rate at 50 mV s^{-1} (see Figure 24.5a). These data indicate that the A-aMEGO/polymer combination with higher specific capacitance allows for better electro-actuation.

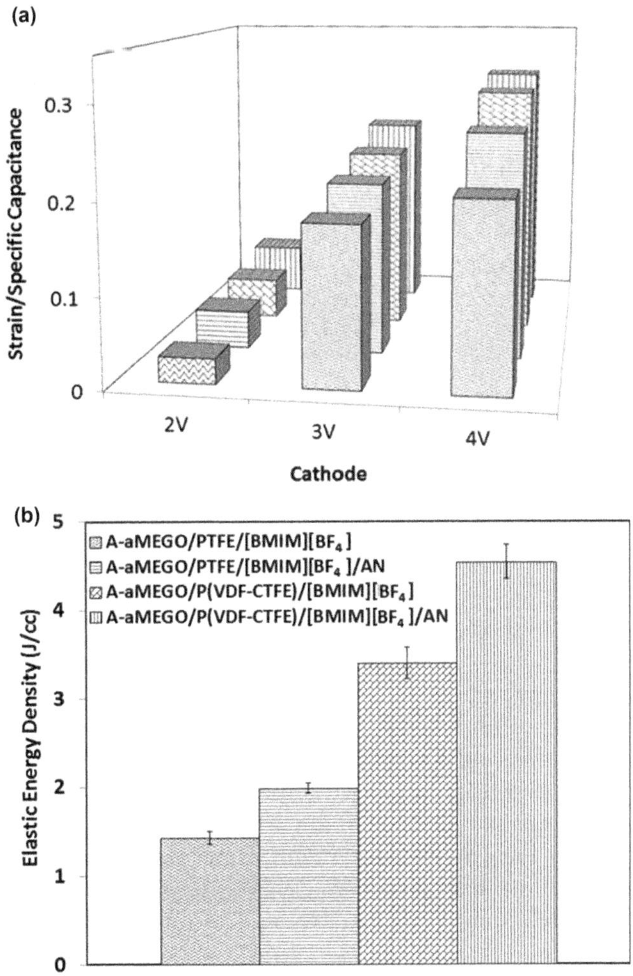

Figure 24.5 (a) Specific capacitances of the varied electrolyte and binder combinations at a peak voltage of 4 V and a scan rate of 50 mV s^{-1}. (b) Elastic energy density of the different nanocomposite samples at a peak voltage of 4 V and a scan rate of 50 mV s^{-1}.

Hence, this A-aMEGO system is affected differently depending on the properties of the electrolytes and binder materials. The P(VDF-CTFE) polymer has already demonstrated large strain, but further tailoring of the polymer concentration and electrolyte may further enhance the electroactuation of the graphene actuator.

The ability of the device to generate strain is not the only important parameter of the actuator. The elastic energy density is also another critical parameter for evaluation. The elastic energy density is used to characterize an actuator's ability to generate strain and stress.[1–3,25] The value of strain generated is proportional to the elastic modulus according to Hooke's law.

To evaluate the generated stress characteristics, some force can be used to generate the strain from the initial thickness, l_0, to the final thickness, l. This force can be described as:

$$F = \sigma A \tag{24.2}$$

where σ is the stress and A is the area occupied by the nanocomposite. The area does not change with the applied voltage due to the high elastic anisotropy of the A-aMEGO system. The mechanical work, W, in this system can be calculated using:

$$W = \int_{l_0}^{l} F\mathrm{d}z' \tag{24.3}$$

For materials that generate large strain, the mechanical stress is the elastic modulus multiplied by the actual strain. This leads to the equation for the elastic energy density, U_m, of eqn (24.4):[2,26,27]

$$U_\mathrm{m} = \int_{l_0}^{l} Y(x)\ln(x)\mathrm{d}x \tag{24.4}$$

The elastic modulus is denoted as Y. By using eqn (24.4), the elastic energy density of the A-aMEGO/polymer nanocomposite actuators can be calculated.

For simplification, an atomic force microscope (AFM) tapping mode method was used to characterize the elastic modulus, Y, for the electrolyte added to the nanocomposite actuators with no external field applied.[26,28] Under an applied potential, the elastic modulus of the actuators was determined by measuring the reduction of the induced strain under a load applied in the z-direction. For this measurement, the change in sample thickness was measured while a load (fixed weight) was applied to the sample. The strain and stress relation allows for Y to be determined *via* eqn (24.5):

$$S = S_0 - \frac{\mathrm{load}}{Y} \tag{24.5}$$

S_0 is the strain generated at 4 V with no load and Y is the elastic modulus at strain S. Once the elastic modulus is determined, the elastic energy density can be calculated.

In the absence of an external field, the elastic modulus was measured to be 67.5 ± 12.5 MPa, which closely resembles the elastic modulus measured in other studies.[7,29] The elastic modulus of samples under an external field was measured by compressing the samples under loads of 0.025, 0.25, and 0.5 MPa. The elastic modulus of the A-aMEGO/P(VDF-CTFE) with the IL/ML electrolyte under an external potential of 4 V (56.6% strain) was 9 MPa. This elastic modulus leads to a high energy density value of 1.5 J cc^{-1}. Thus, a high elastic energy density is obtained since the high elastic modulus is coupled with the high actuator strain.

Successful alignment of graphene sheets containing nanopores was demonstrated in this section. A vacuum-assisted assembly method was utilized to stack the graphene sheets in a layered fashion that allows for large strain. These highly aligned A-aMEGO/polymer nanocomposites exhibited greater than 50% electro-actuation strain with an elastic energy density of 1.5 J cc $^{-1}$. The alignment of the nanopores in the graphene sheets allows for large anisotropy, which results in large strain. Many electro-actuation strain applications require this large anisotropy for unidirectional force and displacement. The operation voltage of these IL-based actuators is relatively low, which offers compatibility for direct integration of these devices into microelectronics.

24.2 Improving the Elastic Energy Density and Electrochemical Conversion Efficiency by Tailoring P(VDF-CTFE) Concentration

Section 24.1 discussed aligned A-aMEGO actuators and the large strain response (>50%) observed with the binders P(VDF-CTFE). The voltage, scan rate, electrolyte type, polymer type, and concentration are several factors that affect the overall electrochemical performance of the actuators. According to Hooke's law, the stress produced is proportional to the elastic modulus of the actuator material. Consequently, the elastic modulus of these actuators should be improved to further the force generation effectiveness of the devices.

In the last section, the elastic modulus of the A-aMEGO with 10 wt% P(VDF-CTFE) was 67.5 MPa without an applied external field, whereas under an applied potential of 4 V, the elastic modulus was reduced to 9 MPa at 56.6% strain. As a result, the actuators showed an electromechanical conversion efficiency of about 1.5%, which is not as high as expected if the large electro-actuation strain and relatively high elastic modulus of the nanocomposites in the non-active state are taken into consideration.

This section investigates the electro-actuation response of the actuators by varying the polymer binder concentration in the A-aMEGO/polymer nanocomposites. The elastic modulus of the A-aMEGO/polymer composition can be improved by increasing the polymer matrix, which reduces the number of voids that were originally occupied by electrolyte in the micro-, meso-, and macro-pores. While the total strain of the actuator will decrease, the elastic energy density and electromechanical efficiency of the device can be effectively increased. In this section, P(VDF-CTFE) concentrations of 10, 20, 35, and 50 wt% were implemented as the binders in A-aMEGO sheets. The electromechanical properties of these different weight compositions were characterized. From these different weight ratios, it is determined that there is an optimum actuator state at 35 wt% P(VDF-CTFE) with an electrochemical conversion efficiency of more than 3.5%, which is high for ionic electroactive polymer actuators.

24.2.1 Polymer Content Adjustment and Characterization

As in Section 24.1, the vacuum-assisted assembly process was used to align the aMEGO sheets with the P(VDF-CTFE) binder.[14–17] The P(VDF-CTFE) at

10, 20, 35, and 50 wt% was added to the aMEGO dispersion in DMF, and the dispersion was then filtered through a porous alumina membrane to align the graphene sheets.

As in the previous section, the samples were cleaved as 2 mm×2 mm square pieces and were about 50 μm thick. These samples were immersed in a 2 M solution of [BMIM][BF_4] in AN and were allowed to soak for 30 min. These actuator samples were characterized using the same CV and electro-actuation testing procedure from the last section. In this case, rather than stainless steel probes, tungsten probes were used due to tungsten's closely matching electrochemical window with the electrolyte.

The Young's moduli of the nanocomposites along the *z*-direction (see Figure 24.1b) were measured without an applied voltage using the Bruker Icon AFM. Next, the elastic moduli of the nanocomposite actuators under different applied voltages were evaluated by applying various loads (0.025, 0.25, and 0.5 MPa) to the nanocomposite samples and measuring the change in electro-actuation strain.

24.2.2 Strain, Elastic Energy Density, and Efficiency Performance

In this configuration, only the electro-actuation strain of the cathode was characterized using a 50 mV s^{-1} scan rate at different applied voltages. The strain magnitude was calculated using eqn (24.1). The strain value in the thickness direction increased almost linearly with the voltage and reached the maximum value at 4 V (see Figure 24.6a). The strain perpendicular to the thickness direction was negligible and had an opposite sign with respect to the thickness direction. The electro-actuation strain in the cathode decreased as polymer content increased. At 10 wt% P(VDF-CTFE), the strain was at a maximum at 56.6%. Under a 4 V potential, the value decreased to 43.2% with 20 wt% P(VDF-CTFE), and the strain value further decreased to a value of 20.3% with 50 wt% P(VDF-CTFE). This trend occurs since the elastic modulus of the nanocomposite increases with an increase in polymer content. By increasing the polymer content, voids originally filled with IL are now filled with polymer, which results in a more robust nanocomposite. In this case, the ion mobility was decreased due to the increase in polymer content, and in turn, this caused a reduction in the mobile ion collection in the electrodes. While the strain response was decreased in the actuators, the elastic modulus was improved, and this allowed for increased force generation capability, elastic energy density, and electromechanical conversion efficiency.

CV experiments at a maximum voltage of 4 V with a scan rate of 50 mV s^{-1} were used to observe the charge and discharge characteristics of the nanocomposites (see Figure 24.6b). The CV curves are nearly rectangular and this indicates that the non-Faradaic charge transfer of the electrodes is similar to the electrode samples in the previous section. This also indicates that an increase in polymer concentration causes the specific capacitance

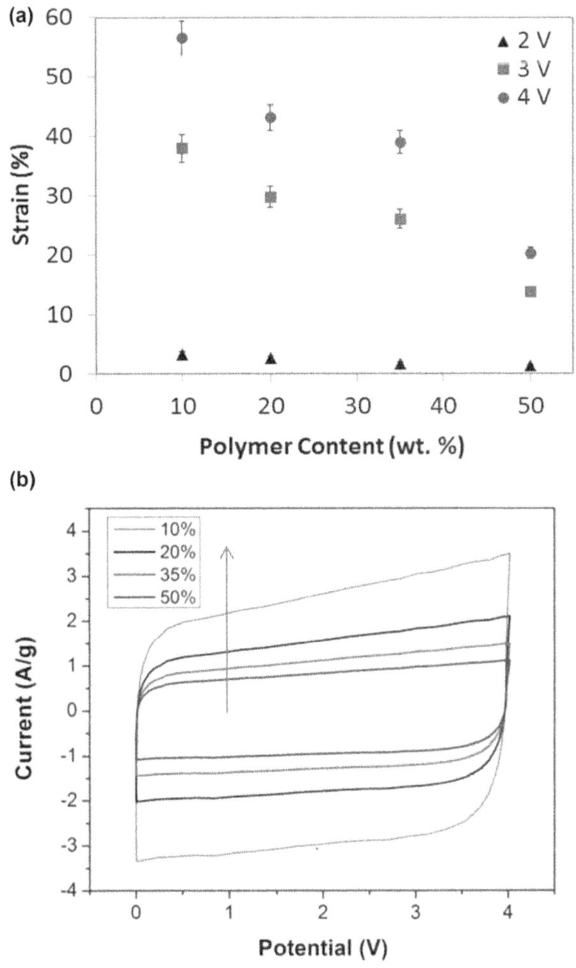

Figure 24.6 (a) Electro-actuation strain on the cathode for varied P(VDF-CTFE) concentrations at a scan rate of 50 mV s^{-1} and applied voltages of 2, 3, and 4 V. (b) CV curves for varied P(VDF-CTFE) concentrations at a 50 mV s^{-1} scan rate.

to decrease. This arises from less mobile ions being stored in the nanocomposites when there is more polymer content present.

Using the method described in the previous section, the elastic moduli of the nanocomposites were characterized. The strain at different fixed weights was used to deduce the elastic moduli by using eqn (24.4). The elastic moduli of the nanocomposites were measured without an applied potential and with an applied potential of 4 V at 50 mV s^{-1} (see Figure 24.7a). The results show that the elastic moduli of the nanocomposites decrease when actuated.

The elastic energy density and efficiency increase with the polymer content up to 35 wt% P(VDF-CTFE) (see Figure 24.7b). At 35 wt%, the elastic energy

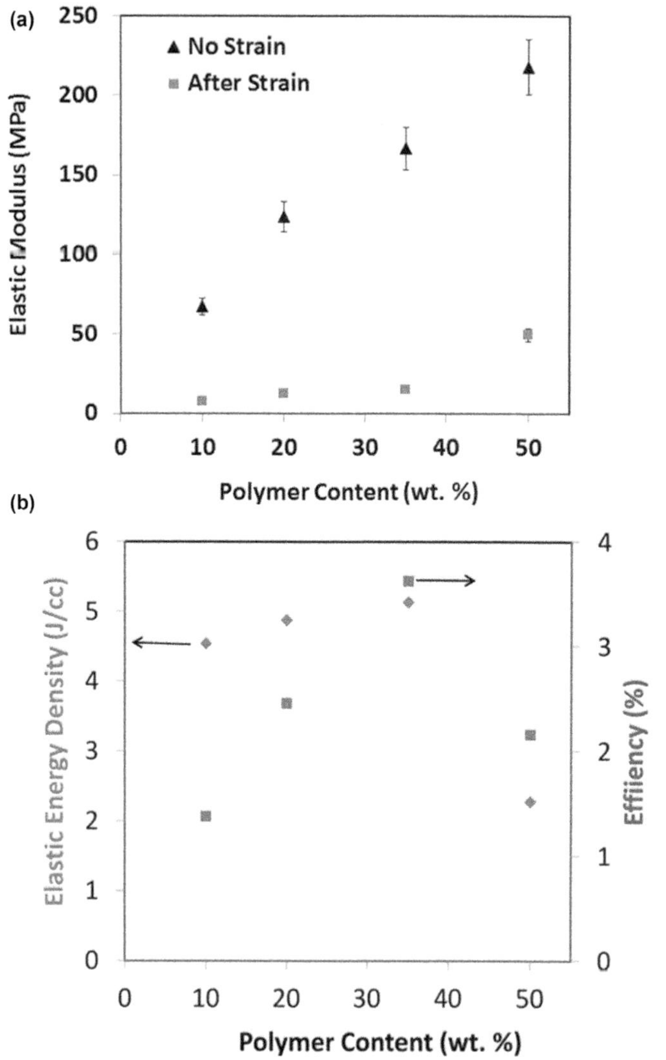

Figure 24.7 (a) Elastic moduli of varied polymer content nanocomposites without an applied potential and with an applied potential of 4 V. (b) Elastic energy density (diamonds) and the efficiency (squares) of the A-aMEGO/P(VDF-CTFE) nanocomposites with varied polymer concentrations.

density reaches a maximum of greater than 5 J cc^{-1}, while at 50 wt%, the elastic energy density is reduced to 2.2 J cc^{-1}, which is due to the reduced strain value. The elastic energy density deduced is the total stored electric energy density in the actuator device, and this is different from what the actuator actually provides to an external load. Under a 0.5 MPa load, the 10 wt% polymer shows a strain of 50%, which represents a work density of

0.25 J cc^{-1}. Hence, the work that is transferred to a load depends on the load condition. The electromechanical conversion efficiency was calculated by using the input electric energy density deduced from the CV curves and the elastic energy density. The ratio of elastic energy density and input electric energy density results in the conversion efficiency. The 35 wt% P(VDF-CTFE) shows the highest efficiency at 3.7%.

As was mentioned, the elastic energy density reaches a maximum of 35 wt% P(VDF-CTFE) and it decreases after this point. This can be attributed to the fact that energy density is dependent on both the strain and elastic modulus. These parameters are typically inversely proportional, which results in the need to optimize these parameters to achieve the best energy density. A similar trend is seen for the highest efficiency 35 wt% P(VDF-CTFE) nanocomposite. This relates to the highest elastic energy density exerted per unit value of the electrical energy consumed. The elastic energy density and efficiency for the A-aMEGO/P(VDF-CTFE) actuators is among the highest for ionic actuators that are performing with less than a 4 V potential.

This section optimized the polymer concentration in the A-aMEGO/P(VDF-CTFE) composites to achieve the best elastic energy density and efficiency. The 10 wt% P(VDF-CTFE) nanocomposites demonstrated the highest strain, but had the lowest elastic modulus of 67.5 MPa under a 4 V potential. The 50 wt% P(VDF-CTFE) had the highest elastic modulus of 218.0 MPa. The optimal nanocomposite sample was the 35 wt% polymer with an elastic energy density of over 5 J cc^{-1}. This nanocomposite concentration resulted in a strain of around 40% with an elastic modulus of 170 MPa and an efficiency of 3.7%. Thus, by fine tuning the polymer content, the energy density of the nanocomposite was maximized.

24.3 Improving Mobile Ion Transport in the A-aMEGO Actuator Electrodes

24.3.1 Background

This section demonstrates improved mobile ion transport in A-aMEGO/PTFE nanocomposite actuators with graphene hierarchical structures that are configured to form quick ion transport channels. Previously, it was demonstrated that A-aMEGO/PTFE nanocomposite actuators with a [BMIM][BF_4]/AN electrolyte achieved a strain of 46.3% under a potential of 4 V. With this earlier experimental setup, long ion transport distances (approximately millimeters) were present in the actuator charge and discharge cycles. This led to slow actuator response times, and hence led to reduced electro-actuation strain responses. The purpose here is to demonstrate that ion transport distances can be reduced in these graphene-based actuators by creating high-speed ion transport channels between graphene hierarchical features in each of the electrodes. One investigation has shown that shortening the ion path length led to an increase in actuation speed.[30] Using hierarchical nanoporous graphene composite features that uphold the high-density nanopores of the graphene and implement micron-sized ion transport channels is one way of addressing this challenge. By incorporating these ion channels, the travel distance of mobile ions can be drastically reduced in the graphene nanocomposite electrodes. Experiments have shown improvements in the ion transport and rate performance of battery electrodes by reducing the ion transport distance.[31–33]

The new experimental configuration still relies on the basic working principle of the aMEGO actuators that is depicted in Figure 24.1. As was mentioned earlier, the A-aMEGO sheets trap the mobile ions between layers, and upon applying an electric potential, a change in the graphene sheet spacing occurs. Ion transport through the nanopores of the aMEGO can be slow, therefore resulting in slow electro-actuation responses if the ions need to be transported over long distances. The cross-sectional views of the previously configured nanoporous graphene actuator *versus* the improved hierarchical nanoporous graphene actuator are illustrated in Figure 24.8. The implemented ion channels allow shorter transport distances for the mobile ions, which results in quicker transport of the ions. Consequently, the electro-actuation response is improved with the hierarchical graphene structures. The width of the ion channels is denoted as w_{gap} and the width of the aMEGO features is denoted as $w_{composite}$.

The proposed method for incorporating these ion channels within hierarchical graphene structures is depicted in Figure 24.9. The challenge associated with this project was patterning the relatively thick A-aMEGO features. By successfully blocking nanopores on an alumina anodisc, selective areas of that porous substrate can be patterned with A-aMEGO. It is demonstrated that poly(methyl methacrylate) (PMMA) can be successfully used to create these ion channels. Thick layers of PMMA have been patterned quickly (~10–20 min) using a light source with a wavelength of 254 nm at doses greater than 216 J.[34] Chain scissions in the PMMA molecular structure occurs when the PMMA is subjected to deep UV light. In turn, the areas of exposed PMMA become soluble in solvent-based developers. If the deep UV light is controlled so that it only is exposed to certain areas of the PMMA, the PMMA can be successfully patterned. Once the PMMA is patterned, aMEGO/PTFE can be deposited

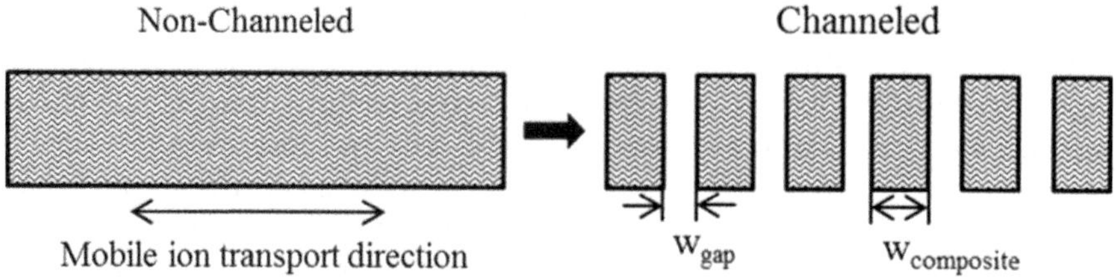

Figure 24.8 Cross-sectional views illustrating implemented ion channels. Non-channeled electrode (left) *versus* channeled electrode (right).

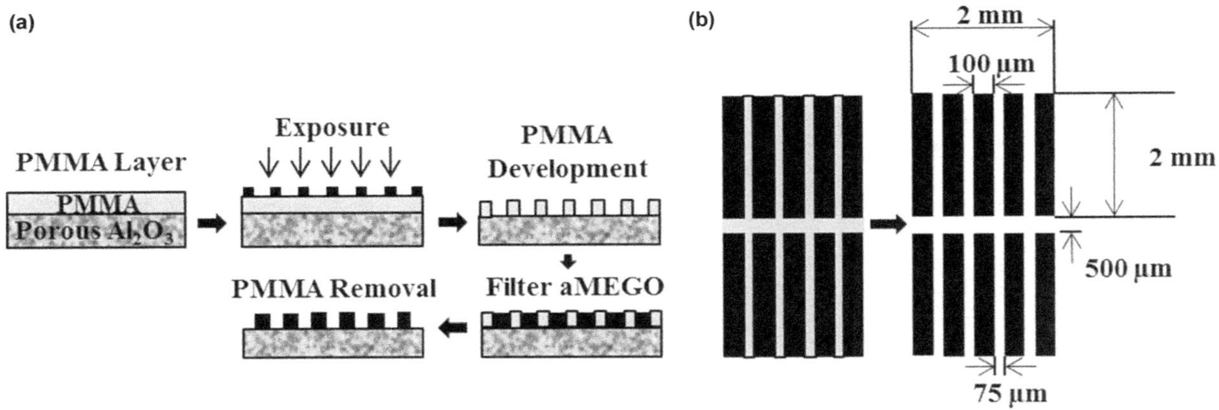

Figure 24.9 (a) Cross-sectional process flow plan for implementing hierarchical aMEGO structures. The ion channels for each electrode are patterned on the porous substrate using a sacrificial poly(methyl methacrylate) (PMMA) layer. (b) Top view illustration of A-aMEGO showing the PMMA pattern design (right) and the ion channel dimensions along with a separation of 500 µm between the anode and cathode.

between the PMMA features using the vacuum-assisted assembly process to form the highly aligned nanoporous graphene structures (A-aMEGO). After the graphene features are patterned, the PMMA can be removed using acetone.

24.3.2 Experimental Modification

The PMMA (2 wt%) was solution cast on a glass slide using *N*-methyl-2-pyrrolidone (NMP) as the dissolving solvent. The PMMA film was set in an oven at 100 °C for 36 h to evaporate the NMP. The thickness of the PMMA was around 20 μm once the NMP was removed. The PMMA was detached from the surface of the glass by submerging it in water. The PMMA film was then heated and attached to the porous alumina under vacuum at 180 °C.

To control the deep UV exposure, a quartz photomask was designed. The dimensions of the actuator are shown in Figure 24.9. The graphene features (areas on the mask where deep UV would pass through) were designed so that $w_{composite} = 100$ μm, and the ion channels (areas on the mask where the deep UV would not pass through) were designed so that $w_{gap} = 75$ μm. These graphene hierarchical features along with the ion channels spanned a width of 2 mm. These features were also designed as 2 mm in length. Each electrode was symmetric to the other electrode with a gap of 500 μm between the electrodes. The photomask was placed in contact with the PMMA layer, and the PMMA was exposed to 363 J of deep UV light. Next, the sample was submerged in methyl isobutyl ketone for 10 min to develop and remove the exposed PMMA. The patterned sample thickness was measured with a profilometer.

As in the previous sections, a vacuum-assisted process was used to align the aMEGO sheets.[14,16,17] Ethanol (20 ml) was used to disperse the nanoporous graphene (2 mg) and the PTFE binder (10 wt% PTFE and 90 wt% graphene). The vacuum-assisted assembly process was used to filter the aMEGO dispersion through the patterned template. The patterned template only allowed fluid flow through the developed regions, which allowed the aMEGO/PTFE dispersion to be deposited between the PMMA features. After aligning the A-aMEGO sheets, acetone was filtered through porous substrate to remove the PMMA, and consequently, the ion channels were formed between the A-aMEGO hierarchical features. The thickness of the hierarchical features was measured using a profilometer. The final nanocomposite structure was annealed in a vacuum oven at 140 °C.

The A-aMEGO/polymer nanocomposite and the alumina substrate were mounted onto a glass sample holder for electro-actuation testing. The nanocomposites were immersed in [BMIM][BF_4]/AN electrolyte for about 30 min. Each of the A-aMEGO composite electrodes was placed gently in contact with tungsten probes for electro-actuation strain testing. The positioning of the tungsten probes was precisely controlled using micrometer adjustments. Scan rates of 50, 100, 250, and 500 mV s^{-1} at 0–4 V were applied to the graphene electrodes using a potentiostat. The strain response of the cathode was characterized by measuring the displacement with the MTI 2100 Fotonic fiber optical sensor. It was observed previously that the cathode has a much higher strain than the anode does with the $[BMIM][BF_4]$/AN electrolyte. The strain of the cathode was calculated using eqn (24.1).

24.3.3 Improved Strain Results due to Ion Channels

Figure 24.10 shows the step-by-step process for the successful fabrication of the A-aMEGO actuator with ion channels. Figure 24.10a presents successful patterning of the PMMA film on the porous alumina substrate. The scanning electron microscope image in Figure 24.10b shows the aMEGO stacked on both sides of one PMMA feature. Figure 24.10c is an optical microscope image showing the hierarchical graphene structures with ion channels remaining between the structures. A profilometer was used to measure the thicknesses of the A-aMEGO hierarchical structures, which are shown in

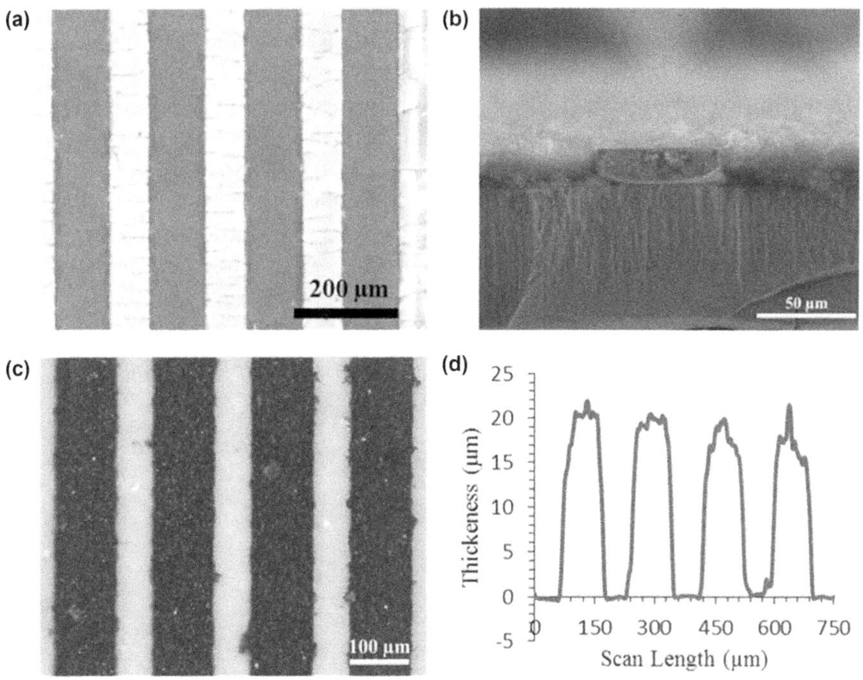

Figure 24.10 (a) Optical microscope image of patterned PMMA on the porous alumina substrate. (b) Cross-section image of a PMMA feature between stacked A-aMEGO/PTFE composites. (c) Optical microscope image of four A-aMEGO features (black regions) with PMMA (white regions) removed. (d) Measured thicknesses of four nanocomposite structures with ion transport channels between each structure.

Figure 24.10d. The images demonstrate the successful process flow shown in Figure 24.9.

Figure 24.11 compares the strain results of the previously synthesized A-aMEGO composites without ion channels, which were described in Section 24.1, to the nanocomposites developed with ion transport channels. A significant increase in strain is observed at the lower scan rates. At 50 mV s^{-1}, the highest strain of 77.5% is observed. Furthermore, at 100 mV s^{-1}, a net strain increase of over 40% is observed. These two scan rates demonstrate the significance of ion transport in the nanoporous graphene sheets. By implementing these ion channels, the mobile ions have better ability to ingress between the graphene sheets. At 250 mV s^{-1}, the strain of the channeled electrodes was 28.7%, which is very comparable to the non-channeled structures at 100 mV s^{-1}. This demonstrates a scan rate performance improvement of 2.5 times with the hierarchical structures. At the higher scan rate of 500 mV s^{-1}, the strain of the device still improved by a few percent, which also indicates that the ILs are better suited for lower scan rates. As was shown earlier, P(VDF-CTFE) was also used as a binder in the previous actuators. This particular binder showed a maximum strain of 56.6% under a potential of 4 V. The implemented ion channels with just PTFE as the binder outperform this previously recorded strain. This also indicates that by incorporating this binder into the hierarchical structures, further improvements in the electro-actuation strain characteristics may be observed. These results demonstrate that improved ion transport results in increased actuator performance. This decreased time for ion transport significantly improves the nanoporous graphene actuator.

This section presented successful patterning of nanoporous graphene and implemented ion channels within the electrodes of the actuator.

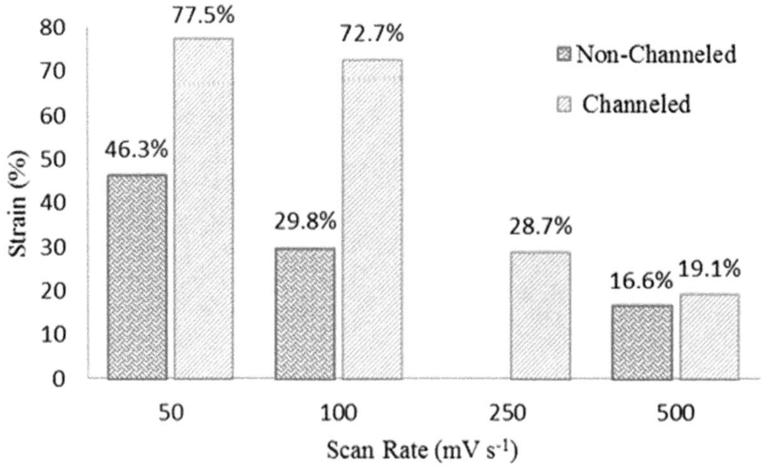

Figure 24.11 Investigation of the electro-actuation strain at 4 V for the previously fabricated non-channeled nanocomposite actuators and the actuators with implemented ion channels.

The implemented ion channels in the electrodes allowed for improved ion transport. Consequently, this allowed for the net transport time of the ions into the graphene sheets to be decreased. In turn, the actuation speed of the nanocomposites was increased, which resulted in better strain responses in the A-aMEGO nanocomposite actuators. Therefore, ion transport distance plays an important role in device characteristics, and specifically in this work, significant improvement in actuation performance was seen with the reduced ion travel distance.

References

1. Y. Bar-Cohen and Q. M. Zhang, *MRS Bull*, 2008, **33**, 173.
2. Q. M. Zhang, V. Bharti and X. Zhao, *Science*, 1998, **280**, 2101.
3. R. Perline, R. Kornbluh, Q. Pei and J. Joseph, *Science*, 2000, **287**, 836.
4. W. Lu, A. G. Fadeev, B. Qi, E. Smela, B. R. Mattes, J. Ding, G. M. Spinks, J. Mazurkiewicz, D. Zhou, G. G. Wallace, D. R. MacFarlane, S. A. Forsyth and M. Forsyth, *Science*, 2002, **297**, 983.
5. M. Shahinpoor, Y. Bar-Cohen, J. O. Simpson and J. Smith, *Smart Mater. Struct.*, 1998, **7**, R15.
6. R. H. Baughman1, C. Cui1, A. A. Zakhidov, Z. Iqbal, J. N. Barisci, G. M. Spinks, G. G. Wallace, A. Mazzoldi, D. De Rossi, A. G. Rinzler, O. Jaschinski, S. Roth and M. Kertesz, *Science*, 1999, **284**, 1340.
7. S. Liu, Y. Liu, H. Cebeci, R. Guzmán de Villoria, J. H. Lin, B. L. Wardle and Q. M. Zhang, *Adv. Funct. Mater.*, 2010, **20**, 3266.
8. T. Thorsen, S. J. Maerkl and S. R. Quarke, *Science*, 2002, **298**, 580.
9. M. D. Lima, N. Li, M. J. de Andrade, S. Fang, J. Oh, G. M. Spinks, M. E. Kozlov, C. S. Haines, D. Suh, J. Foroughi, S. J. Kim, Y. Chen, T. Ware, M. K. Shin, L. D. Machado, A. F. Fonseca, J. D. W. Madden, W. E. Voit, D. S. Galvão and R. H. Baughman, *Science*, 2012, **338**, 928.
10. P. Calvert, *MRS Bull.*, 2008, **33**, 207.
11. M. Knoblauch, G. A. Noll, T. Müller, D. Prüfer, I. Schneider-Hüther, D. Scharner, A. J. E. V. Bel and W. S. Peters, *Nat. Mater.*, 2003, **2**, 600.
12. D. J. Beebe, J. S. Moore, J. M. Bauer, Q. Yu, R. H. Liu, C. Devadoss and B. H. Jo, *Nature*, 2000, **404**, 588.
13. Y. Zhu, S. Murali, M. D. Stoller, K. J. Ganesh, W. Cai, P. J. Ferreira, A. Pirkle, R. M. Wallace, K. A. Cychosz, M. Thommes, D. Su, E. A. Stach and R. S. Ruoff, *Science*, 2011, **332**, 1537.
14. Q. Liang, X. Yao, W. Wang, Y. Liu and C. P. Wong, *ACS Nano*, 2011, **5**, 2392.
15. K. W. Putz, O. C. Compton, C. Segar, Z. An, S. T. Nguyen and L. C. Brinson, *ACS Nano*, 2011, **5**, 6601.
16. K. W. Putz, O. C. Compton, M. J. Palmeri, S. T. Nguyen and L. C. Brinson, *Adv. Funct. Mater.*, 2010, **20**, 3322.
17. D. A. Dikin, S. Stankovich, E. J. Zimney, R. D. Piner, G. H. B. Dommett, G. Evmenenko, S. T. Nguyen and R. S. Ruoff, *Nature*, 2007, **448**, 457.

18. J. D. Madden, N. A. Vandesteeg, P. A. Anquetil, P. G. A. Madden, A. Takshi, R. Z. Pytel, S. R. Lafontaine, P. A. Wieringa and I. W. Hunter, *IEEE J. Oceanic Eng.*, 2004, **29**, 706.
19. B. E. Conway, *Electrochemical Supercapacitors: Scientific Fundamentals and Technological Applications*, Plenum Publishers, New York, 1999.
20. Y. Liu, R. Zhao, M. Ghaffari, J. Lin, S. Liu, H. Cebeci, R. Guzmán de Villoria, R. Montazami, D. Wang, B. L. Wardle, J. R. Heflin and Q. M. Zhang, *Sens. Actuators, A*, 2012, **181**, 70.
21. S. Liu, W. Liu, Y. Liu, J. Lin, X. Zhou, M. J. Janik, R. H. Colby and Q. M. Zhang, *Polym. Int.*, 2010, **59**, 321.
22. A. Jarosik, S. R. Krajewski, A. Lewandowski and P. Radzimski, *J. Mol. Liq.*, 2006, **123**, 43.
23. J. M. Herbert, *Ferroelectric Transducers and Sensors*, Gordon and Breach Science Publishers, New York, 1982, vol. 3.
24. R. E. Newnham, D. P. Skinner and L. E. Cross, *Mater. Res. Bull.*, 1978, **13**, 525.
25. V. Giurgiutiu and C. J. Rogers, *Intell. Mater. Syst. Struct*, 1996, 7, 656.
26. D. Maugis, *Contact, Adhesion and Rupture of Elastic Solid*, Springer, New York, 2000.
27. X. Zhou, X. Zhao, Z. Suo, C. Zou, J. Runt, S. Liu, S. Zhang and Q. M. Zhang, *Appl. Phys. Lett.*, 2009, **94**, 162901.
28. O. Sahin and N. Erina, *Nanotechnology*, 2008, **19**, 445717.
29. Y. Liu, M. Ghaffari, R. Zhao, J. Lin, M. Lin and Q. M. Zhang, *Macromolecules*, 2012, **45**, 5128.
30. J. Lin, Y. Liu and Q. M. Zhang, *Macromolecules*, 2012, **45**, 2050.
31. S. Yin, Y. Zhang, J. Kong, C. Zou, C. M. Li, X. Lu, J. Ma, F. Y. C. Boey and X. Chen, *ACS Nano*, 2011, **5**, 3831.
32. C.-J. Bae, C. K. Erdonmez, J. W. Halloran and Y.-M. Chiang, *Adv. Mater.*, 2013, **25**, 1254.
33. Y.-S. Hu, P. Adelhelm, B. M. Smarsly, S. Hore, M. Antonietti and J. Maier, *Adv. Funct. Mater.*, 2007, **17**, 1873.
34. M. Haiducu, M. Rahbar, I. G. Foulds, R. W. Johnstone, D. Sameoto and M. Parameswaran, *J. Micromech. Microeng.*, 2008, **18**, 115029.

CHAPTER 25

Multiphysics Modeling and Simulation of Dynamics Sensing in Ionic Polymer Metal Composites with Applications to Soft Robotics

YOUSEF BAHRAMZADEH

Mechanical Engineering Department, University of Maine, Orono, ME, USA
Email: yousef.bahramzadeh@maine.edu

25.1 Ionomers and Electrodes in Ionic Polymer Metal Composites

The core part of ionic polymer actuators and sensors is an ion-exchange membrane or ionomer. An ionomer is a polyelectrolyte that contains strong ionic groups such as sulfonic or carboxylic acid attached to the backbone of a stable polymer (with electrically neutral repeating units) such as Teflon.[1] The composition forms highly ionic clusters that selectively facilitate the transport of the mobile ions across the backbone of the polymer. The mobile ions inside the polymer can be transported and accumulated at one side of the polymer by applying a small electric field, which results in ionic electroactivity (Figure 25.1). To this end, the manufacturing of ionic polymeric actuators and sensors starts with compositing an ionic polymeric membrane with two

RSC Smart Materials No. 18
Ionic Polymer Metal Composites (IPMCs): Smart Multi-Functional Materials and Artificial Muscles, Volume 2
Edited by Mohsen Shahinpoor

Published by the Royal Society of Chemistry, www.rsc.org

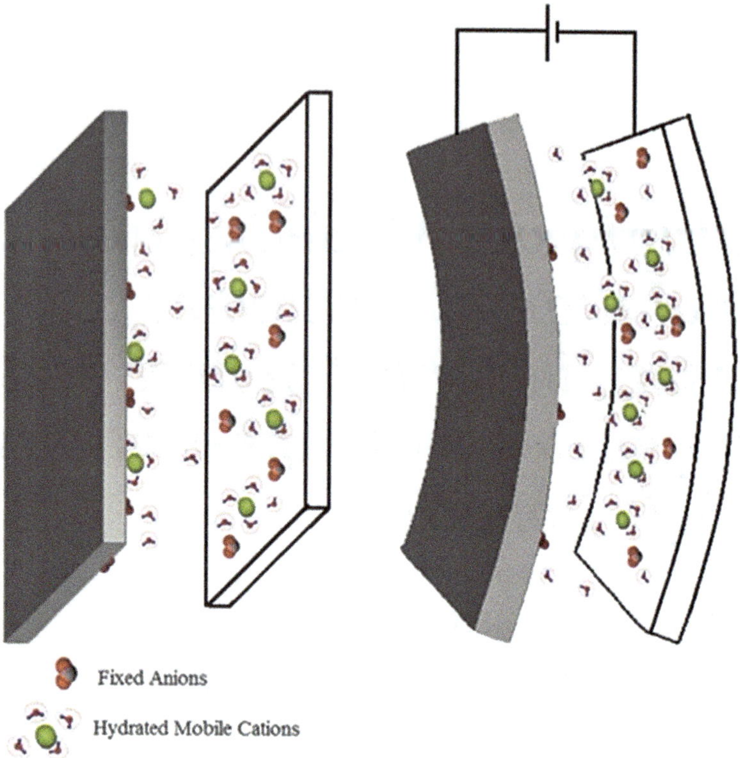

Figure 25.1 Transduction mechanism in an IPMC-based actuator.

conductive thin electrodes on both sides to apply an electric field along the membrane. Other than ionomers and electrodes, the diluent that facilitates the transport of the mobile ions across the membrane is an important factor in the ionic responses of ionic polymer metal composites (IPMCs).[2]

The major role of the ionic polymer is storing ions in the ionic sites and maintaining the required mobility for counterions across the membrane. It also provides the mechanical stress and force required for the actuator. Ionic polymers are considered to be polyelectrolytes that consist of a certain amount of hydrophilic ionic repeat units covalently bound to repeat units of a hydrophobic nonionic polymer. The network of hydrophilic clusters and/or channels provides porosity for charge transport when swollen in the presence of an added diluent. The proton exchange membrane (PEM) and anion exchange membrane are two broad categories of ionic polymers that facilitate the transport of the cations and anions, respectively. Upon applying the electric field to two sides of a PEM, the cationic counterions move toward the cathode while the bound anionic species remain immobile. The acidity or ion-exchange capacity (IEC) of the ionomer is an indication of its capacity for storing the counterions in the ionic sites, and the ion conductivity of the ionomer is an indication of ion mobility across the membrane.

IEC and ion conductivity are the two major characteristics of ionic polymers that directly affect the deformation response and blocked force of IPMCs, as well as their sensing properties. The IEC and ion conductivity of the membrane depend directly on the structure of the membrane;[3] however, the ion conductivity depends on the size and charge of the counterions, as well as the electrolyte type. In most reported PEMs, the proton conductivity is strongly related to the amount of ionic groups (sulfonic or carboxylic acid groups), and membranes with more acid groups typically have higher proton conductivity. On the other hand, the tensile modulus typically scales inversely when ionic conductivity is increased by higher diluent uptake. Clearly, for the device to work efficiently, direct electrical contact between the anode and the cathode must be avoided. Regarding these mentioned properties, the main efforts in this area are to synthesize ionomers with higher ionic conductivity both in hydrated and dry conditions, to improve chemical and thermal stability and to improve mechanical properties. While a considerable number of potential polymer backbone and side-chain combinations have been synthesized to meet these requirements, DuPont's Nafion® continues to hold a place of prominence in the field because of its high proton conductivity and good mechanical properties, and thus it provides a useful standard against which other ionic membranes are often compared. Several other type of ionomers, such as Flemion from Asahi Glass, Aciplex from Asahi Chemical, Aquivion (Hyflon) from Solvay and other synthesized sulfonated aromatic ionomers, have been widely studied as polyelectrolytes for IPMCs.[4] The structures of the majority of these ionomers consist of perfluorinated alkenes with short side chains terminated by ionic groups (typically sulfonate or carboxylate [SO_3^- or COO^-]) for cation exchange or ammonium cations for anion exchange. The large polymer backbones determine their mechanical strength. Short side chains provide ionic groups that interact with water and enable the passage of appropriate ions.

The actuation and sensing of IPMCs occur due to mobile ion transport, so the transduction properties, such as deflection speeds, displacement and degrees of relaxation, are directly affected by the properties of mobile ions such as ion diameter, hydrated ion diameter, charge number and ion mobility.

Following the electroding process, the mobile cation in the membrane can be exchanged for any suitable ion by immersing the membrane in a salt solution of that ion. Alkali metal ions (Li^+, Na^+, K^+, Rb^+ and Cs^+) may be exchanged by soaking them in an appropriate salt solution such as chloride salts (LiCl, NaCl and KCl) at moderate temperatures (30 °C) for 1–3 days.

It has been reported that, in terms of force generation with a given voltage of 1.2 V, Li^+ ions have superior properties:[5] $Li^+ \gg Na^+ > (K^+, Ca^{2+}, Mg^{2+}$ and $Ba^{2+})$. These observations indicate that the hydrated volume of ions and hydration phenomena play important roles in achieving the highest force.

Finally, the electroding process of the ionic polymers for forming a conductor–ionomer composite is a key step in the fabrication of IPMCs and has a significant role in the overall performance of the IPMCs. The electrode layers affect the performance of IPMCs in two major aspects: the electronic conductivity of the electrodes and the electrode–polymer interlocking. The conductor layer should provide a high electronic conductivity for all points of the IPMC so that a consistent electric potential can reach the entire membrane. On the other hand, the large interfacial area between metal particles and the membrane directly enhances the adhesion and capacitance of the IPMC, which leads to larger charge accumulation at the electrode layer and consequently better actuation.

25.2 IPMC Curvature Sensor

IPMC sensing due to bending, twisting and general deformation has been reported by Shahinpoor.[6] The general characteristics of IPMCs as biomimetic sensors, actuators and artificial muscles were investigated by Shahinpoor *et al.*[7] The sensing of IPMCs with respect to different mechanical stimuli, such as the bending deflection of a cantilever, sharp angular bending and distributed pressure loading, have also been investigated in the literature.[8] The feasibility of IPMCs as bending sensors for providing feedback for hand prostheses has been studied by Biddiss *et al.*[9] and it has been shown that bending angles may be accurately measured with 4–5% error. It has been demonstrated that there is a linear relationship between the bending angle of an IPMC and the output voltage.

While several studies have been conducted on the actuation of IPMCs, there are still phenomena in the sensing mode that need to be further investigated in order to characterize IPMC sensors for specific applications such as curvature sensing. For instance, there is a rate dependency of sensor output signals, as well a phase that must be considered for characterizing the IPMC sensors. Figure 25.2 shows a general response of IPMCs to a high-frequency vibration followed by low-speed excitation at the same amplitude. It is observed that bending the IPMC to the same amplitude at different rates generates signals with different amplitudes.

Calibrating the output electric potential signal of a sensor to the curvature variation, rather than the sharp bending angle or tip bending deflection, is of particular interest. It is also important to investigate the effects of curvature variation rates, especially at low rates, both experimentally and theoretically. In order to accomplish this, the experimental test setup should provide accurate control over the curvature and curvature rate of change. The periodic step response of an IPMC sensor is depicted in Figure 25.3. It is observed that there is a fast response region followed by a recovery period in which the sensor output undergoes a relaxation toward the initial static voltage. This behavior of IPMC sensors is similar to that of piezo-resistive materials. The mechanism of this phenomenon can be explained as part of the ionic behavior of IPMC sensors, in that ions migrate across the thickness of the membrane to reach a more stable level of energy. The recovery voltage causes nonlinearity in the sensor behavior; however, it can be observed that there is a nearly constant peak voltage of 2.5 mV in every period. The average

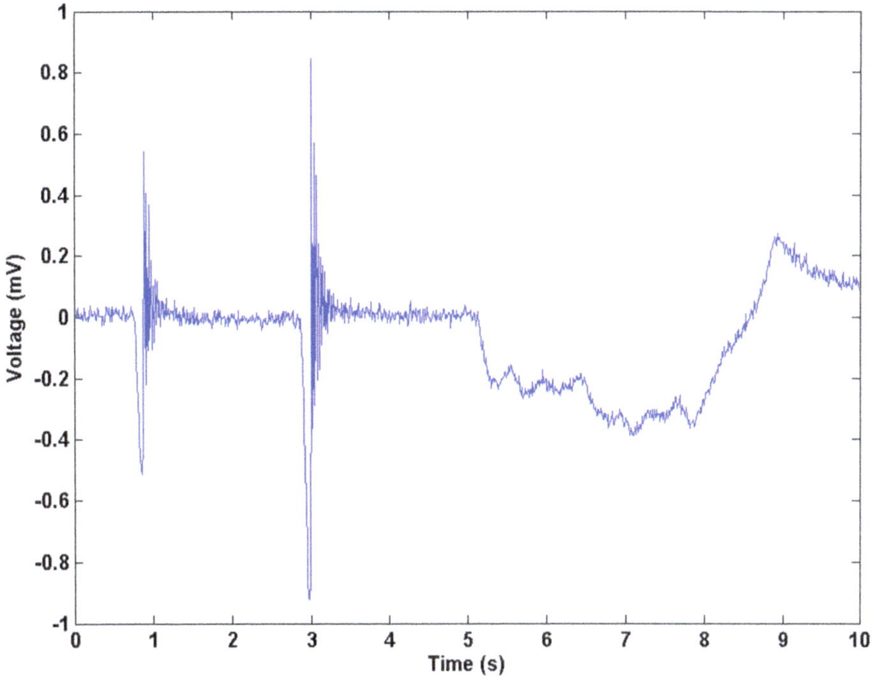

Figure 25.2 General response of an IPMC sensor to high-frequency excitations followed by slow bending accompanied by high-frequency noise.

deviation from 2.5 mV is less than 5%. The plots in Figure 25.3 shows the high repeatability of the sensor during the periodic step deflection.

The influence of voltage recovery on the output of IPMC sensor voltage is more significant at very slow rates. As illustrated in Figure 25.4, the cantilever beam experienced a 10 mm deflection of ramp function for the durations of 10, 20, 40 and 60 s.

It is observed that at high bending rates (10 and 20 s), the linear behavior of the sensor is more significant. At lower rates, the sensor still shows linear behavior, even though the ion drift results in more noisy behavior of the signal.

It is concluded that at the slow bending rates, the output of IPMC sensors is coupled with the rate of bending, which implies that the sensor calibration has to be performed in terms of the specific curvature rate.

25.3 IPMC Curvature Actuators as Soft Robots for Biomedical Instrumentation

The unique properties of IPMCs, such as flexibility and large deflection amplitudes, have opened the way for a multitude of new applications for IPMCs, including in biomedical instrumentation. For instance, there are numerous surgical procedures that require flexible soft actuators with a low actuation voltage. In the area of minimally invasive procedures, advances in

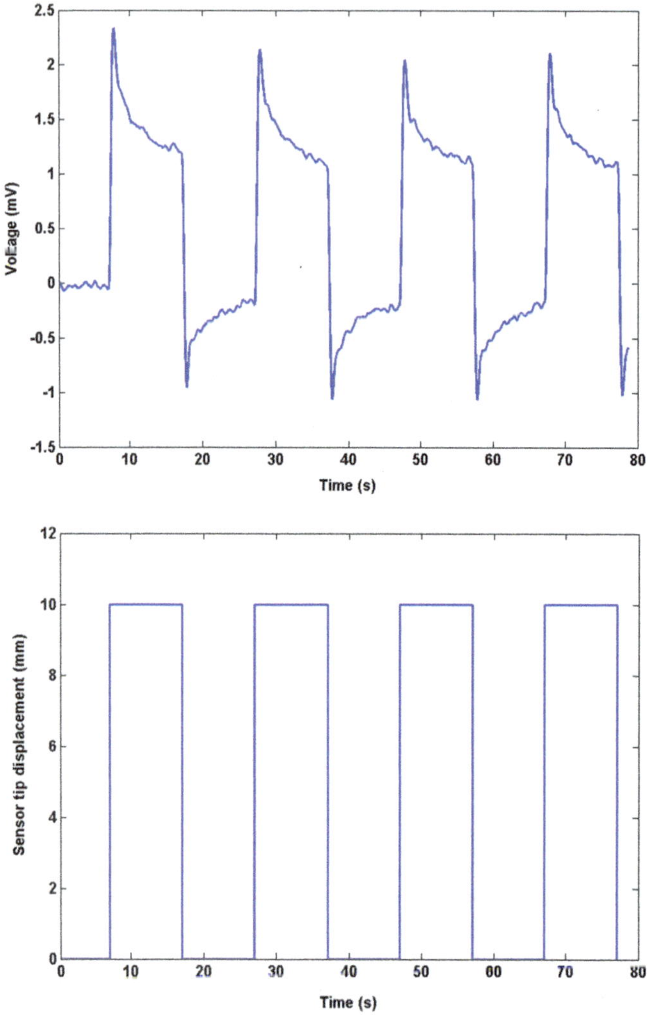

Figure 25.3 Voltage response of an IPMC sensor to a periodic step deflection (top). The bottom plot corresponds to the applied displacements function at the tip of IPMC sensor with respect to time.

developing new minimally invasive methods rely on developing new materials that are biocompatible and meet biomedical requirements. Novel soft and miniaturized actuators capable of operating in small cavities, such as inside the arteries and veins, can extend the efficiency and development of new methods in minimally invasive surgery methods.

The flexibility of an IPMC makes it possible for it to be applied in both small- and large-deflection applications. Successive photographs of an IPMC strip are shown in Figure 25.5 that demonstrates very large deformation in the presence of low voltages.

The underlying principles of the IPMC actuation and sensing capabilities can be described by the standard Onsager formula using linear irreversible

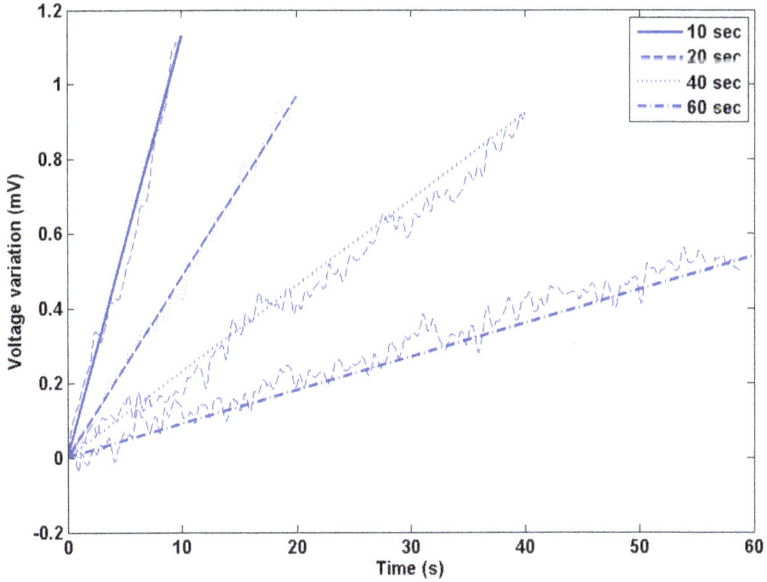

Figure 25.4 Ramp response of an IPMC sensor at four different curvature rates of 10, 20, 40 and 60 s. The beam was bent by 10 mm, which is equivalent to a 500 mm radius of curvature.

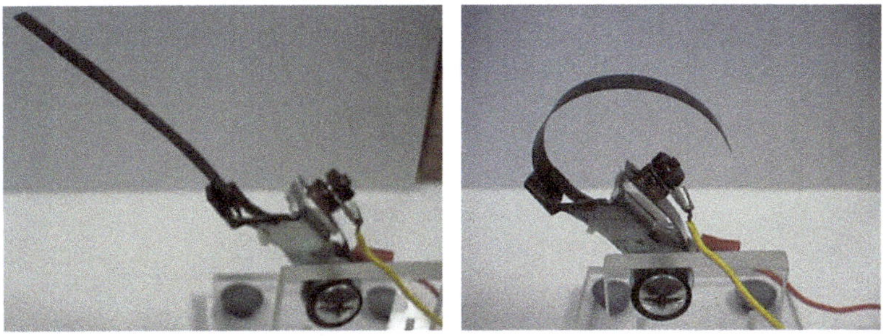

Figure 25.5 Successive photographs of an IPMC strip before actuation (left) and after actuation (right).[10]

thermodynamics. When static conditions are imposed, a simple description of the mechanoelectric effect is possible based upon two forms of transport: ion transport (with a current density, J, normal to the material) and solvent transport (with a flux, Q, which we can assume is water flux).

The conjugate forces include the electric field, E, and the pressure gradient, ∇_p. The resulting equation has the concise form of:[11]

$$\boldsymbol{J}(x, y, z, t) = \sigma \boldsymbol{E}(x, y, z, t) - L_{12}\boldsymbol{\nabla}_p(x, y, z, t) \tag{25.1}$$

$$\boldsymbol{Q}(x, y, z, t) = L_{21}\boldsymbol{E}(x, y, z, t) - K\boldsymbol{\nabla}_p(x, y, z, t) \tag{25.2}$$

where σ and K are the material electric conductance and the Darcy permeability, respectively. A cross-coefficient is usually $L = L_{12} = L_{21}$.

The simplicity of the above equations provides a compact view of the underlying principles of the actuation, transduction and sensing of IPMCs.

Catheter insertion is a minimally invasive surgical process in which a thin wire is inserted into the body through a small incision in the groin or the arm to reach a femoral or brachial artery. The distal tip of a wire is steered or stirred through blood vessels to reach the desired location. IPMCs are capable of being steered or stirred for directional change within the vasculature or for mixing purposes. Currently, various surgical operations apply catheter insertion techniques for different types of surgeries. However, guidance of a catheter inside the complex channels of the blood vessels or vasculature is a complicated task that is currently performed manually in a trial-and-error process. Correct insertion depends on the surgeon's skill to a great extent. A catheter with one or two controllable degrees of freedom for bending the distal tip will give surgeons more dexterity in maneuvering the wire inside the body. Consider the tip of a catheter that is equipped with an IPMC artificial muscle as a steerer/stirrer. A schematic of an attached IPMC steerer/stirrer to the tip of a catheter is depicted in Figure 25.6.[12]

Bi-directional bending of the catheter, along with the manual twisting motion of the wire, enables 3D orientation control of the active catheter. Successful actuation is achieved in a fluidic environment, as shown in Figure 25.7. The tip of the actuator is easily bent by about 90°, which is sufficient to maneuver through the endovascular branches. By applying a voltage of 0.2–3 V on an IPMC film, bending towards the anode occurs.

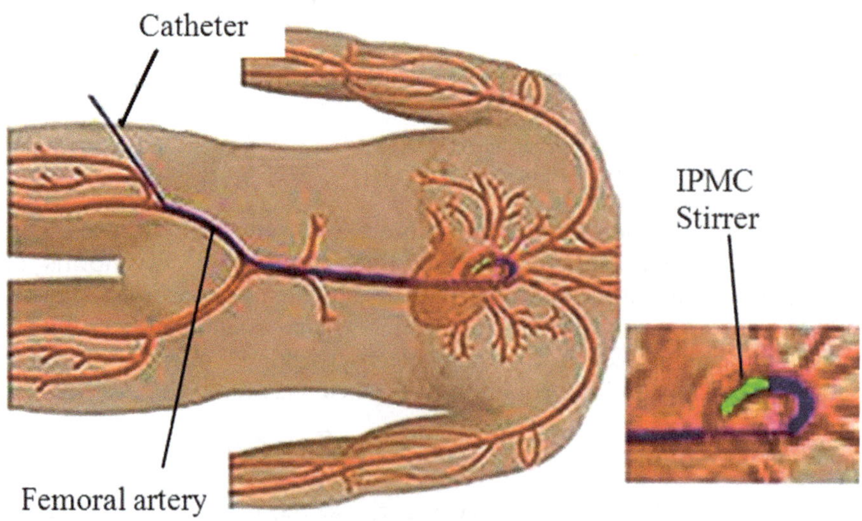

Figure 25.6 Schematic of an IPMC micro catheter steerer/stirrer.

Figure 25.7 IPMC steerer/stirrer attached to the tip of a catheter and actuated in a saline fluidic environment resembling blood.

An increase in voltage level (up to 6 or 7 V) causes larger bending displacement, along with nonlinear saturation in displacement. IPMCs also work very well in water or blood environments.

In addition, with a customized design, IPMC actuators maintain the required dexterity for the 2D bending of a robotic distal tip. The overall design of the robot could be considered to be a hybrid robot with the combination of rigid robotic links and a flexible IPMC actuator with two degrees of freedom.

Figure 25.8 demonstrates a design in which the tip is bent in two directions using a 2D IPMC actuator. The flexible distal tip is attached to the rigid part of the instrument. The instrument is positioned by the manipulator of a surgical robot, such that the system can be considered to be a hybrid robot composed of a rigid robotic console and a flexible polymeric distal tip. The ion-exchange pellets of Nafion can be cast to have a rectangular or circular cross-section and metalized with gold as a highly conductive electrode around the surface. The electrodes at four sides must be then separated so that four distinct electrodes are used to control the ion transport.

In order to have larger actuation and a minimal moment of inertia, the circular cross-section is the optimum choice for undulation. In Figure 25.9, a schematic of the chemo-electro-mechanical bending of a 2D IPMC is demonstrated. The tubular IPMC acts as a continuum robot at the tip of the instrument and antagonist forces are applied by induced strain due to the

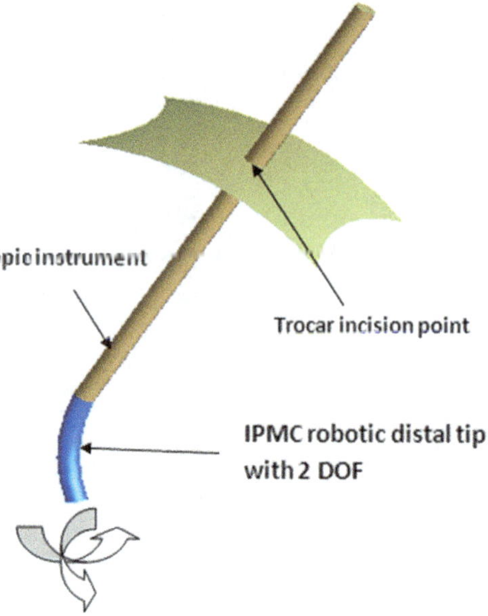

Figure 25.8 Schematic of a surgical instrument equipped with an IPMC actuator.

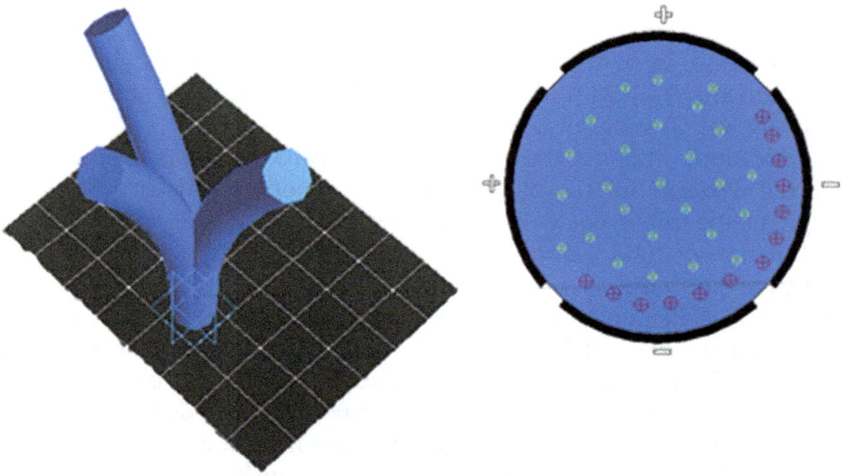

Figure 25.9 (Left) Schematic of 2D bending of an IPMC actuator. (Right) Cross-section of a tubular IPMC. The pressure gradient induced by the accumulation of cations at each side results in bending.

accumulation of ions at a desired side of the cross-section. The mechanism of bending is the same as that of IPMC membranes, except that mobile cations can transport in two dimensions.

By applying an electric potential to each pair of electrodes, rotation about any axis can be achieved. The magnitude and direction of rotation depend on the direction and magnitude of the applied electric potential. The nature of the actuation is comparable to the antagonistic muscles in which the bending direction is controlled by the amount of applied strain at each side of the cylindrical organ.

25.4 Multiphysics Modeling of Ionic Electroactivity in IPMCs

In order to explore the dynamic ionic electroactivity of IPMCs, herein we review a dynamic model for finding the dynamic response between the applied mechanical bending stress and the generated electric potential. Following the methods of Shahinpoor and Kim,[13] Nemat-Nasser and Li,[14] Porfiri,[15] Chen *et al.*[16] and Bahramzadeh and Shahinpoor,[17] the basic governing equations for charge dynamics in ionic polymers are:[18]

1. The constitutive equation of Nernst–Planck;
2. The continuity equation of ions;
3. The equilibrium equation of Nernst–Planck;
4. Poisson's equation.

The Poisson–Nernst–Planck equations for ion diffusion across the IPMC membrane can be then written as:

$$\frac{\delta c^+}{\delta t} = D\frac{\delta}{\delta x}\left[\frac{\delta c^+}{\delta x} + \frac{F}{RT}c^+\frac{\delta V}{\delta x}\right] \quad 0<x<h \tag{25.3}$$

$$\nabla^2 V + \frac{\rho}{\varepsilon} = 0 \tag{25.4}$$

in which $c^+$is the concentration of cations (mol m^{-3}), ρ is the charge density (Colomb m^{-3}), V is the electric potential field (volts), D is the diffusion coefficient (m^2 s^{-1}), F is Faraday's constant (96 487 C mol^{-1}), R is the universal gas constant (8.3143 J [mol K]$^{-1}$), T is thetemperature (K) and x is in normal direction to the IPMC thickness (m).

The relation between charge density and ion concentration is as follows:

$$\rho = F(c^+ - c^-) \tag{25.5}$$

Placing the electric potential term from eqn (25.2) into eqn (25.1) and using eqn (25.3), the modified Poisson–Nernst–Planck equations for kinetic of charge density will be derived as:

$$\frac{\delta \rho}{\delta t} = D\frac{\partial^2 \rho}{\partial x^2} - \alpha\rho \tag{25.6}$$

$$\alpha = \frac{DF^2}{RT\varepsilon}c^- \tag{25.7}$$

Here, we made the following assumptions:

1. The diffusion coefficient D is constant over the thickness of the membrane;
2. The ion diffusion is dominant over the thickness of the membrane;
3. The mobile ions contributing to diffusion are cations.

This partial differential equation in eqn (25.6) represents the kinetics of charge density in terms of time and the thickness of the ionomer. The ionic properties of the ionomer are explicitly represented in this equation by the two constants of D and α.

25.5 Conclusion

In this chapter, first a review was given of the various types of ionic polymers, electrodes and diluents that have been studied in the literature and of the effects of IEC, the importance of electrodes and the different types of mobile ions were discussed. Next,the characteristics of IPMC sensors for curvature measurement of structures were investigated and it was shown that there is a linear relationship between the curvature of an IPMC sensor strip and the output voltage of the IPMC sensor. Curvature–voltage diagrams were constructed for a periodic step function and ramp function at different rates. According to various conducted tests, it can be concluded that the IPMC sensor maintains important characteristics of sensors, including linearity, sensitivity and repeatability for the curvature sensing of structures. The next part of the chapter was dedicated to the application of IPMC curvature actuation in biomedical instrumentation. The design and actuation mechanisms of single degree of freedom (DOF) and two DOF bio-inspired ionic actuators for bio-robotic applications were presented based on custom fabrication of IPMCs. A multiphysics model based on Poisson–Nernst–Planck equations was presented to explain the observed ionic electroactivity and dynamic sensing/actuation of IPMCs.

References

1. K. A. Mauritz and R. B. Moore, State of understanding of Nafion, *Chem. Rev.*, 2004, **104**(10), 4535–4586.
2. Y. Bahramzadeh and M. Shahinpoor, A Review of Ionic Polymeric Soft Actuators and Sensors, *Soft Robotics*, 2014, **1**(1), 38–52.
3. H. L. Yeager and A. Steck, Cation and water diffusion in Nafion ion exchange membranes: influence of polymer structure, *J. Electrochem. Soc.*, 1981, **126**(9), 1880–1884.
4. Y. Bahramzadeh and M. Shahinpoor, A Review of Ionic Polymeric Soft Actuators and Sensors, *Soft Robotics*, 2014, **1**(1), 38–52.
5. K. J. Kim and M. Shahinpoor, Ionic polymer–metal composites: II. Manufacturing techniques, *Smart Mater. Struct.*, 2003, **12**(1), 65.

6. M. Shahinpoor and K. J. Kim, Ionic polymer-metal composites: I. Fundamentals, *Smart Mater. Struct.*, 2001, **10**(4), 819.
7. M. Shahinpoor *et al.*, Ionic polymer-metal composites (IPMCs) as biomimetic sensors, actuators and artificial muscles-a review, *Smart Mater. Struct.*, 1998, 7(6), R15.
8. M. Shahinpoor, K. J. Kim and M. Mojarrad, *Artificial Muscles: Applications of Advanced Polymeric Nanocomposites*, CRC Press, 2010.
9. E. Biddiss and T. Chau, Electroactive polymeric sensors in hand prostheses: bending response of an ionic polymer metal composite, *Med. Eng. Phys.*, 2006, **28**, 568–578.
10. M. Shahinpoor and K. J. Kim, Ionic polymer–metal composites: III. Modeling and simulation as biomimetic sensors, actuators, transducers, and artificial muscles, *Smart Mater. Struct.*, 2004, **13**(6), 1362.
11. P. G. De Gennes *et al.*, Mechanoelectric effects in ionic gels, *Europhys. Lett.*, 2000, **50**(4), 513.
12. L. Rasmussen, *Electroactivity in Polymeric Materials*, Springer, 2012.
13. M. Shahinpoor and K. J. Kim, Mass transfer induced hydraulic actuation in ionic polymer-metal composites, *J. Intell. Mater. Syst. Struct.*, 2002, **13**(6), 369–376.
14. S. Nemat-Nasser and J. Yu Li, Electromechanical response of ionic polymer-metal composites, *J. Appl. Phys.*, 2000, **87**(7), 3321–3331.
15. M. Porfiri, Charge dynamics in ionic polymer metal composites, *J. Appl. Phys.*, 2008, **104**(10), 104915.
16. Z. Chen *et al.*, A dynamic model for ionic polymer–metal composite sensors, *Smart Mater. Struct.*, 2007, **16**(4), 1477.
17. Y. Bahramzadeh and M. Shahinpoor, Charge modeling of ionic polymer-metal composites for dynamic curvature sensing, SPIE Smart Structures and Materials + Nondestructive Evaluation and Health Monitoring. International Society for Optics and Photonics, 2011.
18. Y. Bahramzadeh and M. Shahinpoor, Dynamic curvature sensing employing ionic-polymer–metal composite sensors, *Smart Mater. Struct.*, 2011, **20**(9), 094011.

CHAPTER 26

A Comprehensive Review of Electroactive Paper Actuators

JAEHWAN KIM,* SEONGCHEOL MUN, HYUN-U KO, LINDONG ZHAI, SEUNG-KI MIN AND HYUN CHAN KIM

Inha University, Department of Mechanical Engineering, 253 Yonghyun-Dong, Nam-Ku, Incheon 402-751, Republic of Korea
*Email: jaehwan@inha.ac.kr

26.1 Introduction

Cellulose has a yearly estimated biomass production of 1.5 trillion tons and is an almost inexhaustible raw polymer material with a fascinating structure and properties.[1] Since cellulose is biodegradable, renewable and biocompatible, its derivatives are used for many applications, including the immobilization of proteins and antibodies, coatings, laminates, optical films, pharmaceuticals, textiles and foodstuffs, as well as the formation of cellulose composites with synthetic polymers and biopolymers. Natural cellulose has two morphologies: nanocrystal and amorphous domains. Figure 26.1 shows the hierarchical structure of cellulose from wood.[2] Cell walls of wood are made with macrofibers of cellulose, hemicellulose and lignin, which form a cellulose fiber composite. The macrofibers are composed of microfibrils, which are formed from nanofibrils of cellulose. Interestingly, nanofibrils of cellulose have crystal parts and amorphous parts in a row. The crystal part of cellulose cannot be broken due to the strong

RSC Smart Materials No. 18
Ionic Polymer Metal Composites (IPMCs): Smart Multi-Functional Materials and Artificial Muscles, Volume 2
Edited by Mohsen Shahinpoor

Published by the Royal Society of Chemistry, www.rsc.org

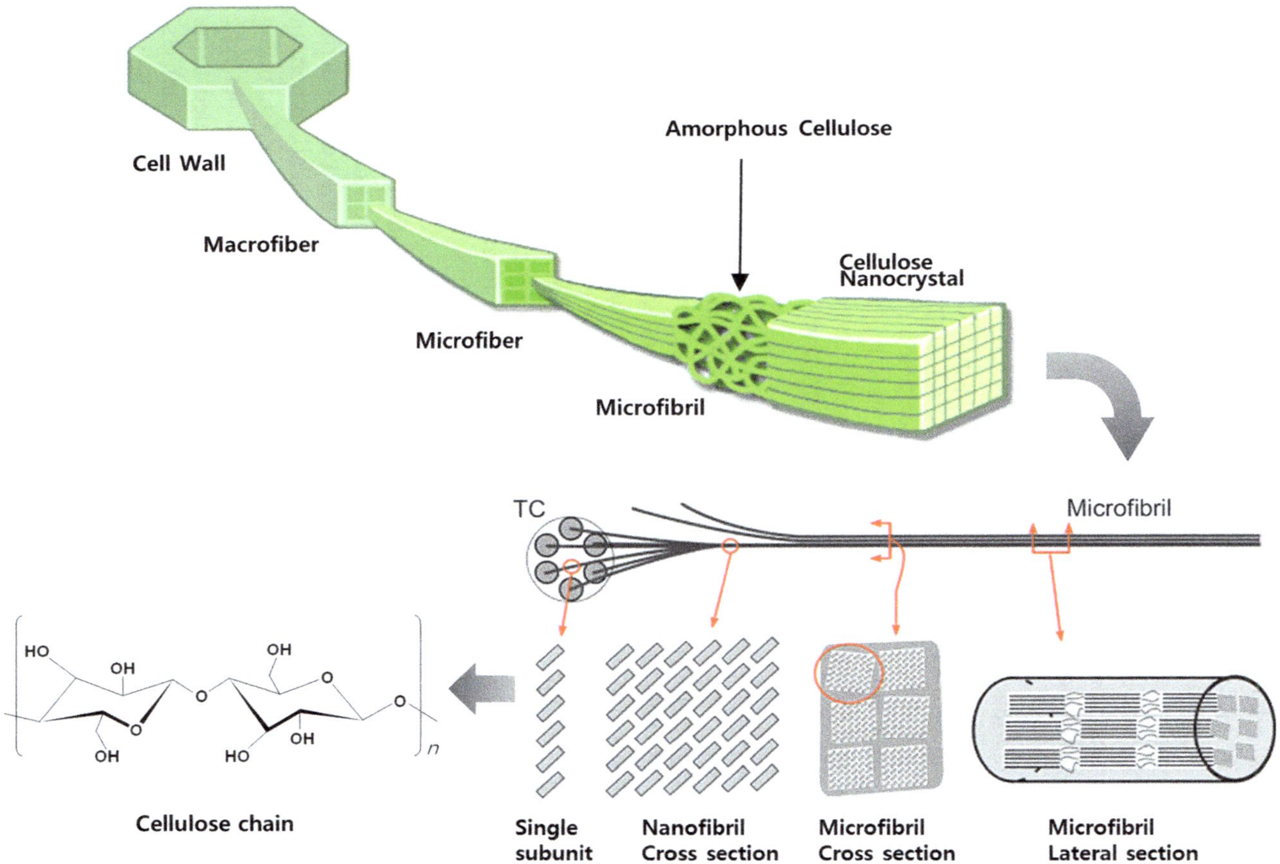

Figure 26.1 Hierarchical structure of cellulose. TC: terminal enzyme complexes.

hydrogen bond of the hydroxyl groups in cellulose. Crystalline cellulose has several polymorphs: cellulose I, II, III and IV. Cellulose I is the crystalline cellulose that is naturally produced by a variety of organisms, which is sometimes referred to as natural cellulose. Its structure is thermodynamically metastable and can be converted to either cellulose II or III. Cellulose II is the most stable crystalline structure and can be produced by regeneration and mercerization. There are two kind of nanocellulose: cellulose nanofiber (CNF) and cellulose nanocrystal (CNC). There are various extraction methods for obtaining CNC or CNF from cellulose microfibrils, which include pre-treatment, disintegration or deconstruction processes, such as chemical pulping and bleaching, mechanical grinding, high-pressure homogenization, acid hydrolysis, enzyme treatment and solvent treatment, as mentioned earlier.[3] Nanofibrillated cellulose, CNC and CNF have unique properties, including a high Young's modulus, dimensional stability, a low thermal expansion coefficient, outstanding reinforcing potential and transparency.[2,3] They are very highly crystalline with a very high Young's modulus of ~150 GPa, making them very strong. Figure 26.2 shows a comparison of the specific strengths and Young's moduli of various materials.

Dipolar orientation and trapped charge are two major phenomena that contribute to the pyro-, piezo- and ferro-electricity of polymers. Shear piezoelectricity in polymers of biological origin such as cellulose and collagen was discovered in the 1950s.[4,5] Piezoelectricity was observed in the uniaxially oriented systems of crystallites of cellulose and elongated films of optical synthetic polymers. In cellulose-based biopolymers such as wood, ramie, chitin, amylase and starch, the shear piezoelectricity is very comparable to that of quartz crystal.[6] The discovery of tensile piezoelectricity in stretched polyvinylidene fluoride (PVDF) triggered research into PVDF and its copolymers in later years.[7] Shear piezoelectricity in wood depends largely on the type of wood, orientation and environmental conditions. Despite these early inroads, there had been very few investigations on the potential of cellulose as a smart material until Kim discovered an interesting actuation mechanism in cellulose paper. This smart cellulose was termed electroactive paper (EAPap).[8,9] Figure 26.3 shows the concept behind EAPap. When an electric voltage was applied to the electrodes, EAPap produced bending displacement depending on actuation voltage, frequency, host paper type and adhesion. The working principle of cellulose EAPap is claimed to be a combination of the piezoelectric effect and ionic migration effects, associated with the dipole moment of the cellulose paper ingredients. The electromechanical coupling and mechanical properties of cellulose EAPap has been reported to be very similar to those of piezopolymer.[10] This extends its possibilities for strain sensors, self-powered vibration sensors and energy scavenging transducers.[11] Once cellulose EAPap acquires the requirements of smart materials, various applications are possible, including wirelessly controlled EAPap actuators, flexible speakers, flying

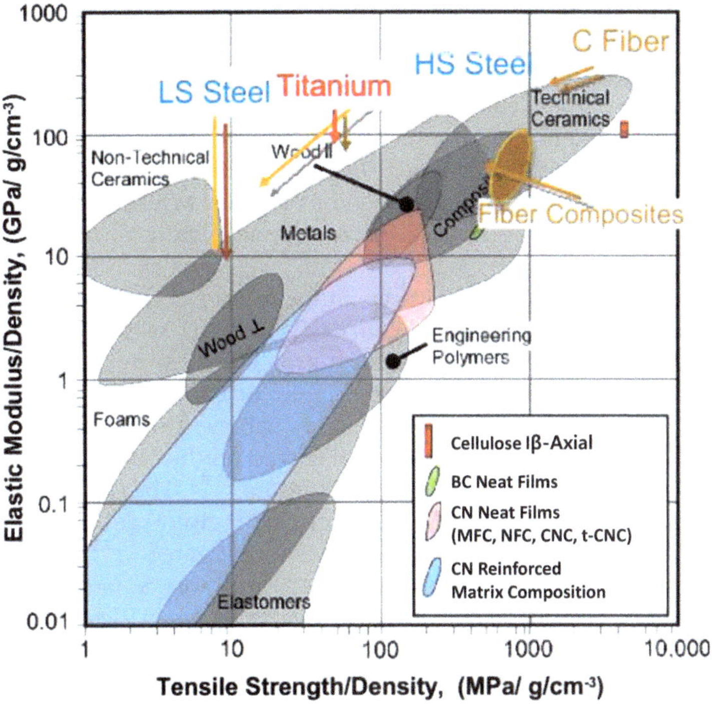

Figure 26.2 Comparison of the specific strengths and Young's moduli of various materials. LS: low strength, HS: high strength, C fiber: carbon fiber, CN: cellulose nanoparticle, MFC: microfiber cellulose, NFC: nanofiber cellulose, CNC: cellulose nanocrystal, t-CNC: tunicate cellulose nanocrystals.

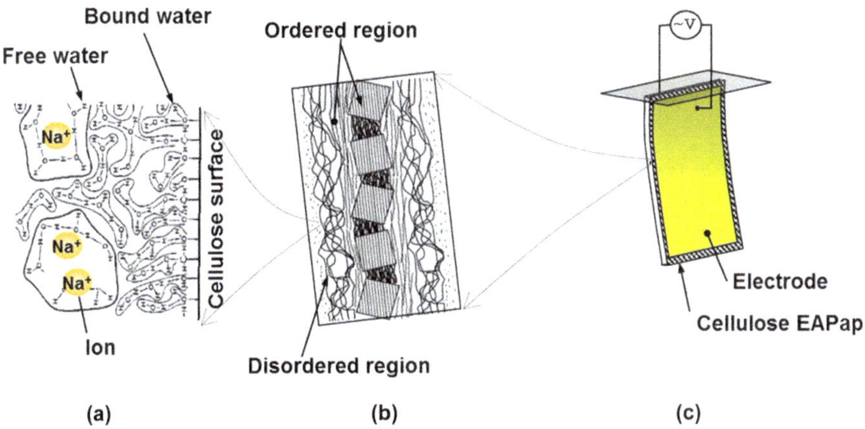

Figure 26.3 Schematic of an EAPap actuator: (a) surface of cellulose fiber, (b) ordered and disordered regions of cellulose, (c) cellulose EAPap.

objects and micro electromechanical systems (MEMS)/nano electromechanical systems (NEMS) devices.[12–14] To overcome the drawbacks of low output force and low actuation frequency, multi-walled CNTs (MWNTs) were coated on the cellulose EAPap.[15] When MWNTs were chemically bonded to cellulose in order to maintain a uniform dispersion of MWNTs in cellulose, its actuator performance exhibited a dramatic improvement in terms of force and actuation frequency.[16,17] This MWNT–cellulose hybrid EAPap can be tuned in terms of its electric conductivity so as to make a paper transistor.[18,19]

Cellulose also has an ionic effect. Since cellulose has a lot of hydroxyl groups around its chains, they can easily interact with ions and water molecules, which results in an ion migration effect. By maximizing the ion migration effect in cellulose EAPap, largely deformable actuators can be made. Many attempts have been made to achieve this. Conducting polypyrrole (PPy) or polyaniline (PANI) were coated on the cellulose film, which provided good results for bending actuators.[20,21] The influence of the addition of poly(ethylene oxide) –poly(ethylene glycol) (PEO–PEG) on the actuation behavior of cellulose EAPap was investigated.[22,23] The increased displacement output and decreased electrical power consumption of the actuator might be due to the improved polymer chain flexibility and ion mobility. Cellulose and chitosan blended EAPap was introduced to solve the issue of degradation with time and sensitivity to humidity that affects the performance of EAPap actuators.[24,25] This chitosan–cellulose composite can be used for biomedical applications.[26] Ionic liquids (ILs) were nanocoated onto cellulose to improve the ionic effect of cellulose EAPap by incorporating a conducting polymers (CPs).[27–33] Bacterial cellulose (BC) composites are unique and promising for medical implants and scaffolds in tissue engineering.[34–39] Recently, BC was used for making bending actuators.[40] Hybrid cellulose nanocomposites with enhanced material properties have been studied for biosensor, chemical disposable sensor, energy conversion and various other applications.[41–48] With the addition of metal oxide into a cellulose matrix, the chemical stability, mechanical properties, conductivity and photosensitivity can be enhanced for the use of cellulose in bioelectronics applications.[49]

This chapter reviews recent advances in cellulose EAPap. The actuation mechanism, physical properties and piezoelectric behavior of cellulose EAPap are summarized and ionic EAPap is introduced, which incorporate CPs, chitosan, PEO–PEG and ILs. To further improve the functionality of EAPap, hybrid nanocomposites are introduced by hybridizing CNTs and metal oxide with cellulose. The recent progress and development in cellulose-based bending actuators, vibration sensors, biosensors and chemical disposable sensors, as well as their applications in acoustic and bioelectronics, are critically discussed. Finally, the potential of cellulose EAPap and its hybrid composites, as well as challenges in this area, are addressed.

26.2 Cellulose EAPap

26.2.1 Fabrication of EAPap

Cellulose has a crystal part and an amorphous part. The crystal part of cellulose cannot be broken due to the strong hydrogen bonds of the hydroxyl groups in cellulose. Cellulose can be extracted from a variety of natural resources (*e.g.*, wood, plants, tunicates, algae and bacteria). The amorphous part of cellulose can be easily broken by strong mechanical force or special solvents, such as lithium chloride/dimethylacetamide (LiCl/DMAc).[3] The CNF is fiber-like and is generally of microns in length with a 4–20 nm cross-section. It has good flexibility and consists of amorphous and crystalline regions. The CNC has a needle-shaped structure generally of 50–500 nm in length and ~3–10 nm wide.[2]

To fabricate cellulose EAPap, cellulose pulp should be dissolved and regenerated so as to align CNFs. Figure 26.4 shows the cellulose EAPap fabrication process.[50–52] To dissolve cellulose pulp, a LiCl/DMAc system has been used. The following is a brief explanation of the regenerated cellulose fabrication of cellulose EAPap. Cotton pulp is torn into small pieces and LiCl is heated inside of the oven at 110 °C for 2 h. After LiCl is completely dissolved in the DMAc, dried small pieces of cotton cellulose pulp are mixed with LiCl/DMAc solution in a ratio of 1.5 : 8.5 : 90. The mixture is heated at 150 °C and two steps of heating then cooling to room temperature occur. The cellulose/LiCl/DMAc viscose solution is obtained. Generally, the viscosity of the cellulose solution will be 30 000–40 000 cP. To prepare the regenerated cellulose film, a certain amount of solution is cast on a glass plate with a doctor blade. The cast film is immersed into deionized water/isopropyl alcohol mixture to eliminate solvent from the film.[53] Then,

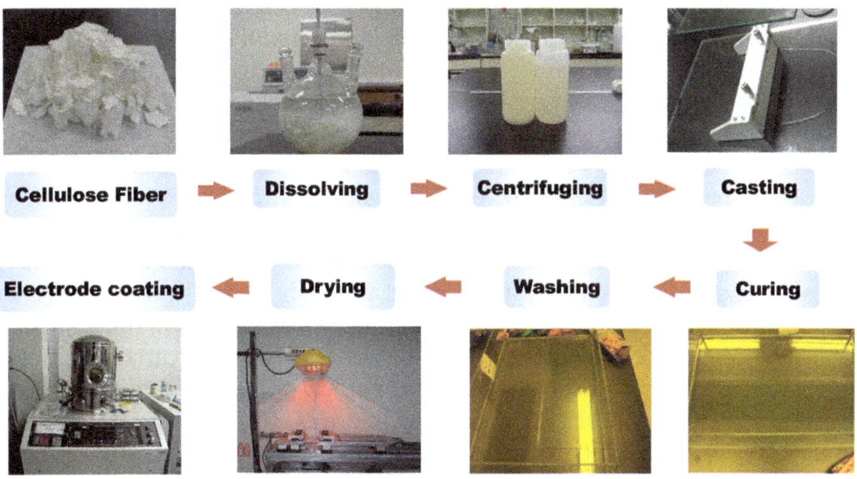

Figure 26.4 Fabrication process of cellulose EAPap.

a wet cellulose film can be obtained. The wet cellulose film is stretched by a stretching machine under a near-infrared heater for 1 h to align cellulose nanofibrils along the stretching direction. The dried and aligned cellulose film with a 20 μm thickness is then obtained. Metal electrodes are coated on both sides of the cellulose film by means of thermally evaporating aluminum. The typical thickness of the metal electrode is in the range of 100–150 nm. Since many hydroxyl groups appear on the surface of the cellulose film, thermally evaporated aluminum is securely adhered onto it.

26.2.2 Actuation Principle

Considering the cellulose structure and processing of the cellulose EAPap, the EAPap actuation principle is due to a combination of two mechanisms: ion migration and the piezoelectric effect.[9] Cellulose EAPap material is a sheet of regenerated cellulose and, morphologically, regenerated cellulose has ordered and disordered regions. The ordered domains are mostly crystalline and disordered molecules retain a preferential direction parallel to the chains in the microfibrils, and they form surface disorder on the microfibrils. A conceptual configuration of cellulose EAPap is depicted in Figure 26.3a. The concept of microfibrils is shown in Figure 26.3b. The EAPap material has large regions of disordered cellulose chains, where water molecules and ions can interact with the hydroxyl groups that appear on the cellulose chains (Figure 26.3c). When an external electric field is applied, these ions can be mobile and migrate to the anode. In addition, the molecular motion of free water in disordered regions cannot be restricted by the cellulose molecules, and the water molecules can interact with ions in the cellulose. In the presence of an electric field, the sodium ions surrounded by free water molecules will move to the anode. Selective ionic and water transport across the polymer under an electric field results in volumetric changes, which in turn lead to bending. When a DC electric field is applied, the cellulose EAPap actuator is bent to the positive electrode, which confirms the actuation mechanism.

On the other hand, cellulose EAPap material is composed of molecular chains with a dipolar nature. In particular, the crystal structure of cellulose II is monoclinic, which is non-centrosymmetric and exhibits piezoelectric and pyroelectric properties. Generally, the polarizability of dielectric materials may be separated into several parts. An electronic contribution arises from a displacement of the electron shell relative to the nucleus and an ionic contribution arises from the displacement of a charged ion with respect to other ions. In cellulose material that possesses molecular groups that have permanent molecular dipole moments, such as water or the hydroxyl and carboxyl groups, these groups will also make a contribution.[54] At low frequency, all of these parts contribute to the polarizability, as will any free ions (space charges) in the material. As the frequency increases, the space charges and permanent dipoles relax out. Space charges are usually the first to relax out, followed by the permanent dipole groups. In the cellulose EAPap, the presence of disordered regions gives rise to localized states associated with the hydrogen bonding of cellulose chains. Since there are many localized states, the release or excitation of the carriers in these states may dominate the charge transfer process. Thus, disordered regions mainly contribute to the dipolar orientation by stabilizing dipoles and leading to a permanent polarization, resulting in piezoelectric behavior.

26.2.3 Physical Properties

For enhancing piezoelectric effects in EAPap, cellulose fiber alignment is very important. The alignment of the cellulose fiber was investigated by taking cross-sectional scanning electron microscope (SEM) and high-voltage electron microscope (HVEM) images, as shown in Figure 26.5.[50] Aligned bundles of cellulose chains in the stretching direction were apparently observed after stretching. In general, regenerated cellulose film consists of crystalline and amorphous regions. Cellulose chains in amorphous regions exist in a randomly oriented formation. However, in this study, the cellulose chains could be aligned along the stretching direction by a stretching process. From the cross-sectional image, it can be seen that the regenerated cellulose exhibits a layered structure. After stretching, the nanofiber formation was well observed in the layered cellulose structure. When the cross-sectional morphology was observed by HVEM, the crystal parts of cellulose could be clearly seen. The diameters of the nanofibers are in the range of 30–150 nm and have a circular shape.

Figure 26.5 shows the XRD patterns of EAPap with different stretching ratios: $D_R = 1.0$, 1.1 and 1.5 and $D_R = 2.0$.[50] Originally, cellulose II has three peaks at 12.1°, 19.8° and 22° assigned to (110), $(1\bar{1}0)$ and (200), respectively. However, the cellulose EAPap showed a more diffused XRD profile compared with cellulose II. The diffused pattern indicates that the cellulose film has considerably more amorphous regions due to slow coagulation. The cellulose sample also shows a diffused XRD pattern; however, the $(1\bar{1}0)$ directional peak intensity gradually increased as the drawing ratio increased from 1.0 to 2.0 (Figure 26.6). Thus, it can be concluded that cellulose EAPap has more ordered cellulose chains by mechanical stretching.

Figure 26.7 shows the strain–stress curves of cellulose films with different stretching ratios.[50] The Young's modulus of cellulose film without stretching

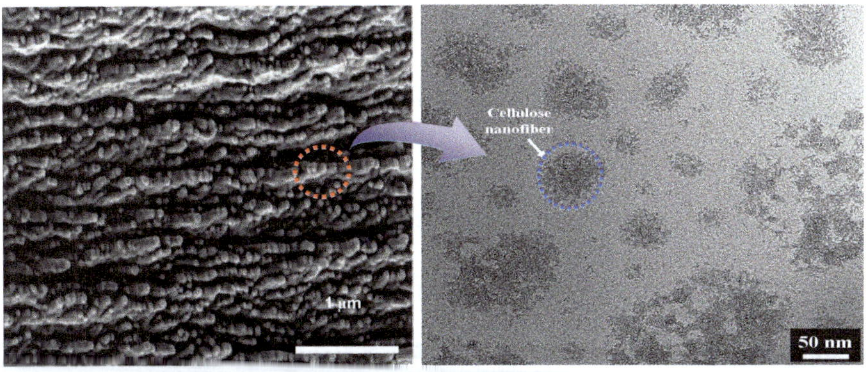

Figure 26.5 SEM and TEM images of cellulose EAPap.[50]

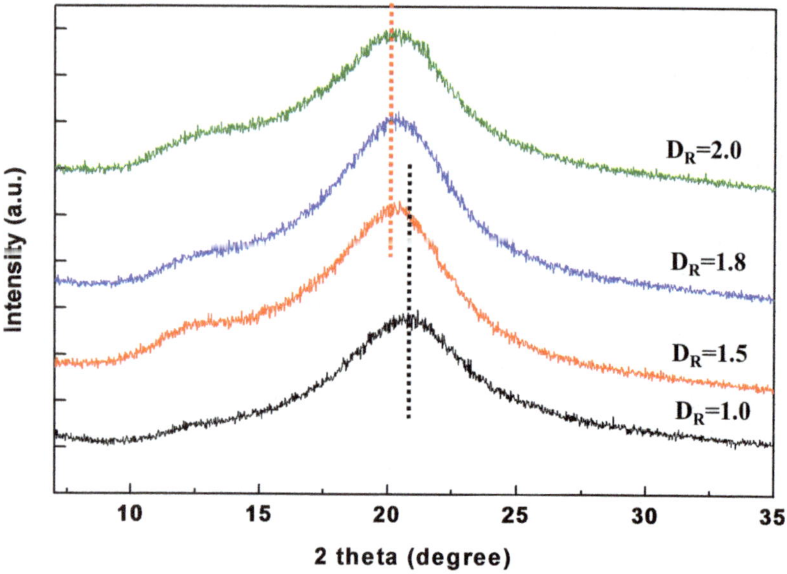

Figure 26.6 XRD patterns of cellulose EAPap with different stretching ratios.[50]

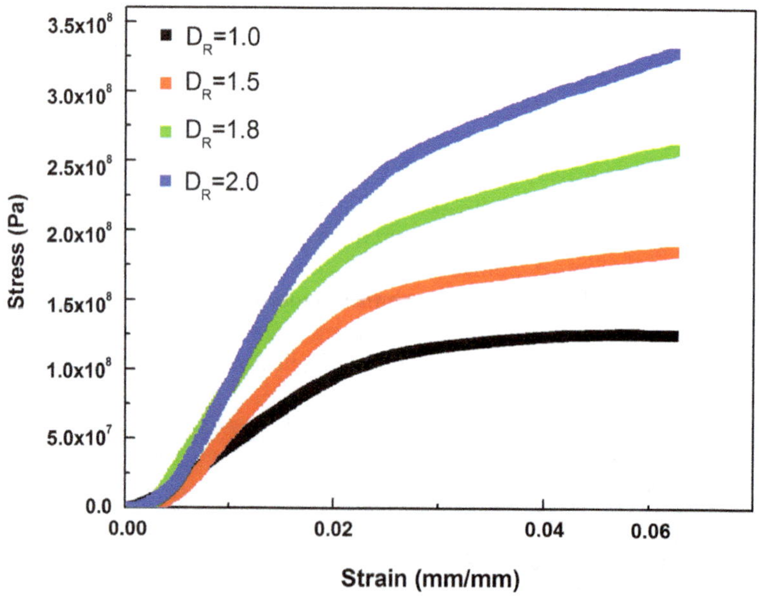

Figure 26.7 Stress–strain curves of cellulose EAPap with different stretching ratios.

($D_R = 1.0$) was 5.3 GPa. When the drawing ratio increased, the Young's modulus also dramatically increased up to 14.1 GPa, which is 270% larger than that of the non-stretched case. The mechanical properties of cellulose film were improved due to the improved alignment of cellulose chains along the stretching direction. However, since cellulose is hydrophilic, its

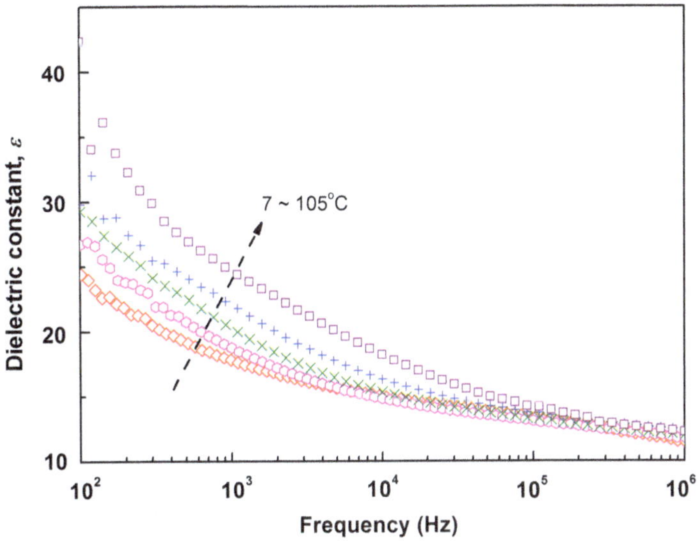

Figure 26.8 Temperature-dependent dielectric behavior of cellulose EAPap from room condition (20 °C) to 105 °C.[52]

mechanical properties are sensitive to environmental humidity conditions. It was found that the Young's modulus of cellulose EAPap decreases as the relative humidity (RH) increases.[51]

The dielectric constant of cellulose EAPap is important for determining its actuation behavior as well as its power consumption. Figure 26.8 shows the measured dielectric constant for cellulose EAPap as a function of frequency at different temperature conditions.[52] This reveals that the dielectric constant of cellulose EAPap is strongly dependant on the temperature as well as the frequency. The dielectric constant increases as the temperature increases. At around 100 °C, near the boiling point of water, the dielectric constant of cellulose EAPap shows its maximum value. However, at over 100 °C, the dielectric constant starts to decrease as the temperature increases. The dielectric behavior at over 100 °C may be related to the vaporized water molecules, which are under higher entropy conditions compared to liquid water. Vaporized water molecules can easily escape from the cellulose surface, resulting in the lower dielectric constant of cellulose EAPap. Nevertheless, the range of the dielectric constant of cellulose EAPap is higher than those of other polymers.

Figure 26.9 shows the optical transparency of cellulose EAPap. The UV-vis value at 450 nm increases from 88 to 90 with the increase in the stretching ratio.

26.2.4 Piezoelectric Properties

The piezoelectric charge constants of different orientations of EAPap samples under quasi-static direct piezoelectricity were measured.[50–52] The crystallinity and alignment of CNFs increase when increasing the stretching ratio. Figure 26.10a shows the sample orientation along the stretching direction and Figure 26.10b shows the induced charge of cellulose EAPap

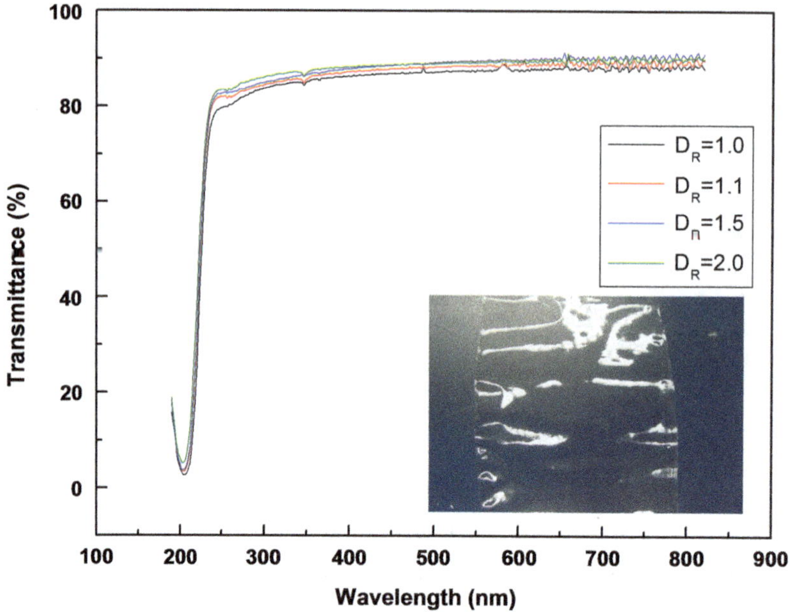

Figure 26.9 UV-vis of cellulose EAPap.

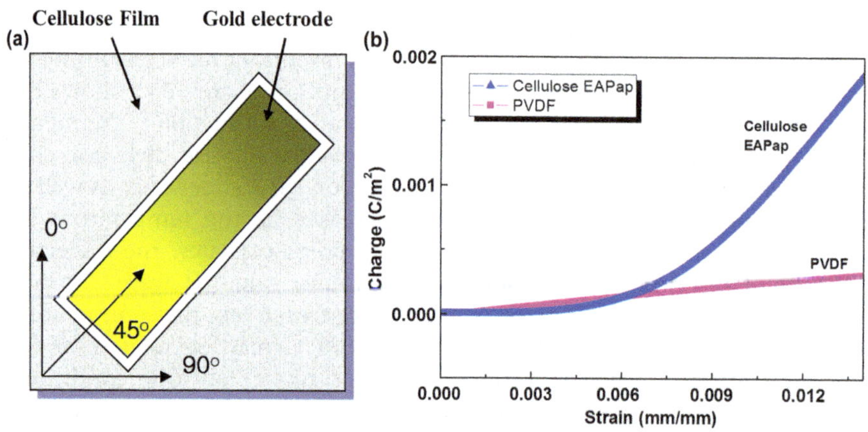

Figure 26.10 (a) Orientation of cellulose EAPap along the stretching direction and (b) induced charge curves of cellulose EAPap and PVDF.

and PVDF film with strain. The measured piezoelectric charge constant of a 45° sample with a 2.0 stretching ratio is 28.2 pC/N, which is larger than other 0° and 90° samples.[50] The measured piezoelectric charge constant of cellulose EAPap is similar to that of PVDF.

Electrically aligned regenerated cellulose films showed higher in-plane piezoelectric constants due to an increased crystallinity index.[53] Heat treatment of EAPap specimens also generates better piezoelectric effects as compared to specimens with no heat treatment.[54]

To understand the actuation phenomenon, thermally stimulated current (TSC) measurements were performed.[9,52] TSC was used to characterize cellulose EAPap under different electric field and temperature conditions. TSC analysis is a sensitive tool for detecting the relaxation phenomena of a material composed molecular chains with a dipolar nature, such as cellulose. From this analysis, the relaxation phenomena can be inferred, which are associated with the orientation polarization of permanent or induced dipoles and real charge injection. Before testing the TSC, the sample was kept at 80 °C in a vacuum oven for 1 day to remove the free water from the cellulose paper sample. The glass transition temperature of the cellulose film was 203 °C. Figure 26.11a shows the depolarized current with temperature under different poling electric fields and Figure 26.11b shows the peak current values as a function of the poling electric field. The depolarization current increases linearly as the poling electric field increases. This behavior is usually indicative of dipole orientation.[9]

26.3 Ionic EAPap

26.3.1 CP-Coated EAPap

To improve the actuation performance of cellulose EAPap, PPy and PANI CPs were coated onto cellulose EAPap and their performances were compared.[20,21,54] CP-coated EAPap (CP-EAPap) was made by electrochemically coating CPs on cellulose EAPap. Two types of actuators were designed: a CP/cellulose bilayer and a CP/cellulose/CP trilayer, as shown in Figure 26.12. ClO_4^- and BF_4^- dopants were examined for improving the actuator performance. The PPy-coated film with BF_4^- dopant with a trilayer configuration exhibited a maximum bending displacement of 10.5 mm. The maximum displacement of the PANI-coated trilayer EAPap actuator with BF_4^- dopant was shown to be 10.1 mm. The trilayer devices show better performance than the bilayer devices. The working principle of the actuator is deeply related to the adsorption or desorption of water molecules and structural changes under applied electrical fields.

26.3.2 PEO–PEG Blended EAPap[22,23]

To improve the actuation performance of cellulose EAPap, PEO–PEG was blended with cellulose solution and cast to form a film. Figure 26.13 shows the association of PEO–PEG with cellulose and its disruption under the excitation of an electric field in high humidity conditions. Cellulose/PEO–PEG blended films were fabricated with varying PEO : PEG ratios from 0.1 : 0.9 to 0.9 : 0.1 *via* traditional solution blending techniques. Although there is some morphological change of cellulose due to the addition of PEO–PEG, both XRD and FT-IR analysis revealed no substantial changes in its structure. The actuator showed a maximum bending displacement of 5.0 mm with very low electrical power consumption (7 mW mm^{-1}) under ambient conditions. The increased displacement output and decreased electrical power consumption

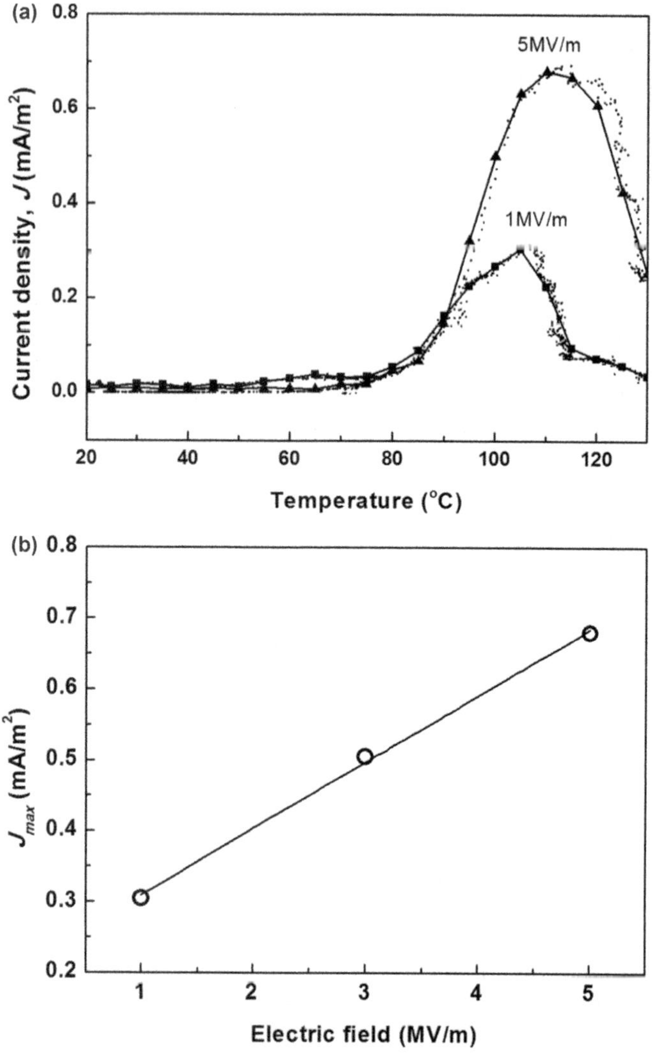

Figure 26.11 Measured TSC data. (a) Current–temperature plot of cellulose EAPap under different external fields. (b) Maximum current versus the electric field.[52]

of the actuator might be due to the improved polymer chain flexibility and ion mobility. The ion migration effect might play a more important role in the actuation principle.

26.3.3 Chitosan Blended EAPap[24,25]

To prepare cellulose EAPap actuators with high performance in ambient humidity conditions, chitosan–cellulose laminated films were used as EAPap actuators. Since the molecular structures of cellulose and chitosan are very similar, there is expected to be high compatibility between cellulose

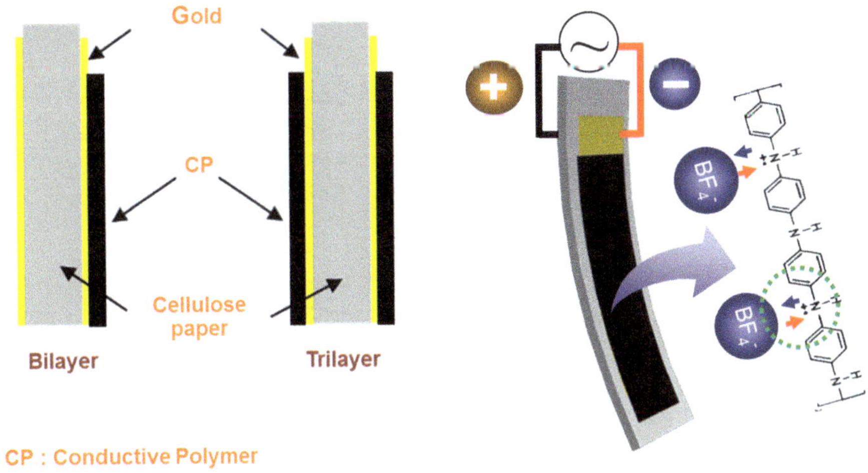

Figure 26.12 Bilayer and trilayer models of CP-EAPap and its activation schematic.

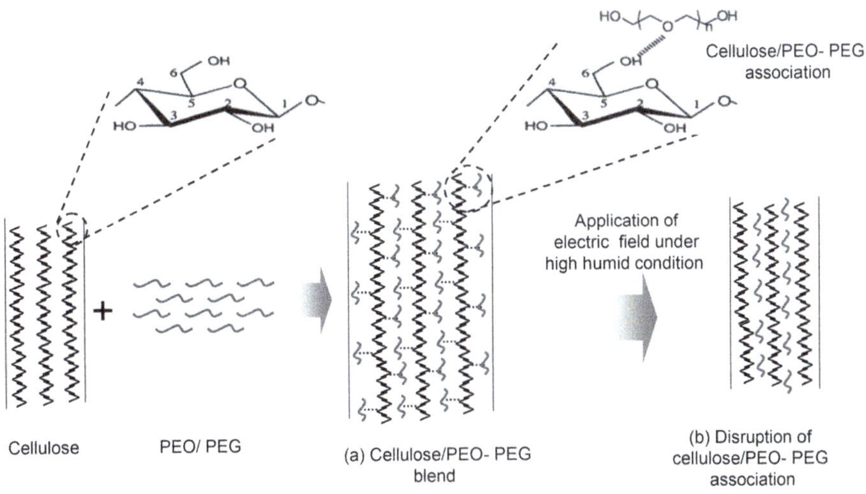

Figure 26.13 Schematic representation of the PEO–PEG association with cellulose and its disruption under the excitation of an electric field in high humidity conditions.

and chitosan. Figure 26.14 shows a schematic diagram of a chitosan and cellulose inter-penetrating polymer network. The chitosan and cellulose blending ratio and humidity effects on actuator performance were investigated. When cellulose–chitosan was blended at a 60 : 40 ratio, the maximum bending displacement was obtained under 50% RH conditions. The actuation principle of cellulose–chitosan blended EAPap actuators might be associated with an ion migration effect associated with Cl^- mobile anions and NH_3^+ fixed cations. When the actuator was activated with an AC voltage under 60% RH conditions, the bending displacement increased with the voltage increase and a maximum bending displacement of 4.1 mm was

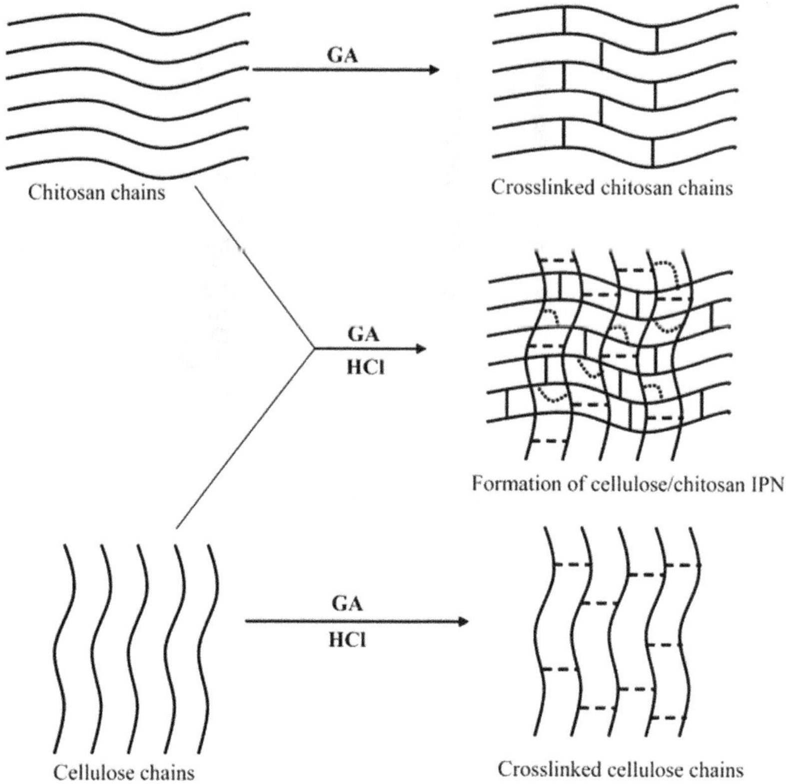

Figure 26.14 Schematic representation of the synthesis of a cellulose–chitosan inter-penetrating polymer network (IPN). GA: glutaraldehyde.

achieved at 6 V and 4 Hz, which is almost the same as the cellulose EAPap actuator at 90% RH. The cellulose–chitosan EAPap actuator is quite durable under more than 9 h of actuation. The electrical power consumption was 17 mW cm^{-2}. The SEM images and XRD patterns of this cellulose–chitosan EAPap actuator did not show any significant damages of electrode surfaces after 9 h of actuation.

26.3.4 IL Dispersed EAPap[27–33]

In spite of its excellent advantages over other electroactive polymers, there are several material drawbacks to cellulose EAPap: the performance of cellulose EAPap is sensitive to the surrounding humidity conditions and there is performance degradation with time. Recently, nanocoating with polyelectrolytes and CPs has been used to modify the surfaces and electrical properties of materials used in the fields of science, engineering and information technology, as well as medical supplies.[55,56] The actuator performance and durability of cellulose EAPap actuators was greatly improved by PPy and IL nanocoating. ILs have interesting properties such as non-volatility, high stability, suitable polarity, high ionic conductivity and easy recyclability. Also, room temperature ILs consume little energy for numerous electrochemical processes. Owing to their ideal properties, ILs are receiving much attention as environmentally benign solvents for organic chemical reactions, separations and for electrochemical applications.[57,58] An aqueous dispersion of IL was made to create an IL nanolayer on the cellulose EAPap surface. To ensure uniform and stable IL nanocoating, nanoscaled PPy intermediate layer was introduced onto the cellulose surface *via* a polymerization-induced adsorption process. Figure 26.15 shows the fabrication process of the cellulose–PPy–IL (CPIL) nanocomposites and its transmission electron microscopy (TEM) image. During the wet impregnation and polymerization process, PPy molecules could penetrate into the bulk and be deposited onto cellulose nanofibrils. Furthermore, activating wet PPy-adsorbed cellulose in IL solution results in the adsorption and uniform deposition of IL over the cellulose–PPy surface by ionic interaction. The thickness of the IL layer was found to be about 8 nm. Three ILs, namely 1-butyl-3-methylimidazolium chloride, 1-butyl-3-methylimidazolium tetrafluoroborate ($BMIBF_4$) and 1-butyl-3-methylimidazolium hexafluorophosphate were tried for the CPIL fabrication, and it was found that $BMIBF_4$ exhibited the best results.

The maximum bending displacement of CPIL actuators is 10 mm under ambient temperature and humidity conditions at 5 V and 3.5 Hz. The power consumption was 30 mW. The enhancement of actuation performance is due to the greater mobility and high ion-transporting ability of IL. Table 26.1 provides a comparison of the actuator performance of CPIL with other reported EAPap actuators. The CPIL actuator showed enhanced durability of up to 6 h in ambient humidity and temperature conditions without any significant performance degradation.

With durable CPIL actuators, a wirelessly driven CPIL actuator was demonstrated.[33] Figure 26.16 shows a photograph of this configuration. The actuator system consists of a dipole rectenna array, a power control circuit and two CPIL actuators. Once the rectenna array receives microwaves and converts them into DC power, the control circuit distributes the power so as to active the CPIL actuators wirelessly. This was the first demonstration of wirelessly driven polymer actuators by using microwaves.

26.4 Hybrid EAPap

26.4.1 CNT Blended EAPap

Polymer–CNT composites have been researched for enhancing their properties and multi-functionality. Cellulose–CNT composites were made for bending actuation with various conditions of CNT. MWNT-mixed cellulose EAPap was fabricated by blending small amounts (0.1 wt%) of MWNTs into a cellulose solution.[17] The MWNT–cellulose EAPap actuator increased its Young's modulus by 76% and its resonance frequency by 35%. As a result,

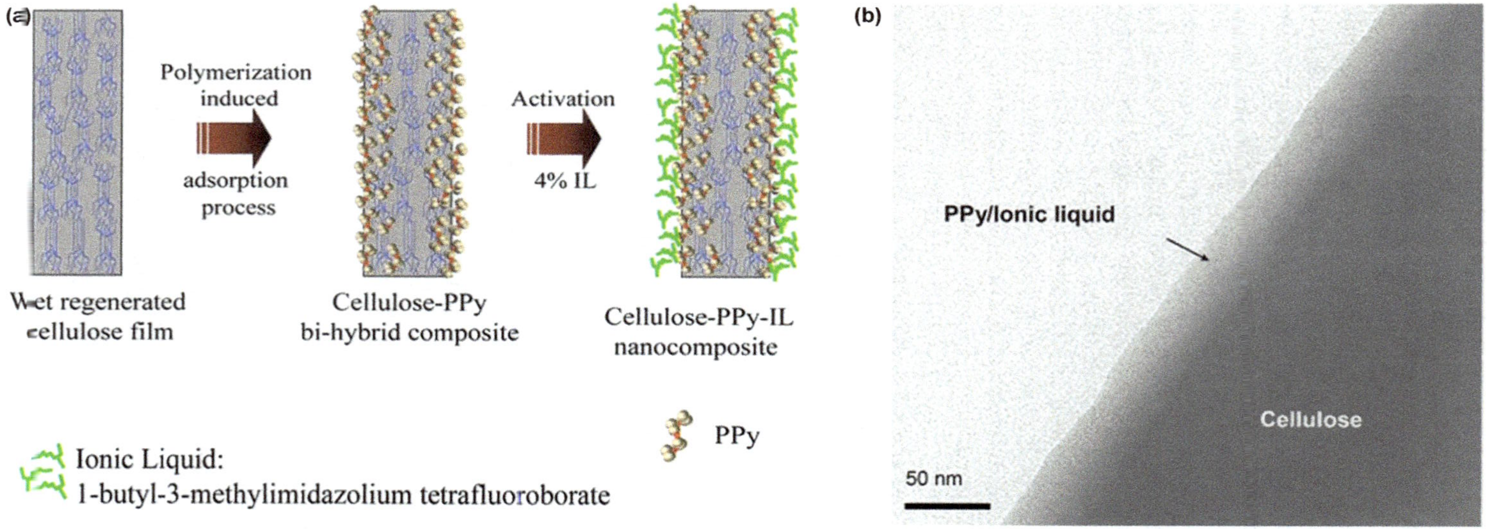

Figure 26.15 (a) Fabrication process of a CPIL nanocomposite and (b) its TEM image.

Table 26.1 Comparison of the performance properties of CPIL actuators and other reported EAPap actuators.

	Maximum displacement	Resonance frequency	Output force	Electrical power consumption
Cellulose	3.5 mm[a]	7.5 Hz	32 μN	55 mW
PPy-coated cellulose	6.7 mm[a]	6 Hz	62 μN	68 mW
Single walled carbon nanotube/ PANI-coated cellulose	3.1 mm[a]	5 Hz	3.0 μN	23.2 mW
CNT/cellulose	4.5 mm[b]	6 Hz	198.4 μN	28 mW
CPIL nanocomposite	10 mm[c]	3.5 Hz	3.0 μN	30 mW

[a]70% RH.
[b]90% RH.
[c]30% RH.

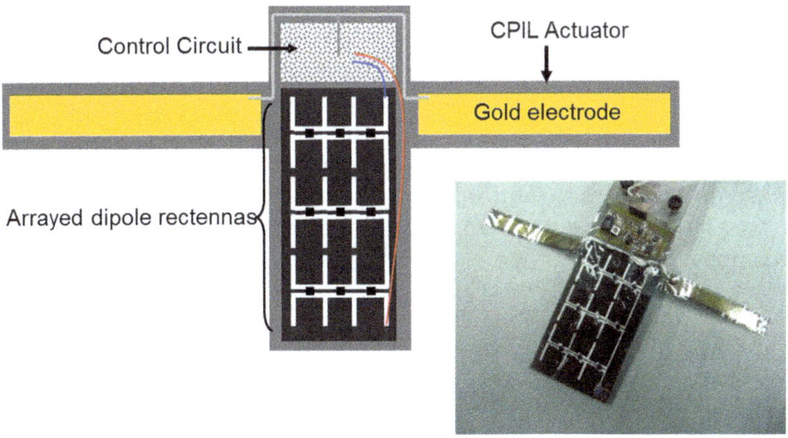

Figure 26.16 Microwave-driven CPIL actuator.

the mechanical power output, which is a significant drawback of cellulose EAPap, was enhanced by up to 65%, increasing its mechanical properties. However, MWNTs were inhomogeneously dispersed in the cellulose solution owing to them aggregating to each other by interaction forces. To overcome this problem, the functionalized MWNT (f-MWNT) was suggested.[59] The well-dispersed f-MWNT is shown in Figure 26.17. By blending 0.15 wt% of f-MWNT with cellulose, the f-MWNT blending improved the output force of the actuator remarkably to over ten times that of the cellulose EAPap without loss of its maximum bending displacement.[21]

26.4.2 TiO_2-Coated EAPap[60–62]

Inorganic nanomaterials have a broad range of functionalities, which are promising for sensors, actuators, electronics and energy devices. TiO_2

has been used as a building block to develop new nanoarchitectures with advanced properties for achieving high-performance sensing and improved photocatalytic power. Since TiO_2 is a nanoparticle, it is hard to make flexible films with it. TiO_2 can be blended with polymers to form films. However, its functionalities are limited, since TiO_2 can be aggregated in polymer matrices. In particular, the cellulose matrix is difficult to bond with TiO_2 nanoparticles. Thus, MWNTs were used to coat TiO_2 so as to enable chemical bonding between MWNTs and cellulose. This incorporation of MWNTs can achieve good dispersion of TiO_2 in a cellulose matrix. TiO_2 and MWNT nanocomposites was synthesized by a hydrothermal method, and a cellulose–TiO_2–MWNT (CTM) hybrid nanocomposite was made by blending the TiO_2/MWNT with cellulose. The CTM hybrid nanocomposite was applied as an ammonia gas sensor at room temperature. The hybrid nanocomposite of MWNT and TiO_2 with cellulose is flexible and environmentally friendly, without sacrificing the functionality of MWNTs and TiO_2 for ammonia gas sensing at room temperature. Figure 26.18 shows the bonding between cellulose, MWNTs and TiO_2 nanoparticles, TEM images of TiO_2–MWNT, the electrode deposition on the CTM hybrid nanocomposite and ammonia gas sensing apparatus, along with the sensitivity results.[60] This ammonia gas sensor works at room temperature, which has the advantage of not requiring heating, as is required by conventional gas sensors. Glucose biosensor and pH sensor properties were also demonstrated by the hybrid nanocomposite.[61,62]

26.4.3 SnO_2-Coated EAPap[63–65]

Synthesizing a very thin and uniform nanocrystalline metal oxide layer on a cellulose film can be advantageous for applying this material to flexible electronic devices, disposable sensors and biosensors. A flexible nanocomposite was developed by coating a regenerated cellulose film with a thin layer of SnO_2 by liquid-phase deposition. SnO_2 is an electrical conductor that is optically transparent in the visible spectrum with a wide band gap of 3.6 eV at room temperature. SnO_2 has been widely used in gas sensors, optical devices and lithium batteries. SnO_2 was crystallized in solution and formed nanocrystal coatings on regenerated cellulose. The nanocrystalline layers did not exfoliate from cellulose. TEM and energy dispersive X-ray spectroscopy suggested that SnO_2 was not only deposited over the cellulose surface, but also nucleated and grew inside the cellulose film.[63] The cellulose–SnO_2 hybrid nanocomposite was used for biodegradable and disposable glucose biosensors and pH sensors. Figure 26.19 shows the fabrication process of this nanocomposite by thermal hydrolysis, a photograph of the sample and the sensitivity curves for urea and glucose biosensors.

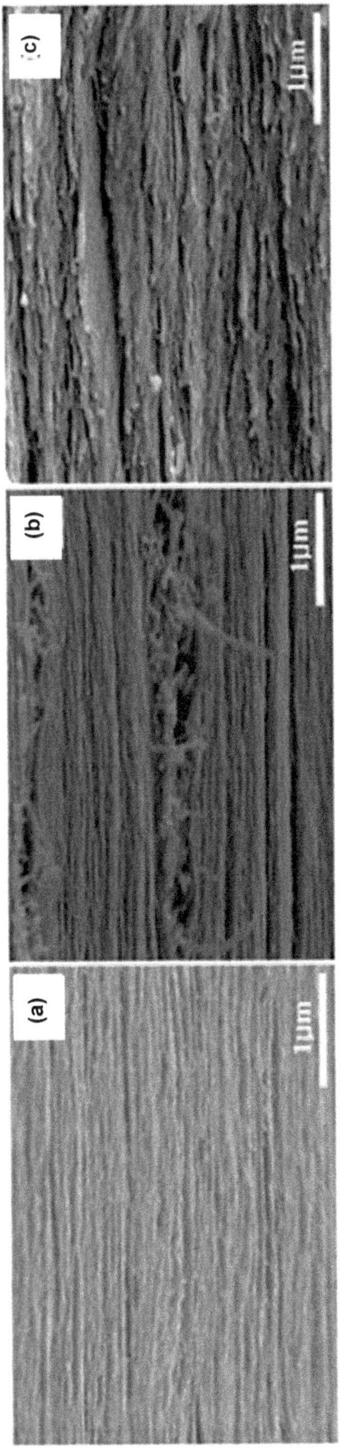

Figure 26.17 SEM cross-sectional images of: (a) cellulose film; (b) MWNT blended cellulose composite; and (c) f-MWNT blended cellulose composite.

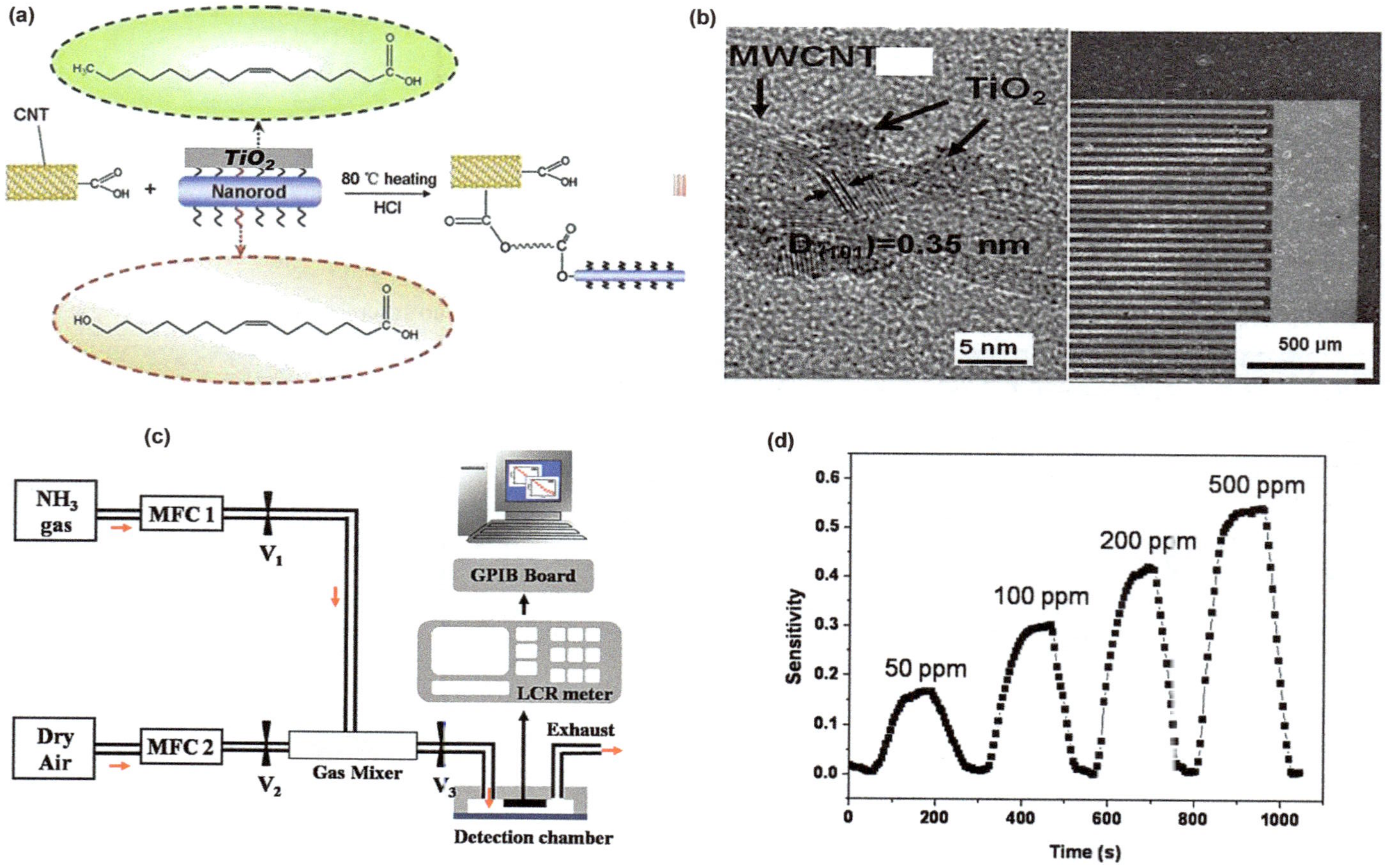

Figure 26.18 Cellulose–TiO_2–MWNT hybrid nanocomposite: (a) TiO_2–MWNT bonding; (b) TEM of TiO_2–MWNT and electrode for a gas sensor; (c) CTO ammonia gas sensor; and (d) sensitivity of the ammonia gas sensor. CTO: cellulose tin oxide, MWCNT: multi walled carbon nanotube, MFC: mass flow controller, GPIB: general purpose interface bus, LCR: inductance (L), capacitance (C), resistance (R).

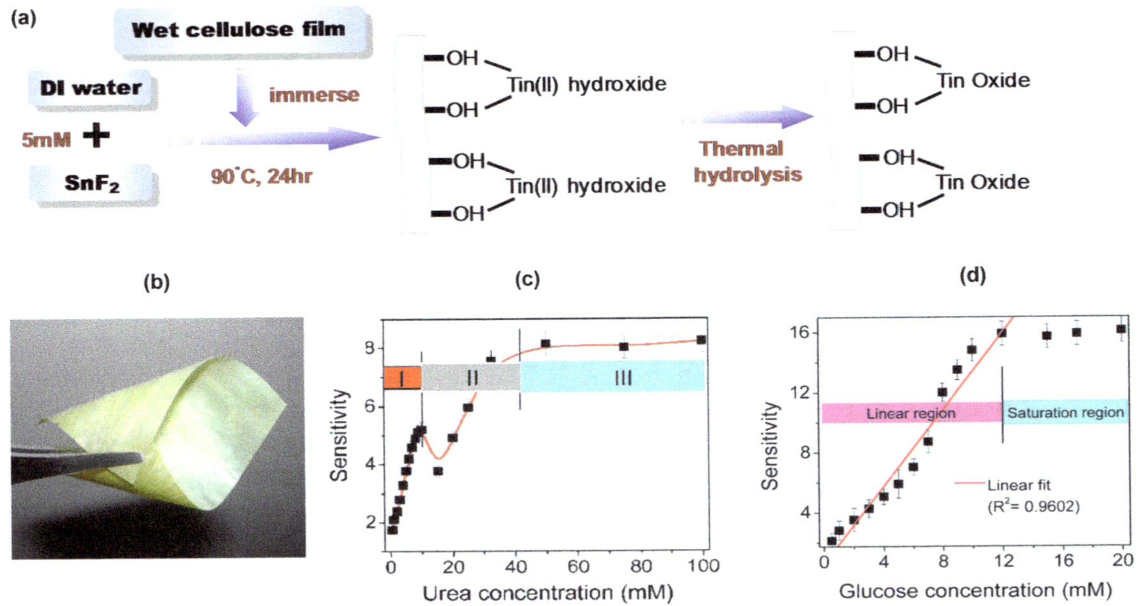

Figure 26.19 Cellulose–SnO_2 hybrid nanocomposite: (a) fabrication process; (b) photograph of the sample; (c) sensitivity of the urea biosensor (I is the first linear region, II is the second linear region and III is saturation region); and (d) sensitivity of the glucose biosensor.

26.5 Conclusions

This chapter reviewed the fabrication and actuation principles of EAPap and its three subareas—piezoelectric EAPap, ionic EAPap and hybrid EAPap—along with their applications. Cellulose EAPap is made by dissolving cellulose followed by casting and stretching so as to maximize the piezoelectricity in the EAPap. The actuation principle is a combination of piezoelectric and ion migration effects. Cellulose EAPap exhibits a high Young's modulus and high dielectric constant, which result in a high piezoelectric constant ($d_{31} = 28.2$ pC/N), which is similar to that of the PVDF piezoelectric polymer.

To enhance the ion migration effect in cellulose, PANI and PPy CP coating, PEO–PEG blending, chitosan blending and IL dispersion approaches were undertaken with cellulose. Aqueous dispersion of IL was done to create an IL nanolayer on the cellulose EAPap surface, and a nanoscaled PPy intermediate layer was introduced onto the cellulose surface *via* polymerization-induced adsorption to ensure uniform and stable IL nanocoating. The cellulose–PPy–IL nanocomposite exhibits three-fold greater displacement output than cellulose EAPap under room temperature and humidity conditions without any significant performance degradation over 6 h.

To further improve the functionality of cellulose, hybrid composites of inorganic functional materials were introduced by incorporating CNTs, TiO_2 and SnO_2 with cellulose. The incorporation of CNTs showed improvements in mechanical properties and force output. The blending/coating of TiO_2 and SnO_2 with cellulose expanded the functionality and application areas to biosensor and chemical sensor devices. Since cellulose is biocompatible, sustainable, biodegradable, capable of broad chemical modification, hydrophilic and has high mechanical strength and stiffness, various cellulose-based devices are possible.

References

1. D. Klemm, B. Heublein, H. P. Fink and A. Bohn, *Angew. Chem., Int. Ed.*, 2005, **44**, 3358.
2. R. J. Moon, A. Martini, J. Nairn, J. Simonsen and J. Youngblood, *Chem. Soc. Rev.*, 2011, **40**, 3941.
3. M. A. Hubbe, O. J. Rojas, L. A. Lucia and M. Sain, *BioResources*, 2008, **3**, 929.
4. E. Fukada, *J. Phys. Soc. Jpn.*, 1995, **10**, 149.
5. V. A. Bazhenov, *Piezoelectric Properties of Woods*, Consultants Bureau, New York, 1961.
6. E. Fukada, *IEEE Trans. Ultrason. Ferroelectr. Freq. Control*, 2000, **47**, 1277.
7. H. Kawai, *Jpn. J. Appl. Phys.*, 1969, **8**, 975.
8. J. Kim and Y. B. Seo, *Smart Mater. Struct.*, 2002, **11**, 355.
9. J. Kim, S. Yun and Z. Ounaies, *Macromolecules*, 2006, **39**, 4202.
10. J. Kim, S. Yun and S. K. Lee, *J. Intell. Mater. Syst. Struct.*, 2008, **19**, 417.

11. K. M. Suresha, S. Y. Yang, M. H. Lee, J. H. Kim and J. Kim, *Comp. Interf.*, 2008, **15**, 679.
12. J. H. Kim, S. Yun, J. H. Kim and J. Kim, *J. Bionic. Eng.*, 2009, **6**, 18.
13. J. Kim, S. H. Bae and H. G. Lim, *Smart Mater. Struct.*, 2006, **15**, 889.
14. S. W. Lee, J. H. Kim, J. Kim and H. S. Kim, *Chin. Sci. Bull.*, 2009, **54**, 2703.
15. S. Yun, J. Kim and Z. Ounaies, *Smart Mater. Struct.*, 2006, **15**, N61.
16. S. Yun and J. Kim, *Carbon*, 2011, **49**, 518.
17. S. Yun and J. Kim, *Sens. Actuators, A*, 2009, **154**, 73.
18. S. Yun, S. D. Jang, G. Y. Yun, J.-H. Kim and J. Kim, *Appl. Phys. Lett.*, 2009, **95**, 104102.
19. J.-H. Kim, S. Yun, H.-U. Ko and J. Kim, *Curr. Appl. Phys.*, 2013, **13**, 897.
20. S. D. Deshpande, J. Kim and S. Yun, *Smart Mater. Struct.*, 2005, **14**, 876.
21. S. D. Deshpande, J. Kim and S. Yun, *Synth. Met.*, 2005, **149**, 53.
22. S. K. Mahadeva and J. Kim, *Polym. Eng. Sci.*, 2010, **50**, 1199.
23. S. K. Mahadeva, J. Kim, K. S. Kang, H. S. Kim and J. M. Park, *J. Appl. Polym. Sci.*, 2009, **114**, 847.
24. Z. Cai and J. Kim, *J. Appl. Polym. Sci.*, 2010, **115**, 2044.
25. Z. Cai and J. Kim, *J. Appl. Polym. Sci.*, 2009, **114**, 288.
26. J. Kim, Z. Cai, H. S. Lee, G. S. Choi, D. H. Lee and C. Jo, *J. Polym. Res.*, 2011, **18**, 739.
27. S. K. Mahadeva, K.-J. Yun, J.-H. Kim and J. Kim, *J. Nanosci. Nanotechnol.*, 2011, **11**, 270.
28. S. K. Mahadeva, J. Kim and C. Jo, *Int. J. Precis. Eng. Manuf.*, 2011, **12**, 47.
29. S. K. Mahadeva, S. Y. Yang and J. Kim, *IEEE Trans. Nanotechnol.*, 2011, **10**, 445.
30. K.-B. Kim and J. Kim, *Proc. Inst. Mech. Eng., Part C*, 2013, **227**, 2665.
31. S. K. Mahadeva and J. Kim, *Fibers Polym.*, 2012, **13**, 289.
32. S. K. Mahadeva, Y. Chen and J. Kim, *Ionics*, 2011, **17**, 41.
33. S. Y. Yang, S. K. Mahadeva and J. Kim, *Smart Mater. Struct.*, 2010, **19**, 105026.
34. J. H. Jeon, I. K. Oh, C. D. Kee and S. J. Kim, *Sens. Actuators, B*, 2010, **146**, 307.
35. N. Petersen and P. Gatenholm, *Appl. Microbiol. Biotechnol.*, 2011, **91**, 1277.
36. D. Ciechanska, *Fibres Text. East. Eur.*, 2004, **12**, 69.
37. W. C. Lin, C. C. Lien, H. J. Yeh, C. M. Yu and S. H. Hsu, *Carbohydr. Polym.*, 2013, **94**, 603.
38. A. Svensson, E. Nicklasson, T. Harrah, B. Panilaitis, D. L. Kaplan, M. Brittberg and P. Gatenholm, *Biomaterials*, 2005, **26**, 419.
39. O. M. Alvarez, M. Patel, J. Booker and L. Markowitz, *Wounds*, 2004, **16**, 224.
40. J. Kim, Z. Cai and Y. Chen, *J. Nanotechnol. Eng. Med*, 2010, **1**, 011006.
41. S. Kalia, S. Boufi, A. Celli and S. Kango, *Colloid Polym. Sci.*, 2014, **292**, 5.
42. W. Gindl and J. Keckes, *Polymer*, 2005, **46**, 10221.
43. I. Siro and D. Plackett, *Cellulose*, 2010, **17**, 459.

44. A. Kumar, H. Gullapalli, K. Balakrishnan, A. Botello-Mendez, R. Vajtai, M. Terrones and P. M. Ajayan, *Small*, 2011, 7, 2173.
45. A. S. Zakirov, S. U. Yuldashev, H. D. Cho, J. C. Lee, T. W. Kang, J. J. Khamdamov and A. T. Mamadalimov, *J. Korean Phys. Soc.*, 2012, **60**, 1526.
46. J. H. Kim, S. Mun, H. U. Ko, G. Y. Yun and J. Kim, *Nanotechnology*, 2014, **25**, 092001.
47. Y. Zhou, F. H. Canek, M. K. Telha, J. C. Liu, J. Hsu, J. W. Shim, A. Dindar, P. Y. Jeffrey, R. J. Moon and B. Kippelen, *Sci. Rep.*, 2013, **3**, 1536.
48. K. K. Sadasivuni, K. Abdullahil, L. Zhai, H.-U. Ko and S. Mun, *Small*, 2014, DOI: 10.1002/small.201402109.
49. S. K. Mahadeva and J. Kim, *Sci. Technol. Adv. Mater.*, 2011, **12**, 055006.
50. C. Yang, J. H. Kim, J.-H. Kim, J. Kim and H. S. Kim, *Sens. Actuators, A*, 2009, **154**, 117.
51. J.-H. Kim, K. J. Yun, J. H. Kim and J. Kim, *Smart Mater. Struct.*, 2009, **18**, 055005.
52. G.-Y. Yun, J.-H. Kim and J. Kim, *J. Phys. D: Appl. Phys.*, 2009, **42**, 082003.
53. S. Yun, S. D. Jang, G. Y. Yun and J. Kim, *Smart Mater. Struct.*, 2009, **18**, 117001.
54. J. Kim, S. D. Deshpande, S. Yun and Q. Li, *Polym. J.*, 2006, **38**, 659.
55. M. Agarwal, Y. Lvov and V. Kody, *Nanotechnology*, 2006, **17**, 5319.
56. A. Sufen, H. Qiang, T. Cheng, S. Zheng and J. Li, *Macromol. Rapid Commun.*, 2005, **26**, 1965.
57. S. K. Mahadeva, J. Kim and C. Jo, *Int. J. Precis. Eng. Manuf.*, 2011, **12**, 47.
58. L. Wen, A. G. Fadeev, Q. Baohua, E. Smela, B. R. Mattes, J. Ding, G. M. Spinks, M. Jakub, D. Zhou, G. W. Gordon, R. M. Douglas, A. F. Stewart and F. Maria, *Science*, 2002, **297**, 983.
59. S. Yun, J. Kim and K. S. Kang, *Polym. Int.*, 2010, **59**, 1071.
60. S. Mun, Y. Chen and J. Kim, *Sens. Actuators, B*, 2012, **171–172**, 1186.
61. M. Maniruzzaman, S. D. Jang and J. Kim, *Mater. Sci. Eng. B*, 2012, **177**, 844.
62. Y. Chen, S. Mun and J. Kim, *IEEE Sens. J.*, 2013, **13**, 4157.
63. S. K. Mahadeva and J. Kim, *Sci. Technol. Adv. Mater.*, 2011, **12**, 055006.
64. S. K. Mahadeva and J. Kim, *Sens. Actuators, B*, 2011, **157**, 177.
65. S. K. Mahadeva, H.-U. Ko and J. Kim, *Z. Phys. Chem.*, 2013, **227**, 419.

Subject Index